Chaotic Electronics in TELECOMMUNICATIONS

Chaotic Electronics in TELECOMMUNICATIONS

Edited by

Michael Peter Kennedy

Riccardo Rovatti

Gianluca Setti

CRC Press

Boca Raton London New York Washington, D.C.

Library of Congress Cataloging-in-Publication Data

Chaotic electronics in telecommunications / edited by Michael P. Kennedy, Riccardo Rovatti, Gianluca Setti.

p. cm.

Includes bibliographical references and index.

ISBN 0-8493-2348-7 (alk. paper)

1. Digital communications—Mathematics. 2. Chaotic behavior in systems. 3. Integrated circuits—Design and construction. 4. Telecommunication systems. I. Kennedy, Michael P. II. Rovatti, Riccardo. III. Setti, Gianluca. IV. Series.

TK5103.7 .C534 2000
621.382—dc21 00-039814
CIP

Direct all inquiries to CRC Press LLC, 2000 N.W. Corporate Blvd., Boca Raton, Florida 33431.

International Standard Book Number 0-8493-2348-7
Library of Congress Card Number 00-039814
Printed in the United States of America 1 2 3 4 5 6 7 8 9 0
Printed on acid-free paper

Preface

In recent years chaos engineering has attracted much interest in the academic and industrial worlds. In fact, the continuous pursuit of both methodological and technological innovation has led to the realization that common linear models of real systems suffer from severe limitations. In particular, they preclude the exploitation of phenomena whose complexity may be an intrinsic advantage in cryptography, spread spectrum communication, electro-magnetic interference reduction, noise generation, digital image watermarking, etc.

Today the potential of chaos engineering is recognized worldwide with research groups actively working on this topic in every part of the globe. Large-scale research projects are being funded by the European Community as well as by the United States Department of Defense.

Historically, at least three achievements were fundamental for the acceptance of chaos engineering as a field worthy of attention and exploitation. The first was the implementation and characterization in the early eighties of several electronic circuits exhibiting chaotic behavior. This brought chaotic systems from mathematical abstraction into the core of electronic engineering.

The second historical event in the path leading to chaos exploitation was the observation of Pecora and Carroll in the early nineties that two chaotic systems can synchronize under suitable conditions if one of them is driven by (at least) one component of the first one. This suggested that chaotic signals could be used for communication, where their noise-like broadband nature could improve disturbance rejection as well as security.

A third, and fundamental, step was the awareness of the engineering community that chaotic systems enjoy a mixed deterministic/stochastic nature. This had been known to mathematicians since at least the early seventies, and advanced methods from that theory have been recently incorporated in the tools of chaos engineering. These tools helped to clarify that the tasks which are most likely to benefit from chaos-based techniques are those where the statistical properties of the signals are the dominant factor. The same tools were also of paramount importance in developing the quantitative models needed to effectively design chaotic systems following engineering specifications.

Following this evolution, the aim of this book is to present the state of the art in the application of chaos in digital communication applications where a statistical approach is adopted. The material is organized according to the level of abstraction at which chaos is applied: from coding-theory to circuit design.

This volume is intended for applied mathematicians and information engineers in general who specialize in electronics as well as telecommunications or systems science. It is meant for readers from an academic environment who are interested in the latest developments in chaos applications and those from an industrial environment who are interested in a first but in-depth look at the improvements that chaos-based techniques can bring to existing systems.

To reach such a qualified and diversified audience, a pool of internationally renowned scientists has graciously agreed to contribute chapters treating the aspects of chaos engineering that they have mastered. All of the contributors lead or are involved in large-scale projects, funded by national and international agencies, and the editors wish to thank them warmly for devoting their time to the preparation of this book. Finally, special thanks to Prof. L.C. Jain, D. Mesa, C. Andreasen, and all the CRC staff who helped and supported the editors' work.

Michael P. Kennedy
Riccardo Rovatti
Gianluca Setti

THE EDITORS

Michael Peter Kennedy received a B.E. degree in electronics from the National University of Ireland in 1984, and the M.S. and Ph.D. degrees from the University of California at Berkeley in 1987 and 1991, respectively. From 1992 to 1999, he was with the Department of Electronic and Electrical Engineering, University College Dublin. In 2000, he was appointed Professor and Head of the Department of Microelectronic Engineering at University College Cork. He has written more than 180 articles in the area of nonlinear circuits and systems and has taught courses on nonlinear dynamics and chaos in England, Switzerland, Italy, and Hungary. He has held visiting research positions at the EPFL, AGH Krakow, TU Budapest, and UC Berkeley. His research interests are in the simulation, design, and analysis of nonlinear dynamical systems for applications in communications and signal processing. He is also engaged in research into algorithms for mixed-signal testing. Dr. Kennedy received the 1991 Best Paper Award from the *International Journal of Circuit Theory and Applications*, delivered the 88th Kelvin Lecture to the Institution of Electrical Engineers in 1997, and received the 1999 Best Paper Award at the European Conference on Circuit Theory and Design in 1999. He served as Associate Editor of the *IEEE Transactions on Circuits and Systems* (Parts I and II), as Guest Editor of Special Issues on "Chaos Synchronization and Control," "Advances in Nonlinear Electronic Circuits," and "Noncoherent Chaotic Communications," and was Chair of the IEEE TECHNICAL COMMITTEE ON NONLINEAR CIRCUITS AND SYSTEMS and the IEE Circuits and Systems Professional Group. Dr. Kennedy was elected a Fellow of the IEEE in 1998 "for contributions to the theory of neural networks and nonlinear dynamics, and for leadership in nonlinear circuits research and education." In 2000, he was awarded a Golden Jubilee Medal by the IEEE Circuits and Systems Society and an IEEE Third Millenium Medal.

Riccardo Rovatti received a Dr. Eng. degree (with honors) in Electronic Engineering and a Ph.D. degree in Electronic Engineering and Computer Science from the University of Bologna, in 1992 and 1996, respectively. Since 1997 he has been a lecturer of Digital Electronics at the University of Bologna. He has authored or co-authored more than 100 international scientific publications. His research interests include fuzzy theory foundations, learning and CAD algorithms for fuzzy and neural systems, statistical pattern recognition, function approximation, non-linear system theory and identification as well as theory and applications of chaotic systems.

Gianluca Setti received a Dr. Eng. degree (with honors) in Electronic Engineering and a Ph.D. degree in Electronic Engineering and Computer Science from the University of Bologna, in 1992 and in 1997, respectively. His focus was on the study of neural networks and chaotic systems. From May 1994 to July 1995 he was a visiting research assistant with the Circuits & Systems Group of the Swiss Federal Institute of Technology. In 1997 he was named Lecturer at the University of Ferrara. Since 1998 he has been Assistant Professor of Analog Electronics at the University of Ferrara. Dr. Setti received the 1998 Caianiello Prize for the best Italian Ph.D. thesis on neural networks. He is currently Secretary of the IEEE TECHNICAL COMMITTEE ON NONLINEAR CIRCUITS AND SYSTEMS and also serves as Associate Editor of the *IEEE Transactions on Circuits and Systems - Part I*, for the area of Nonlinear Circuits and Systems. His research interests include nonlinear circuit theory, recurrent neural networks, and design and implementation of chaotic circuits and systems as well as their applications to electronics and signal processing.

Contents

Chapter 1

Introduction

Michael Peter Kennedy
Department of Microelectronic Engineering
University College Cork
Cork, Ireland
Peter.Kennedy@ucc.ie

1.1 Motivation

In recent years, there has been explosive growth in personal communications, the aim of which is to guarantee the availability of voice and/or data services between mobile communications terminals. In order to provide these services, radio links are required for a large number of compact terminals in densely populated areas. As a result, there is a need to provide high-frequency, low-power, low-voltage circuitry. The huge demand for telecommunications results in a large number of users; therefore, today's telecommunications systems are limited primarily by interference from other users. In some applications, the efficient use of available bandwidth is extremely important, but in other applications, where the exploitation of communication channels is relatively low, a wideband communication technique having limited bandwidth efficiency can also be used.

Often, many users must be provided with simultaneous access to the same or neighboring frequency bands. The optimum strategy in this situation, where every user appears as interference to every other user, is for each communicator's signal to look like Gaussian noise which is as wideband as possible [1].

There are two ways in which a communicator's signal can be made to appear like wideband noise: by spreading each symbol using a pseudorandom sequence to increase the bandwidth of the transmitted signal, or by representing each symbol by a piece of "noiselike" waveform.

The conventional solution to this problem is the first approach: to use a synchronizable pseudorandom sequence to spread the transmitted signal, and to use a conventional modulation scheme based on Phase Shift Keying (PSK) or Frequency Shift Keying (FSK).

Such Direct Sequence (DS) Spread Spectrum (SS) schemes have processing gain associated with despreading at the receiver, and the possibility to provide multiple access by assigning mutually orthogonal sequences to different users. This is the basis of Code Division Multiple Access (CDMA) communications systems. Limitations are imposed by the periodic nature of the spreading sequences, the limited number of available orthogonal sequences, and the periodic nature of the carrier. One further problem is that the orthogonality of the spreading sequences requires the synchronization of all spreading sequences used in the same frequency band, i.e., the whole system must be synchronized. Due to different propagation times for different users, perfect synchronization can never be achieved in real systems.

The use of chaotic spreading sequences in DS systems offers the potential for both asynchronous CDMA capability and a large number of orthogonal codes.

An alternative approach to making a transmission "noiselike," without spreading, is to represent the transmitted symbols not as weighted sums of periodic basis functions but as chaotic basis functions.

In this book, we consider both chaotic spreading codes for DS-CDMA and chaotic basis functions for digital modulation.

1.2 What is Chaos?

Deterministic dynamical systems are those whose states change with time in a deterministic way. They may be described mathematically by differential or difference equations, depending on whether they evolve in continuous or discrete time. Deterministic dynamical systems can produce a number of different steady-state behaviors including DC, periodic, and chaotic solutions [2].

DC is a non-oscillatory state. Periodic behavior is the simplest type of steady-state oscillatory motion. Sinusoidal signals, which are universally used as carriers in analog and digital communications systems, are periodic solutions of continuous-time deterministic dynamical systems. The time-domain waveform and spectrum of a periodic signal are shown in Figure 1.1(a).

Deterministic dynamical systems also admit a class of non-periodic signals which are characterized by a continuous "noiselike" broad power spectrum; this is chaos [3]. In the time domain, chaotic signals appear "random," as shown in Figure 1.1(b). Chaotic systems are characterized by "sensitive dependence on initial conditions"; a small perturbation eventually causes a large change in the state of the system. Equivalently, chaotic signals decorrelate rapidly with themselves. The autocorrelation function of a chaotic signal has a large peak at zero and decays rapidly.

Thus, while chaotic systems share many of the properties of stochastic processes, they also possess a deterministic structure which makes it possible to generate "noiselike" chaotic signals in a theoretically reproducible manner.

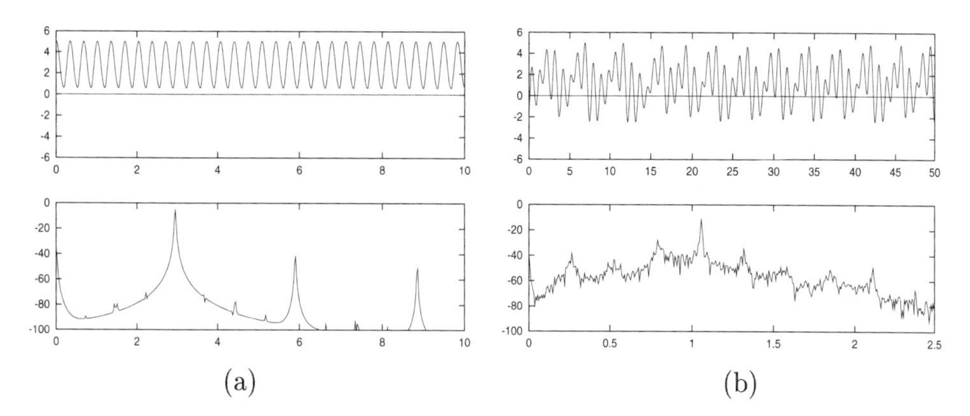

Figure 1.1: (a) Periodic and (b) chaotic signals in the time (upper) and frequency (lower) domains.

1.3 Outline of the Book

Over the past decade, there has been significant interest in exploiting chaotic dynamics in communications. Topics which have been explored include chaotic encryption for security, chaotic spreading codes for multi-user access in spread-spectrum systems, and chaotic modulation for the transmission of analog and digital information.

In this book, we consider applications of chaotic electronics in telecommunications at three levels. At code level, discrete-time chaotic systems can be used to generate spreading codes for DS-SS systems. Part I of the book, consisting of three chapters, focuses on chaos at code level. Chapter 2 provides an introduction to spread spectrum communications, highlighting the key performance issues for DS-CDMA. Chapter 3 deals with chaos-based asynchronous DS-CDMA systems. Chapter 4 discusses information sources using chaotic dynamics, also addressing briefly the use of chaos in encryption.

At signal level, continuous-time chaotic systems can be used to generate wideband carriers for digital modulation schemes. Part II of the book, consisting of five chapters, is concerned with chaos at signal level. Chapter 5 provides an overview of digital communications. Chapter 6 focuses on chaotic modulation schemes, highlighting the problems associated with coherent detection of chaotic transmissions. The best digital chaotic modulation scheme in the literature is FM-DCSK. Chapter 7 analyzes the performance of slow- and fast-frequency-hopping variants of FM-DCSK in multipath environments. Chapter 8 discusses noise filtering in chaos-based communications. Chapter 9 addresses statistical analysis and design strategies for chaotic systems.

The three chapters in Part III are concerned with chaos at the hardware level. Chapter 10 provides an introduction to applications and architectures for chaotic integrated circuits. Chapter 11 focuses on chaos-based noise generation in silicon. Chapter 12 is concerned with robustness issues for chaos generators.

References

[1] A. J. Viterbi. Wireless digital communication: A view based on three lessons learned. *IEEE Communications Magazine*, pages 33–36, September 1991.

[2] M. P. Kennedy. Basic concepts of nonlinear dynamics and chaos. In C. Toumazou, editor, *Circuits and Systems Tutorials*, pages 289–313. IEEE Press, London, 1994.

[3] M. P. Kennedy. Bifurcation and chaos. In W. K. Chen, editor, *The Circuits and Filters Handbook*, pages 1089–1164. CRC Press, Boca Raton, 1995.

Part I

CHAOS AT CODE LEVEL

Chapter 2

Introduction to DS-CDMA

Gianluca Mazzini
DI
University of Ferrara
via Saragat 1, 44100 Ferrara, Italy
g.mazzini@chaos.cc

2.1 Introduction

Since the very first developments of communication systems by Morse, Bell and Marconi, it was clear that a natural way to increase communication efficiency and implementability as well as to allow multiple links, was to keep the bandwidth of a single link as small as possible.

Great efforts have been made in this direction. One of the most powerful results obtained in the field of information theory was Shannon's discovery of the maximum capacity that a medium can offer when a given ratio between power of the signal and of the white noise is considered.

While new digital modulation schemes and channel coding systems were developed to approach the Shannon limit, other techniques were investigated from a completely different point of view. These new techniques were indicated as *spread spectrum* [1] techniques since the signals used for the transmissions are characterized by a bandwidth much greater than the one necessary for transmitting the same signals with conventional techniques. Hence, spread spectrum techniques are band-wise inefficient though they feature some properties that made them first interesting for military applications [1, 2], i.e., jamming immunity, intrinsic robustness to multipath effect, signal hiding under noise or under other transmissions, message intrinsic privacy.

In this chapter a short review of spread spectrum techniques will be given: starting from the more intuitive frequency and time hopping approaches to the more recent and promising *Direct Sequence - Code Division Multiple Access* (DS-CDMA). Actually, the importance of this technique has grown in the last decade also in consumer applications mainly thanks to the development of VLSI modules providing the computational resources needed for modulation and demodulation at low cost.

This review will also include brief sketches of recent open problems and applications, giving a slightly deeper insight into those topics which will not be covered by the subsequent chapters.

2.2 Spread Spectrum Concept

Many different techniques have been proposed to spread the spectrum of digital signals. Three main "spaces" may be used to this aim: frequency, time, and code [2–4].

2.2.1 Frequency Hopping

In the frequency space the spread spectrum communication is called *Frequency Hopping* (FH) [3, 4]. The channel bandwidth is subdivided into a number of continuous frequency slots and in each information symbol time the transmitting signal is modulated using a carrier that falls in a different set of frequency slots. The simplest scheme actually uses only one slot per symbol.

The selection of the frequency slots depends on *Hopping Sequences* (HS). At the receiver, a synchronized replica of the HS of the transmitter is needed for signal demodulation. If more than one slot is used for each symbol, the signals recovered are combined in order to obtain a more reliable estimation of the transmitted symbol.

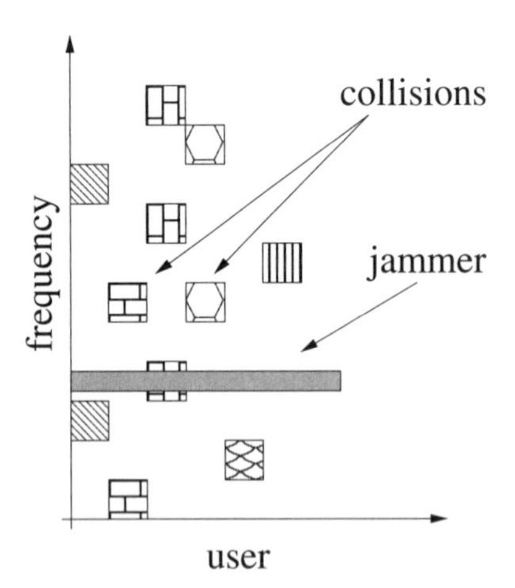

Figure 2.1: Frequency hopping with collisions and one jamming signal.

An intentional jamming signal that does not know the HS is unable to systematically disturb the useful signal and the jamming effects decrease as the bandwidth increases. This mechanism also protects the useful link from interference from other transmissions systems and can be considered a multiple access scheme if the HSs are designed in order to reduce the collision on the same frequency slots. In Figure 2.1 an example of a frequency hopping multiple access scheme is reported in the frequency versus user space, by also including the possibility to have frequency slots collisions and a jamming signal attacking one frequency slot.

When multipath propagation is present, the delayed path of the same signals (usually a consequence of reflection, refraction, and diffraction by scattering objects in the environment) combine at the receiver and are equivalent to a selective channel strongly attenuating certain frequencies.

Attenuated frequencies actually vary in time as each delayed ray is characterized by a random phase, arrival time and amplitude, depending on terminal mobility and on mutating environment.

A simple example of multipath propagation is the two-ray model. If $i(t)$ is the complex enveloped of the transmitted signal, and the second ray is characterized by an instantaneous amplitude a, a delay τ and phase ϕ; the envelope at the channel output is $i(t) + ai(t-\tau)e^{\mathbf{i}\,\phi}$ and the modulus of frequency response is $\sqrt{1 + a^2 + 2a\cos(\phi - 2\pi f_0 \tau)}$ where f_0 is the carrier frequency.

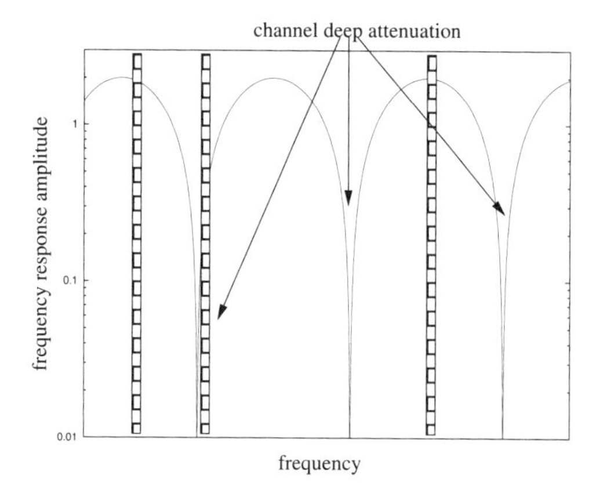

Figure 2.2: Two rays selective fading channel and signal frequency hopping.

The demodulation of a narrowband signal whose carrier falls near a channel null is then impaired by stronger relative levels of noise and interference.

On the contrary, when FH is considered, the narrow band signal is affected by the channel null only in certain slots and the larger the global allocated bandwidth the less likely the event of a complete demodulation failure. In Figure 2.2 the case with $a = 1$ and a selective channel with two rays is reported with the indication of possible hopping frequencies inside the system bandwidth.

Finally, with FH the privacy depends on the knowledge of the HS, the complexity of the latter being strictly related to the number of channel observations needed by an eavesdropper to identify the sequence of slots used by the target communication.

The FH has actually few but relevant applications; in particular, it may be optionally used on the GSM system and it has been selected by IEEE 802.11 for the wireless local area network (WLAN) systems.

2.2.2 Time Hopping

In the time space the spread spectrum communication is called *Time Hopping* (TH) [3,4]. The information symbol time is subdivided into many smaller intervals and the actual signal is transmitted only in some of these time slots chosen by means of an HS. This baseband signal is then modulated. Obviously, the receiver must have knowledge of the HS in order to retrieve the information.

The main problems with this bursty transmission technique are the nonlinearities of the power amplifier in the transmitter and the strict timing synchronization which is required in the receiver. In strict analogy with FH, this system achieves security and multiple access as it uses HS but it has not yet found interesting practical implementations.

2.2.3 Direct Sequence

In the code space the spread spectrum communication is called *Direct Sequence* (DS) [3,4]. While FH and TH mechanisms (especially in their multi-access aspects) are quite straightforward generalizations of classical FDM and TDM multiplexing techniques, DS techniques do not differentiate signals either in frequency or time domain.

To understand the DS concept let us consider a narrowband signal and multiply it for a PAM signal in which the symbol time is N time shorter than the information symbols. We call this PAM signal the *spreading signal* and the sequence of its symbols (*chips*) is called *spreading sequence* (SS).

The spectrum of the resulting signal is the convolution of narrow spectrum of the original signal and the spectrum of the spreading signal. If N (called *spreading factor*) is large enough its bandwidth is the same as the spreading one.

At the receiver the incoming signal is re-multiplied with a replica of the spreading signal, whose symbols have been chosen so that their squares value is always 1. A lowpass filter or a correlator is then enough to re-extract the narrowband information signal.

Though identical, the two blocks comprising the spreading signal generator and the multiplier are given different names at the transmitter and the receiver side, i.e., *spreader* and *despreader*. In general, it is clear that with this mechanism a signal is spread each time it passes an odd number of these units and it is narrowed when it passes an even number of the same.

The jamming rejection is then due to the fact that jamming is usually inserted in the channel and its effect is spread by the same block that actually narrows the information signal: rejection is then of the same order of the spreading factor.

The spreading and despreading mechanisms are clarified in Figure 2.3, where a jamming carrier is also reported and the spectra of the signal at several points of the transmission chain are depicted.

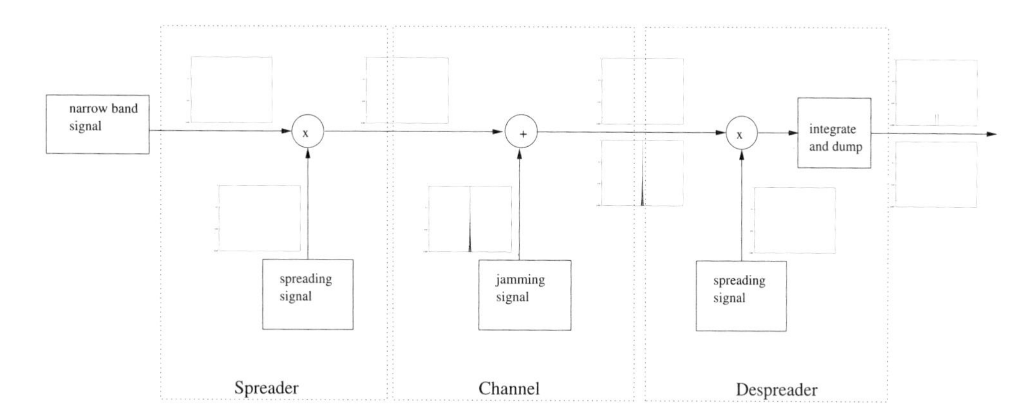

Figure 2.3: Spreading and despreading concept with jamming signals.

Multiple access is also achieved by DS techniques by considering "orthogonal" SSs. In this way all the users use the same bandwidth at the same time, without interfering with each other. This multiple access scheme is called DS-CDMA and its simple basic scheme is reported in Figure 2.4.

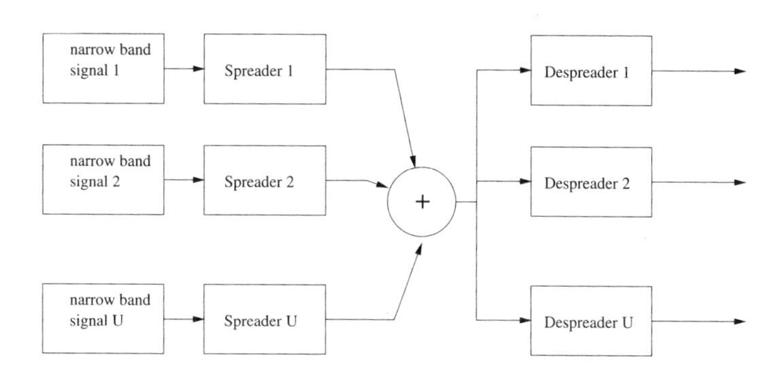

Figure 2.4: DS-CDMA basic scheme.

The DS technique also features an intrinsic immunity to multipath-related channel fading. In fact, if the SSs are chosen so that they have a rapidly vanishing autocorrelation, delayed replica of the information signal have almost negligible effect once they are fed into the final correlation block.

The security of communication is given by the use of spreading sequences but also from the fact that the performance does not decrease as the spreading factor increases, i.e., as the power density of the information is reduced. Hence it is possible to hide the spread signal under the noise level.

The DS approach is the most diffuse one, first selected from the IS-95 standard for the cellular system, by the IEEE 802.11 for WLANs and finally from the third generation of cellular systems. The following chapter is devoted to the application of chaos-related techniques to the generation of suitable SSs.

2.2.4 Final Remarks

Let us conclude this section observing some points common to all the described techniques.

First of all HS or SS do improve communication privacy but require a further synchronization between the transmitter and the receiver.

Hence, with respect to the classical narrowband communication systems, where carrier (for coherent demodulation) and symbol timing recovery is needed, in the spread spectrum communications it is also necessary to have the new HS or SS synchronization level.

Finally, note that the three techniques are not mutually exclusive so that hybrid systems can be considered in which different spreading techniques are merged.

2.3 Spreading Codes and Environments

The first main problem in designing a DS-CDMA system is the generation of a set of SSs. Such a problem cannot be effectively tackled without first clarifying the environment in which the system operates.

Let us consider as an example a cellular system [5]: the devices involved in radio links are the mobiles and the base stations (ground or satellite). For each cell, two different links must be considered, those from the base station to a mobile terminal, called *downlink*, and those from mobile to base stations, called *uplink*.

In the downlink all the signals addressed to the system users are generated at the same physical point (the base station) and are therefore perfectly synchronous. On the contrary, in the uplink the signals are generated by different sources (the mobiles) which do not share a common time scale and can be expected to be at random distance from the base station.

Channel variability further impairs any attempt to achieve synchronization between the mobiles so that downlinks and uplinks have a completely different nature, the first being inherently synchronous while the second is inherently asynchronous. An asynchronous system with random channel is shown in Figure 2.5 and the subsequent chapter is devoted to this class of systems.

The spreading sequence selection is then a function of the environment and in the synchronous one, where no carrier phases or channel delays are present between users, orthogonal spreading sequence may be used to achieve multiple access. On the other hand, for asynchronous environment a classical choice is that of pseudo-random codes [4], where the spreading sequences are chosen to approximate independent and identically distributed sequences. A discussion about generators able to approach these sequences is in Chapter 4. The main motivation to select pseudo-random sequences is that when no information on channel delays and phases is known, i.e., when all is random, random spreading sequences may be intuitively thought to give the best performance.

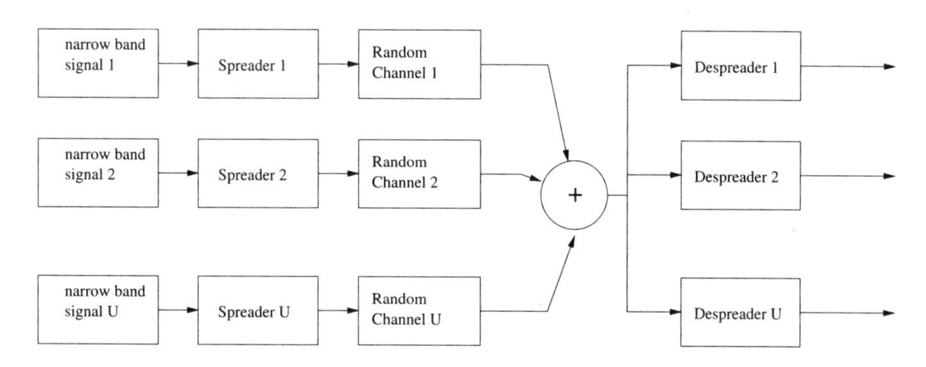

Figure 2.5: Asynchronous DS-CDMA systems.

Though it will be proved in Chapter 3 that this is not in general true, pseudo-random sequences remain in general an appealing choice and are often used as a reference case.

Let us also note that in synchronous environment the number of orthogonal codes (which have to be a basis of the sequence vector space) is limited by the sequence space dimension, i.e., by the spreading factor.

On the other hand, the stronger constraint of pseudo-randomness needed to generate sequences for asynchronous systems may suggest the adoption of shift-register based sequences such that the m-sequences [4, 6] whose number is quite limited even for relatively large spreading factors.

When pseudo-randomness is slightly relaxed, more choices are available even if we stick to shift-register generation. The construction of Gold sequences [7, 8] is one of these choices but a more general approach is also available.

In fact, we may think of at least two different solutions: the so called *short sequences* and *long sequences*. The two approaches are distinguished by the relationship between the sequence period P (which is related to the length of the generating shift register) and the spreading factor N. Short sequences are those for which $N = P$, long sequences are those for which $N > P$.

2.3.1 Short and Long Sequences

To understand the relationship between short and long sequence [9, 10] we introduce the simple performance index which accounts for the interfering on useful power ratio per interfering users, $\mathcal{R}$.

Let $S_k^u \in \{-1, 1\}$ be the k-th binary information symbols of the u-th user, $Y_k^u \in \{-1, 1\}$ be the binary k-th chip of the SS associated to the u-th user.

We will consider pseudo-random sequences of period P which are multiplied by a chip waveform $g(t)$ which is assumed to be rectangular of unit amplitude and extension equal to the chip time, T_{chip}. The thermal noise is neglected by assuming that the main cause of performance degradation is the interference. The system serves U users whose power is assumed to be equalized at the receiver.

If the information source is ergodic we may analyze, without loss of generality, the recovery of bit S_0^v. At the receiver, the integrate-and-dump correlates the incoming signal and its local replica of the v-th SS, in the interval $[0, NT_{\text{chip}}]$. The contribution of the u-th transmitter to the output of the v-th receiver is

$$\mathcal{O}^{uv} = \sum_k \sum_j \sum_n S_k^u Y_j^u Y_n^v \mathcal{F}(\tau^{uv}, k, j, n) \cos \theta^{uv}$$

where τ^{uv} and θ^{uv} are the random channel delay and phase, modeled as uniform random variables in $[0, NT_{\text{chip}}]$ and $[0, 2\pi]$, respectively. The function $\mathcal{F}$ is defined as:

$$\begin{aligned}
\mathcal{F}(\tau, k, j, n) &= \sum_{i=0}^{N-1} \delta_{kN+i,j} \int_0^T g(t - \tau - jT) g(t - nT) dt \\
&= \sum_{i=0}^{N-1} \delta_{kN+i,j} \sum_{w=0}^{N-1} \delta_{n,w} \int_{wT}^{(w+1)T} g(t - \tau - jT) dt
\end{aligned}$$

where $\delta_{\cdot,\cdot}$ is equal to one if its indexes are equal and zero otherwise. Note how the definition of the index w limits the range of the index n for which $\mathcal{F}$ is non-vanishing. By splitting the delay τ such that $\tau = (D + M)T$ with $D \in [0, 1[$ and M integer one obtains

$$\mathcal{F}(\tau, k, j, n) = T[(1 - D)\delta_{n,j+M} + D\delta_{n-1,j+M}] \sum_{i=0}^{N-1} \delta_{kN+i,j}$$

As we are looking for interfering power terms, the key point in the performance evaluation is the computation of $E[(\mathcal{O}^{uv})^2]$ for $u \neq v$. To this aim, assume that bits, chips, delays, and phases are independent. A first consequence of this is that as long as $u \neq v$ the value of $E[(\mathcal{O}^{uv})^2]$ does not depend on the particular transmitter-receiver pair. Hence, we may compute the expectation as a function of M and D to obtain

$$\begin{aligned}
&E[(\mathcal{O}^{uv})^2] = \\
&= \frac{T_{\text{chip}}^2}{2} \sum_k \sum_j \sum_{j'} \sum_{n=0}^{N-1} \sum_{n'=0}^{N-1} E[Y_j^u Y_{j'}^u] E[Y_n^v Y_{n'}^v] \sum_{i=0}^{N-1} \sum_{i'=0}^{N-1} \delta_{kN+i,j} \delta_{kN+i',j'} \\
&\quad \frac{1}{N} \int_0^1 \sum_{M=0}^{N-1} [(1 - D)^2 \delta_{n,j+M} \delta_{n',j'+M} + D^2 \delta_{n-1,j+M} \delta_{n'-1,j'+M} \\
&\quad + D(1 - D)\delta_{n,j+M} \delta_{n'-1,j'+M} + D(1 - D)\delta_{n-1,j+M} \delta_{n',j'+M}] dD
\end{aligned}$$

By taking into account the spreading code periodicity $E[Y_j^u Y_{j'}^u] = \sum_\rho \delta_{j,j'+\rho P}$, we have

$$E[(\mathcal{O}^{uv})^2] =$$

$$= \frac{T_{\text{chip}}^2}{6N} \sum_k \sum_j \sum_{j'} \sum_{n=0}^{N-1} \sum_{n'=0}^{N-1} \sum_\rho \sum_{\rho'} \delta_{j,j'+\rho P} \delta_{n,n'+\rho' P} \sum_{i=0}^{N-1} \sum_{i'=0}^{N-1} \delta_{kN+i,j} \delta_{kN+i',j'}$$

$$\sum_{M=0}^{N-1} \delta_{n,j+M} \delta_{n',j'+M} + \delta_{n-1,j+M} \delta_{n'-1,j'+M} + \frac{1}{2} \delta_{n,j+M} \delta_{n'-1,j'+M} + \frac{1}{2} \delta_{n-1,j+M} \delta_{n',j'+M}$$

Due to the presence of the term $\delta_{kN+i,j}\delta_{kN+i',j'}$, the two sums in the indexes j and j' give no contribution but for one pair of indexes. Furthermore, terms with coefficient $1/2$ vanish for all indexes combinations when $P \neq 1$. Hence, the power term can be rewritten as

$$\begin{aligned} E[(\mathcal{O}^{uv})^2] & \\ = & \frac{T_{\text{chip}}^2}{3N} \sum_k \sum_{v=0}^{N-1} \sum_{v'=0}^{N-1} \sum_{n=0}^{N-1} \sum_{n'=0}^{N-1} \sum_\rho \sum_{\rho'} \delta_{v,v'+\rho P} \delta_{n,n'+\rho' P} \delta_{n,k+v} \delta_{n',k+v'} \\ = & \frac{T_{\text{chip}}^2}{3N} \sum_k \sum_{a=\max[k,0]}^{\min[N-1+k,N-1]} \sum_{b=\max[k,0]}^{\min[N-1+k,N-1]} \sum_\rho \delta_{a,b+\rho P} \\ = & \frac{T_{\text{chip}}^2}{3N} \left[Q(N-1) + 2 \sum_{z=0}^{N-2} Q(z) \right] \end{aligned}$$

where in the second line the substitutions $a = i + k$ and $b = i' + k$ have been used, while the auxiliary function $Q(A) = \sum_{x=0}^{A} \sum_{y=0}^{A} \sum_\rho \delta_{x,y+\rho P}$ has been introduced in the last line.

Such an auxiliary function can be given a closed form expression considering for any integer A the decomposition $A = BP + C$, with B, C integer and $C < P$. With this we obtain that, as the indexes x and y in the definition of $Q(A)$ are such that $x - y \in \{-A, .., A\}$, then

$$Q(A) = \sum_{x=0}^{A} \sum_{y=0}^{A} \sum_{\rho=-B}^{B} \delta_{x,y+\rho P} = PB^2 + (C+1)(2B+1)$$

Now, let H be a further integer decomposed as in $H + 1 = ZP + G$, with Z, G integer and $G < P$. The term $\sum_{A=0}^{H} Q(A)$ can be also given a closed form expression

$$\begin{aligned}
\sum_{A=0}^{H} Q(A) &= \sum_{B=0}^{Z-1}\sum_{C=0}^{P-1} Q(BP+C) + \sum_{C=0}^{G-1} Q(ZP+C) \\
&= \frac{P^2}{6}(Z-1)Z(2Z-1) + Z^2\frac{P}{2}(1+P) + \\
& \quad PZ^2G + (2Z+1)\frac{G}{2}(1+G).
\end{aligned}$$

It should be emphasized that this relation is true for all H; in fact, when $H+1 < P$ we have $Z = 0$ and it is possible to check that the sum with index B gives a null contribution. With these expressions we may set $H = N-2$ and obtain

$$\begin{aligned}
& E[(\mathcal{O}^{uv})^2] = \\
&= \frac{T_{\text{chip}}^2}{3N}\left[Z^2P(2+P+2G) + (2Z+1)(G+1)^2 + \frac{P^2}{3}(Z-1)Z(2Z-1)\right]
\end{aligned}$$

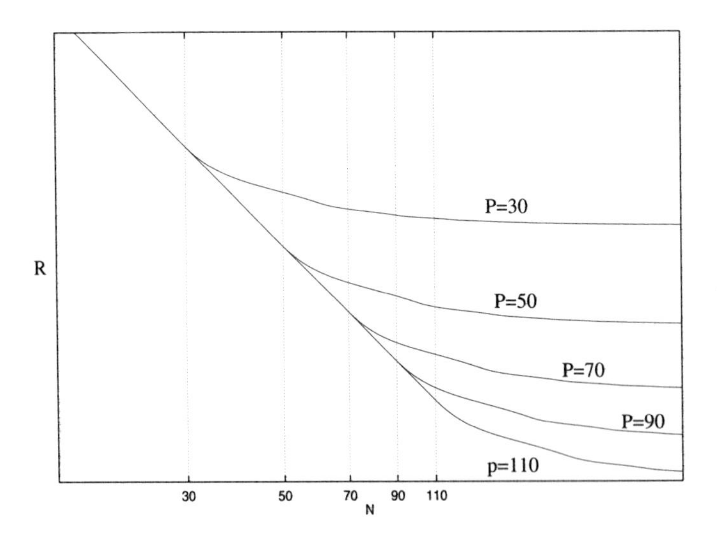

Figure 2.6: Performance index $\mathcal{R}$ versus spreading factor N by varying P.

As the receiver is synchronized with the link it has to decode, the useful term is such that $\theta_{vv} = 0$ and $\tau_{vv} = 0$, i.e., $\mathcal{O}^{vv} = NT_{\text{chip}}S_0^v$. The useful power is therefore $(\mathcal{O}^{vv})^2 = N^2T_{\text{chip}}^2$. Recalling the definition of $\mathcal{R}$ we finally obtain [10]

$$\begin{aligned}
& \mathcal{R} = \frac{E[(\mathcal{O}^{uv})^2]}{(\mathcal{O}^{vv})^2/2} = \\
&= \frac{4N^3 + 2NP^2 - 4G^3 - 12G^2 + 6G^2P + 12GP - 2G^2P - 12G - 2P^2 + 6P - 4}{9PN^3}
\end{aligned}$$

where $G = N - 1 - \left\lfloor \frac{N-1}{P} \right\rfloor P$.

This closed form emphasizes two key points: first $\mathcal{R} = 2/(3N)$ for $N \leq P$, then $\lim_{N\to\infty} \mathcal{R} = 4/(9P) \neq 0$. Hence, performance increases linearly with the spreading factor for short sequences but saturated to a level depending on the period of the sequences.

The performance index $\mathcal{R}$ is reported in Figure 2.6 as a function of N for different values of P.

From the saturating behavior we get that it is better to choose $N = P$ to maximize performance without wasting bandwidth. This is actually the case that will be discussed in Chapter 3 in which the pseudo-random behavior and $\mathcal{R} = 2/(3N)$ will be taken as a reference case.

The total signal to interference ratio (SIR), in considering the U users in the system, is

$$\rho = \frac{1}{\mathcal{R}(U-1)}$$

2.3.2 Chip Waveform

In this subsection we relax the assumption that $g(t)$ is rectangular and analyze the effect of different chip waveforms on the performance when short pseudo-random SSs are adopted and, again, perfect power equalization is assumed [11–13].

By defining $\hat{g}(t) = \sum_m g(t - mNT_{\text{chip}})$ as the periodic repetition of the chip waveform g, with period NT_{chip}, the component due to the useful bit is

$$\mathcal{O}^{vv} = \sum_k S_k^v \sum_{j=0}^{N-1} \sum_{n=0}^{N-1} Y_j^v Y_n^v \mathcal{F}(0, k, j, n)$$

where the function $\mathcal{F}$ must now be generalized as in

$$\mathcal{F}(\tau, k, j, n) = \int_0^{NT_{\text{chip}}} g(t - \tau - jT_{\text{chip}} - kNT_{\text{chip}})\hat{g}(t - nT_{\text{chip}})dt$$

The sum in the expression of $\mathcal{O}^{vv}$ can be decomposed to obtain that the useful component is made of three contributions

$$
\begin{aligned}
\mathcal{O}_1^{vv} &= S_0^v \sum_{j=0}^{N-1} \mathcal{F}(0,0,j,j) \\
\mathcal{O}_2^{vv} &= S_0^v \sum_{j=0}^{N-1} \sum_{\substack{n=0 \\ n \neq v}}^{N-1} Y_j^v Y_n^v \mathcal{F}(0,0,j,n) \\
\mathcal{O}_3^{vv} &= \sum_{k \neq 0} S_k^v \sum_{j=0}^{N-1} \sum_{n=0}^{N-1} Y_j^v Y_n^v \mathcal{F}(0,k,j,n)
\end{aligned}
$$

where $\mathcal{O}_1^{vv}$ is the term not taking into account inter-chip interference, $\mathcal{O}_2^{vv}$ is the term due to interfering chips around S_0^v, and $\mathcal{O}_3^{vv}$ accounts for the interference from other bits.

As far as the interference due to other users is concerned, if the u-th interfering user $u \neq v$ is considered we have

$$
\mathcal{O}^{uv} = \sum_k S_k^u \sum_{j=0}^{N-1} \sum_{n=0}^{N-1} Y_j^u Y_n^v \mathcal{F}(\tau^{uv}, k, j, n) \cos\theta^{uv}
$$

The total signal to interference ratio (SIR) can be computed by considering the power of $\mathcal{O}_1^{vv}$ with respect to the sum of the second-order moments of the terms $\mathcal{O}_2^{vv}$, $\mathcal{O}_3^{vv}$ and $\mathcal{O}^{uv}$, which can be considered independent and zero mean disturbances.

For $E[(\mathcal{O}_2^{vv})^2]$ we have

$$
\begin{aligned}
E[(\mathcal{O}_2^{vv})^2] &= \sum_{j=0}^{N-1} \sum_{\substack{n=0 \\ n \neq j}}^{N-1} \sum_{\tilde{\jmath}=0}^{N-1} \sum_{\substack{\tilde{n}=0 \\ \tilde{m} \neq \tilde{\jmath}}}^{N-1} E[Y_j^v Y_n^v Y_{\tilde{\jmath}}^v Y_{\tilde{n}}^v] \mathcal{F}(0,0,j,n) \mathcal{F}(0,0,\tilde{\jmath},\tilde{n}) \\
&= \sum_{j=0}^{N-1} \sum_{\substack{n=0 \\ n \neq j}}^{N-1} \mathcal{F}^2(0,0,j,n) + \mathcal{F}(0,0,j,n)\mathcal{F}(0,0,n,j)
\end{aligned}
$$

where it has been observed that $E[Y_j^v Y_n^v Y_{\tilde{\jmath}}^v Y_{\tilde{n}}^v] = \delta_{jn}\delta_{\tilde{\jmath}\tilde{n}}\overline{\delta}_{jn\tilde{\jmath}\tilde{n}} + \delta_{j\tilde{\jmath}}\delta_{n\tilde{n}}\overline{\delta}_{jn\tilde{\jmath}\tilde{n}} + \delta_{j\tilde{n}}\delta_{n\tilde{\jmath}}\overline{\delta}_{jn\tilde{\jmath}\tilde{n}} + \delta_{jn\tilde{\jmath}\tilde{n}}$ where $\overline{\delta}_{\cdot,\cdot} = 1 - \delta_{\cdot,\cdot}$.

The second-order moment of $\mathcal{O}_3^{vv}$ is

$$
\begin{aligned}
&E[(\mathcal{O}_3^{vv})^2] = \\
&= \sum_{k\neq 0}\sum_{\tilde{k}\neq 0} E[S_0^v S_{\tilde{k}}^v] \sum_{j=0}^{N-1}\sum_{n=0}^{N-1}\sum_{\tilde{j}=0}^{N-1}\sum_{\tilde{n}=0}^{N-1} E[Y_j^v Y_n^v Y_{\tilde{j}}^v Y_{\tilde{n}}^v]\mathcal{F}(0,k,n)\mathcal{F}(0,\tilde{k},\tilde{j},\tilde{n}) \\
&= \sum_{k\neq 0}\sum_{j=0}^{N-1} \mathcal{F}^2(0,k,j,j) \times \\
&\quad \sum_{\substack{n=0\\ n\neq j}}^{N-1} \mathcal{F}(0,k,j,j)\mathcal{F}(0,k,n,n) + \mathcal{F}^2(0,k,j,n) + \mathcal{F}(0,k,j,n)\mathcal{F}(0,k,n,j)
\end{aligned}
$$

where the incorrelation of the information bits $E[S_k^v S_{\tilde{k}}^v] = \delta_{k\tilde{k}}$ has been used.

The cross-interference term has power

$$
\begin{aligned}
&E[(\mathcal{O}^{uv})^2] = \\
&= \sum_{k}\sum_{\tilde{k}} E[S_k^u S_{\tilde{k}}^u] \sum_{j=0}^{N-1}\sum_{n=0}^{N-1}\sum_{\tilde{j}=0}^{N-1}\sum_{\tilde{n}=0}^{N-1} E[Y_j^u Y_n^v Y_{\tilde{j}}^u Y_{\tilde{n}}^v] \times \\
&\qquad E[\mathcal{F}(\tau^{uv},k,j,n)\mathcal{F}(\tau^{uv},\tilde{k},\tilde{j},\tilde{n})]E[\cos^2\theta^{uv}] \\
&= \frac{1}{2}\sum_{k}\sum_{j=0}^{N-1}\sum_{n=0}^{N-1} E[\mathcal{F}^2(\tau^{uv},k,j,n)] = \frac{1}{2T_{\text{chip}}}\sum_{n=0}^{N-1}\int_{-\infty}^{\infty}\mathcal{F}^2(t,0,0,n)dt
\end{aligned}
$$

where in order to select the non-zero components, the positions $j = \tilde{j}$ and $m = \tilde{m}$ have been considered.

Finally, the useful power results

$$
(\mathcal{O}_1^{vv})^2 = \left[\sum_{j=0}^{N-1} \mathcal{F}(0,0,j,j)\right]^2
$$

With these expressions, the system performance can be estimated defining a self-interference and cross-interference power as in

$$
\begin{aligned}
\mathcal{R}_a &= \frac{E[(\mathcal{O}_2^{vv})^2] + E[(\mathcal{O}_3^{vv})^2]}{(\mathcal{O}_1^{vv})^2/2} \\
\mathcal{R} &= \frac{E[(\mathcal{O}^{uv})^2]}{(\mathcal{O}_1^{vv})^2/2}
\end{aligned}
$$

and setting the total SIR as in

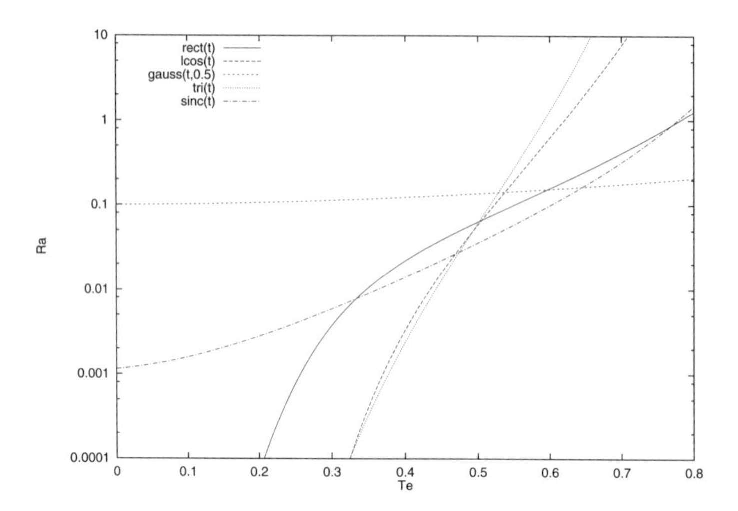

Figure 2.7: $\mathcal{R}_a$ as function of timing error T_e by varying the spreading and despreading waveforms.

$$\rho = \frac{1}{\mathcal{R}(U-1) + \mathcal{R}_a}$$

The analytical considerations developed so far can be concretized considering the waveforms

$$\text{rect}(t) = \begin{cases} 1 & 0 \le t \le 1 \\ 0 & \text{otherwise} \end{cases} \qquad \text{tri}(t) = \begin{cases} 4t & 0 \le t \le 1/2 \\ 4(1-t) & 1/2 < t \le 1 \\ 0 & \text{otherwise} \end{cases}$$

$$\text{gauss}(t, \sigma^2) = \frac{1}{\sqrt{2\pi\sigma^2}} e^{-\frac{(t-1/2)^2}{2\sigma^2}} \qquad \text{sinc}(t) = \frac{\sin(\pi(t-1/2))}{\pi(t-1/2)}$$

$$\text{lcos}(t) = \begin{cases} \frac{2}{\pi}\cos(t) & 0 \le t \le 1 \\ 0 & \text{otherwise} \end{cases}$$

and evaluate the performance as a function of error in the chip timing synchronization, T_e, that plays the role of a time offset in the transmitting waveform. As an example the two merit factors are reported in Figures 2.7 and 2.8 for $N = 31$.

2.3.3 Near Far Problem, Soft Degradation and Voice Activity Factor

In the previous subsections we assumed that all the received powers are the same [14]. This assumption is far from being trivially satisfied in real systems

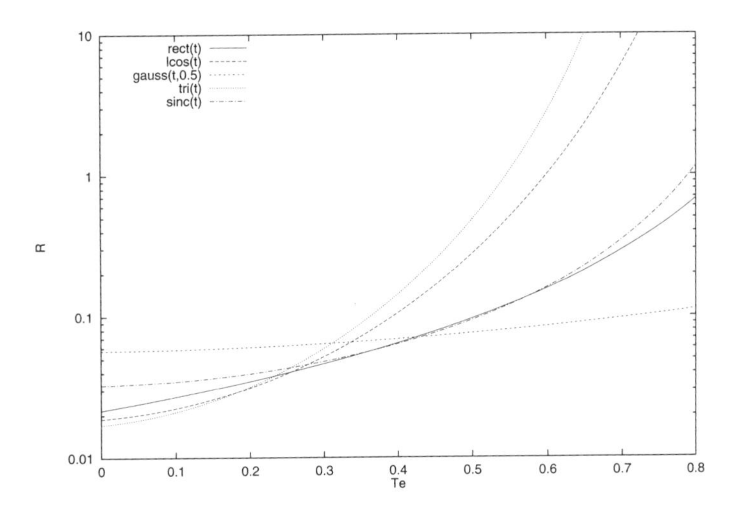

Figure 2.8: $\mathcal{R}$ as function of timing error T_e by varying the spreading and despreading waveforms.

as it depends on the implementation of a feedback from the receiver to the transmitter.

A more realistic model would take into account the different amplitudes V^{uv} with which the signal from the u-th user arrives at the v-th receiver. With this, the expression of the total SIR at the v-th receiver may be rearranged as follows

$$\rho^v = \frac{(V^{vv}\mathcal{O}^{vv})^2/2}{\sum_{\substack{u=1\\u\neq v}}^{U}(V^{uv})^2 E[(\mathcal{O}^{uv})^2]} = \frac{1}{\mathcal{R}}\frac{(V_{vv})^2}{\sum_{\substack{u=1\\u\neq v}}^{U}(V^{uv})^2}$$

If all the users but the w-th feature the same received power while the w-th features a received power different by a factor η, it follows that:

$$\begin{aligned} \rho^w &= \frac{\eta}{\mathcal{R}(U-1)} \\ \rho^{v\neq w} &= \frac{1}{\mathcal{R}(U-2+\eta)} \end{aligned}$$

where if $\eta > 1$ the w-th user benefits from an improved SIR but all the other $U-1$ suffer from an increased interference and thus experience a degraded performance. For these $U-1$ users an increase in η is equivalent to adding more users in the system. This phenomenon is called *Near Far Problem*, as terminals close to the receiver, if not perfectly compensated in power, blind the receiver with respect to the far ones.

If the set of SS is large enough it is possible to continue to add users without any drastic failure but a smooth decrease of the experienced SIR. This behavior is very different with respect to classical TDMA and FDMA systems, where, when all slots (in time or frequency) are busy, no other user may access the system. This characteristic is called *soft degradation* and it is a very attractive feature because it gives to the provider the possibility to accommodate more users when particular congestion occurs.

As a final remark, note how the DS-CDMA approach allows an easy integration between the access policy and the source characterization. As an example, we may consider a vocal source whose output may be characterized by a mixture of activity and idle periods. The ratio between the active and frame times is called *Voice Activity Factor* Ω. A strategy to reduce the power consumption of the transmission devices is to use a codec that transmits only when the source is active. No *a priori* agreement is needed to incorporate this in a DS-CDMA system which will automatically benefit from the non-continuous transmission also in term of improved performance. In fact, the average total SIR becomes

$$\rho = \frac{1}{\Omega \mathcal{R}(U-1)}$$

which can be significantly better than the normal one as some measurements show that Ω can be as low as 3/8.

2.3.4 From SIR to Bit Error Probability: The Standard Gaussian Approximation

The quality of a digital communication system can be often evaluated by considering the bit error probability as merit figure. Usually this evaluation is performed by using the so-called *Standard Gaussian Approximation* (SGA) [15] [16] [17] .

It is well known that, if we consider a large number of independent random variables, then by the Central Limit Theorem their sum is a random variable whose density function approaches a Gaussian. Obviously, this convergence increases with the number of random variables involved and depends on the starting density functions of the variable and on their relative absolute ratio.

In DS-CDMA systems with ten users or so, the total interference can be easily considered Gaussian due to the presence of the independent information symbols S_k^u.

The interference is then characterized by a variance that depends on the channel parameters (delays and phases) and on the particular user selected for demodulation.

If the channel is not stationary, e.g., when a mobile scenario is taken into account, the delays and phases may be fast varying, with a speed of the same order of information symbols. In this case these random variables increase the signal randomness and increase the quality of the approximation.

In this case the interference depends only on the useful user as channel variations are averaged in the variance computation. This explicit dependence from the decoded user may be finally eliminated if a power control is assumed and all the spreading sequences have the same statistical behaviors.

All these hypotheses are usually reasonable when analyzing DS-CDMA systems in asynchronous environments. Hence, the bit error probability may be directly evaluated from the SIR recalling standard results on the BPSK modulation and writing

$$P_{\text{err}} = \frac{1}{2}\text{erfc}\sqrt{\rho}$$

2.4 Synchronization

In DS-CDMA receivers three different synchronization levels must be taken into account: the carrier recovery for coherent demodulation, the chip time recovery and the spreading sequence phase alignment for correct signal despreading.

Usually the synchronization procedure is divided into two steps: first the *acquisition* and then the *tracking* [4]. Even if a joint procedure for simultaneous synchronization at several levels may be possible, especially when tracking slow changes in the timing of the received signal, here we concentrate on the recovery of the phase of SSs which is the one that is peculiar to DS-CDMA systems. Furthermore, to simplify the description, tracking is considered as a restart of the acquisition procedure when synchronization is lost [18]. Two basic schemes may be used.

2.4.1 Serial Search

The incoming signal is despread using a local replica of the SS with a random starting phase. If we assume chip synchronization the relative phase between the incoming and local SS is an integer number of chip times.

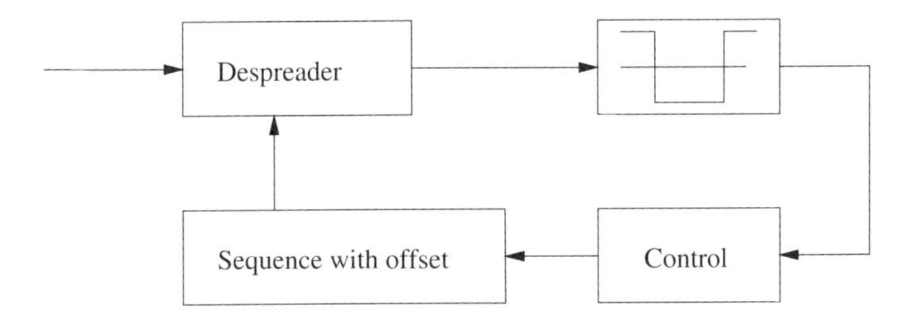

Figure 2.9: Serial Search Synchronizer.

A threshold detector is inserted as in Figure 2.9. Such a detector is triggered by outputs of the correlator which are large enough in absolute value. This allows the system to look for the peak of spreading sequence autocorrelation independently of the transmitted symbol. If the output of the correlator is found to be large enough for a certain number of subsequent symbol durations, then synchronization is claimed, otherwise another shift of the local SS is tried. Synchronization may be revoked if the magnitude of the output of the correlator falls below threshold for a certain number of symbol durations. In this case the search for a new shift is started.

This algorithm is normally used for short SSs. When long SSs are employed, which may exhibit more than one autocorrelation peak, synchronization may be assisted by a signaling channel over which a short sequence is transmitted.

Two main causes of error affect the serial synchronization procedure: *false alarms* and *detection missing.* They both depend on the noise and interference contributions at the output of the correlator.

A false alarm occurs when the correlator output is above threshold even if synchronization is not achieved. A detection missing occurs when the correlator output is below threshold even if the local SS is synchronized with the incoming one.

If all the possible shifts are tried in circular order, a false alarm simply delays synchronization by one information symbol time while missing a detection may mean retrying all the possible shifts.

As new correlator outputs are available at a rate of one every NT_{chip}, the average time to achieve synchronization starting by a random point is $N^2T_{\text{chip}}/2$, when the search proceeds without false alarms or detection missing events. In order to reduce this time it is possible to search subsequence of length $\alpha < N$ instead of operating on the complete sequence [19]. This approach reduces the time to obtain a sample from NT to αT, but increases the possibility of both false alarms and detection missing.

2.4.2 Parallel Search

The second acquisition scheme is called *Parallel Search*, where N parallel despreaders are fed with N local replicas of the SS with all the possible offsets. The offset corresponding to the maximum output is assumed to be the synchronization offset. In this way the search time is reduced by a factor N but circuit complexity is increased by the same factor. The scheme of a parallel search is reported in Figure 2.10.

Note how both schemes rely on the detection of the peak of autocorrelation of the SS which is hidden by interference and noise. Hence, it may seem desirable to have autocorrelation peaks as high as possible. Actually, this behavior is in contrast with the need of having low crosscorrelation to obtain good performance in terms of bit error probability. A tradeoff between these two different trends may be found on the Welch bound.

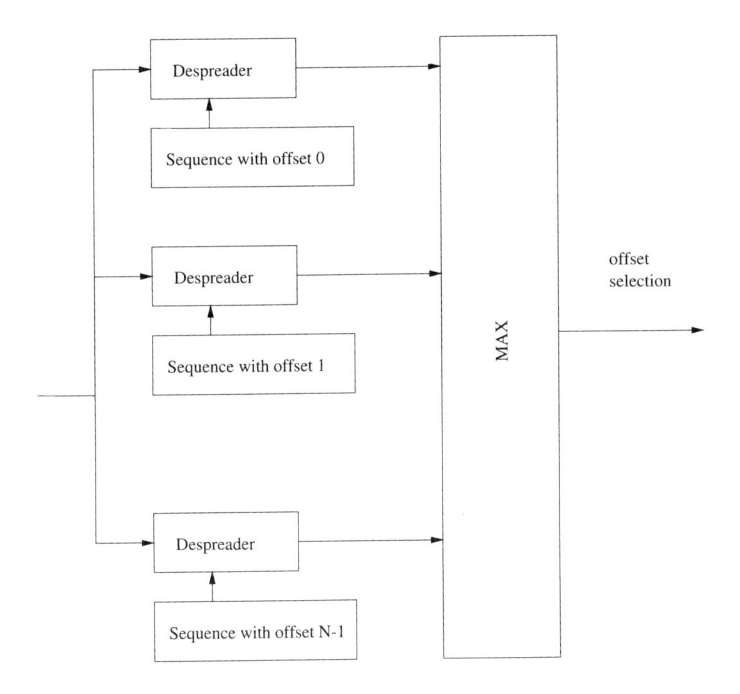

Figure 2.10: Parallel Search Synchronizer.

2.5 Advanced Topics on DS-CDMA

The last decade of investigations on DS-CDMA has focused on increasing performance and thus system capacity in terms of users. Most efforts have been spent along three directions: spreading sequence design, multiuser detection and multipath rejection [20].

Chapter 4 is devoted to spreading sequence design by means of chaotic systems for non-selective and selective channels and no further detail will be given here on that topic. Multiuser detection, on the contrary, will not be addressed and the main related issues will be sketched in the following for the sake of completeness.

The basic idea of optimal decoding for multiple access systems with interference starts with the work of Verdù [21] and stems from the fact that in many systems all the receivers are in the same physical position (think of a base station of a cellular system which receives all the uplinks) and thus cooperation between receivers can improve overall performance.

Ideally, the signal received by the multiuser receiver is a waveform depending on all the transmitted bits. Hence, a set of waveforms could be constructed and optimal decoding strategies can be adopted to match every element in the set with the incoming signal to obtain the best joint estimation of all the information.

Though this can be done by resorting to Viterbi-like implementation of matched filters, the global decoding algorithm features an exponential complexity which makes it unrealistic for a reasonable number of users.

Hence, most of the research stimulated by the work of Verdù has concentrated on the development of sub-optimal multiuser decoding procedures.

Many possible schemes have been proposed. Here, two basic examples are briefly reviewed: the serial and the parallel canceller. Due to the intrinsic complexity of the topic, discussion will concentrate on key issues and neglect technical subtleties as, for example, the problem of signal re-timing after cancellation in each stage, which may dominate practical implementations.

2.5.1 Serial Canceller

Serial joint-detection systems or *serial cancellers* attempt to first decode the more reliable users, i.e., the users with higher received power. They then use these decoded bits to cancel the contribution of these users from the global signal. This procedure is iteratively applied to all users up to the weakest received. With this, near-far resistance is intrinsic in the system as it is in all the multiuser detectors, which, in principle, relaxes the requirement of a strong power control.

A scheme of the serial canceller is reported in Figure 2.11.

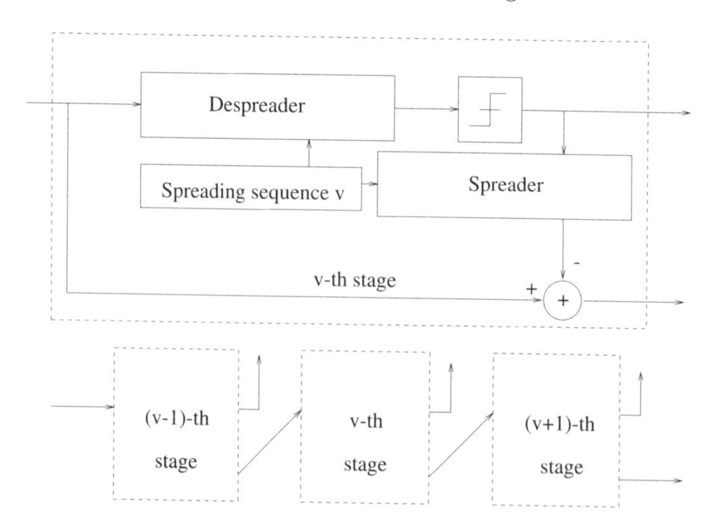

Figure 2.11: Cancellation stage scheme.

To briefly analyze this system [22], let us first recall that the performance of a conventional system, which assumes $V^{v1} = V^{v2} = \cdots = V^{vU} = V^v$ and in which we also introduce a thermal noise with spectral density N_o, depends on a total signal to noise plus interference ratio (SNIR) defined as

$$\rho^v = \frac{W^v}{1 + \mathcal{R}\sum_{\substack{u=1 \\ u\neq v}}^{U} W^u}$$

where $W^v = (V^v)^2 N T_{\text{chip}}/2N_o$ is the signal to noise ration (SNR) of the system.

Refer now to Figure 2.11 and note that, obviously, only the contributions from non-cancelled users affect the decoding of each user. Consider now the v-th stage which is in charge of the decoding the v-th user. If the bit in the u-th stage, with $u < v$, has been correctly recovered, then the signal due to the u-th user no longer affects the v-th stage and any of the following, giving an appreciable SNR improvement. On the contrary, if the bit in the u-th has been mistaken, the power of the interfering contribution of the u-th is multiplied by 4 as the amplitude of the sum of two in-phase signals doubles. With this, the SNIR at the v-th stage becomes

$$\rho^v = \frac{W^v}{1 + \mathcal{R}\left[4\sum_{u=1}^{v-1} P_{\text{err}}^u W^u + \sum_{u=v+1}^{U} W^u\right]}$$

where the term P_{err}^u is the expected bit error probability at the u-th stage. Under SGA with BPSK modulation we have have $P_{\text{err}}^v = 1/2 \operatorname{erfc}\sqrt{\rho^v}$.

Let us now assume to look for a system with equal bit error probability in each link, so that $P_{\text{err}}^v = P_{\text{err}}$ and $\rho^v = \rho$. With this the previous equation may be re-written as

$$W^v = \rho + 4\mathcal{R}P_{\text{err}}\rho \sum_{u=1}^{v-1} W^u + \mathcal{R}\rho \sum_{u=v+1}^{U} W^u$$

for all v. By expliciting the W^1 SNR, the generic W^v may be written as a function of W^u for $u = 1, \ldots, v-1$ and few algebraic manipulations allow unrolling of the recursion to obtain

$$\begin{aligned} W^1 &= \frac{\rho}{1 - \mathcal{R}\rho\frac{\Xi - \Xi^U}{1-\Xi}} \\ W^v &= \Xi^{v-1} W_1 \\ \Xi &= \frac{1 + 4P_{\text{err}}\mathcal{R}\rho}{1 + \mathcal{R}\rho} \end{aligned}$$

Note that, when no cancellation is present and ρ is equally given, we must have

$$W = \frac{\rho}{1 - \mathcal{R}\rho(U-1)}$$

which allows us to compare the serial canceller and the conventional receiver from the point of view of the total received power for a given communication quality.

The gain of the serial canceller, i.e., the ratio between the two overall received powers, is reported in Figure 2.12 for $U = 10$ as a function of the spreading factor for different desired SNIR ρ.

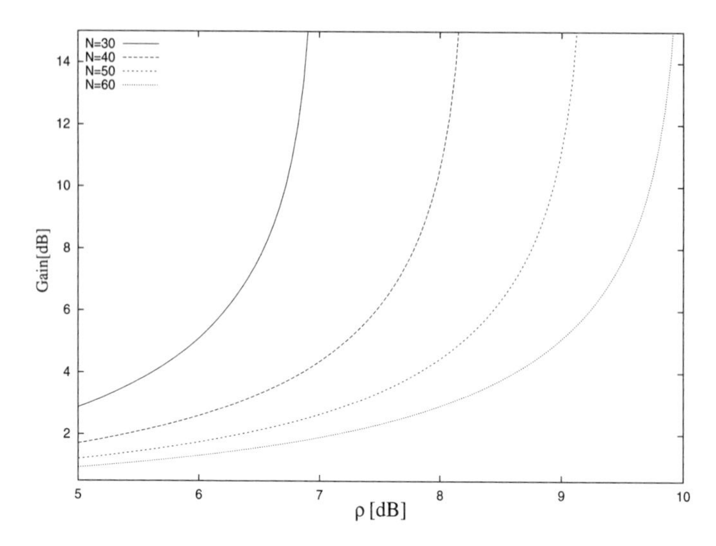

Figure 2.12: Total power gain as a function of the ρ.

2.5.2 Parallel Canceller

The parallel joint detection mechanism or *parallel canceller* proceeds with a multi-stage approach as reported in Figure 2.13. In the stage 0 all the signals are decoded by U parallel conventional receivers.

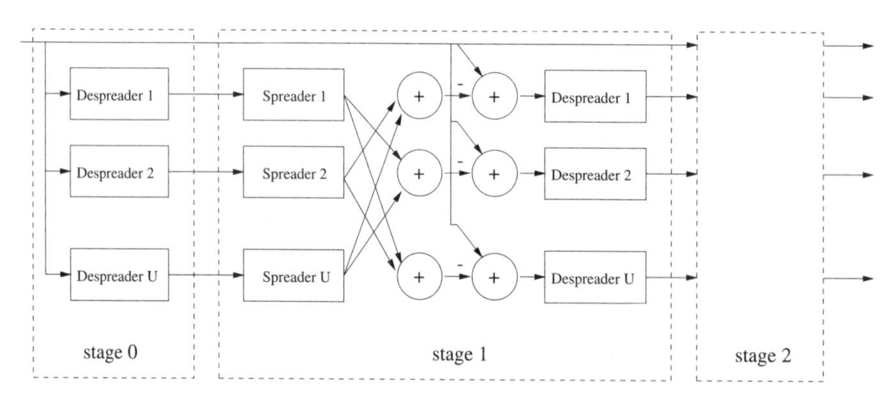

Figure 2.13: Parallel canceller.

In the subsequent stages the outputs estimated by the previous stage are re-modulated and re-spread. These signals are then composed so that the new v-th receiver is fed with a signal in which all the $u \neq v$ contributions are subtracted.

Several stages may be cascaded to increase the decoding reliability. At each stage the despreader may operate by decoding the bit (hard decision) or simply by propagating the output (soft decision) and leaving the bit decision to the last stage.

The system complexity is obviously increased with respect to a serial canceller. As far as the performance is concerned note that for a system with hard decision the SNIR may be computed following the same path used for the serial canceller. We obtain

$$\rho^v(0) = \frac{W^v}{1+\mathcal{R}\sum_{\substack{u=1\\u\neq v}}^{U} W^u}$$

$$\rho^v(s) = \frac{W^v}{1+4\mathcal{R}\sum_{\substack{u=1\\u\neq v}}^{U} P_{\text{err}}^u(s-1)W^u}$$

where the parenthesized index indicates the stage at which SNIR and error probability are measured.

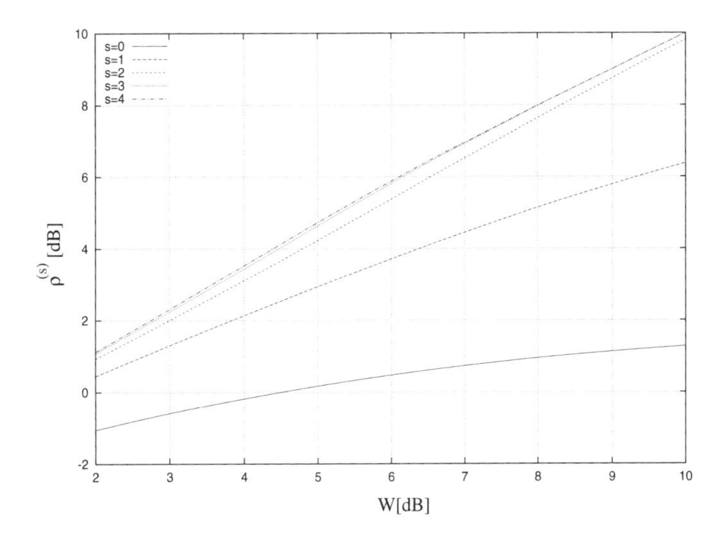

Figure 2.14: Parallel canceller performance.

If all the received powers are the same, we have $W^v = W$, $P_{\text{err}}^v(s-1) = P_{\text{err}}(s-1)$ and $\rho^v(s) = \rho(s)$ and the SNIR can be rewritten as

$$\rho(0) = \frac{W}{1+\mathcal{R}(U-1)W}$$

$$\rho(s) = \frac{W}{1+4\mathcal{R}(U-1)P_{\text{err}}(s-1)W}$$

which can be easily seen to converge to $\rho(s) = W$ when s and W are large enough. Hence, as thermal noise remains the only cause of error, interference is asymptotically cancelled. In Figure 2.14 the SNIR $\rho(s)$ is reported for $s =$

$0, \ldots, 4$ as a function of the SNR W for $U = N = 31$. It is interesting to observe that $\rho(3) \simeq \rho(4) \simeq W$ so that even with so many users competing for the channel three stages are sufficient to obtain complete interference cancellation.

References

[1] Scholtz, R. A., "The Origin of Spread Spectrum," *IEEE Transactions on Communications*, pp. 822-854, May 1982.

[2] Scholtz, R. A., "The Spread Spectrum Concept," *IEEE Transactions on Communications*, pp. 748-755, Aug. 1977.

[3] Dixon, R. C., "Spread Spectrum Systems," John Wiley & Sons, New York, 1976.

[4] Proakis, J. G., "Digital Communications," McGraw-Hill Int. Ed., New York, 1995.

[5] Lee, W. C. Y., "Overview of Cellular CDMA," *IEEE Transactions on Vehicular Technology*, pp. 291-302, May 1991.

[6] Golomb, S. W., "Shift Register Sequences," Holden-Day Inc., 1967.

[7] Gold, R., "Optimum Binary Sequences for Spread Spectrum Multiplexing," *IEEE Transactions on Information Theory*, pp. 619-621, Oct. 1967.

[8] Gold, R., "Maximal Recursive Sequences with 3-Value Recursive Crosscorrelation Functions," *IEEE Transactions on Information Theory*, pp. 154-156, Jan. 1968.

[9] Giubilei, R., "Code Period Effects in DS-CDMA Systems," *Electronics Letters*, 30 March 1995, pp. 521-522.

[10] Mazzini, G., "Closed Form Performance Evaluation for Code Period versus Spreading Factor on Asynchronous DS-CDMA Systems," Proceedings of IEEE ISSSTA'98, pp. 806-808, Sun City, South Africa.

[11] Asano, Y., Daido, Y., and Holtzman, J. M., "Performance Evaluation for Band-Limited DS-CDMA Communication System," Proceedings of IEEE VTC'93, pp. 464-468, Chicago.

[12] Ikegami, T. and Pursley, M. B., "The Effect of Transmitter Filtering on the Performance of Direct Sequence Spread Spectrum Multiple Access Systems," Proceedings of ISSTA'94, pp. 569-537, Houlu, Finland.

[13] Mazzini, G. and Volta, A., "On the Inter Chip Interference in DS-CDMA Systems," Proceedings of IEEE ISSSTA'96, pp. 702-707, Mainz, Germany.

[14] Mokhtar, M. A. and Gupta, S. C., "Power Control Considerations for DS/CDMA Personal Communication Systems," *IEEE Transactions on Vehicular Technology*, pp. 479-487, Nov. 1992.

[15] Pursley, M. B., "Performance Evaluation for Phase-Coded Spread-Spectrum Multiple-Access Communication-part I: System Analysis," *IEEE Transactions on Communications*, pp. 795-799, Aug. 1977.

[16] Morrow, R. K. and Lehnert, J. S., "Bit-to-Bit Error Dependence in Slotted DS/SSMA Packet System with Random Signature Sequences," *IEEE Transactions on Communications*, pp. 1052-1061, Oct. 1989.

[17] Mazzini, G., "DS-CDMA Systems using q-Level m Sequences: Coding Map Theory," *IEEE Transactions on Communications*, pp. 1304-1313, Oct. 1997.

[18] Mazzini, G., "Analytical formulation for the synchronization performance of Q-level M-sequences in DS-CDMA," Proceedings of IEEE CTMC GLOBECOM'94, pp. 44-50, San Francisco, CA.

[19] Wainberger, S. and Wolf, J. K., "Subsequences of Pseudorandom Sequences," *IEEE Transactions on Communication Technology*, pp. 606-612, Oct. 1970.

[20] Mazzini, G. and Tralli, V., "Performance Evaluation of DS-CDMA Systems on Multipath Propagation Channels with Multilevel Spreading Sequences and Rake Receivers," *IEEE Transactions on Communications*, pp. 244-257, Feb. 1998.

[21] Verdù, S., "Minimum Probability of Error for Asynchronous Gaussian Multiple-Access Channels," *IEEE Transactions on Information Theory*, pp. 85-96, Jan. 1986.

[22] Mazzini, G., "Equal BER with Successive Interference Cancellation DS-CDMA Systems on AWGN and Ricean Channels," Proceedings of IEEE PIMRC'95, pp. 727-731, Toronto, Canada.

Chapter 3

Chaos-Based Asynchronous DS-CDMA Systems

Gianluca Mazzini
DI
University of Ferrara
via Saragat 1, 44100 Ferrara, Italy
`g.mazzini@chaos.cc`

Riccardo Rovatti
DEIS
University of Bologna
viale Risorgimento 2, 40136 Bologna, Italy
`r.rovatti@chaos.cc`

Gianluca Setti
DI
University of Ferrara
via Saragat 1, 44100 Ferrara, Italy
`g.setti@chaos.cc`

3.1 Introduction[1]

In the last decade, increasing research efforts have been devoted to DS-CDMA as a possible multiple access transmission methodology able to achieve both high spectrum efficiency and relevant performance in several scenarios. For example, the IS95 supported by QUALCOMM is one of the first commercial standards based on DS-CDMA for cellular communication and this same technique has been recently assumed as the base for the next generation of cellular network in the UMTS and IMT2000 systems.

[1]This research has been partially funded by the EU through ESPRIT Project 31103, INSPECT.

This interest is actually due to some positive features intrinsic to DS-CDMA such as the ease of cell planning, soft degradation of performance, and ability to support the variable bit-rate access needed by multimedia applications.

DS-CDMA systems can also be applied in the wireless local area network to obey the constraint that ITU put on the use of the non-license ISM band. In fact, the standard prescribes that, to avoid interference with other services, every link should appear as noise from the point of view of the other links.

This chapter is devoted to the use of one-dimensional chaotic maps to generate the finite-length discrete-time sequence of symbols which is periodically repeated to obtain the infinite one needed in DS-CDMA systems to spread the spectrum of information signals. Such a system is indicated as a Chaos-Based DS-CDMA (CBDS-CDMA) and, apart from the introduction of different spreading sequences, is *identical* to the simplest conventional DS-CDMA system. As a by-product of this choice, any improvement due to chaos-based spreading can be zero-cost achieved in a conventional system by simply storing different sequences in the transmitter and receiver memories. Even if the basic idea can be dated back to [1], it is now a topic of thorough investigation as it has been proved that this may result in telecommunication systems outperforming the classical ones [2–6].

To perform a theoretical investigation of the expected performance of a CBDS-CDMA system, two parallel paths have been followed. On the one hand, a model of a DS-CDMA system has been developed isolating some performance indexes and linking them with some design-dependent quantities, i.e., the correlation properties of the spreading sequences. On the other hand, the class of maps adopted for spreading sequence generation has been thoroughly characterized to link the map structure to the correlation properties of the resulting sequences.

The starting point for the last development is the observation that a statistical approach may greatly benefit the study of causal systems that exhibit a chaotic behavior. In this case, in fact, critical dependence on initial condition prevents the study of single trajectories from giving information which is globally valid. On the contrary, considering a set of trajectories of non-vanishing measure allows tracking of the mixing mechanism causing the loss of information and characterizing it quantitatively.

This approach has recently found many promising applications in fields like communication [2, 3, 7], data encryption [8], noise generation [9] or suppression [10].

Yet, realistic, complex signal processing tasks involve blocks with memory in which the signals are non-linearly combined with their delayed versions. The higher the non-linearity of the block (e.g., the higher the number of significant terms in its multi-dimensional Taylor expansion), the higher the order of the signal correlations that are needed to compute even the simple variance of its output. Such a situation is common in real systems as non-linear equalization, noise filtering, correlation receivers coping with complex channel models, etc.

Hence, the topic of computing higher-order statistics of trajectories generated by a chaotic dynamics is emerging as one of the points to tackle to fully exploit the potentiality of chaos-based techniques in real applications.

Some pioneering work in this field can be traced back to [11–13]. In [12], non-quantized trajectories produced by piecewise-affine fully stretching maps are considered and any-order correlation function involving polynomial observables are given an analytical expression. In [13] a generalization to n-dimensional map is proposed, although the main aim is to address implementation issues.

The aim of this chapter is to show that by chaining the results of these two paths, one arrives at design criteria which, depending on the environment in which the communication system is set and on the performance index considered, allows the optimization of the CBDS-CDMA system so that it results in better than conventional DS-CDMA systems.

The chapter is structured as follows. First, the modeling of a classical DS-CDMA system correlation receiver when asynchronous multi-user access is considered over an exponentially vanishing multipath channel and in the case where selective fading can be neglected. In any case, to highlight the effect of different spreading strategies the receiver is kept to minimum complexity. The useful user transmitter and receiver are assumed to be synchronized and the channel parameters are assumed to vary fast enough that the average bit error probability is a meaningful performance index. The Standard Gaussian Assumption, which is widely employed in a similar investigation, is used here to relate such an index to the powers of some signals and thus to their correlation properties.

Section 3.3 contains the formal development of the mathematical tools needed to link the features of a chaotic map to the correlation properties of the signals it generates.

The starting point for this development is the statistical approach to dynamical system theory which has been specialized and enhanced to cover the phenomena of interest in this investigation.

A proper tensor formalism is introduced to cope with the complexity of some calculation. The fundamental results achieved and needed in the subsequent derivation are the analytical computation of any-order correlation terms of quantized chaotic trajectories generated by Piecewise Affine Markov (PWAM) maps and the establishment of a formal link between such correlation terms and the map design parameters whenever a particular family of function, called (n, t)-tailed shifts, is used for sequences generation.

Section 3.4 analyzes the case in which the selective effect of the channel is negligible, i.e., a non-multipath environment.

In this case, the general performance model can be specialized to show that the bit error probability depends only on the interference from the other users (cross-interference) and thus on the second-order auto-correlation properties of the spreading sequences through a simple and analytically tractable expression. Exploiting that expression it is proven that an auto-correlation profile exists that results in the minimum achievable bit error probability and thus that, in the non-multipath case, spreading sequences with that auto-correlation profile are the optimum choice. It is finally shown that (n, t)-tailed shifts produce spreading sequences that are so close to the optimum ones to give rise to a final performance which is indistinguishable from the global optimum for any realistic system.

For a given quality of the communication link (i.e., for a given level of bit error probability), the adoption of optimum sequences allows the allocation of $\approx$15% more users on the same bandwidth. As this is the expected gain over an infinite number of users, a trivial optimization procedure has been set up to explore the availability of finite sets of spreading sequences resulting in even better performance. The results of this optimization show that an increase of up to 20% in the number of users is feasible.

In Section 3.5, the more general case of multipath channel is considered, where an additional cause of error appears in the form of a self-interference between the useful synchronized signals and its delayed version. With this, auto-correlations of second as well as fourth order turn out to affect the final performance is that the correlation shaping optimization described in the previous section no longer applies. Yet, the mathematical tools developed support the analytical calculation of all the key quantities for the class of maps achieving optimum performance over non-selective channels.

With this, the trade off between cross-interference and self-interference can be addressed by numerically optimizing the analytical expression of the bit error probability. The results of such an optimization show that, when a given communication quality is set, CBDS-CDMA allow the allocation of a number of users which is from 5% to 15% higher than what is achieved with conventional spreading sequences. Maximum improvement is achieved over channels in which either multipath is almost negligible (as the optimum non-selective performance is retrieved) or a lot of secondary rays share the same non-negligible power. Lower improvement is achieved when only few (on the order of one or two) secondary rays carry most of the multipath power.

Finally, Section 3.6 reports some conclusions on our investigation.

3.2 Channels and System Model

In this section a baseband equivalent model of a DS-CDMA system is introduced when transmission is accomplished over a selective fading channel in which multipath propagation effects are taken into account. Furthermore, a simplified model for transmission over a non-selective channel is also derived as a particular, although significant, case of the previous one.

3.2.1 Transmission Over a Selective Fading Channel

The propagation medium is modeled as a linear time-varying multipath channel. Let us consider U users, the v-th being the useful one. The channel transmits the signal of the generic u-th user as described by the following low-pass impulse response

$$h^u(t) = \beta_0\delta(t) + \sum_{\tau=1}^{\infty} \beta_\tau a_\tau^u e^{-\mathbf{i}\,\phi_\tau^u}\delta\left(t - \tau\frac{T}{N}\right)$$

where $\mathbf{i}$ is the imaginary unit, $\delta(t)$ is the Dirac's generalized function, T is the information symbol duration, $\beta_\tau^2(a_\tau^u)^2$ is the instantaneous random power attenuation of the τ-th ray while ϕ_τ^u and $\tau T/N$ are its phase and delay respectively. The random nature of the secondary rays, assumed independent of each other, is modeled by the Rayleigh distributed real random variables a_τ^u which are such that $\mathbf{E}[(a_\tau^u)^2] = 1$ and by considering ϕ_τ^u as a further random variable uniformly distributed in $[0, 2\pi]$. As thoroughly discussed in [14], considering ray delays multiple of the chip time T/N leads to no loss of generality.

This channel model is usually referred to as a specular multipath channel model and is able to take into account the effects of a fixed dominant component. Even if a more general model may, for example, take into account that the first ray is composed by a fixed number of random components, we here adopt a slightly simplified model to ease analytical investigation and focus on the impact of chaos-based spreading.

Indicate with K the Rice factor, i.e., the ratio between the power of the first ray and the remaining rays, such that

$$\beta_0 = \sqrt{K\sum_{\tau=1}^{\infty}\beta_\tau^2}$$

The power attenuation will be considered exponential so that two constants A and B exist such that $\beta_\tau = Ae^{-B\tau}$ for $\tau \geq 1$. With this, to ensure that the power flowing through the channel is constant, we must set

$$\int_0^\infty \mathbf{E}[|h^u(t)|^2]dt = (1+K)\sum_{\tau=1}^{\infty}\beta_\tau^2 = 1 \tag{3.1}$$

so that A and β_0 are linked to B and to the Rice factor by

$$A^2 = \frac{e^{2B}-1}{1+K} \quad \beta_0^2 = \frac{K}{1+K}$$

The modeled selective fading channel adds up the contributions of all the users so that the baseband receiver is presented with a signal which is the sum of the convolutions of the signal transmitted by each user and the corresponding channel impulse response. All the random variables describing the channel contributions will be considered independent. The U sources producing the information symbols will also be assumed to be independent in time and from user to user.

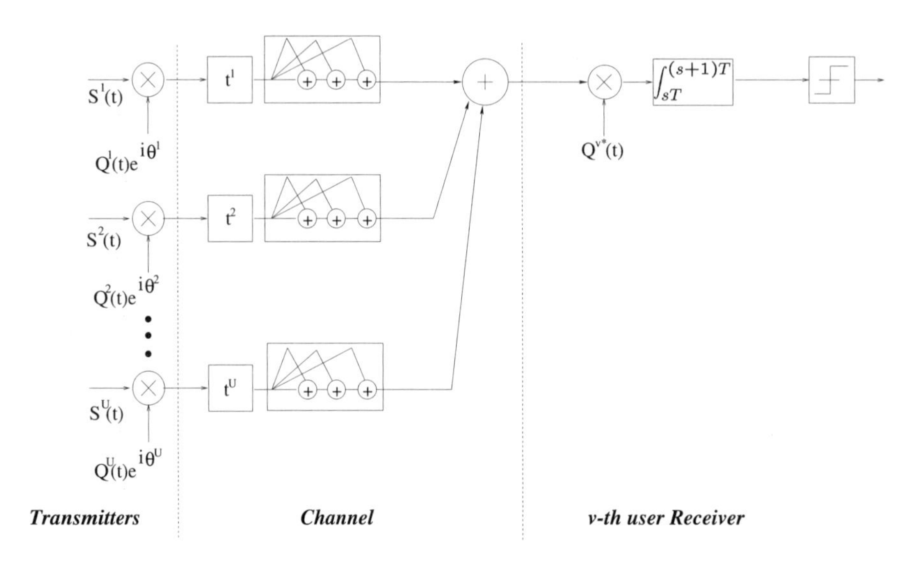

Figure 3.1: Lowpass equivalent of a DS-CDMA system with a multipath channel.

As far as the system model is concerned, a baseband equivalent of a DS-CDMA system is reported in Figure 3.1. The generic u-th users information signal is

$$S^u(t) = \sum_{s=-\infty}^{\infty} S_s^u g_T(t - sT)$$

where $S_s^u = \pm 1$ are the information symbols and $g_T(t)$ is the rectangular pulse which is 1 within $[0, T]$ and vanishes otherwise.

The spreading signal depends on sequences of symbols x_s^u from the alphabet X which are mapped into the set of the L complex roots of the unity by the function Q, and combined to form the complex signal

$$Q^u(t) = \sum_{s=-\infty}^{\infty} Q(x_s^u) g_{T/N}\left(t - s\frac{T}{N}\right)$$

The signal $Q^u(t)$ is then multiplied by $S^u(t)$ and transmitted along the (equivalent) channel together with the spread-spectrum signals from the other users.

The receiver is made of a multiplier which combines the incoming signal with a synchronized replica of the spreading sequence of the v-th user and of an integrate-and-dump stage in charge of extracting the information symbol by correlation. Even when thermal noise can be neglected, symbol recovery is hindered by the presence of the current and delayed signals of the other users and by the delayed version of the useful signal.

The system performance depends on how much these spurious signals affect the integrate-and-dump output, i.e., on how they are correlated with the spreading sequence of the useful user.

As correlation with a fixed sequence is a linear operation, error causes add at the integrate-and-dump output Υ_s^v which can be decomposed into three main terms:

- the useful component carrying the information to be retrieved, Ω_s^v;
- the disturbance due to the secondary rays carrying the previous symbols of the useful user, Ξ_s^v;
- the disturbance due to the primary and secondary rays carrying the information transmitted by the other users, Ψ_s^v.

These three terms are such that

$$\Upsilon_s^v = \Omega_s^v + \Xi_s^v + \Psi_s^v$$

To express each term we assume that a function $\Theta_s^{uv}(a, \Delta\theta, \Delta t)$ can be defined giving the contribution to the correlation with the v-th spreading sequence of a signal coming from the u-th users with amplitude a relative phase $\Delta\theta$ and relative delay Δt. With this we may write

$$\begin{aligned}
\Omega_s^v &= \frac{1}{2} S_s^v \beta_0 \\
\Xi_s^v &= \frac{1}{2} \sum_{\tau=1}^{\infty} \Theta_s^{vv} \left(\beta_\tau a_\tau^v, -\phi_\tau^v, \tau \frac{T}{N} \right) \\
\Psi_s^v &= \frac{1}{2} \sum_{u \neq v} \left[\Theta_s^{uv}(\beta_0, \theta^{uv}, t^{uv}) + \sum_{\tau=1}^{\infty} \Theta_s^{uv} \left(\beta_\tau a_\tau^u, \theta^{uv} - \phi_\tau^u, t^{uv} - \tau \frac{T}{N} \right) \right]
\end{aligned}$$

where S_s^u is the symbol transmitted by the u-th user at the s-th time step, θ^{uv} is the relative phase between the u-th and v-th users after demodulation and t^{uv} is the relative delay [2]. Assuming that the receiver is synchronized with the signal it is supposed to decode, we also have $\theta^{vv} = 0$ and $t^{vv} = 0$.

The quantity $\Theta_s^{uv}(a, \Delta\theta, \Delta t)$ can be given an expression depending on the two symbols from the u-th transmitter which are overlapping with the s-th symbol from the v-th one [2], i.e.,

$$\Theta_s^{uv}(a, \Delta\theta, \Delta t) = \frac{a}{N} \left\{ S_{s-\lfloor \Delta t/T \rfloor}^u \operatorname{Re}[e^{\mathrm{i}\,\Delta\theta} X^{uv}(\Delta t)] + S_{s-\lfloor \Delta t/T \rfloor - 1}^u \operatorname{Re}[e^{\mathrm{i}\,\Delta\theta} Y^{uv}(\Delta t)] \right\}$$

where $\lfloor \cdot \rfloor$ gives the largest integer not greater than its argument and the two functions $X^{uv}(\Delta t)$ and $Y^{uv}(\Delta t)$ account for the symbols overlapping. They are defined to be periodic of period T as further delay is simply accounted for skipping to the next couple of symbols. Their expression is

$$\begin{aligned} X^{uv}(\Delta t) &= X^{uv}(\Delta t + kT) = \left(1 - \frac{N\Delta t}{T} + \left\lfloor \frac{N\Delta t}{T} \right\rfloor\right) \Gamma_{N,\lfloor \frac{N\Delta t}{T} \rfloor}(\underline{x}^u, \underline{x}^v) \\ &+ \left(\frac{N\Delta t}{T} - \left\lfloor \frac{N\Delta t}{T} \right\rfloor\right) \Gamma_{N,\lfloor \frac{N\Delta t}{T} \rfloor + 1}(\underline{x}^u, \underline{x}^v)) \\ Y^{uv}(\Delta t) &= Y^{uv}(\Delta t + kT) = \left(1 - \frac{N\Delta t}{T} + \left\lfloor \frac{N\Delta t}{T} \right\rfloor\right) \Gamma_{N,\lfloor \frac{N\Delta t}{T} \rfloor - N}(\underline{x}^u, \underline{x}^v) \\ &+ \left(\frac{N\Delta t}{T} - \left\lfloor \frac{N\Delta t}{T} \right\rfloor\right) \Gamma_{N,\lfloor \frac{N\Delta t}{T} \rfloor + 1 - N}(\underline{x}^u, \underline{x}^v) \end{aligned} \quad (3.2)$$

where $\underline{x}^u$ is the periodic sequence of symbols generating the spreading sequence of the u-th user and the partial correlation function is defined as [2]

$$\Gamma_{N,\tau}(\underline{x}^u, \underline{x}^v) = \begin{cases} \sum_{k=0}^{N-\tau-1} Q(x_k^u) Q^*(x_{k+\tau}^v) & \tau = 0, 1, \ldots, N-1 \\ \Gamma_{N,-\tau}^*(\underline{x}^v, \underline{x}^u) & \tau = -1, -2, \ldots, -N+1 \\ 0 & |\tau| \geq N. \end{cases}$$

where $\cdot^*$ stands for complex conjugation.

3.2.2 System with Non-Selective Channel

In the case of non-selective channel, the channel response can be trivially rewritten as $h^u(t) = \delta(t)$, and the system model is simplified as reported in Figure 3.2. By assuming the same meaning and characterization, for the variables of Figure 3.2 with respect to the multipath fading channel case, the integrate-and-dump output Υ_s^v can be decomposed into the useful component Ω_s^v and the disturbance due to the other users, Ψ_s^v, so that

$$\Upsilon_s^v = \Omega_s^v + \Psi_s^v$$

which, by using the previously defined functions, can be expressed as

$$\Omega_s^v = \frac{1}{2} S_s^v \quad \Psi_s^v = \frac{1}{2} \sum_{u \neq v} \Theta_s^{uv}(1, \theta^{uv}, t^{uv})$$

3.3 Sequences Generation with Chaotic Maps

Generally speaking, any conceivable procedure for spreading sequence generation can be considered composed of at least two steps. First, one needs to generate a time-series $\underline{x} = \{x_k\}$ in a given domain X, then one must identify a suitable quantization function $Q : X \mapsto Z$ mapping each symbol x_k into one of the set Z of L complex symbols.

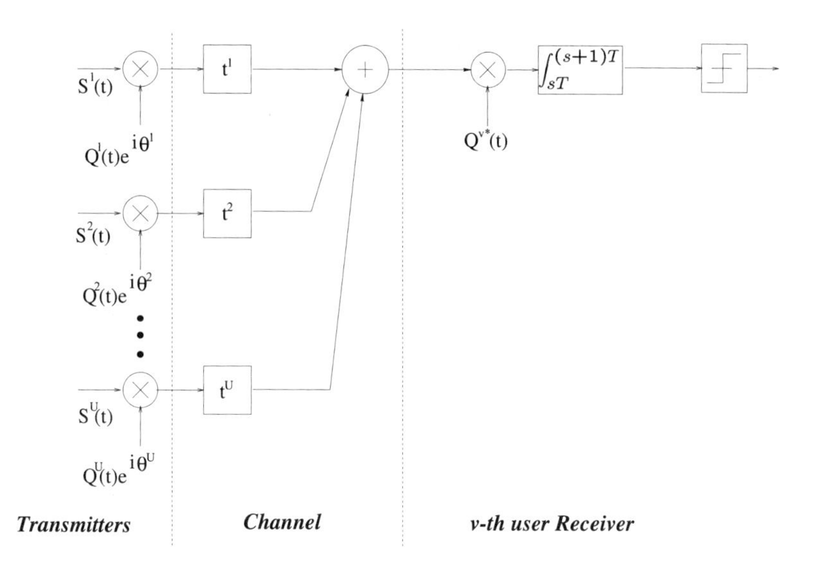

Figure 3.2: Baseband equivalent of a DS-CDMA system with a non-selective channel.

To exploit chaotic time-series we now set $X = [0, 1]$ and take a piecewise-continuous map $M : X \mapsto X$ to construct the periodic time series such that $x_{k-lN} = x_k$ for any integer l and $x_{k+1} = M(x_k)$ for $k = 0, 1, \ldots, N-2$ while x_0 is a degree of freedom. It can intuitively be accepted that, for non-degenerate functions Q, the statistical properties of the spreading sequence depend on the statistical features of the time series $\underline{x}$. Actually, such features can be studied by using techniques from the statistical approach to non-linear dynamical system theory. In fact, whenever a dynamical system exhibits a chaotic behavior, because of the well-known sensible dependence on initial conditions, the study of single trajectories brings little or no information on the general properties of the system. On the contrary, a better characterization of chaotic systems is in terms of the statistical evolution of significant sets of trajectories [15].

3.3.1 The Perron-Frobenius Operator: A Tool for Studying Chaos with Densities

To follow the statistical dynamical system theory approach, let M^k be the k-th iterate of M and for any set $Y \subseteq X$ indicate with $M^k(Y)$ the set $\{y \in X |\ y = M^k(x) \text{ and } x \in Y\}$, while $M^{-k}(Y)$ is the set $\{x \in X |\ y = M^k(x) \text{ and } y \in Y\}$.

Let $\mathbb{D}(X)$ be the set of probability densities defined on X and assume that a point $x_{k-1} \in X$ is randomly drawn according to $\rho_k \in \mathbb{D}(X)$. Then, the probability density regulating the distribution of the random point $x_k = M(x_{k-1})$ is given by $\rho_k = \mathbf{P}\rho_{k-1}$ where $\mathbf{P}$ is implicitly defined by the following probability conservation constraint

$$\mathbf{P}\rho_{k-1}(\xi) = \frac{\partial}{\partial \xi} \int_{M^{-1}[0,\xi]} \rho_{k-1}(x)dx \tag{3.3}$$

which holds for any $[0, \xi] \subseteq X$. If $\mathbb{L}_1$ is the space of all the Lebesgue integrable functions $f : X \mapsto \mathbb{C}$, then the operator $\mathbf{P} : \mathbb{L}_1 \mapsto \mathbb{L}_1$ satisfying (3.3) is called the Perron-Frobenius operator corresponding to M. A few properties can be deduced from (3.3). First, $\mathbf{P}$ is a *linear* infinite-dimensional operator. Then, if one defines $\mathbb{L}_\infty$ as the subspace of all the bounded elements of $\mathbb{L}_1$ and takes $f \in \mathbb{L}_1$ and $g \in \mathbb{L}_\infty$, the following equality holds [15, chap. 3]

$$\int_X \mathbf{P}f(x)g(x)dx = \int_X f(x)g(M(x))dx \tag{3.4}$$

From the dynamical system point of view, the iteration of the Perron-Frobenius operator accounts for the evolution of the set of trajectories of the discrete-time system $x_k = M(x_{k-1}) = M^k(x_0)$, whose initial conditions x_0 are drawn according to $\rho_0 \in \mathbb{D}(X)$.

Further important characteristics are related to the features of the dynamical system in terms of behavior complexity. More specifically, we will refer to *ergodic*, *mixing*, and *exact* maps [15, chap. 4]. To this aim, indicate with $\mu(Y)$ the normalized Borel measure of any measurable set $Y \subseteq X$. A map M is *ergodic*, if for any subset $Y \subseteq X$ which is invariant with respect to M, namely $M^{-1}(Y) = Y$, one has that either $\mu(Y) = 0$ or $\mu(X \setminus Y) = 0$. In other terms, any ergodic map must be studied on the entire space X. Moreover, it can be shown that if M is ergodic, then there is at most a density $\bar{\rho} \in \mathbb{D}(X)$ which is *invariant* under the Perron-Frobenius operator, namely it is such that $\bar{\rho} = \mathbf{P}\bar{\rho}$.

Though ergodicity ensures that trajectories almost always traverse the whole domain it says little about how they do this. In purely ergodic maps this wandering need not to be anything highly complex: nearby trajectories may remain reasonably close while they explore the map domain (see [15, Example (4.3.3)] for an enlightening two-dimensional case). Since we are interested in the generation of sequences with decaying cross-correlations, ergodicity is obviously not enough. To be sure that the correlation between any two different trajectories vanishes in time, maps which are at least mixing should be adopted. Yet, since checking exactness is easier than checking the mixing property, and it can be proved [15, chap. 4] that exactness implies mixing and ergodicity, we will here concentrate on exact maps.

We say that a map M is *exact* if, for any $Y \subseteq X$ such that $\mu(Y) > 0$, we have $\mu(\lim_{k\to\infty} M^k(Y)) = \mu(X) = 1$ [15, chap. 4]. The above condition has a very simple interpretation. If one starts with a set Y of initial conditions of nonzero measure, then after a large number of iterations of an exact transformation, the points have spread and completely filled the whole space X.

Exact maps possess interesting properties in terms of probability densities evolution. To highlight these features, recall that the variation var $\cdot$ of a function

f over X is defined as

$$\operatorname{var} f = \sup_{m \geq 1} \left\{ \sum_{j=1}^{m} |f(\xi_j) - f(\xi_{j-1})| : \ 0 \leq \xi_0 < \xi_1 < \cdots < \xi_m \leq 1 \right\}$$

Then, define $\mathbb{L}_{\mathrm{BV}} = \{f \in \mathbb{L}_1 : \ \operatorname{var} f < \infty\}$, namely the space of integrable function of bounded variation on X [16]. If M is exact it can be shown that for any $\rho_0 \in \mathbb{D}(X) \bigcap \mathbb{L}_{\mathrm{BV}}$, one has $\lim_{p \to \infty} \mathbf{P}^p \rho_0 = \bar{\rho}$ [16].

To complete the set of tools needed to obtain closed-form expressions for m-th order correlations of quantized chaotic maps trajectories, we need to generalize the definition (3.3) of the Perron-Frobenius operator.

Consider the family of operators $\mathbf{P}^{p_1,\ldots,p_m} : \mathbb{D}(X) \mapsto \mathbb{D}(X^m)$ defined in the following.

Definition 1. *For any $\rho \in \mathbb{D}(X)$ and any m integers $0 < p_1 < p_2 < \cdots < p_m$ set $\boldsymbol{\xi} = (\boldsymbol{\xi}_1, \ldots, \boldsymbol{\xi}_m)$ and define*

$$\mathbf{P}^{p_1,\ldots,p_m}[\rho](\boldsymbol{\xi}) = \frac{\partial^m}{\partial \boldsymbol{\xi}_1 \ldots \partial \boldsymbol{\xi}_m} \int_{\bigcap_{i=1}^{m} M^{-p_i}([0,\boldsymbol{\xi}_i])} \rho(x) dx$$

From Definition 1 we get that $\mathbf{P}^{p_1,\ldots,p_m}$ can be applied to the probability density ρ_k regulating the distribution of x_k to obtain the joint probability density $\mathbf{P}^{p_1,\ldots,p_m}[\rho_k]$ of the random vector $(x_{k+p_1}, \ldots, x_{k+p_m})$. From Definition 1 it is also evident that $\mathbf{P}^1$ is the usual Perron-Frobenius operator (3.3).

The operators in Definition 1 allow for a complete statistical description of the chaotic systems. In fact, for any integrable function $f : X^m \mapsto \mathbb{C}$, where $\mathbb{C}$ is the set of complex numbers,

$$\begin{aligned} \mathbf{E}[f(x_{k+p_1}, \ldots, x_{k+p_m})] &= \int_{X^m} f(\boldsymbol{\xi}) \mathbf{P}^{p_1,\ldots,p_m}[\rho_k](\boldsymbol{\xi}) d\boldsymbol{\xi} \\ &= \int_X f(M^{p_1}(x_k), \ldots, M^{p_m}(x_k)) \rho_k(x_k) dx_k \end{aligned} \tag{3.5}$$

The last equality, which is due to the causality of the system, is the analogous of the duality condition (3.4) extended to more than one time steps.

A further link between the operator in Definition 1 and the Perron-Frobenius operator (3.3) is given by the following

Property 1. *The operator in Definition 1 admits the following alternative expression*

$$\mathbf{P}^{p_1,\ldots,p_m}[\rho_k](\boldsymbol{\xi}) = \mathbf{P}^{p_1}[\rho_k](\boldsymbol{\xi}_1) \prod_{i=2}^{m} \delta(\boldsymbol{\xi}_i - M^{p_i - p_1}(\boldsymbol{\xi}_1))$$

where $\delta(\cdot)$ is the Dirac's generalized function.

Figure 3.3: The support of $\mathbf{P}^{0,2,3}[1](\boldsymbol{\xi}_1, \boldsymbol{\xi}_2, \boldsymbol{\xi}_3)$ and its projection on the coordinate planes when M is the tent map.

Proof. Note first that $\mathbf{P}^{p_1}[\rho_k]$ is the p_1-th iteration of the Perron-Frobenius operator so that the point $\boldsymbol{\xi}_1$ distributes according to $\mathbf{P}^{p_1}[\rho_k] = \rho_{k+p_1}$. Yet, from $\boldsymbol{\xi}_1$ on, the points are linked by the causal nature of the map. Hence, the joint probability of the random vector collecting samples of the trajectory at different time steps is vanishing whenever such a link is not fulfilled (i.e., whenever $\boldsymbol{\xi}_i \neq M^{p_i - p_1}(\boldsymbol{\xi}_1)$), and depends only on the distribution of the first point $\boldsymbol{\xi}_1$. □

Figure 3.3 shows the support (i.e., set of points in which it is non-null) of $\mathbf{P}^{0,2,3}[1](\boldsymbol{\xi}_1, \boldsymbol{\xi}_2, \boldsymbol{\xi}_3)$ when M is the well-known tent-map $M(x) = 1 - 2|x - 1/2|$. Note how the causal link between the subsequent map states is encoded in the projections of such a support on the coordinate planes whose profile is the shape of the iterate linking the state at the two corresponding time steps.

From the Definition we can also easily derive that the operator $\mathbf{P}^{p_1,\ldots,p_m}[\rho](\boldsymbol{\xi})$ is linear with respect to ρ.

In the following we will assume that M is exact and we will deduce some properties highlighting stationarity features of the chaotic dynamics. To help the discussion let us define a shortcut operator.

Definition 2.

$$\bar{\mathbf{P}}^{p_2,\ldots,p_m} = \mathbf{P}^{0,p_2,\ldots,p_m}$$

The following properties hold.

Property 2. *As long as* $p_1 < p_2 < \cdots < p_m$

$$\lim_{p_1 \to \infty} \mathbf{P}^{p_1,\ldots,p_m}[\rho_k] = \lim_{p_1 \to \infty} \bar{\mathbf{P}}^{p_2-p_1,\ldots,p_m-p_1}[\bar{\rho}]$$

Proof. Exploit Property 1 and $\lim_{p_1\to\infty} \mathbf{P}^{p_1}\rho_k = \bar{\rho}$ to get $\lim_{p_1\to\infty} \mathbf{P}^{p_1,\ldots,p_m}[\rho_k] = \bar{\rho}(\boldsymbol{\xi}_1)\lim_{p_1\to\infty}\prod_{i=2}^m \delta(\boldsymbol{\xi}_i - M^{p_i-p_1}(\boldsymbol{\xi}_1))$ which can be brought back to $\mathbf{P}^{q'_1,\ldots,q'_m}$ by means of Property 1 with $q'_1 = 0$ and $q'_i = p_i - p_1$ for $i = 2,\ldots,m$. □

Note how the absence of the initial condition ρ_k in the right-hand side of the thesis of Property 2 means that exact maps are asymptotically stationary. Property 2 can be specialized in two cases which are the most commonly encountered. Note first that, if $\lim_{p_1\to\infty} p_i - p_1 = \Delta p_i$ then

$$\lim_{p_1\to\infty} \mathbf{P}^{p_2,\ldots,p_m}[\rho_k] = \bar{\mathbf{P}}^{\Delta p_2,\ldots,\Delta p_m}[\bar{\rho}]$$

where convergence holds almost at each point. Moreover, if $\lim_{p_1\to\infty} p_i - p_1 = \infty$ then

$$\lim_{p_1\to\infty} \mathbf{P}^{p_1,\ldots,p_m}[\rho_k](\boldsymbol{\xi}) = \prod_{i=1}^{m} \bar{\rho}(\boldsymbol{\xi}_i)$$

where convergence holds in the weak Cesaro sense [15, chap. 2].

Beyond these asymptotic considerations, a noteworthy simplification holds even for finite p_i. In fact, the following Property is analogous to Property 2 and follows from the invariance of $\bar{\rho}$ and from Property 1.

Property 3.

$$\mathbf{P}^{p_1,\ldots,p_m}[\bar{\rho}] = \bar{\mathbf{P}}^{p_2-p_1,\ldots,p_m-p_1}[\bar{\rho}]$$

Hence, the adoption of initial condition distributed according to $\bar{\rho}$ makes all the characteristic features of the resulting trajectories appear stationary even though the underlying system is causal.

When this stationarity condition holds, the transition mechanism of the map state variable can be assimilated to that of a continuous-state Markov chain as indicated by the following

Property 4.

$$\bar{\mathbf{P}}^{p_2,\ldots,p_m}[\bar{\rho}](\boldsymbol{\xi}) = \bar{\mathbf{P}}^{p_2,\ldots,p_{m-1}}[\bar{\rho}](\boldsymbol{\xi}_1,\ldots,\boldsymbol{\xi}_{m-1})\frac{\bar{\mathbf{P}}^{p_m-p_{m-1}}[\bar{\rho}](\boldsymbol{\xi}_{m-1},\boldsymbol{\xi}_m)}{\bar{\rho}(\boldsymbol{\xi}_{m-1})}$$

Proof. Use Definition 2, Property 1 and the invariance of $\bar{\rho}$ to express the product in the right-hand side as

$$\bar{\mathbf{P}}^{p_2,\ldots,p_{m-1}}[\bar{\rho}](\boldsymbol{\xi}_1,\ldots,\boldsymbol{\xi}_{m-1})\bar{\mathbf{P}}^{p_m-p_{m-1}}[\bar{\rho}](\boldsymbol{\xi}_{m-1},\boldsymbol{\xi}_m) =$$
$$\bar{\rho}(\boldsymbol{\xi}_1)\prod_{i=2}^{m-1}\delta(\boldsymbol{\xi}_i - M^{p_i}(\boldsymbol{\xi}_1))\bar{\rho}(\boldsymbol{\xi}_{m-1})\delta(\boldsymbol{\xi}_m - M^{p_m-p_{m-1}}(\boldsymbol{\xi}_{m-1}))$$

Note then that

$$\delta(\boldsymbol{\xi}_m - M^{p_m-p_{m-1}}(\boldsymbol{\xi}_{m-1})) = \delta(\boldsymbol{\xi}_m - M^{p_m}(\boldsymbol{\xi}_1))$$

to obtain the thesis. □

Note how the quantity $\bar{\mathbf{P}}^{p_m-p_{m-1}}[\bar{\rho}](\boldsymbol{\xi}_{m-1},\boldsymbol{\xi}_m)/\bar{\rho}(\boldsymbol{\xi}_{m-1})$ plays the role of a conditioned probability density quantifying the probability of moving from $\boldsymbol{\xi}_{m-1}$ to $\boldsymbol{\xi}_m$ given the density of $\boldsymbol{\xi}_{m-1}$.

Exploiting the consequences of the above definition we are now able to give some interesting results on the correlation functions of quantized trajectories and, consequently, of spreading sequences generated by exact maps.

3.3.2 Correlations of Quantized Chaotic Trajectories

From now on we will consider trajectories generated with $\rho_0 = \bar{\rho}$ so to exploit all the stationarity properties highlighted so far. We also consider the quantized version of the system trajectories that are obtained applying to the state variable at time p_i the function $f_i : X \mapsto \mathbb{C}$ which is constant in each of n quantization intervals $X_1, \ldots, X_n$ partitioning X. The n-dimensional vector $\mathbf{f}_i = (\mathbf{f}_{i1}, \ldots, \mathbf{f}_{in})$ is defined so that $f_i(X_j) = \{\mathbf{f}_{ij}\}$.

The m-th order correlation of the quantized trajectories can be obtained from (3.5). More formally, assuming from now on $p_1 = 0$, we have

$$\begin{aligned}\mathbf{E}\left[\prod_{i=1}^m f_i(x_{p_i})\right] &= \int_{X^m} \prod_{i=1}^m f_i(\boldsymbol{\xi}_i)\bar{\mathbf{P}}^{p_2,\ldots,p_m}[\bar{\rho}](\boldsymbol{\xi})d\boldsymbol{\xi} \\ &= \sum_{j_1=1}^n \cdots \sum_{j_m=1}^n \left[\prod_{i=1}^m \mathbf{f}_{ij_i} \int_{X_{j_1}\times\cdots\times X_{j_m}} \bar{\mathbf{P}}^{p_2,\ldots,p_m}[\bar{\rho}](\boldsymbol{\xi})d\boldsymbol{\xi}\right]\end{aligned}$$

in which the integral term is nothing but the probability that a trajectory starting in X_{j_1} falls in X_{j_2} at time step p_2, in X_{j_3} at time step p_3, etc. We can collect all the values of such a term in a m-dimensional tensor $\mathcal{H}^{p_2,\ldots,p_m}$

$$\mathcal{H}^{p_2,\ldots,p_m}_{j_1,\ldots,j_m} = \int_{X_{j_1}\times\cdots\times X_{j_m}} \bar{\mathbf{P}}^{p_2,\ldots,p_m}[\bar{\rho}](\boldsymbol{\xi})d\boldsymbol{\xi}$$

which will be indicated as the *symbolic dynamic tracking tensor* (SDTT) as it tracks the symbolic evolution [17] of the systems at prescribed time steps.

Since multi-index quantities like the SDTT will be useful in the subsequent discussion some useful basic operators acting on tensors are recalled in the Appendix. By means of this operators, the terms $\prod_{i=1}^m \mathbf{f}_{ij_i}$ for $j_1, \ldots, j_m$ spanning the range $1, \ldots, n$ can be compounded in the m-dimensional tensor $\mathcal{F} = \mathbf{f}_1 \otimes \cdots \otimes \mathbf{f}_m$, and that

$$\mathbf{E}\left[\prod_{i=1}^m f_i(x_{p_i})\right] = \langle \mathcal{F}, \mathcal{H}^{p_2,\ldots,p_m}\rangle \tag{3.6}$$

Such an expression is redundant whenever the f_i are such that if $f_{i'}(X_{j'}) = f_{i'}(X_{j''})$ then $f_{i''}(X_{j'}) = f_{i''}(X_{j''})$ for any i', i'', j', j'' and their codomains contain only $L < n$ different complex symbols $\hat{\mathbf{f}}_{i1}, \ldots, \hat{\mathbf{f}}_{iL}$. In fact, an expression analogous to (3.6) exists entailing $L \times \cdots \times L$ tensors instead of $n \times \cdots \times n$ tensors.

To obtain such an expression construct the $n \times L$ matrix $\mathcal{R}$, depending on n, L and the f_i as

$$\mathcal{R}_{kj} = \begin{cases} 1 & \text{if } f_i(X_j) = \{\hat{\mathbf{f}}_{ik}\} \\ 0 & \text{otherwise} \end{cases}$$

as well as an m-dimensional tensor $\hat{\mathcal{F}} = \bigotimes_{i=1}^{m} \hat{\mathbf{f}}_i$ which are such that $\mathcal{F} = \hat{\mathcal{F}}\big| \mathcal{R}$. With this, and with the aid of Property 16 we easily get

$$\langle \mathcal{F}, \mathcal{H}^{p_2,\ldots,p_m} \rangle = \left\langle \hat{\mathcal{F}} \middle| \mathcal{R}, \mathcal{H}^{p_2,\ldots,p_m} \right\rangle = \left\langle \hat{\mathcal{F}}, \mathcal{H}^{p_2,\ldots,p_m} \middle| \mathcal{R}^T \right\rangle = \left\langle \hat{\mathcal{F}}, \hat{\mathcal{H}}^{p_2,\ldots,p_m} \right\rangle$$

where the last equality implicitly defines the SDTT related to the symbolic dynamics of the largest pluri-intervals $Y_{k_i} = \bigcup_{\mathcal{R}_{k_i j_i}=1} X_{j_i}$ in which the f_i are constant.

3.3.3 Specialization to Piecewise Affine Markov Maps

In the following we will assume that the map M is a PWAM map with respect to the quantization intervals, i.e., that M restricted to any X_j is affine and that, for any X_{j_1} and X_{j_2}, either $M(X_{j_1}) \cap X_{j_2} = \emptyset$ or $X_{j_2} \subseteq M(X_{j_1})$. The most interesting properties of such kind of maps depend on a very informative square matrix characterizing the statistical behavior of trajectories. It is defined component-wise as

$$\mathcal{K}_{j_1 j_2} = \frac{\mu(X_{j_1} \cap M^{-1}(X_{j_2}))}{\mu(X_{j_1})} \tag{3.7}$$

and is often referred to as the *kneading matrix* $\mathcal{K}$, since its entry in the j_1-th row and j_2-th column records the fraction of X_{j_1} that is mapped into X_{j_2}. From this it easily follows that the sum of the row entries of any kneading matrix is always 1.

A first useful result achieved by using $\mathcal{K}$ allows one to easily check exactness for PAWM maps as it is stated in the following property.

Property 5. *If M is a piecewise affine Markov map and an integer l exists such that $\mathcal{K}^l$ has no null entry, then M is exact (and thus mixing and ergodic).*

Proof. Recall that a non-negative matrix $\mathcal{K}$ such that $\mathcal{K}^q$ has no null entry for a certain integer q is said to be primitive [18]. Let also $Y_1, \ldots, Y_p$ be any collection of intervals whose interiors are disjoint and whose union covers the whole X and let $\mathcal{K}_Y$ be the corresponding kneading matrix.

From [19, Theorem 2.3] we know that if $\mathcal{K}$ is primitive, then $\mathcal{K}_Y$ is also primitive if

$$\alpha_1 \alpha_2 \ldots \alpha_n \left[1 + \frac{\min_{j_i} \mu(X_{j_i})}{\max_{j_i} \mu(X_{j_i})} \right] \geq 1,$$

where α_{j_i} is the ratio between the minimum of $|(d/dx)M|$ on X_{j_i} and the slope of its local affine approximation, i.e., 1 for PWAM maps. Thus, $\mathcal{K}_Y$ is primitive for any collection of intervals $Y_1, \ldots, Y_p$.

Let now a $Y \subseteq X$ be given such that $\mu(Y) > 0$. Let n be the number of quantization intervals and construct the intervals $Y_1, \ldots, Y_p$ for p multiple of n dividing each X_{j_i} into p/n equal, adjacent intervals. Assume, without loss of generality that p can be increased so that $Y \supseteq Y_1$.

Since M is piecewise-affine and from the construction of $Y_1, \ldots, Y_p$ we know that, for any j_m, j_n either $Y_{j_m} \subseteq M(Y_{j_n})$ or $Y_{j_m} \cap M(Y_{j_n}) = \emptyset$. Hence, the kneading matrix of M^l with respect to the intervals $Y_1, \ldots, Y_p$ is nothing but the l-th power of the kneading matrix of M.

Yet, an integer q exists such that $\mathcal{K}_Y^q$ has no null entry, i.e., such that $M^q(Y_1) = X$. For that integer $\mu(M^q(Y)) \geq \mu(M^q(Y_1)) = 1$ which matches the exactness definition and proves the thesis. □

Additionally, it can be proved [3] that $\mathcal{K}$ is the restriction of the Perron-Frobenius operator to the space of functions which are constant in each X_{j_1}. As $\bar{\rho}$ can also be seen to belong to that space it can be written as $\bar{\rho} = \sum_{i=1}^{n} \mathbf{e}_i \chi_{X_i}$ where the vector $\mathbf{e} = (\mathbf{e}_1, \ldots, \mathbf{e}_n)$ is the unique eigenvector of $\mathcal{K}$ corresponding to a unit eigenvalue, i.e., $\mathbf{e} = \mathbf{e}\mathcal{K}$, and χ_{X_i} is the characteristic function of X_i. From this we also get that, if $\bar{\rho} = 1$, then $\mathbf{e}_i = 1$ and also the sum of the row entries of $\mathcal{K}$ is always 1, i.e., $\mathcal{K}$ is a *doubly stochastic* matrix.

The adoption of PWAM maps allows also to prove the following property.

Property 6. *If M is piecewise affine Markov with respect to the quantization intervals then*

$$\int_{X_{j_1}\times\cdots\times X_{j_m}} \bar{\mathbf{P}}^{p_2,\ldots,p_m}[\bar{\rho}](\boldsymbol{\xi})d\boldsymbol{\xi} = \int_{X_{j_1}\times\cdots\times X_{j_{m-1}}} \bar{\mathbf{P}}^{p_2,\ldots,p_{m-1}}[\bar{\rho}](\boldsymbol{\xi}_1,\ldots,\boldsymbol{\xi}_{m-1})d\boldsymbol{\xi}_1\ldots d\boldsymbol{\xi}_{m-1} \times$$
$$\times \frac{1}{\bar{\mu}(X_{j_{m-1}})} \int_{X_{j_{m-1}}\times X_{j_m}} \bar{\mathbf{P}}^{p_m-p_{m-1}}[\bar{\rho}](\boldsymbol{\xi}_{m-1},\boldsymbol{\xi}_m)d\boldsymbol{\xi}_{m-1}d\boldsymbol{\xi}_m$$

where the invariant measure is defined as $\bar{\mu}(X) = \int_X \bar{\rho}(x)dx$.

Proof. Consider the mapping mechanism of a piecewise affine Markov map as depicted in Figure 3.4. From that figure we get that, as $\int_{X_{j_1}\times\cdots\times X_{j_{m-1}}} \bar{\mathbf{P}}^{p_2,\ldots,p_{m-1}}[\bar{\rho}]$ $(\boldsymbol{\xi}_1,\ldots,\boldsymbol{\xi}_{m-1})d\boldsymbol{\xi}_1\ldots d\boldsymbol{\xi}_{m-1}$ is the probability that the map state belongs to X_{j_i} at time p_i for $i = 1, \ldots, m-1$, it is the invariant measure $\bar{\mu}$ of the intersection of all the counterimages $M^{-p_i}(X_{j_i})$. If the further condition $x_{p_m} \in X_{j_m}$ is added, such an intersection is restricted to have an invariant measure equal to $\int_{X_{j_1}\times\cdots\times X_{j_m}} \bar{\mathbf{P}}^{p_2,\ldots,p_m}[\bar{\rho}](\boldsymbol{\xi})d\boldsymbol{\xi}$. Yet, from the piecewise affine nature of the map we know that $\bar{\rho}$ is constant in each X_j and from the Markov property, the intersection with the additional counterimage $M^{-p_m}(X_{j_m})$ divides the previous intersection in two parts proportional to the ratio of points in $X_{j_{m-1}}$ that are mapped into

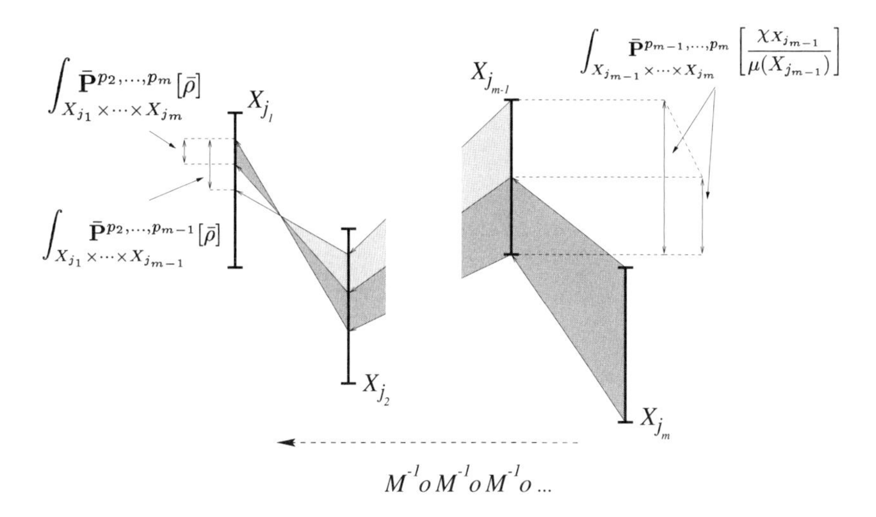

Figure 3.4: Interval mapping mechanism for piecewise affine Markov maps and its relationship with the SDTT.

X_{j_m}. With this we may write

$$\frac{\int_{X_{j_1}\times\cdots\times X_{j_m}} \bar{\mathbf{P}}^{p_2,\ldots,p_m}[\bar{\rho}](\boldsymbol{\xi})d\boldsymbol{\xi}}{\int_{X_{j_1}\times\cdots\times X_{j_{m-1}}} \bar{\mathbf{P}}^{p_2,\ldots,p_{m-1}}[\bar{\rho}](\boldsymbol{\xi}_1,\ldots,\boldsymbol{\xi}_{m-1})d\boldsymbol{\xi}_1\ldots d\boldsymbol{\xi}_{m-1}} = \frac{\mu(X_{j_{m-1}}\cap M^{-p_m+p_{m-1}}(X_{j_m}))}{\mu(X_{j_{m-1}})} \tag{3.8}$$

Yet, by definition, the last term is nothing but the probability that a trajectory starts with a point uniformly distributed in $X_{j_{m-1}}$ and is in X_{j_m} at time $p_m - p_{m-1}$, i.e.,

$$\int_{X_{j_{m-1}}\times X_{j_m}} \bar{\mathbf{P}}^{p_m-p_{m-1}}\left[\frac{\chi_{X_{j_{m-1}}}}{\mu(X_{j_{m-1}})}\right](\boldsymbol{\xi}_{m-1},\boldsymbol{\xi}_m)d\boldsymbol{\xi}_{m-1}d\boldsymbol{\xi}_m$$

As $\bar{\rho}$ is constant in each X_j we may rewrite the last expression as

$$\int_{X_{j_{m-1}}\times X_{j_m}} \bar{\mathbf{P}}^{p_m-p_{m-1}}\left[\frac{\bar{\rho}\chi_{X_{j_{m-1}}}}{\bar{\rho}\mu(X_{j_{m-1}})}\right](\boldsymbol{\xi}_{m-1},\boldsymbol{\xi}_m)d\boldsymbol{\xi}_{m-1}d\boldsymbol{\xi}_m$$

which can be substituted in (3.8). Recalling the linearity of $\bar{\mathbf{P}}$, and noting that $\bar{\rho}(\boldsymbol{\xi}_{m-1})\chi_{X_{j_{m-1}}}(\boldsymbol{\xi}_{m-1}) = \bar{\rho}(\boldsymbol{\xi}_{m-1})$ and $\bar{\rho}(\boldsymbol{\xi}_{m-1})\mu(X_{j_{m-1}}) = \bar{\mu}(X_{j_{m-1}})$ for every $\boldsymbol{\xi}_{m-1} \in X_{j_{m-1}}$, we finally get the thesis. □

Note how Property 6 is the analogous of Property 4 for the symbolic dynamics instead of the real dynamics. Note also that Property 4 allows the factorization of multi-indexes $\bar{\mathbf{P}}$ operators into one-index operators and that Property 6 allows the factorization of multi-dimensional SDTT into two-dimensional SDTTs.

In fact, recalling the definition and the associativity of the chain product we may rewrite the SDTT as

$$\mathcal{H}^{p_2,\ldots,p_m} = \mathcal{H}^{p_2} \circ \bigcirc_{i=3}^{m} \frac{\mathcal{H}^{p_i - p_{i-1}}}{\bar{\mu}(X_{j_{i-1}})}$$

Note that this factorization property is a key point distinguishing $\mathcal{H}^{p_2,\ldots,p_m}$ from $\hat{\mathcal{H}}^{p_2,\ldots,p_m} = \mathcal{H}^{p_2,\ldots,p_m}\Big| \mathcal{R}^T$. In fact, the operator in Definition 7 cannot be, in general, factored.

In the following, we will concentrate on piecewise affine Markov maps with $\bar{\rho} = 1$. In this case, the two-dimensional tensor $\mathcal{H}^q$ can be related to $\mathcal{K}$ as stated in the following property.

Property 7. *If $\mu(X_j) = 1/n$ for $j = 1, \ldots, n$ then*

$$\mathcal{H}^p = \frac{1}{n}\mathcal{K}^p$$

Proof. Let us first prove that $(\mathcal{K}^p)_{j_1 j_2} = \mu(X_{j_1} \cap M^{-p}(X_{j_2}))/\mu(X_{j_1})$, where $(\mathcal{K}^p)_{j_1 j_2}$ is the $j_1 j_2$-th entry of the p-th power of the kneading matrix. To see this proceed by induction, the property being true by definition for $p = 1$. Note then, that from the fact that either $X_i \subseteq M(X_j)$ or $X_i \cap M(X_j) = \emptyset$ and from the piecewise affinity of M we have

$$\frac{\mu(X_{j_1} \cap M^{-p}(X_{j_2}))}{\mu(X_{j_1})} = \sum_{i=1}^{n} \frac{\mu(X_{j_1} \cap M^{-(p-1)}(X_i))}{\mu(X_{j_1})} \frac{\mu(X_i \cap M^{-1}(X_{j_2}))}{\mu(X_i)} = \sum_{i=1}^{n} (\mathcal{K}^{p-1})_{j_1 i} \mathcal{K}_{i j_2}$$

which is nothing but the matrix product $\mathcal{K}^{p-1}\mathcal{K} = \mathcal{K}^p$. As far as $\mathcal{H}^p$ is concerned, note that it may be written as

$$\begin{aligned}
\mathcal{H}^p_{j_1 j_2} &= \int_{X_{j_1} \times X_{j_2}} \bar{\mathbf{P}}^p[\bar{\rho}](\boldsymbol{\xi}_1, \boldsymbol{\xi}_2) d\boldsymbol{\xi}_1 d\boldsymbol{\xi}_2 = \int_{X_{j_1} \times X_{j_2}} \bar{\rho}(\boldsymbol{\xi}_1)\delta(\boldsymbol{\xi}_2 - M^p(\boldsymbol{\xi}_1)) d\boldsymbol{\xi}_1 d\boldsymbol{\xi}_2 \\
&= \mu(X_{j_1} \cap M^{-p}(X_{j_2})) = \frac{1}{n}(\mathcal{K}^p)_{j_1 j_2}
\end{aligned}$$

□

Note that, Property 7 is a further step in the direction of Property 6. In fact, while the latter allows factorization of multi-indexes SDTT into single-index SDTTs, the former states that single-index SDTTs are actually powers of a matrix which is characteristic of the map.

In the light of all the previously stated properties, the expression of the m-order correlation of the quantized trajectories becomes

$$\mathbf{E}\left[\prod_{i=1}^{m} f_i(x_{p_i})\right] = \frac{1}{n}\left\langle \mathcal{F}, \bigcirc_{i=2}^{m} \mathcal{K}^{p_i - p_{i-1}} \right\rangle \tag{3.9}$$

where $p_1 = 0$ and which can be effectively exploited noting the following property.

Property 8. *If M is a piecewise affine Markov map such that $\mathcal{K}$ is a primitive matrix a family of matrixes $\mathcal{A}(p)$ exists such that*

$$\mathcal{K}^p = \Im + \mathcal{A}(p)$$

with $\mathcal{A}(p)$ exponentially vanishing as $p \to \infty$ and $\Im = (1/n)\mathbf{1} \otimes \mathbf{1}$ with $\mathbf{1} = (1, \ldots, 1)$.

Proof. As $\mathcal{K}$ is primitive and doubly stochastic it enjoys special spectral properties [18]. In fact, only $\mathbf{1}$ is an eigenvector with an unit eigenvalue and all the other eigenvalues λ are such that $|\lambda| < 1$. Hence, for any vector $\mathbf{e}$, $\mathcal{K}^p \mathbf{e}$ exponentially approaches $\mathbf{1}/n$ as $p \to \infty$ so that $\mathcal{K}^p$ exponentially approaches $\Im$. □

Using Property 8, the expression in (3.9) can be often simplified if the following considerations are taken into account.

Property 9. *If $\mathcal{A}'$ is an m'-dimensional tensor and $\mathcal{A}''$ is an m''-dimensional tensor such that $m = m' + m''$, then*

$$\langle \mathcal{F}, \mathcal{A}' \circ \Im \circ \mathcal{A}'' \rangle = \frac{1}{n} \langle \mathcal{F}', \mathcal{A}' \rangle \langle \mathcal{F}'', \mathcal{A}'' \rangle$$

where $\mathcal{F}' = \mathbf{f}_1 \otimes \cdots \otimes \mathbf{f}_{m'}$ and $\mathcal{F}'' = \mathbf{f}_{m'+1} \otimes \cdots \otimes \mathbf{f}_m$.

Proof. Recall that $\Im = (1/n)\mathbf{1} \otimes \mathbf{1}$. Hence, by Property 16 reported in Appendix we have $\mathcal{A}' \circ \Im \circ \mathcal{A}'' = (1/n)\mathcal{A}' \circ \mathbf{1} \otimes \mathbf{1} \circ \mathcal{A}'' = (1/n)\mathcal{A}' \otimes \mathcal{A}''$ and, from $\mathcal{F} = \mathcal{F}' \otimes \mathcal{F}''$, $(1/n) \langle \mathcal{F}, \mathcal{A}' \otimes \mathcal{A}'' \rangle = (1/n) \langle \mathcal{F}' \otimes \mathcal{F}'', \mathcal{A}' \otimes \mathcal{A}'' \rangle = (1/n) \langle \mathcal{F}', \mathcal{A}' \rangle \langle \mathcal{F}'', \mathcal{A}'' \rangle$. □

Property 10. *For any two pairs of tensors $\mathcal{F}', \mathcal{A}'$ and $\mathcal{F}'', \mathcal{A}''$ with the same dimension*

$$\begin{aligned}
\langle \mathbf{f}_1 \otimes \mathcal{F}', \Im \circ \mathcal{A}' \rangle &= \langle \mathcal{F}', \mathcal{A}' \rangle \frac{\sum_{j=1}^n \mathbf{f}_{1j}}{n} \\
\langle \mathcal{F}' \otimes \mathbf{f}_m, \mathcal{A}' \circ \Im \rangle &= \langle \mathcal{F}', \mathcal{A}' \rangle \frac{\sum_{j=1}^n \mathbf{f}_{mj}}{n} \\
\langle \mathcal{F}' \otimes \mathbf{f}_i \otimes \mathcal{F}'', \mathcal{A}' \circ \Im \circ \Im \circ \mathcal{A}'' \rangle &= \langle \mathcal{F}', \mathcal{A}' \rangle \langle \mathcal{F}'', \mathcal{A}'' \rangle \frac{\sum_{j=1}^n \mathbf{f}_{ij}}{n^2}
\end{aligned}$$

Proof. The thesis simply follows noting that in all the three left-hand side expressions the right term of the inner product is independent, respectively, from the first, m-th and i-th index. □

The above property can be exploited in a particular but common case to substantially reduce the number of different terms entailed in the computation of (3.9).

Property 11. *If $\sum_{j=1}^n \mathbf{f}_{ij} = 0$ for any i and $\mu(X_j) = 1/n$ for any j, then*

$$\mathbf{E}\left[\prod_{i=1}^m f_i(x_{p_i})\right] = \frac{1}{n} \sum_{\mathbf{b} \in B} \langle \mathcal{F}, \bigcirc_{i=1}^m \mathcal{B}(i) \rangle$$

where

$$\mathcal{B}(i) = \begin{cases} \mathcal{A}(p_i - p_{i-1}) & \textit{if } \mathbf{b}_i = 1 \\ \Im & \textit{otherwise} \end{cases}$$

$p_1 = 0$ and $B \subset \{0,1\}^m$ contains all the binary vectors whose first and last entry are 1 and in which no two subsequent entries are both null.

Proof. Substitute Property 8 in (3.9) and exploit the linearity of $\circ$ and of the inner product to obtain

$$\frac{1}{n} \sum_{\mathbf{b}\in\{0,1\}^m} \langle \mathcal{F}, \bigcirc_{i=1}^m \mathcal{B}(i) \rangle$$

With this, the thesis follows noting that all the terms corresponding to a binary vector $\mathbf{b}$ not belonging to B may be factored as in Property 10 to obtain quantities proportional to a vanishing sum. □

Note that all the terms in the sum of Property 11 are of the kind

$$\langle \mathcal{F}, \mathcal{A}' \circ \Im \circ \mathcal{A}'' \circ \Im \circ \ldots \rangle$$

to which Property 9 applies giving products like $\langle \mathcal{F}', \mathcal{A}' \rangle (1/n) \langle \mathcal{F}'', \mathcal{A}'' \rangle (1/n) \cdots$ where $\mathcal{A}', \mathcal{A}'', \ldots$ are chain products of $\mathcal{B}(i)$ for $\mathbf{b}_i = 1$.

To clarify some possible uses of the previous results we give here examples of the computation of a second-, third- and fourth-order correlation under conditions whose significance will be clarified in the final applicative section.

Example 1. *If $\sum_{j=1}^n \mathbf{f}_{1j} = 0$ then*

$$\mathbf{E}[f_1(x_0) f_2(x_{p_2})] = \frac{1}{n} \langle \mathcal{F}, \Im + \mathcal{A}(p_2) \rangle = \frac{1}{n} \langle \mathcal{F}, \mathcal{A}(p_2) \rangle$$

Example 2. *If $\sum_{j=1}^n \mathbf{f}_{ij} = 0$ for $i = 1$ and $i = 3$ then*

$$\mathbf{E}[f_1(x_0) f_2(x_{p_2}) f_3(x_{p_3})] = \frac{1}{n} \langle \mathcal{F}, (\Im + \mathcal{A}(p_2)) \circ (\Im + \mathcal{A}(p_3 - p_2)) \rangle$$

$$= \frac{1}{n} \Big[\langle \mathcal{F}, \Im \circ \Im \rangle + \langle \mathcal{F}, \Im \circ \mathcal{A}(p_3 - p_2) \rangle + \langle \mathcal{F}, \mathcal{A}(p_2) \circ \Im \rangle + \langle \mathcal{F}, \mathcal{A}(p_2) \circ \mathcal{A}(p_3 - p_2) \rangle \Big]$$

$$= \frac{1}{n} \langle \mathcal{F}, \mathcal{A}(p_2) \circ \mathcal{A}(p_3 - p_2) \rangle$$

Example 3. *If $\sum_{j=1}^n \mathbf{f}_{ij} = 0$ for $i = 1, \ldots, 4$ then*

$$\mathbf{E}[f_1(x_0) f_2(x_{p_2}) f_3(x_{p_3}) f_4(x_{p_4})] =$$

$$= \frac{1}{n} \langle \mathcal{F}, (\Im + \mathcal{A}(p_2)) \circ (\Im + \mathcal{A}(p_3 - p_2)) \circ (\Im + \mathcal{A}(p_4 - p_3)) \rangle$$

$$= \frac{1}{n} \Big[\langle \mathcal{F}, \Im \circ \Im \circ \Im \rangle + \langle \mathcal{F}, \Im \circ \Im \circ \mathcal{A}(p_4 - p_3) \rangle + \langle \mathcal{F}, \Im \circ \mathcal{A}(p_3 - p_2) \circ \Im \rangle$$

$$+ \langle \mathcal{F}, \mathcal{A}(p_2) \circ \Im \circ \Im \rangle + \langle \mathcal{F}, \mathcal{A}(p_2) \circ \Im \circ \mathcal{A}(p_4 - p_3) \rangle + \langle \mathcal{F}, \mathcal{A}(p_2) \circ \mathcal{A}(p_3 - p_2) \circ \Im \rangle$$

$$+ \langle \mathcal{F}, \Im \circ \mathcal{A}(p_3 - p_2) \circ \mathcal{A}(p_4 - p_3) \rangle + \langle \mathcal{F}, \mathcal{A}(p_2) \circ \mathcal{A}(p_3 - p_2) \circ \mathcal{A}(p_4 - p_3) \rangle \Big]$$

$$= \frac{1}{n} \Big[\frac{1}{n} \langle \mathcal{F}', \mathcal{A}(p_2) \rangle \langle \mathcal{F}'', \mathcal{A}(p_4 - p_3) \rangle + \langle \mathcal{F}, \mathcal{A}(p_2) \circ \mathcal{A}(p_3 - p_2) \circ \mathcal{A}(p_4 - p_3) \rangle \Big]$$

3.3.4 The Case of (n,t)-Tailed Shifts

In this section a particular family of piecewise affine Markov maps is considered. They will be called (n,t)-tailed shifts and are defined as

$$M(x)=\begin{cases}(n-t)x \pmod{\frac{n-t}{n}}+\frac{t}{n} & \text{if } 0\le x<\frac{n-t}{n}\\ t\left(x-\frac{n-t}{n}\right) \pmod{\frac{t}{n}} & \text{otherwise}\end{cases}$$

for $t<n/2$. The family of (n,t)-tailed shifts includes some of the maps already analyzed in the chaos-based DS-CDMA framework [2] like the n-way Bernoulli shift (which is a $(n,0)$-tailed shift) and the n-way tailed shift (which is a $(n,1)$-tailed shift). Figure 3.5 shows the graphics of (n,t)-tailed shift for $n=10$ and $t=0,1,2,3$. For these maps $X_j=[(j-1)/n, j/n]$ and the kneading matrix can be written as

$$\mathcal{K}=\left(\begin{array}{c|c}0 & (n-t)^{-1}\\ \hline t^{-1} & 0\end{array}\right)$$

where the size of upper left block is $(n-t)\times t$. From the fact that $\mathcal{K}^3$ has no null entry and from Property 5 we get that they are exact.

Shifts from the (n,t)-tailed family enjoy the particular structure for the matrix $\mathcal{A}(p)$ considered in Property 8, as stated in the following property.

Property 12. *If M is an (n,t)-tailed shift, then for $p>0$*

$$\mathcal{A}(p)=-\frac{r^p}{n}\left(\begin{array}{c|c}r^{-1} & 1\\ \hline r^{-2} & r^{-1}\end{array}\right)$$

where $r=-t/(n-t)$ and the size of the upper left block is $(n-t)\times t$.

Proof. In order to compute the powers of $\mathcal{K}$, let us consider its Jordan decomposition $\mathcal{K}=\mathcal{T}^{-1}\mathcal{D}\mathcal{T}$, so that $\mathcal{K}^p=\mathcal{T}^{-1}\mathcal{D}^p\mathcal{T}$ which, in this case, is extremely simple to compute.

In fact, note that r is the unique non-vanishing eigenvalue of $\mathcal{K}$ whose modulus is less than 1, so that

$$\mathcal{D}=\begin{pmatrix}r & 0 & \dots & 0 & 0\\ 0 & 0 & \dots & 0 & 0\\ \vdots & \vdots & \ddots & \vdots & \vdots\\ 0 & 0 & \dots & 0 & 0\\ 0 & 0 & \dots & 0 & 1\end{pmatrix}\quad \mathcal{T}=\left(\begin{array}{ccccc|ccccc}
r^{-1} & \dots & r^{-1} & & & 1 & \dots & \dots & \dots & 1\\ \hline
 & & & & & -1 & 1 & 0 & \dots & 0\\
 & & 0 & & & -1 & 0 & 1 & \dots & 0\\
 & & & & & \vdots & \vdots & \vdots & \ddots & \vdots\\
 & & & & & -1 & 0 & 0 & \dots & 1\\ \hline
-1 & 1 & 0 & \dots & 0 & & & & & \\
-1 & 0 & 1 & \dots & 0 & & & 0 & & \\
\vdots & \vdots & \vdots & \ddots & \vdots & & & & & \\
-1 & 0 & 0 & \dots & 1 & & & & & \\ \hline
1 & \dots & \dots & \dots & \dots & \dots & \dots & \dots & \dots & 1
\end{array}\right)$$

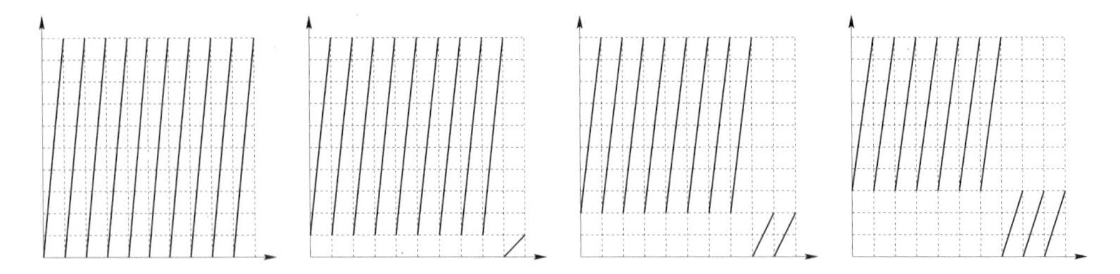

Figure 3.5: From left to right, graphics of a $(10,0)$-tailed shift, $(10,1)$-tailed shift, $(10,2)$-tailed shift, and $(10,3)$-tailed shift.

where the first row contains t entries equal to $1/r$, the size of the upper non-null rectangular block is $(t-1)\times t$ and the size of the lower one is $(n-t-1)\times(n-t)$.

Let now $\mathbf{e}' = (1,0,\ldots,0)$ and $\mathbf{e}'' = (0,\ldots,0,1)$ and note that $\mathcal{D}^p = r^p\mathbf{e}'\otimes\mathbf{e}' + \mathbf{e}''\otimes\mathbf{e}''$. We may now recall Property 8 and get

$$\mathcal{A}(p) = \mathcal{K}^p - \Im = \mathcal{T}^{-1}(r^p\mathbf{e}'\otimes\mathbf{e}')\mathcal{T} + \mathcal{T}^{-1}(\mathbf{e}''\otimes\mathbf{e}'')\mathcal{T} - \Im$$

Yet, as $\mathcal{T}^{-1}(\mathbf{e}''\otimes\mathbf{e}'')\mathcal{T} = \Im$ we have $\mathcal{A}(p) = r^p(\mathcal{T}^{-1}\mathbf{e}')\otimes(\mathbf{e}'\mathcal{T})$, i.e., r^p times the outer product of the first column of $\mathcal{T}^{-1}$ and the first row of $\mathcal{T}$. The first column of $\mathcal{T}^{-1}$ can be easily computed to be $-1/n(1,\ldots,1,r^{-1},\ldots,r^{-1})$ with $n-t$ unit entries. The thesis easily follows. □

From the above property it is also clear that another characteristic of the family of (n,t)-tailed shifts is their ability of featuring a kneading matrix whose second largest eigenvalue modulus can be made arbitrarily close to any real value in $[0,1]$. In fact given a desired modulus $\hat{r}$, if t is the integer closest to $n\hat{r}/(1+\hat{r})$, then $|r| = t/(n-t)$ is an approximation of $\hat{r}$ whose accuracy increases when $n\to\infty$.

As the importance of $|r|$ in selecting maps with optimal cross-interference has been highlighted in [4], the availability of a family of maps whose $|r|$ densely spans the whole $[0,1]$ is of great interest.

Moreover, the following property gives a general expression for the building blocks of any-order correlations of trajectories generated by an (n,t)-tailed shift.

Property 13.

$$\begin{aligned}
&\langle \mathcal{F}, \mathcal{A}(p_2)\circ\cdots\circ\mathcal{A}(p_m)\rangle = \\
&r^{\sum_{i=2}^m p_i}(-nr)^{1-m}\left[r\alpha_1(1,n) + (1-r)\alpha_1(1,n-t+1,n)\right]\times \\
&\prod_{i=2}^{m-1}\left[\alpha_i(1,n) - (1-r)\alpha_i(t+1,n-t)\right]\left[r^{-1}\alpha_m(1,n) + (1-r^{-1})\alpha_m(t+1,n)\right]
\end{aligned}$$

where $\alpha_i(k',k'') = \sum_{j=k'}^{k''}\mathbf{f}_{ij}$.

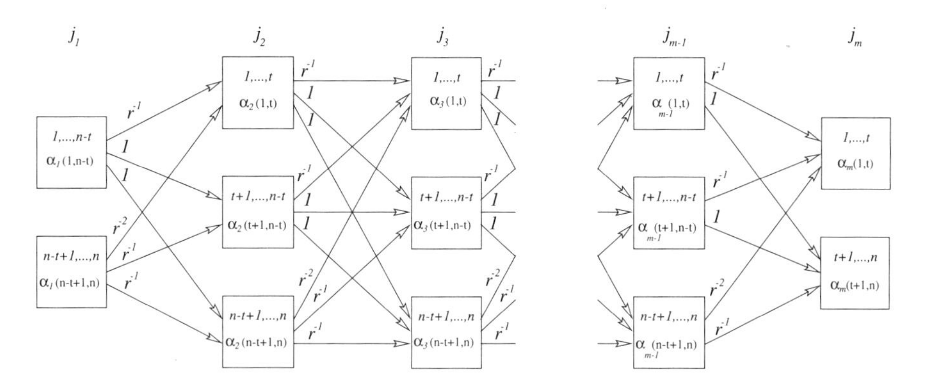

Figure 3.6: Graph representation of the block product terms making $\langle \mathcal{F}, \mathcal{A}(p_2) \circ \cdots \circ \mathcal{A}(p_m) \rangle$.

Proof. To simplify the derivation consider the matrix $\mathcal{A}'(p)$ which takes into account the block structure of $\mathcal{A}(p)$ and it is defined assuming that $\mathcal{A}(p) = -\frac{r^p}{n}\mathcal{A}'(p)$ (see Property 12). The linearity of the inner product immediately allows to write

$$\langle \mathcal{F}, \mathcal{A}(p_2) \circ \cdots \circ \mathcal{A}(p_m) \rangle = r^{\sum_{i=2}^{m} p_i} \frac{(-1)^{m-1}}{n^{m-1}} \langle \mathcal{F}, \mathcal{A}'(p_2) \circ \cdots \circ \mathcal{A}'(p_m) \rangle \quad (3.10)$$

As the generic $\mathcal{A}'(p)$ is constant in each of its four blocks, also $\mathcal{A}'(p_2) \circ \cdots \circ \mathcal{A}'(p_m)$ can be partitioned into a number of blocks in which it is constant. Once these blocks are identified, the value of $\mathcal{A}'(p_2) \circ \cdots \circ \mathcal{A}'(p_m)$ can be collected in the contribution to the global inner product so to leave a product of sums of components of $\mathbf{f}$.

From the block structure of $\mathcal{A}'(p_2)$ we get that the value of the final chain product in (3.10) will be different if either $j_1 = 1, \ldots, n-t$ or $j_1 = n-t+1, \ldots, n$. As every index j_i for $i = 2, \ldots, m-1$ is the second index of $\mathcal{A}'(p_i)$ and the first index of $\mathcal{A}'(p_{i+1})$ the value of the final chain product in (3.10) will be different if either $j_i = 1, \ldots, t$, if $j_i = t+1, \ldots, n-t$ or $j_i = n-t+1, \ldots, n$. From the block structure of $\mathcal{A}'(p_m)$ we get that the value of the final chain product in (3.10) will be different if either $j_m = 1, \ldots, t$ or $j_m = t+1, \ldots, n$.

To take all the possible combinations of the above condition into account we introduce the graph representation in Figure 3.6. Each of the possible path beginning from one of the two leftmost nodes in the graph and ending in one of the two rightmost ones correspond to one of the possible contribution to the final inner product. Along the path, quantities associated to edges multiply up to the value of the tensor in the corresponding block while quantities associated with nodes are the result of collecting such a value and extracting partial sums of the components of $\mathbf{f}$.

From inspection of Figure 3.6 it is easy to see that, if γ_0 indicates the whole inner product (and thus the sum of all the possible product paths) in (3.10) then $\gamma_0 = \left[1 + \frac{r\alpha_1(1,n-t)}{\alpha_1(n-t+1,n)}\right] \gamma_1$ where γ_1 is the contribution to the inner product in (3.10) of all the product paths beginning with the bottom leftmost node (i.e. that corresponding to the condition $j_1 = n-t+1, \ldots, n$). Noting that, in general $\alpha_i(k', k'') = \alpha_i(1, k'') - \alpha_i(1, k'-1)$, the above expression can be simplified to give

$$\gamma_0 = \frac{r\alpha_1(1,n) + (1-r)\alpha_1(n-t+1,n)}{\alpha_1(n-t+1,n)}\gamma_1$$

Further inspection of the column related to j_2 reveals that

$$\begin{aligned}
\gamma_1 &= \alpha_1(n-t+1,n)\times \\
&\left[r^{-2}\left(r\frac{\alpha_2(1,t)}{\alpha_2(n-t+1,n)}\gamma_2\right) + r^{-1}\left(r\frac{\alpha_2(t+1,n-t)}{\alpha_2(n-t+1,n)}\gamma_2\right) + r^{-1}\gamma_2\right] = \\
&= r^{-1}\alpha_1(n-t+1,n)\frac{\alpha_2(1,n) - (1-r)\alpha_2(t+1,n-t)}{\alpha_2(n-t+1,n)}\gamma_2
\end{aligned}$$

where γ_2 is the contribution to the inner product in (3.10) of all the product paths starting from the second bottom node of the second column of the graph (i.e. such that $j_2 = n-t+1, \ldots, n$ and considering $j_2, j_3, \ldots, j_m$).

Let us now define γ_i to be the contribution to the inner product in (3.10) of all the product paths starting from the bottom note of the i-th column of the graph (i.e. such that $j_i = n-t+1$ and considering $j_i, j_{i+1}, \ldots, j_m$). Again, direct graph inspection reveals that, for $i = 1, \ldots, m-2$,

$$\gamma_i = r^{-1}\alpha_1(n-t+1,n)\frac{\alpha_{i+1}(1,n) - (1-r)\alpha_{i+1}(t+1,n-t)}{\alpha_{i+1}(n-t+1,n)}\gamma_{i+1}$$

A final graph inspection easily gives that

$$\begin{aligned}
\gamma_{m-1} &= \alpha_{m-1}(n-t+1,n)\left[r^{-2}\alpha_m(1,t) + r^{-1}\alpha_m(t+1,n)\right] = \\
&= r^{-1}\alpha_{m-1}(n-t+1,n)\left[r^{-1}\alpha_m(1,n) + (1-r^{-1})\alpha_m(t+1,n)\right]
\end{aligned}$$

so that the required result is then given by trivial back-substitution of the γ_i. □

Property 13 has a straightforward specialization to the case in which $\alpha_i(1,n) = \sum_{j=1}^{n} \mathbf{f}_{ij} = 0$ for every i. In this case the statement becomes

$$\begin{aligned}
\langle \mathcal{F}, \mathcal{A}(p_2)\circ\cdots\circ\mathcal{A}(p_m)\rangle &= r^{\sum_{i=2}^m p_i}(-nr)^{1-m}(-1)^{m-2}(1-r)^{m-1}(1-r^{-1})\times \\
&\alpha_1(n-t+1,n)\prod_{i=2}^{m-1}\alpha_i(t+1,n-t)\alpha_m(t+1,n)
\end{aligned}$$

and, noting that $(1-r^{-1}) = n/t$,

$$\begin{aligned}
\langle \mathcal{F}, \mathcal{A}(p_2)\circ\cdots\circ\mathcal{A}(p_m)\rangle &= r^{\sum_{i=2}^m p_i}\frac{(-1)^m n}{t^m}\times \\
&\alpha_1(n-t+1,n)\prod_{i=2}^{m-1}\alpha_i(t+1,n-t)\alpha_m(t+1,n)
\end{aligned}$$

3.3.5 Application to DS-CDMA Systems

To obtain the spreading sequences for a CBDS-CDMA system we will assume in the following that the quantization function $Q : X \mapsto Z$ is constant in each interval $X_1, \ldots, X_n$ and it is defined by means of the n-dimensional symbols vector $\mathbf{q} = (\mathbf{q}_1, \ldots, \mathbf{q}_n)$ so that $Q(X_j) = \{\mathbf{q}_j\}$ and $\sum_{j=1}^{n} \mathbf{q}_j = 0$.

The partial correlation properties of the spreading sequences and their periodic nature are taken into account in terms as (3.2) by means of $\Gamma_{N,\tau}(\underline{x}^u, \underline{x}^v)$ where $\underline{x}^u$ represents here the slice of chaotic trajectory generating the spreading sequence of the u-th user.

As reported in Section 3.2 two kinds of effects are present in a DS-CDMA communication systems affected by multipath propagation.

The first disturbance Ψ_s^v, due to cross-interference, i.e., to the presence of users other than the useful one whose spreading sequence are partially correlated with the useful one, can be thought to enter the final symbol reconstruction stage as a random Gaussian disturbance term [2]. As it will be clarified in the following sections, the variance (and thus the impact) of such a disturbance depends on terms of the kind

$$
\begin{aligned}
&\mathbf{E}^{u \neq v}_{\underline{x}^u, \underline{x}^v} \left[\Gamma_{N,\tau}(\underline{x}^u, \underline{x}^v) \Gamma^*_{N,\tau+l}(\underline{x}^u, \underline{x}^v) \right] = \\
&\quad = \sum_{j=0}^{N-\tau-1} \sum_{k=0}^{N-\tau-l-1} \mathbf{E}^{u \neq v}_{\underline{x}^u, \underline{x}^v} \left[Q(x_j^u) Q^*(x_{j+\tau}^v) Q^*(x_k^u) Q(x_{k+\tau+l}^v) \right] \\
&\quad = \sum_{j=0}^{N-\tau-1} \sum_{k=0}^{N-\tau-l-1} \mathbf{E}_{\underline{x}^u} \left[Q(x_j^u) Q^*(x_k^u) \right] \mathbf{E}_{\underline{x}^v} \left[Q^*(x_{j+\tau}^v) Q(x_{k+\tau+l}^v) \right]
\end{aligned}
\tag{3.11}
$$

where the expectation is taken over all the possible pairs of spreading sequences of two different users and l is either 0 or 1.

The other disturbance Ξ_s^v, due to self-interference, i.e., to the presence of delayed versions of the useful signal which add up at the receiver, gives rise to correlation between delayed versions of the same spreading sequence. This effect is then responsible for a further transmission degradation which depends on terms of the kind

$$
\mathbf{E}_{\underline{x}^v} \left[|\Gamma_{N,\tau}(\underline{x}^v, \underline{x}^v)|^2 \right] = \sum_{j=0}^{N-\tau-1} \sum_{k=0}^{N-\tau-1} \mathbf{E}_{\underline{x}^v} [Q(x_j^v) Q^*(x_{j+\tau}^v) Q^*(x_k^v) Q(x_{k+\tau}^v)] \tag{3.12}
$$

where the expectation is taken over all the possible spreading sequences and $\tau > 0$.

Hence, the quantities involved in the performance of chaos-based DS-CDMA systems affected by multipath depend on second-, third-, and fourth-order correlations of the spreading sequences. Such quantities can be computed by means of the previous results when (n, t)-tailed shifts are adopted to generate spreading sequences with $\rho_0 = \bar{\rho} = 1$.

As far as the cross-interference term is concerned, recall (3.11) along with the stationarity highlighted by Property 3 and Equation (3.5) to write

$$\begin{aligned}\mathbf{E}^{u\neq v}_{\underline{x}^u,\underline{x}^v}\left[\Gamma_{N,\tau}(\underline{x}^u,\underline{x}^v)\Gamma^*_{N,\tau+l}(\underline{x}^u,\underline{x}^v)\right] = \\ = (l+1)(N-\tau-l)\,\mathbf{E}_{\underline{x}^u}[|Q(x^u_0)|^2]\,\mathbf{E}_{\underline{x}^v}[Q^*(x^v_0)Q(x^v_l)] \\ +2\sum_{k=1}^{N-\tau-l-1}(N-\tau-l-k)\,\mathbf{E}_{\underline{x}^u}[Q(x^u_0)Q^*(x^u_k)]\,\mathbf{E}_{\underline{x}^v}[Q^*(x^v_0)Q(x^v_{k+l})]\end{aligned} \tag{3.13}$$

which holds for $l = 0$ or $l = 1$.

In the above expression a statistical average and a second-order correlation appear. The average can be easily computed to give

$$C_0 = \mathbf{E}_{\underline{x}^u}[|Q(x^u_0)|^2] = \frac{1}{n}\sum_{j=1}^{n}|\mathbf{q}_j|^2 \tag{3.14}$$

and the second-order correlation can be derived from Example 1 and Property 13 so that setting

$$C_1 = \frac{1}{t^2}\sum_{j=n-t+1}^{n}\mathbf{q}_j\sum_{j=t+1}^{n}\mathbf{q}^*_j$$

one gets

$$\mathbf{E}_{\underline{x}^v}\left[Q(x^v_0)Q^*(x^v_{p_2})\right] = C_1 r^{p_2}$$

The two expressions for the average and the second-order correlation can be substituted into Equation (3.13) to obtain a mixed polynomial-exponential sum that can be closed by means of [20]

$$\sum_{k=1}^{N-\tau-1}(N-\tau-k)r^k = \frac{r}{(1-r)^2}\left[(N-\tau-1)-(N-\tau)r+r^{N-\tau}\right]$$

to obtain the generic term of the cross-interference disturbance as

$$\begin{aligned}\mathbf{E}^{u\neq v}_{\underline{x}^u,\underline{x}^v}\left[\Gamma_{N,\tau}(\underline{x}^u,\underline{x}^v)\Gamma^*_{N,\tau+l}(\underline{x}^u,\underline{x}^v)\right] = (l+1)(N-\tau-l)C^*_l C_0 r^l \\ +2|C_1|^2\frac{r^{2+l}}{(1-r^2)^2}\left[(1-r^2)(N-\tau-l)+r^{2(N-\tau-l)}-1\right]\end{aligned} \tag{3.15}$$

Almost the same path can be followed for the self-interference term. In this case the stationarity highlighted by Property 3 allow us to write

$$\begin{aligned}\mathbf{E}_{\underline{x}^v}[|\Gamma_{N,\tau}(\underline{x}^v,\underline{x}^v)|^2] = (N-\tau)\,\mathbf{E}_{\underline{x}^v}[|Q(x^v_0)|^2|Q(x^v_\tau)|^2] \\ +2\sum_{k=1}^{N-\tau-1}(N-\tau-k)\,\mathrm{Re}\left[\mathbf{E}_{\underline{x}^v}\left[Q(x^v_0)Q^*(x^v_\tau)Q^*(x^v_k)Q(x^v_{\tau+k})\right]\right]\end{aligned} \tag{3.16}$$

where second- and fourth-order correlation appear. A deeper look at (3.16) reveals that if $\tau \leq (N-1)/2$ and when the index k of the sum equals τ, the third-order correlation $\mathbf{E}_{\underline{x}^v}\left[Q(x_0^v)\left(Q^*(x_\tau^v)\right)^2 Q(x_{2\tau}^v)\right]$ is also present.

Again Example 1 and Property 13 help in deriving an expression for the second-order correlation as

$$\mathbf{E}_{\underline{x}^v}\left[|Q(x_0^v)|^2|Q(x_{p_2}^v)|^2\right] = C_0^2 + C_2 r^{p_2}$$

with

$$C_2 = -\frac{1}{n^2}\left[\sum_{j=1}^{n}|\mathbf{q}_j|^2 + (r^{-1}-1)\sum_{j=n-t+1}^{n}|\mathbf{q}_j|^2\right] \times \left[r^{-1}\sum_{j=1}^{n}|\mathbf{q}_j|^2 + (1-r^{-1})\sum_{j=t+1}^{n}|\mathbf{q}_j|^2\right]$$

Then, Example 2 and Property 13 give an expression for the third-order correlation that may appear when $\tau \leq (N-1)/2$ as

$$\mathbf{E}_{\underline{x}^v}\left[Q(x_0^v)(Q^*(x_{p_2}^v))^2 Q(x_{p_3}^v)\right] = C_3 r^{p_3}$$

with

$$C_3 = -\frac{1}{nt^2}\sum_{j=n-t+1}^{n}\mathbf{q}_j\sum_{j=t+1}^{n}\mathbf{q}_j\left[r^{-1}\sum_{j=1}^{n}(\mathbf{q}_j^*)^2 - (r^{-1}-1)\sum_{j=t+1}^{n-t}(\mathbf{q}_j^*)^2\right]$$

Finally, Example 3 and Property 13, may be used to give an expression for the fourth-order correlations

$$\mathbf{E}_{\underline{x}^v}\left[Q(x_0^v)Q^*(x_{p_2}^v)Q^*(x_{p_3}^v)Q(x_{p_4}^v)\right] = |C_1|^2 r^{p_2-p_3+p_4} + C_4 r^{p_4}$$

with

$$C_4 = \frac{1}{t^4}\sum_{j=n-t+1}^{n}\mathbf{q}_j\sum_{j=t+1}^{n}\mathbf{q}_j\left(\sum_{j=t+1}^{n-t}\mathbf{q}_j^*\right)^2$$

In the light of these results a final closed form for the generic term of the self-interference disturbance can be obtained distinguishing the two cases $\tau \leq (N-1)/2$ and $\tau > (N-1)/2$. For $\tau \leq (N-1)/2$ we obtain

$$\begin{aligned}\mathbf{E}_{\underline{x}^v}[|\Gamma_{N,\tau}(\underline{x}^v,\underline{x}^v)|^2] &= (N-\tau)(C_0^2 + C_2 r^\tau) \\ &+\left\{|C_1|^2[(N-\tau)^2-(N-\tau)] + 2(N-2\tau)[\mathrm{Re}[C_3-C_4]-|C_1|^2]\right\}r^{2\tau} \\ &+\frac{2\,\mathrm{Re}[C_4]r}{(1-r)^2}\left[(N-\tau-1)r^\tau-(N-\tau)r^{\tau+1}+r^N\right]\end{aligned} \tag{3.17}$$

while for $\tau > (N-1)/2$ we obtain

$$\begin{aligned}\mathbf{E}_{\underline{x}^v}[|\Gamma_{N,\tau}(\underline{x}^v,\underline{x}^v)|^2] &= (N-\tau)(C_0^2 + C_2 r^\tau) + |C_1|^2[(N-\tau)^2 \\ &-(N-\tau)]r^{2\tau} + \frac{2\,\mathrm{Re}[C_4]r}{(1-r)^2}\left[(N-\tau-1)r^\tau-(N-\tau)r^{\tau+1}+r^N\right]\end{aligned} \tag{3.18}$$

3.4 Performance Over a Non-Selective Channel

Our investigation on performance is based on the Standard Gaussian Approximation (SGA), i.e., sums of independent random contributions are considered to be Gaussian random variables, characterized only by means of their average and variance. Let us note that Ψ_s^v is a zero-mean and if we further assume that the speed of variation of the channel is comparable with the speed of variation of the transmitted signal, expectations on both channel-related and transmission-related random variables can be taken at the same time. This assumption makes it possible to consider both information symbols and channel delays and phases at the same level of investigation. These are the conditions under which the error probability is the most sensible merit figure. Other performance indicators, like the *outage probability*, i.e., the probability that the bit error probability increases above a certain threshold, which may be of help when such a stationarity does not hold, are out of the scope of this analysis.

In order to estimate the bit error probability, we need to compute the variance $(\sigma_\Psi^v)^2$ of the cross-interference term. As reported in [2, 3] this quantity can be expressed as

$$(\sigma_\Psi^v)^2 = \frac{1}{8N^2}\sum_{u\neq v}\mathbf{E}\left[|X^{uv}(t^{uv})|^2 + |Y^{uv}(t^{uv})|^2\right]$$

where, considering the periodic nature of the functions X^{uv} and Y^{uv}, the dependence of the expectations on τ has been suppressed assuming to subsequently take the expectation over t^{uv} uniformly distributed over $[0,T]$. With this, and taking the expectations over the information symbols S_s^u, and the relative phases θ^{uv} and delays t^{uv}, we have

$$(\sigma_\Psi^v)^2 = \frac{1}{24N^3}\sum_{u\neq v}\sum_{\tau=1-N}^{N-1}\left[2|\Gamma_{N,\tau}(\underline{x}^u,\underline{x}^v)|^2 + \mathrm{Re}[\Gamma_{N,\tau}(\underline{x}^u,\underline{x}^v)\Gamma^*_{N,\tau+1}(\underline{x}^u,\underline{x}^v)]\right]$$

Note how the interference is summed over all the possible interfering users. Hence, we may use SGA again to average over all the possible pairs users, and obtain a unique $(\sigma_\Psi)^2$ (accounting for the average contribution to the error of the users interfering with the generic receiver) by simple multiplication by $U-1$.

$$(\sigma_\Psi)^2 = \mathbf{E}_{\underline{x}^v}[(\sigma_\Psi^v)^2] = \frac{U-1}{24N^3}\Bigg\{2\,\mathbf{E}^{u\neq v}_{\underline{x}^u,\underline{x}^v}[|\Gamma_{N,0}(\underline{x}^u,\underline{x}^v)|^2]$$

$$+\mathrm{Re}\left[\mathbf{E}^{u\neq v}_{\underline{x}^u,\underline{x}^v}[\Gamma_{N,0}(\underline{x}^u,\underline{x}^v)\Gamma^*_{N,1}(\underline{x}^u,\underline{x}^v)]\right] + \sum_{\tau=1}^{N-1}\Bigg[4\,\mathbf{E}^{u\neq v}_{\underline{x}^u,\underline{x}^v}[|\Gamma_{N,\tau}(\underline{x}^u,\underline{x}^v)|^2]$$

$$+\mathrm{Re}\left[\mathbf{E}^{u\neq v}_{\underline{x}^u,\underline{x}^v}[\Gamma_{N,\tau}(\underline{x}^u,\underline{x}^v)\Gamma^*_{N,\tau+1}(\underline{x}^u,\underline{x}^v)]\right] + \mathrm{Re}\left[\mathbf{E}^{u\neq v}_{\underline{x}^u,\underline{x}^v}[\Gamma^*_{N,\tau}(\underline{x}^u,\underline{x}^v)\Gamma_{N,\tau-1}(\underline{x}^u,\underline{x}^v)]\right]\Bigg]\Bigg\}$$

in which, changing the index in the third term of the sum for $\tau = 1$ to $N-1$

we obtain

$$(\sigma_\Psi)^2 = \mathbf{E}_{\underline{x}^v}[(\sigma_\Psi^v)^2] = \tag{3.19}$$

$$= \frac{U-1}{12N^3}\left\{\mathbf{E}^{u\neq v}_{\underline{x}^u,\underline{x}^v}[|\Gamma_{N,0}(\underline{x}^u,\underline{x}^v)|^2] + \mathrm{Re}\left[\mathbf{E}^{u\neq v}_{\underline{x}^u,\underline{x}^v}[\Gamma_{N,0}(\underline{x}^u,\underline{x}^v)\Gamma^*_{N,1}(\underline{x}^u,\underline{x}^v)]\right]\right.$$

$$\left.+\sum_{\tau=1}^{N-1}\left[2\,\mathbf{E}^{u\neq v}_{\underline{x}^u,\underline{x}^v}[|\Gamma_{N,\tau}(\underline{x}^u,\underline{x}^v)|^2] + \mathrm{Re}\left[\mathbf{E}^{u\neq v}_{\underline{x}^u,\underline{x}^v}[\Gamma_{N,\tau}(\underline{x}^u,\underline{x}^v)\Gamma^*_{N,\tau+1}(\underline{x}^u,\underline{x}^v)]\right]\right]\right\}$$

We will now specialize the results of Section 3.3 to a particular quantization which is often used in telecommunication literature (see, e.g., [21]). More specifically, we suppose that the codomain of Q contains the L symbols $\hat{\mathbf{q}}_j = e^{\mathrm{i}\,j2\pi/L}$ $j = 0,\ldots,L-1$, and take Q such that $Q(X_j) = \{\hat{\mathbf{q}}_{(\lfloor jL/n\rfloor+\Delta)(\mathrm{mod}\ L)}\}$ for some integer Δ. We will further assume that n/L is integer and greater than t so that the quantization partitions the set of X_j into n/L subsets each containing L adjacent intervals and maps every point in any of the intervals in the same subset to the same complex root of 1. The offset Δ rotates the association between interval subsets and the symbol constellation, while the assumption $n/L > t$ means that the intervals corresponding to the tails of the map are always in the same subset.

With this, from (3.14) it readily follows that $C_0 = 1$. Additionally, by using (3.15) and by defining the auxiliary functions $F_0^M(x) = \sum_{l=1}^{M-1} x^l$, $(\sigma_\Psi)^2$ can be expressed in closed form as

$$(\sigma_\Psi)^2 = \frac{U-1}{12N^3}\left\{N^2 + rN(N-1)\,\mathrm{Re}[C_1] + \frac{2|C_1|^2r^2}{(1-r^2)^2}\left[(1-r^2)\times\right.\right. \tag{3.20}$$

$$\left.\left.\left(N^2 + r\frac{N(N-1)}{2}\right) - N(r+2) + r^{2N} + r^{2N-1} + 1 + F_0^N(r^{-2})\left(2r^{2N}+r^{2N-1}\right)\right]\right\}$$

where C_1 is can be computed as stated by the following property.

Property 14. *The constant C_1 assumes the value:*

$$C_1 = -e^{-\mathrm{i}\,\frac{2\pi}{L}}$$

which is independent of the rotation Δ of the constellation.

Proof. Few calculations exploiting the assumptions on Q give that

$$\sum_{j=n-t+1}^{n} \mathbf{q}_j = t\hat{\mathbf{q}}_{(-1+\Delta)(\mathrm{mod}\ L)}$$

$$\sum_{j=t+1}^{n} \mathbf{q}_j = -t\hat{\mathbf{q}}_{\Delta(\mathrm{mod}\ L)}$$

Then $C_1 = -\hat{\mathbf{q}}_{(-1+\Delta)(\mathrm{mod}\ L)}\hat{\mathbf{q}}_{-\Delta(\mathrm{mod}\ L)} = -\hat{\mathbf{q}}_{-1}$ where we exploited the fact the the L-th complex roots of 1 are isomorphic to the classical field of integers modulus L.

□

As a particular case consider the limit $r \to 0$ which can be achieved by an $(n,0)$-tailed shift. One may recall Example 1 and Property 13 to note that

$$\mathbf{E}_{\underline{x}^v}[Q(x_0^v)Q^*(x_p^v)] = C_1 r^p$$

which indicates that adopting the n-way Bernoulli shift the resulting sequences are made of uncorrelated symbols, i.e., they conform to the requirements for ideal random sequences that are often taken as a reference case and that are well approximated by classical maximum-length sequences.

Once the average variance $(\sigma_\Psi)^2$ has been computed, the bit error probability P_{err} can be expressed as [21]

$$P_{\text{err}} = \frac{1}{2}\,\text{erfc}\,\sqrt{\rho}$$

where ρ is a global signal-to-interference ratio defined as

$$\rho = \frac{1}{8\,(\sigma_\Psi)^2} = \frac{1}{(U-1)R} \tag{3.21}$$

It is worthwhile to notice that the quantity

$$R = \frac{2}{3N^3}\left\{\mathbf{E}^{u\neq v}_{\underline{x}^u,\underline{x}^v}[|\Gamma_{N,0}(\underline{x}^u,\underline{x}^v)|^2] + \text{Re}\left[\mathbf{E}^{u\neq v}_{\underline{x}^u,\underline{x}^v}[\Gamma_{N,0}(\underline{x}^u,\underline{x}^v)\Gamma^*_{N,1}(\underline{x}^u,\underline{x}^v)]\right]\right. \tag{3.22}$$
$$\left. + \sum_{\tau=1}^{N-1}\left[2\,\mathbf{E}^{u\neq v}_{\underline{x}^u,\underline{x}^v}[|\Gamma_{N,\tau}(\underline{x}^u,\underline{x}^v)|^2] + \text{Re}\left[\mathbf{E}^{u\neq v}_{\underline{x}^u,\underline{x}^v}[\Gamma_{N,\tau}(\underline{x}^u,\underline{x}^v)\Gamma^*_{N,\tau+1}(\underline{x}^u,\underline{x}^v)]\right]\right]\right\}$$

defined by (3.21) and (3.19) assumes the significance of the expected *interference-to-signal ratio per interfering user*. In other terms, this is the expected degradation in system performance when a new user is added and can be used as merit figure to compare standard and chaos-based DS-CDMA systems [2, 21].

3.4.1 Numerical Results

Let us consider a spreading factor $N = 70$ and spreading sequences generated by means of an (n,t)-tailed shifts.

In Figure 3.7 the normalized cross-interference term, $(\sigma_\Psi)^2/(U-1)$, is reported as a function of r with $L = 2$ and $L = 5$. Extrapolation for $r > 0$, which is not achievable with (n,t)-tailed shifts is reported to clarify the trends.

Simulations points are reported for $n = 10$ and $t = 0,1,2,3$ for $L = 2$ and $t = 0,1,2$ for $L = 5$ ($t = 3$ does not satisfy the assumption that $n/L > t$). The variance of the cross-interference term due to Gold and maximum-length sequences is also reported. Trivial visual inspection is enough to verify the match between theoretical computation and actual data.

Note how multi-level chaos-based spreading cannot improve performance with respect to classical solutions. On the contrary, binary spreading may benefit from chaos-based techniques, and sequences generated by a $(10,2)$-tailed shift are reasonably close to the optimal performance achievable with (n,t)-tailed shifts and $L = 2$.

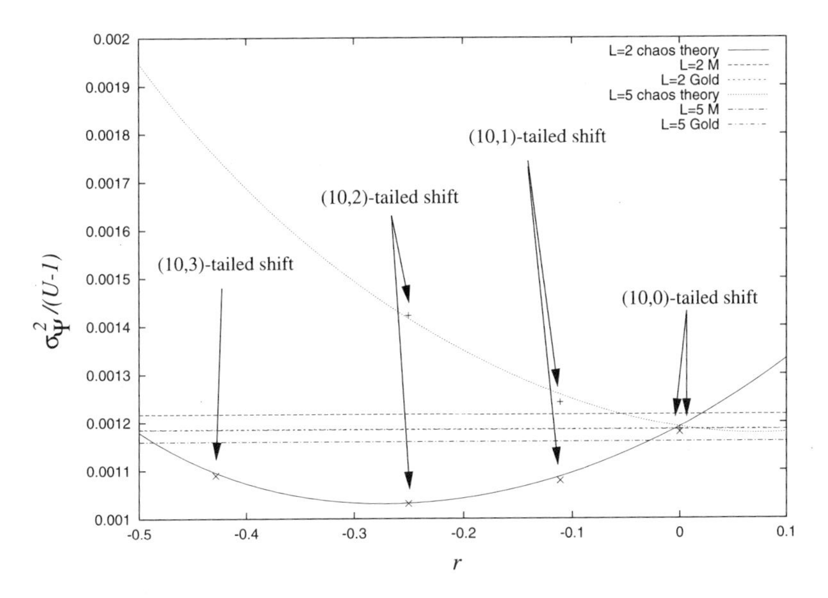

Figure 3.7: $(\sigma_\Psi)^2$ as a function of r for $L = 2$ and $L = 5$. Gold sequences produce almost identical results in the two cases.

3.4.2 Nearly Optimal Performance Over a Non-Selective Channel

The expression (3.22) of R depends in general on the partial cross-correlation functions of the spreading sequences. Yet, if all the sequences can be thought of as independent and if they satisfy a mild form of second-order stationarity so that $\mathbf{E}_{\underline{x}^v}[Q(x_m^v)Q^*(x_n^v)] = \mathbf{E}_{\underline{x}^v}[Q(x_0^v)Q^*(x_{|n-m|}^v)]$, few algebraic manipulations arrive at an alternative expression depending only on the auto-correlation function $A_k = \mathbf{E}_{\underline{x}^v}[Q(x_0^v)Q^*(x_k^v)]$, namely [4]

$$R = \frac{2}{3N} + \frac{4}{3N^3}\sum_{k=1}^{N-1}\left[(N-k)^2\left((A_k^R)^2 + (A_k^I)^2\right) + \frac{(N-k+1)(N-k)}{2}\left(A_{k-1}^R A_k^R + A_{k-1}^I A_k^I\right)\right] \tag{3.23}$$

where $A_k^R = \mathrm{Re}(A_k)$ and $A_k^I = \mathrm{Im}(A_k)$ and where $A_0 = \mathbf{E}_{\underline{x}^v}[|Q(x_0^v)|^2] = 1$.

The expression of R is thus the sum of a linear function and a positive definite quadratic form whose unique minimum is achieved when

$$\frac{\partial R}{\partial A_k^R} = 0 \quad \text{and} \quad \frac{\partial R}{\partial A_k^I} = 0$$

for $k = 1, \ldots, N-1$. Indicate with Y_k either A_k^R or A_k^I, and note how the structure of (3.23) is such that

$$\frac{\partial R}{\partial Y_k} = 2(N-k)^2 Y_k + \frac{(N-k+1)(N-k)}{2} Y_{k-1} + \frac{(N-k)(N-k-1)}{2} Y_{k+1} = 0 \tag{3.24}$$

for $k = 1, \ldots, N-1$ with $Y_N = 0$. This set of simultaneous equations (3.24) can be solved setting

$$A_k^R = \frac{\alpha_k (N-1) A_1^R - \alpha_{k-1} N}{N-k} \quad \text{and} \quad A_k^I = \frac{\alpha_k (N-1) A_1^I}{N-k} \tag{3.25}$$

where the α_k are real coefficients such that $\alpha_{-1} = -1$, $\alpha_0 = 0$ and, by substituting (3.25) in (3.24), $\alpha_{k+1} = -4\alpha_k - \alpha_{k-1}$ for $k = 0, \ldots, N-1$. With this the two equations

$$\frac{\partial R}{\partial A_{N-1}^R} = 2A_{N-1}^R + A_{N-2}^R = 0 \quad \text{and} \quad \frac{\partial R}{\partial A_{N-1}^I} = 2A_{N-1}^I + A_{N-2}^I = 0$$

can be solved to obtain

$$A_1^R = \frac{N}{N-1} \frac{\alpha_{N-1}}{\alpha_N} \quad \text{and} \quad A_1^I = 0$$

so that, respectively

$$A_k^R = \frac{N}{N-k} \frac{\alpha_k \alpha_{N-1} - \alpha_{k-1} \alpha_N}{\alpha_N} \quad \text{and} \quad A_k^I = 0$$

for $k = 2, \ldots, N-1$. This proves that *the optimal auto-correlation is real.*

Furthermore, the Fibonacci-like recursion of the α_k can be unrolled to give

$$\alpha_k = (-1)^{k+1} (h^{-k} - h^k) / (h^{-1} - h)$$

with $h = 2 - \sqrt{3}$ and thus

$$A_k = A_k^R = (-1)^k \frac{N}{N-k} \frac{h^{k-N} - h^{N-k}}{h^{-N} - h^N} \tag{3.26}$$

which is the auto-correlation guaranteeing the minimum R. Note how the A_k are not only real but also *with alternating signs* so that the positive contributions to R due to $|A_k|^2$ are counterbalanced by the negative contributions of $A_{k-1}A_k$.

By substituting (3.26) into (3.23), the minimum interference is

$$R = \frac{\sqrt{3}}{3N} \frac{h^{-2N} - h^{2N}}{h^{-2N} + h^{2N} - 2}$$

which, for large N, tends to $\sqrt{3}/(3N)$ (actually $|R - \sqrt{3}/(3N)| < 10^{-12}$ for $N \geq 10$) which is less than the classical $R = 2/(3N)$ obtained for ideal random sequences where $A_k = 0$ for $k > 0$.

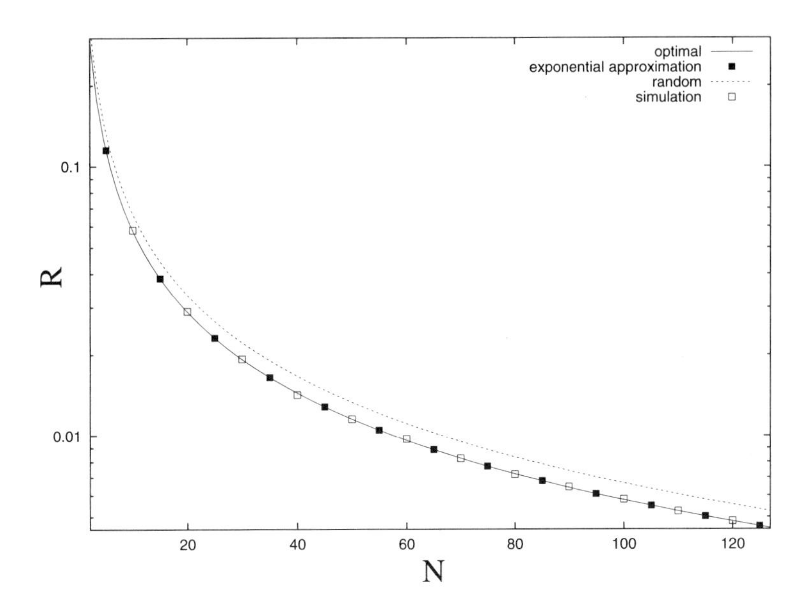

Figure 3.8: Performance of chaos-based spreading compared with optimal and random spreading.

The most significant values of A_k are those higher in modulus, i.e., those with $k \ll N$. In these cases (3.26) reduces to $A_k \simeq (-h)^k$ and, from earlier results on sequences with exponentially vanishing auto-correlations [4], one may reasonably expect that chaos-based spreading can be designed to suitably approximate the optimal performance.

Let us consider (n,t)-tailed shifts as sequence generators and, since the auto-correlation function defining the optimal profile is real-valued, we will refer to binary spreading sequences only, assuming $Q(X_j) = -1$ for $j \leq n/2$ and $Q(X_j) = +1$ for $j > n/2$.

With this, we may recall Example 1, Property 13, and Property 14 with $L = 2$ to note that

$$A_k = \mathbf{E}_{\underline{x}^v}[Q(x_0^v)Q^*(x_v^k)] = r^k$$

where $r = -t/(n-t)$. For any given n, this trend approximates the optimal one choosing t as the integer closest to $nh/(1+h)$. The accuracy of this approximation obviously increases as $n \to \infty$. Since $h \simeq 0.2679$, for $n = 10$ we have $t = 2$ and the approximation is with $r = -0.25$. Note how the above considerations represent therefore a theoretical ground for the numerical results reported in Section 3.4.1.

To further verify this result, simulations of the resulting sequences for $U = 10$ users are reported in Figure 3.8 in which the theoretical optimum, the theoretical performance of the exponential approximation, and the theoretical performance of purely random sequences are reported for $2 \leq N \leq 128$.

U	$(10,1)$ tailed shift	(10,2) tailed shift
10	1.21	1.23
20	1.17	1.18
30	1.15	1.17

Table 3.1: Maximum value of $(2/3N)/R$ over ten thousand trials.

Note how the exponential auto-correlation of chaos-based spreading results in a performance improvement with respect to purely random spreading. Such an improvement leads to an extremely good approximation of the maximum achievable performance of any system which employs second-order stationary sequences and for which the SGA is valid.

Finally, let us remark that the improvement of $2/\sqrt{3} = 1.154$ is an average improvement by considering a large set of users and sequences and without any selection of the starting points x_0^v. On the other hand, it is possible to select a particular set of sequences by means of an optimization procedure that is strongly dependent on the active user number in the system. In this case better performance can be found for low user numbers as reported in Table 3.1. Of course, the adoption of a different set of sequences for any number of users in the system implies a renegotiation between terminals whenever the load changes.

3.5 Performance Over a Selective Fading Channel

Our investigation of performance is again based on SGA. Obviously, Ξ_s^v in addition to Ψ_s^v is zero-mean terms. Once more we will assume that the speed of variation of the channel is comparable with the speed of variation of the transmitted signal.

Let us first consider the variance $(\sigma_{\Xi}^v)^2$ of the self-interference term as

$$(\sigma_{\Xi}^v)^2 = \frac{1}{8N^2}\sum_{\tau=1}^{\infty}\beta_\tau^2\left[\left|X^{vv}\left(\tau\frac{T}{N}\right)\right|^2 + \left|Y^{vv}\left(\tau\frac{T}{N}\right)\right|^2\right]$$

which, recalling the periodicity of the functions X^{uv} and Y^{uv} and their definition in terms of correlation between spreading sequences, can be rewritten as

$$(\sigma_{\Xi}^v)^2 = \frac{1}{8N^2}\left\{N^2\sum_{k=1}^{\infty}\beta_{kN}^2 + \sum_{\tau=1}^{N-1}\left[|\Gamma_{N,\tau-N}(\underline{x}^v,\underline{x}^v)|^2 + |\Gamma_{N,\tau}(\underline{x}^v,\underline{x}^v)|^2\right]\sum_{k=0}^{\infty}\beta_{kN+\tau}^2\right\}$$

Due to the exponential nature of the power attenuation, all the series in the previous expression can be given a closed form. If we also recall that by its very definition $|\Gamma_{N,\tau-N}(\underline{x}^v,\underline{x}^v)|^2 = |\Gamma_{N,N-\tau}(\underline{x}^v,\underline{x}^v)|^2$, we finally obtain

$$(\sigma_{\Xi}^v)^2 = \frac{A^2}{8N^2(1-e^{-2BN})}\sum_{\tau=1}^{N}\left[|\Gamma_{N,N-\tau}(\underline{x}^v,\underline{x}^v)|^2 + |\Gamma_{N,\tau}(\underline{x}^v,\underline{x}^v)|^2\right]e^{-2B\tau}$$

Then, the SGA allows us to average over all the possible users to define a unique $(\sigma_\Xi)^2$ accounting for the average contribution of self-interference to the error on the generic receiver

$$(\sigma_\Xi)^2 = \mathbf{E}_{\underline{x}^v}[(\sigma_\Xi^v)^2] = \frac{A^2}{8N^2(1-e^{-2BN})} \times \tag{3.27}$$
$$\times \sum_{\tau=1}^{N} \left[\mathbf{E}_{\underline{x}^v}[|\Gamma_{N,N-\tau}(\underline{x}^v,\underline{x}^v)|^2] + \mathbf{E}_{\underline{x}^v}[|\Gamma_{N,\tau}(\underline{x}^v,\underline{x}^v)|^2] \right] e^{-2B\tau}$$

By employing the tensor approach of Section 3.3 with the same mapping assumption as in Section 3.4 and using the expression (3.17) for $\tau \leq \lfloor (N-1)/2 \rfloor$ and (3.18) for $\tau > \lfloor (N-1)/2 \rfloor$, $(\sigma_\Xi)^2$ can be given a closed for expression as

$$(\sigma_\Xi)^2 = \frac{A^2}{8N^2(1-e^{-2BN})} \times \tag{3.28}$$
$$\times \left\{ N^2 e^{-2BN} + N F_0^N(e^{-2B}) + |C_1|^2 [\bar{F}_2^N(r^2 e^{-2B}) - \bar{F}_1^N(r^2 e^{-2B}) + r^{2N} F_2^N(r^{-2} e^{-2B}) \right.$$
$$-r^{2N} F_1^N(r^{-2} e^{-2B})] + \frac{2\,\mathrm{Re}[C_4] r}{(1-r)^2} \left[2 r^N F_0^N(e^{-2B}) - F_0^N(r e^{-2B}) - r^N F_0^N(r^{-1} e^{-2B}) \right.$$
$$\left. + \bar{F}_1^N(r e^{-2B})(1-r) + F_1^N(r^{-1} e^{-2B})(1-r) r^N \right] + 2[\mathrm{Re}[C_3 - C_4] - |C_1|^2] \left\{ e^{-2BN} \times \right.$$
$$\left.\left. \times \left[N F_0^{\lfloor \frac{N-1}{2} \rfloor}(r^2 e^{2B}) - 2 F_1^{\lfloor \frac{N-1}{2} \rfloor}(r^2 e^{2B}) \right] + N F_0^{\lfloor \frac{N-1}{2} \rfloor}(r^2 e^{-2B}) - 2 F_1^{\lfloor \frac{N-1}{2} \rfloor}(r^2 e^{-2B}) \right\} \right\}$$

where F_q^M and $\bar{F}_q^M$ are auxiliary functions defined as $F_q^M(x) = \sum_{l=1}^{M-1} l^q x^l$ and $\bar{F}_q^M(x) = \sum_{l=1}^{M-1} (M-l)^q x^l$.

Moreover with the special quantization considered in Section 3.4 we can also prove the following property.

Property 15. *The two constant C_3 and C_4 assume the values:*

$$C_3 = \begin{cases} 1 & \textit{if } L = 2 \\ 2\,\mathrm{Re}[C_1] = -2\cos\frac{2\pi}{L} & \textit{otherwise} \end{cases}$$
$$C_4 = -2(1 - \mathrm{Re}[C_1]) = -2\left(1 + \cos\frac{2\pi}{L}\right)$$

which are independent of the rotation Δ of the constellation.

Proof. Few calculations exploiting the assumptions on Q give that

$$\sum_{j=t+1}^{n-t} (\mathbf{q}_j^*)^2 = \begin{cases} n - 2t & \text{if } L = 2 \\ -t\left[(\mathbf{q}^*_{\Delta (\mathrm{mod}\ L)})^2 + (\mathbf{q}^*_{-1+\Delta(\mathrm{mod}\ L)})^2 \right] & \text{otherwise} \end{cases}$$
$$\sum_{j=t+1}^{n-t} \mathbf{q}_j^* = -t\left[\hat{\mathbf{q}}_{-\Delta(\mathrm{mod}\ L)} + \hat{\mathbf{q}}_{(1-\Delta)(\mathrm{mod}\ L)} \right]$$

With this, recalling also 14, and noting that $(1 - r^{-1}) = n/t$, we obtain

$$C_3 = \hat{\mathbf{q}}_{\Delta (\mathrm{mod}\ L)} \hat{\mathbf{q}}_{(-1+\Delta)(\mathrm{mod}\ L)} \left[\hat{\mathbf{q}}_{-2\Delta (\mathrm{mod}\ L)} + \hat{\mathbf{q}}_{-2(-1+\Delta)(\mathrm{mod}\ L)} \right] = \\ \hat{\mathbf{q}}_{-\Delta (\mathrm{mod}\ L)} \hat{\mathbf{q}}_{(-1+\Delta)(\mathrm{mod}\ L)} + \hat{\mathbf{q}}_{\Delta (\mathrm{mod}\ L)} \hat{\mathbf{q}}_{(1-\Delta)(\mathrm{mod}\ L)} = \hat{\mathbf{q}}_{-1(\mathrm{mod}\ L)} + \hat{\mathbf{q}}_1 = 2\,\mathrm{Re}[C_1]$$

for $L > 2$ and $C_3 = -(1/n)\left(r^{-1}n - (r^{-1} - 1)(n - 2t)\right) = 1$ for $L = 2$.

Finally, as far as C_4 is concerned, we have

$$C_4 = -\hat{\mathbf{q}}_{\Delta (\mathrm{mod}\ L)} \hat{\mathbf{q}}_{(-1+\Delta)(\mathrm{mod}\ L)} \left[\hat{\mathbf{q}}_{-\Delta (\mathrm{mod}\ L)} + \hat{\mathbf{q}}_{(1-\Delta)(\mathrm{mod}\ L)} \right]^2 = \\ -\hat{\mathbf{q}}_{-\Delta (\mathrm{mod}\ L)} \hat{\mathbf{q}}_{(-1+\Delta)(\mathrm{mod}\ L)} + \hat{\mathbf{q}}_{\Delta (\mathrm{mod}\ L)} - \hat{\mathbf{q}}_{(1-\Delta)(\mathrm{mod}\ L)} - 2 = -2(1 - \mathrm{Re}[C_1])$$

□

As far as the variance $\left(\sigma_\Psi^v\right)^2$ of the cross-interference term is concerned we have

$$\left(\sigma_\Psi^v\right)^2 = \frac{1}{8N^2} \sum_{u \neq v} \sum_{\tau=0}^{\infty} \beta_\tau^2\, \mathbf{E}\left[\left| X^{uv}\left(\tau\frac{T}{N} + t^{uv}\right)\right|^2 + \left| Y^{uv}\left(\tau\frac{T}{N} + t^{uv}\right)\right|^2 \right]$$

where, considering once more the periodic nature of the functions X^{uv} and Y^{uv}, one may suppress the dependence on τ to subsequently take the expectation over t^{uv} uniformly distributed over $[0, T]$. Moreover, the channel power normalization constraint sets $\sum_{\tau=0}^{\infty} \beta_\tau^2 = 1$ so that we also obtain

$$\left(\sigma_\Psi^v\right)^2 = \frac{1}{8N^2} \sum_{u \neq v} \mathbf{E}\left[|X^{uv}(t^{uv})|^2 + |Y^{uv}(t^{uv})|^2 \right]$$

which is nothing but the term already considered in the non-selective channel case.

Once that the average variances of both disturbing contributions are known, we may define a global signal-to-disturbance ratio

$$\rho = \frac{K}{8(1+K)\left(\left(\sigma_\Xi\right)^2 + \left(\sigma_\Psi\right)^2\right)} \tag{3.29}$$

so that the resulting P_{err} is

$$P_{\mathrm{err}} = \frac{1}{2} \operatorname{erfc} \sqrt{\rho} \tag{3.30}$$

3.5.1 Numerical Results

Two different kinds of numerical results are reported in this section. All of them deal with a system with spreading factor $N = 70$, whose spreading sequences are generated by means of (n, t)-tailed shifts.

The first group of numerical evaluations is devoted to showing the accuracy of the theory in predicting $(\sigma_\Xi)^2$ as a function of the channel, mapping, and quantization parameters. Under the SGA, these two quantities are the building blocks allowing performance evaluation, and the availability of analytical expression for them results in a straightforward design methodology to optimize the communication system.

Such a methodology is the topic of the second group of numerical evaluation in which maximization of the system performance is pursued under a different condition of channel dispersion and network users load.

Figures 3.9(a) and (b) show the self-interference term $(\sigma_\Xi)^2$ as a function of the channel dispersion B with $K = 1$, respectively, for $L = 2$ and $L = 5$. The self-interference grows for maps with increasing $|r|$ so that the best choice is given by $r = 0$, which can be obtained with $(10, 0)$-tailed shift and is approximately equal to what is given by maximum-length sequences.

Note that the channel dispersion decreases as B increases, and that maps with higher $|r|$ are characterized by higher correlation between symbols. The trends in Figure 3.9(a) and (b) are therefore a consequence of the inability of high-$|r|$ maps to cope with self-interference concentrated in the immediate neighborhood of the main ray.

Figures 3.10(a) and (b) show the self-interference term $(\sigma_\Xi)^2$ as a function of the Rice factor K with $B = 1$, respectively, for $L = 2$ and $L = 5$. As before, maps characterized by an higher $|r|$ give rise to higher self-interference which nevertheless decreases as K increases, $K \to \infty$ being the non-multipath limit.

Note in general the match between theoretical results and numerical estimation and, as a final overall remark, how increasing the number of levels negatively affects cross-interference while it slightly reduces self-interference.

The availability of analytical exact expressions for the key quantities involved in performance evaluation allows an optimization of the system once the channel characteristics and the number of users is known.

The results of such an optimization are the topic of this second group of numerical evaluations. They are obtained by numerically maximizing the signal-to-interference ratio that, thanks to (3.29), (3.28), and (3.20), can be expressed as a function of r. This gives minimum bit error rate in (3.30).

As shown in Section 3.3, for the r achieving best performance, an (n, t)-tailed shift can be constructed so that the generated sequences are arbitrarily close to the optimum ones. So, the curves that follow are the graphical representation of optimal design criteria that give the best sequences for any possible set of channel parameters.

To compare such optimal solutions with the common reference case of purely random sequences (which is approximately equivalent to the adoption of classical

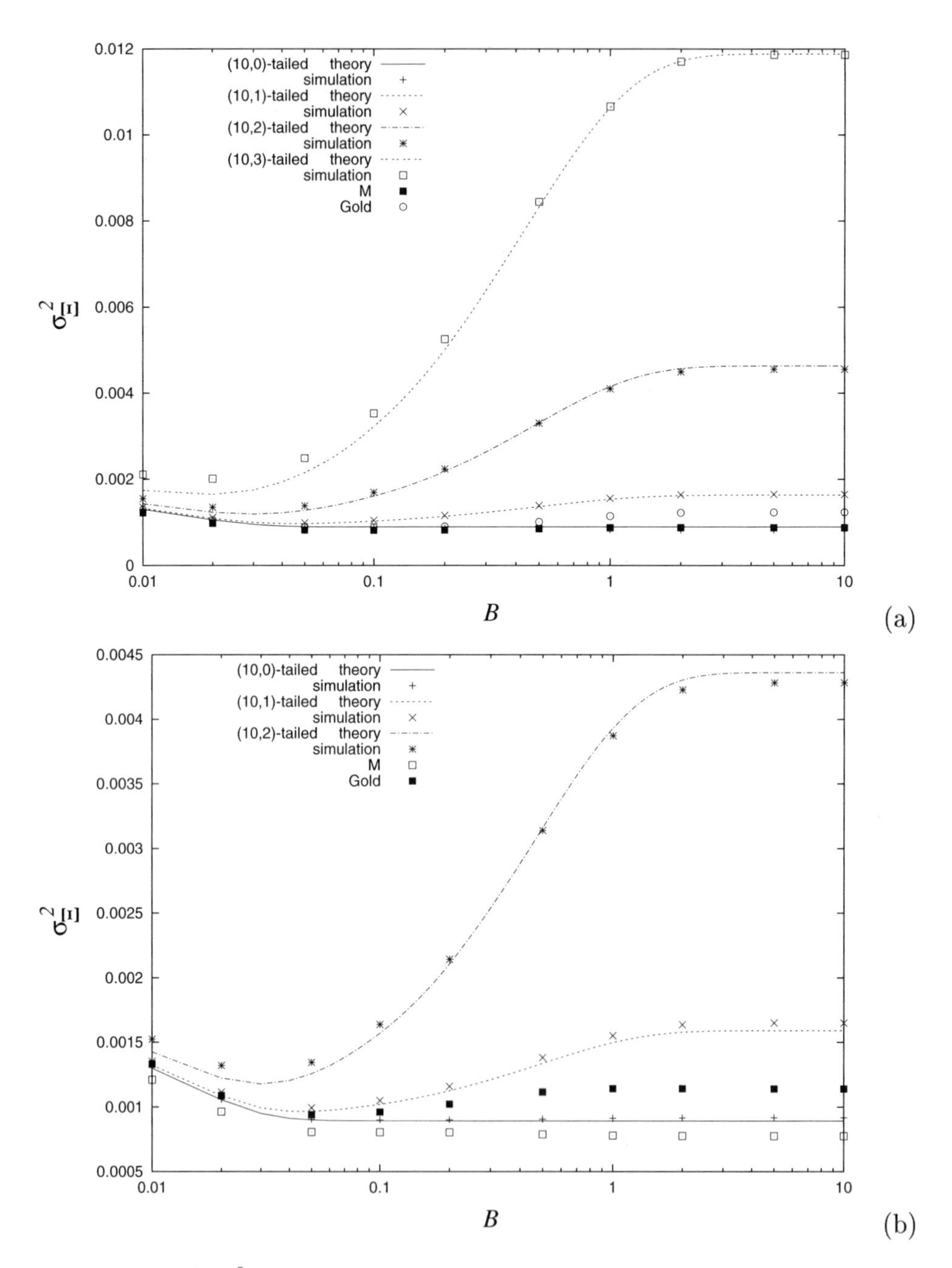

Figure 3.9: $(\sigma_\Xi)^2$ as a function of B for $K = 1$ and (a) $L = 2$ or (b) $L = 5$.

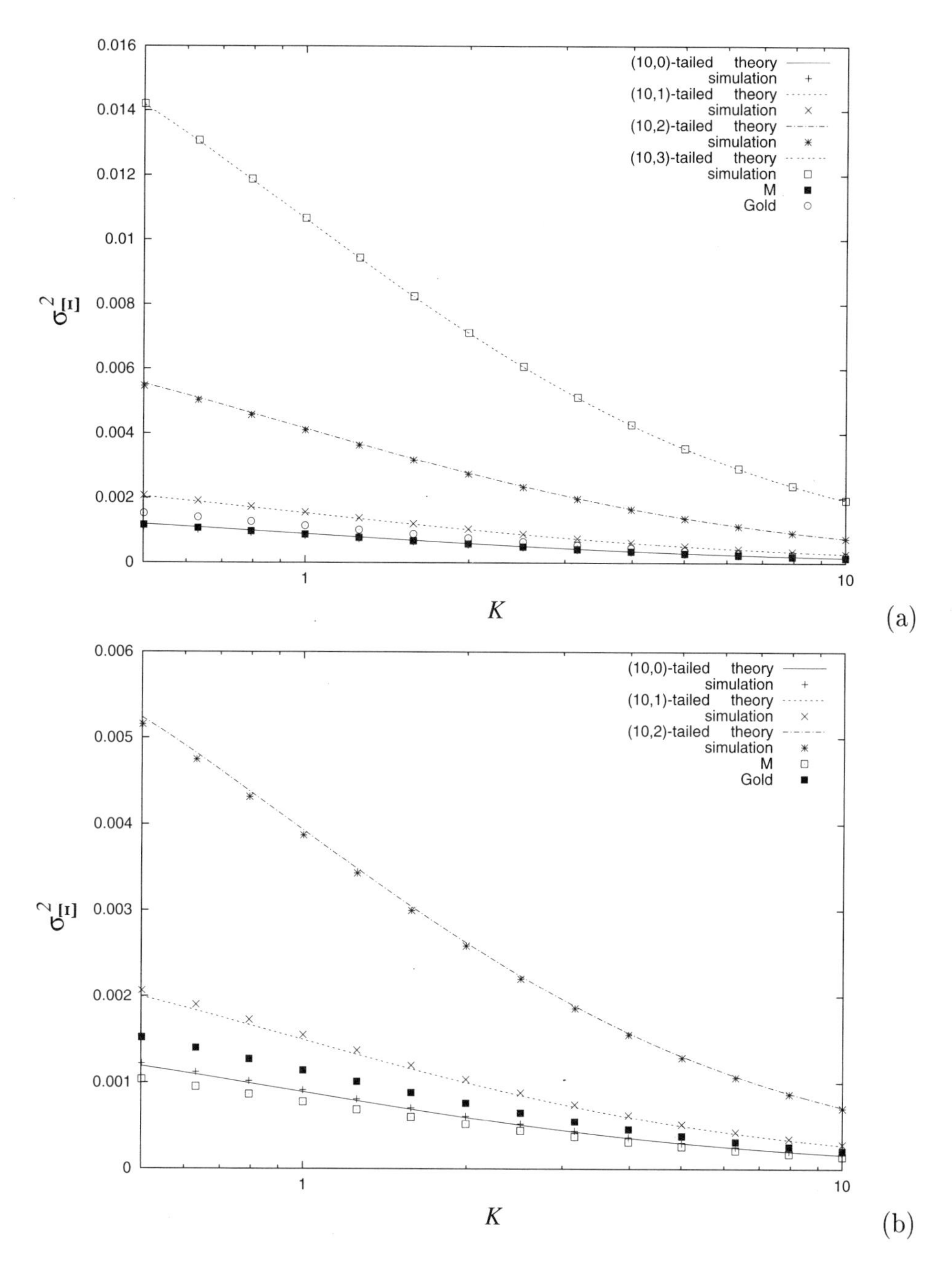

Figure 3.10: $(\sigma_{\Xi})^2$ as a function of K for $B = 1$ and (a) $L = 2$ or (b) $L = 5$.

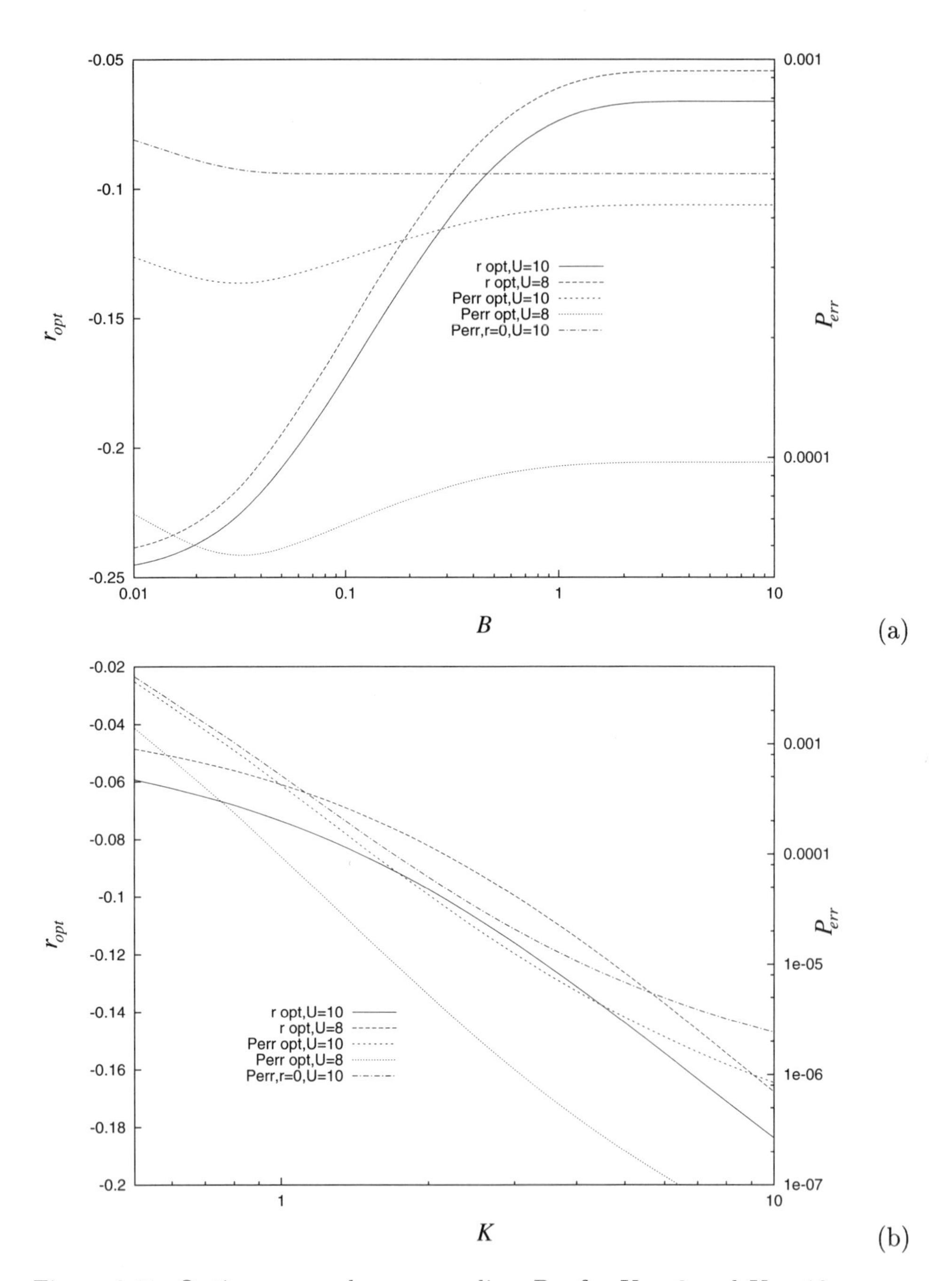

Figure 3.11: Optimum r and corresponding P_{err} for $U = 8$ and $U = 10$ users as a function of (a) B or (b) K.

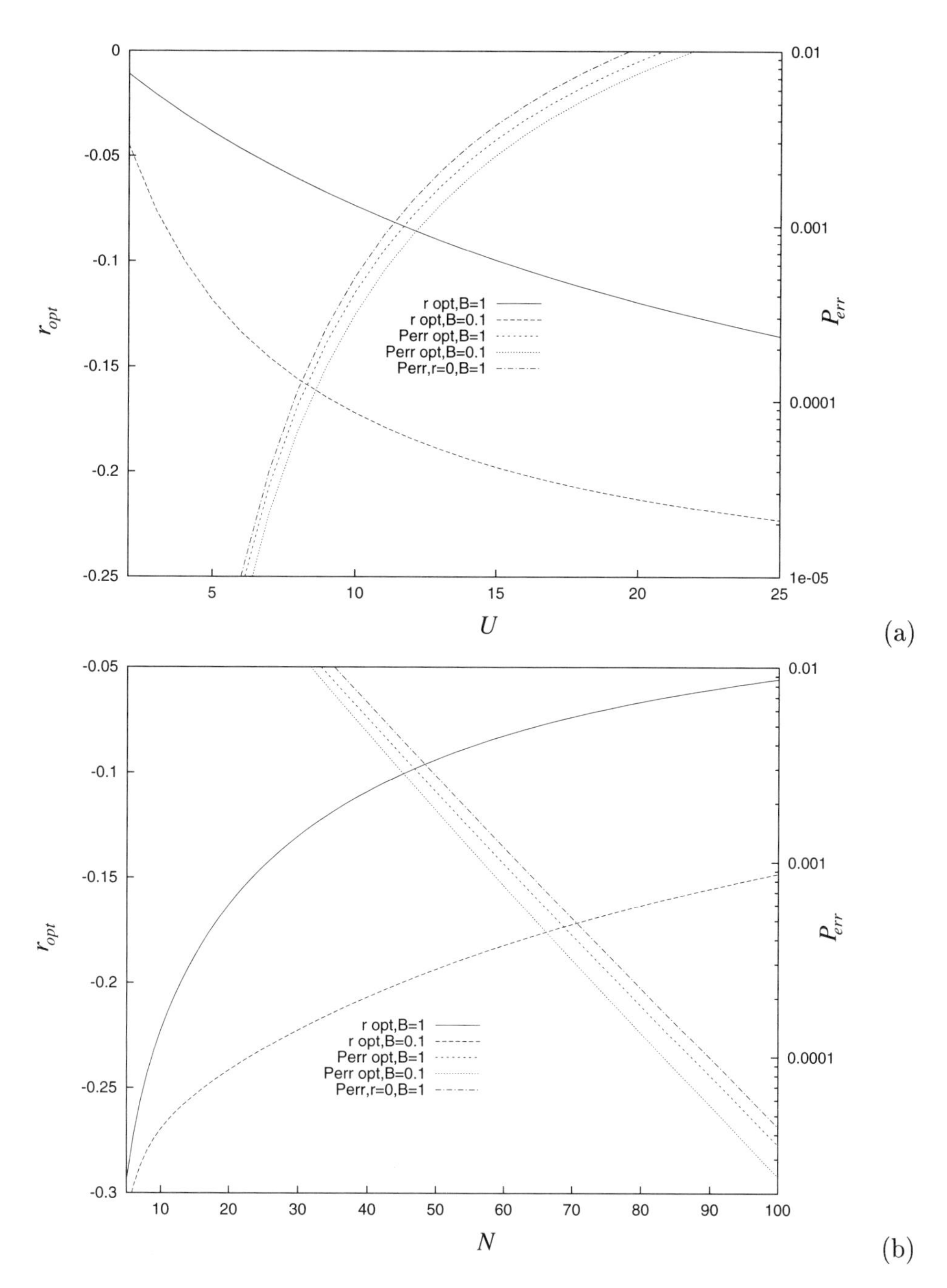

Figure 3.12: Optimum r and corresponding P_{err} (a) as a function of U for $B = 1$ and $K = 0.1$ or $K = 1$ and (b) as a function of N for $U = 10$, $K = 1$ and $B = 0.1$ or $B = 1$.

maximum-length) we also report the performance for $r \to 0$ (which is also what a generalized Bernoulli map would give).

Even if the theory previously developed applies to any possible number L of spreading symbols in the following, we focus the attention on $L = 2$.

In Figure 3.11(a) the optimum r and resulting performance is reported as a function of the channel dispersion B for two different number of users, $U = 8$ and $U = 10$. The optimum $|r|$ decreases as B increase, i.e., as the self-interference concentrates on the second ray, optimization chooses a map which is increasingly good in breaking correlation as soon as possible.

In Figure 3.11(b) the optimum r and resulting performance is reported as a function of the Rice factor K for two different number of users, $U = 8$ and $U = 10$. The greater the power of the secondary rays (i.e., the smaller the K) the lower the optimum $|r|$ to break the self-interference effects.

As K increases, the gain over the reference case increases due to prevailing contribute of cross-interference in almost non-multipath channels ($K \to \infty$).

As far as the effect of the network load is concerned, results are reported in Figure 3.12(a) for two channel configurations with $B = 1$ and $K = 0.1$ or $K = 1$. When U increases, the cross-interference becomes the dominant term in the signal-to-interference ratio. As such a term is minimized for $\sqrt{3} - 2 \simeq -0.27$ the optimum r localizes in its neighborhood for high loads, independently of the channel characteristics.

Finally, let us report the trends of the optimal r and P_{err} when the spreading factor N changes. They are shown in Figure 3.12(b). The optimum $|r|$ decreases when the spreading factor increases. This is due to the fact that, in a system with a fixed load, increasing the bandwidth drastically reduces cross-interference while self-interference still depends on the auto-correlation features of the spreading sequences and becomes the main error cause.

As a general remark, note that optimization always yields a P_{err} which is lower than what can be achieved with the adoption of classical sequences.

3.6 Conclusion

The aim of this chapter has been the theoretical investigation of the achievable performance of a *conventional* DS-CDMA system when chaos-based spreading sequences are substituted to classical Gold and m-sequences.

To perform such an investigation two parallel paths have been followed. On one side, the class of maps adopted for spreading sequence generation has been thoroughly characterized to link the map structure to the correlation properties of the resulting sequences. On the other side, a model of the communication systems under investigation has been developed isolating some synthetic performance indexes and linking them with some design-dependent quantities, i.e., the correlation properties of the spreading sequences.

As far as the first aspect is concerned, some new mathematical tools for the statistical characterization of correlation of the spreading sequence have been introduced. With this, for a quite general family of PWAM maps called (n, t)-tailed shifts it has been possible to compute a closed-form general expression for the building blocks of *any-order correlations* of quantized trajectories. Additionally, it has been proven that the auto-correlation function of binaries sequences obtained quantizing a chaotic time series can be expressed as $A_k = r^k$, where r is a *design* parameter depending on the feature of the map, which can be made arbitrarily close to any real value in $[0, 1]$.

As far as the communication system modeling is concerned, the three main aspects which affect code-level design of a DS-CDMA system have been analyzed, namely co-channel interference and co-channel self-interference in additive Gaussian white noise and multipath channels.

When only co-channel interference is present, the general performance model can be specialized to show that the bit error probability depends only on the interference from the other users (cross-interference) and thus on the second-order auto-correlation properties of the spreading sequences through a simple and analytically tractable expression. Exploiting that expression it is has been shown that an auto-correlation profile exists that results in the minimum achievable bit error probability and thus that, in the non-multipath case, spreading sequences with that auto-correlation profile are the optimum choice. With this it has been shown that the (n, t)-tailed shifts exist producing spreading sequences that are so close to the optimum ones to produce a final performance which is indistinguishable from the global optimum for any realistic system. Finally, for a given level of bit error probability, the adoption of optimum sequences allows *the allocation of ≈15% more users on the same bandwidth, at no additional cost.* As this is the expected gain over an infinite number of users, a trivial optimization procedure has been set up to explore the availability of finite sets of spreading sequences resulting in even better performance. The results of this optimization show that an increase of up to 20% in the number of users is feasible.

Next, the more general case of multipath channel has been considered. In this case, an additional cause of error appears in the form of a self-interference between the useful synchronized signals and its delayed version. With this, auto-correlations of second as well as fourth order turn out to affect the final performance so that the correlation shaping optimization described in the previous case no longer applies. Yet, the mathematical tools developed support the analytical calculation of all the key quantities for the class of maps achieving optimum performance over non-selective channels. With this, the trade off between cross-interference and self-interference can be addressed by numerically optimizing the analytical expression of the bit error probability. The results of such an optimization show that, when a given communication quality is set, *CBDS-CDMA allows the allocation of a number of users which is from 5% to 15% higher than what is achieved with conventional spreading sequences.* Maximum improvement is achieved over channels in which either multipath is almost negligible (as the optimum non-selective performance is retrieved) or a lot of secondary rays share the same non-negligible power. Lower improvement is achieved when only few (on the order of one or two) secondary rays carry most of the multipath power.

Appendix: Basic Operators on Tensors

Here we define some basic operators acting on tensors along with few related properties. Although other more general operators can be defined, only those specialized forms that are most useful for the material developed in this chapter will be considered.

Definition 3. *Given two m-dimensional tensors $\mathcal{A}$ and $\mathcal{B}$ with identical index ranges, their* inner product *is a scalar defined as*

$$\langle \mathcal{A}, \mathcal{B} \rangle = \sum_{j_1,\ldots,j_m} \mathcal{A}_{j_1,\ldots,j_m} \, \mathcal{B}_{j_1,\ldots,j_m}$$

where every index sweeps all its range.

The inner product is a commutative bilinear operator.

Definition 4. *Given an m'-dimensional tensor $\mathcal{A}$ and an m''-dimensional tensor $\mathcal{B}$, their* outer product *is an $(m' + m'')$-dimensional tensor $\mathcal{C} = \mathcal{A} \otimes \mathcal{B}$ such that*

$$\mathcal{C}_{j_1,\ldots,j_{m'+m''}} = \mathcal{A}_{j_1,\ldots,j_{m'}} \, \mathcal{B}_{j_{m'+1},\ldots,j_{m'+m''}}$$

where every index sweeps all its range.

The outer product is a non-commutative, associative, and bilinear operator.

Definition 5. *Given the m'-dimensional tensor $\mathcal{A}$ and the m''-dimensional tensor $\mathcal{B}$ such that the range of the last index of $\mathcal{A}$ is the same of the first index of $\mathcal{B}$, their* product *is an $(m' + m'' - 2)$-dimensional tensor $\mathcal{C} = \mathcal{A}\mathcal{B}$ such that*

$$\mathcal{C}_{j_1,\ldots,j_{m'+m''-2}} = \sum_i \mathcal{A}_{j_1,\ldots,j_{m'-1},i} \, \mathcal{B}_{i,j_{m'},\ldots,j_{m'+m''-2}}$$

where i sweeps all the common range.

The product is a non-commutative, associative, and bilinear operator.

Definition 6. *Given the m'-dimensional tensor $\mathcal{A}$ and the m''-dimensional tensor $\mathcal{B}$ such that the range of the last index of $\mathcal{A}$ is the same of the first index of $\mathcal{B}$, their* chain product *is an $(m' + m'' - 1)$-dimensional tensor $\mathcal{C} = \mathcal{A} \circ \mathcal{B}$ such that*

$$\mathcal{C}_{j_1,\ldots,j_{m'+m''-1}} = \mathcal{A}_{j_1,\ldots,j_{m'}} \, \mathcal{B}_{j_{m'},\ldots,j_{m'+m''-1}}$$

The chain product is a non-commutative associative bilinear operator.

Definition 7. *Given the hyper-cubic m-dimensional tensor $\mathcal{A}$ and the matrix $\mathcal{B}$ whose first index has the same range as the indexes of $\mathcal{A}$, the hyper-cubic m-dimensional tensor $\mathcal{C} = \mathcal{A}\big|\,\mathcal{B}$ is defined as*

$$\mathcal{C}_{j_1,\ldots,j_m} = \sum_{i_1,\ldots,i_m} \mathcal{A}_{i_1,\ldots,i_m} \mathcal{B}_{i_1 j_1} \ldots \mathcal{B}_{i_m j_m}$$

where all the index $i_1, \ldots, i_m$ span their whole range.

The tensor products defined above enjoy few *distributive* properties which are recalled without proof in the following.

Property 16. *For any three tensors $\mathcal{A}$, $\mathcal{B}$, and $\mathcal{C}$ we have*

$$\mathcal{A} \otimes (\mathcal{B} \circ \mathcal{C}) = (\mathcal{A} \otimes \mathcal{B}) \circ \mathcal{C} \quad \text{and} \quad \mathcal{A} \circ (\mathcal{B} \otimes \mathcal{C}) = (\mathcal{A} \circ \mathcal{B}) \otimes \mathcal{C}$$

For any three tensors $\mathcal{A}$, $\mathcal{B}$, and $\mathcal{C}$ with compatible index ranges we have

$$\mathcal{A}(\mathcal{B} \otimes \mathcal{C}) = (\mathcal{A}\mathcal{B}) \otimes \mathcal{C} \quad \text{and} \quad \mathcal{A} \otimes (\mathcal{B}\mathcal{C}) = (\mathcal{A} \otimes \mathcal{B})\mathcal{C}$$

For any four tensors $\mathcal{A}$, $\mathcal{B}$, $\mathcal{C}$, and $\mathcal{D}$ with compatible index ranges we have

$$\langle \mathcal{A} \otimes \mathcal{B}, \mathcal{C} \otimes \mathcal{D} \rangle = \langle \mathcal{A}, \mathcal{C} \rangle \langle \mathcal{B}, \mathcal{D} \rangle$$

For any two tensors $\mathcal{A}$ and $\mathcal{B}$ and any compatible matrix $\mathcal{C}$

$$\left\langle \mathcal{A}\big|\,\mathcal{C}, \mathcal{B} \right\rangle = \left\langle \mathcal{A}, \mathcal{B}\big|\,\mathcal{C}^T \right\rangle$$

where $\cdot^T$ denotes matrix transposition.

References

[1] Heidari-Bateni, G. and McGillem, C.D., "A Chaotic Direct Sequence Spread Spectrum Communication System," *IEEE Transactions on Commununications*, vol. 42, pp. 1524-1527, 1994.

[2] Mazzini, G., Setti, G., and Rovatti, R., "Chaotic Complex Spreading Sequences for Asynchronous CDMA - Part I: System Modelling and Results," *IEEE Transactions on Circuits and Systems - Part I*, vol. 44, n. 10, pp. 937–947, 1997.

[3] Rovatti, R., Setti, G., and Mazzini, G., "Chaotic Complex Spreading Sequences for Asynchronous CDMA - Part II: Some Theoretical Performance Bounds," *IEEE Transactions on Circuits and Systems - Part I*, vol. 45, n. 4, pp. 496–506, 1998.

[4] Rovatti, R. and Mazzini, G., "Interference in DS-CDMA systems with exponentially vanishing auto-correlations: Chaos-based spreading is optimal," *Electronics Letters*, vol. 34, pp. 1911-1913, 1998.

[5] Mazzini, G., Rovatti, R., and Setti, G., "Interference Minimization by Autocorrelation Shaping in Asynchronous DS-CDMA Systems: Chaos-Based Spreading is Nearly Optimal," *Electronics Letters*, vol. 35, pp. 1054-1055, 1999.

[6] Setti, G., Mazzini, G., and Rovatti, R., "Chaos-Based Spreading Sequence Optimization for DS-CDMA Synchronization," *IEICE Transactions on Fundamentals of Electronics, Communications and Computer Sciences*, vol. E82-A, n. 9, pp. 1737–1746, 1999

[7] Kolumbán, G., Kennedy, M. P., and Chua, L. O., "The Role o Synchronization in Digital Communication Using Chaos—Part I: Fundamentals of Digital Communications," *IEEE Transactions on Circuits and Systems - Part I*, vol. 44, pp. 927-936, 1997.

[8] Götz, M., Kelber, K., and Schwarz, W., "Discrete-Time Chaotic Encryption Systems - Part I: Statistical Design Approach," *IEEE Transactions on Circuits and Systems - Part I*, vol. 44, pp. 963-970, 1997.

[9] Kohda, T. and Tsuneda, A., "Statistics of Chaotic Binary Sequences," *IEEE Transactions on Information Theory*, vol. 43, pp. 104-112, 1997.

[10] Dedieu, H. and Ogorzalek, M.J., "Overview of Nonlinear Noise Reduction Algorithms for Systems with Known Dynamics," in Proc. of International Symposium on Nonlinear Theory and Its Applications, pp. 1297-1300, 1998.

[11] Kohda, T. and Tsuneda, A., "Explicit Evaluations of Correlation Functions of Chebyshev Binary and Bit Sequences Based on Perron-Frobenius Operator," *IEICE Transactions on Fundamentals of Electronics, Communications and Computer Sciences*, vol. E77-A, pp. 1794-1800, 1994.

[12] Götz, M. and Schwarz, W., "Spectral Decomposition of the Frobenius-Perron Operator and Higher-Order Correlation Functions of One-Dimensional Chaotic Systems with Fully Stretching Maps," in Proc. of European Conference on Circuit Theory and Design, pp. 1253-1258, 1997.

[13] Götz, M., Falk, T., and Schwarz, W., "Statistical Theory and Canonical Architecture Implementation of N-Dimensional Markov Maps," in *Proc. of International Symposium on Nonlinear Theory and its Applications (NOLTA98)*, pp. 195-198, 1998.

[14] Ochsner, H., "Direct Sequence Spread-Spectrum Receiver for Communication on Frequency Selective Fading Channels," *IEEE JSAC*, vol. 5, pp. 188-193, 1987.

[15] Lasota, A. and Mackey, M. C., *Chaos, Fractals, and Noise*, Springer-Verlag, New York, 1994.

[16] Hofbauer, F. and Keller, G., "Ergodic Properties of Invariant Measures for Piecewise Monotonic Transformations," *Mathematische Zeitschrift*, vol. 180, pp. 119-140, 1982.

[17] Wiggings, S., *Introduction to Applied Nonlinear Dynamical Systems and Chaos*, Springer-Verlag, New York, 1996.

[18] Minc, H., *Nonnegative Matrices*, Wiley-Interscience, 1987.

[19] Friedman, N. and Boyarsky, A., "Irreducibility and Primitivity Using Markov Maps," *Linear Algebra and Its Applications*, vol. 37, pp. 103-117, 1981.

[20] Gradshteyn, I. S. and Ryzhik, I. M., *Table of Integrals, Series, and Products–5th Edition*, ed., A. Jeffrey, Academic Publishers, Boston, 1994.

[21] Mazzini, G., "DS-CDMA Systems using q-Level m Sequences: Coding Map Theory," *IEEE Transactions on Communications*, vol. 45, pp. 1304-1313, 1997.

Chapter 4

Information Sources Using Chaotic Dynamics

Tohru Kohda
Kyushu University
6-10-1 Hakozaki, Higashi-ku, Fukuoka 812-8581, Japan
kohda@csce.kyushu-u.ac.jp

Akio Tsuneda
Kumamoto University
2-39-1 Kurokami, Kumamoto 860-8555, Japan
tsuneda@eecs.kumamoto-u.ac.jp

4.1 Introduction

Computational applications requiring random numbers [1, 2] include computer programming, stochastic simulation, Monte Carlo technique, simulated annealing, and so on. In particular, a sequence of independent and identically distributed (i.i.d.) binary random variables [3–5] has found significant applications in modern digital communication systems such as in spread spectrum (SS) communication systems [6,7] or cryptosystems [8,9]. Although such a binary sequence can be generated in various ways, linear feedback shift register (LFSR) sequences are employed in nearly all the methods [10–12].

It is well known from probability theory [3–5] that coin tossing has always been considered as the symbol of randomness. Randomness itself, however, is very difficult to define. In fact, throwing a die may be seen as both a deterministic and a random event. Both the dyadic map and its associated binary function, called the Rademacher function, have been considered to realize coin tossing [3,5]. Thus, it is an important problem whether any maps other than the dyadic map can realize coin tossing or not. In addition, there is another question: "What are their associated binary functions?"

Many communication systems based on chaotic phenomena [13–17] have been recently proposed. In our opinion, some of these attempts, however, have led many chaos researchers to mistakenly overemphasize the role of chaotic phenomena in communication systems. This in turn has led to the rebuttal of chaotic methods by researchers in existing digital communication systems. One of the main reasons for this is that most of these systems have been based on a chaotic real-valued orbit itself rather than its binary version (namely, *code level*). Another is that in much research on such systems, important statistics such as correlation functions in communication systems have not been theoretically discussed or estimated. On the other hand, sequences of random numbers generated by a Markov chain are discussed as candidates of spreading sequences for CDMA [18], and as running-key sequences in chaotic encryption systems [19]. Moreover, in source coding problems, there have been several attempts to construct a dynamical system with an arbitrarily prescribed Markov information source. In addition, analogies between arithmetic coding and chaotic dynamics are also discussed [20–23]. Study, however, soon informs us that Kalman's 1956 embedding of a Markov chain [24] is to be greatly appreciated.

We have recently given [64] a simple sufficient condition for some class of ergodic maps to produce a sequence of i.i.d. binary random variables and given their applications [55] to two typical communication systems: (1) a stream cipher cryptosystem [61–63] and (2) an image transmission system through quasi-synchronous DS/CDMA channels using spread spectrum sequences of variable-period [59, 60, 66, 67].

In this chapter, we first review how Kalman embedded a prescribed Markov information source into chaotic dynamics of the PLM Markov onto map and review our recently improved Kalman's embedding procedure [69] by using the PLM onto map with the minimum number of subintervals. Secondly, we briefly review that our sufficient condition for some class of ergodic maps and their associated binary functions to produce a sequence of i.i.d. binary random variables consists of two simple symmetric properties: the *equidistributivity property* (or briefly *EDP*) and *constant summation property* (or briefly *CSP*) [68] and furthermore make it clear that the CSP provides us with the design of sequences of p-ary random variables as well as recently defined correlational properties of real-valued chaotic trajectory [70] (*k-th order correlated property* and *d−delay correlated property*). Thirdly, we show the critical roles of such a sequence of i.i.d. binary random variables in the above two communication systems, in terms of cryptographically secure properties and multiple-access interference properties, respectively.

4.2 Information Sources and Markov Chains

4.2.1 A Model for a Communication System

Statistical communication theory or information theory, originated by C.E. Shannon in 1948 [25], is a broad field comprised of methods for the study of the statis-

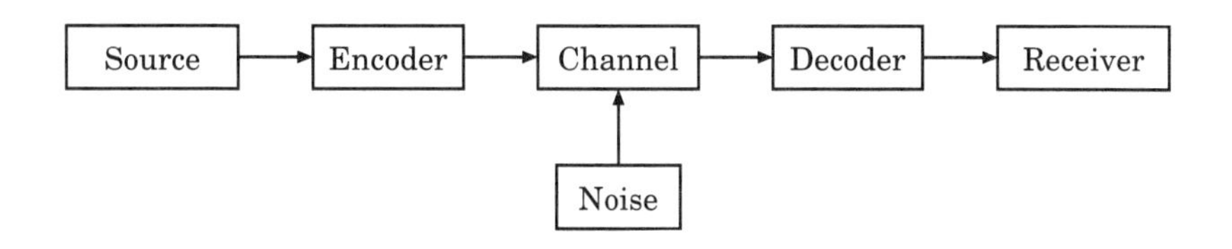

Figure 4.1: A model of a communication system.

tical problems encountered in all types of communications. In fact, information theory intersects physics, mathematics, electrical engineering, and computer science (There are several well-known textbooks on information theory [26–28] and an important reference [29]).

The simplest communication system has three essential parts: (1) source or transmitter, (2) receiver or sink, and (3) channel or transmission network. Since the usual or real communication systems are of a statistical nature, the performance of the system can never be described in a deterministic sense; rather, it is always given in statistical terms. A source is a device that selects and transmits sequences of symbols from a given alphabet. Each selection is made at random, although this selection may be based on some statistical rule. The channel transmits the incoming symbols to the receiver. The performance of the channel is also based on laws of chance. Therefore, in a communication model as shown in Figure 4.1 [26], the source, channel, encoder, decoder, noise source, and receiver must be *statistically* defined (see texts [30, 31] for a more advanced or applied context).

As stated in Reza's textbook [26], a basic study of communication systems requires some knowledge of probability theory; communication theories cannot be adequately studied without having a good background of probability. (Fortunately, an enormous number of introductory textbooks on probability theory are available. Billingsley [5] and Loève [4] have been the strongest influences on our research, and the spirit of Kac's small volume [3] has been very important.)

4.2.2 Kalman's Markov Map and Markov Information Sources

First of all we have to discuss the Kalman's design of Markov information source using chaotic dynamics.

Kalman, the founder of modern control theory, wrote a paper in 1956 [24], which made history with concepts and developments in chaotic dynamics. In the paper, he pointed out that "a nonlinear sampled-data system behaved in a similar way to autonomous nonlinear systems under restricted conditions," and he discussed the equivalence between Markov chains and nonlinear difference equations. Kalman's results were well in advance of Lorenz (1963) [32] and May

(1976) [33] who studied chaotic dynamics of one-dimensional mappings.[1] He also gave a procedure for embedding a Markov chain into the chaotic dynamics of piecewise-linear-monotonic (PLM) onto maps.

Of crucial importance in many fields of engineering is the question of whether irregular sequences observed in physical systems originated from determinism or not. Kalman's latter result is closely related to this question because he gave a *deterministic* procedure which generates a *stochastic* sequence. In fact, Kalman's procedure gave a fundamental answer to the problem of random number generation [1], which is discussed intensively in the field of information theory (e.g., [35,36]). However, few attempts or discussions in the following decades remind us that Kalman's 1956 embedding procedure is to be much appreciated.[2]

Let us begin by reviewing the definition of a Markov chain and the notion of a Markov information source [27]. Given a set of states $S = \{1, 2, \cdots, N_s\}$ and a probability transition matrix $P = \{p_{ij}\}_{i,j=1}^{N_s}$, satisfying $p_{ij} \geq 0$ for all i, j; $\sum_{j=1}^{N_s} p_{ij} = 1$ for all i, we define a sequence of random variables $Z_0, Z_1, \cdots$ taking values in S. If for Z_0 having an arbitrary distribution

$$\text{Prob}\{Z_{n+1} = s_k | Z_0 = s_{i_0}, \cdots, Z_n = s_{i_n}\} = p_{i_n,k}, \tag{4.1}$$

then the sequence of random variables $Z_0, Z_1, \cdots$ is called an N_s-state Markov chain (Prob$\{A\}$ denotes the probability assigned to an event A). Given a Markov chain $Z_0, Z_1, \cdots$ and a function f whose domain is S and whose range is an alphabet set $\Gamma = \{\gamma_1, \cdots, \gamma_{N_a}\}$, assume that the initial state Z_0 is chosen in accordance with a stationary distribution $\mathbf{u} = (u_1, \cdots, u_{N_s})$, that is,

$$\text{Prob}\{Z_0 = s_j\} = u_j \text{ for all states } s_j, \tag{4.2}$$

then the stationary sequence $X_n = f(Z_n)$, $n = 0, 1, 2, \cdots$ is said to be the Markov information source.[3] In this paper, for simplicity, we take $\Gamma = S$, $N_a = N_s$, and f to be the identity function.

Kalman gave a simple procedure for embedding a Markov chain with transition matrix $P = \{p_{ij}\}_{i,j=1}^{N_s}$, satisfying

$$0 < p_{ij} < 1 \quad \text{for all } i, j \tag{4.3}$$

into a PLM onto map $\tau : J = [0, 1] \to J$ with N_s^2 subintervals, defined as

$$\omega_{n+1} = \tau(\omega_n), \quad n = 0, 1, 2, \cdots, \quad \omega_n \in J \tag{4.4}$$

as follows.

[1] Readers interested in the history of concepts and developments in chaotic dynamics should see [34], for example.

[2] Kalman [37] has recently given us a lecture about chaos.

[3] As stated in [28], it is an unfortunate fact that in the communication-theory literature the term 'Markov source' includes 'hidden' Markov chains. In this chapter, however, we speak of 'simple' Markov sources.

Divide first the interval J into N_s subintervals such that

$$J = \bigcup_{i=1}^{N_s} J_i \tag{4.5}$$

where

$$J_i = (d_{i-1}, d_i], \quad d_0 = 0 < d_1 < d_2 < \cdots < d_{N_s} = 1. \tag{4.6}$$

Furthermore, divide the subintervals J_i $(1 \leq i \leq N_s)$ into N_s subintervals such that

$$J_i = \bigcup_{j=1}^{N_s} J_{i,j} \tag{4.7}$$

where

$$J_{i,j} = \begin{cases} (d_{i,j-1}, d_{i,j}] & \text{for } \tau(d_i) = 1, \\ \quad (d_{i,0} = d_{i-1},\ d_{i,N_s} = d_i), & \\ (d_{i,j}, d_{i,j-1}] & \text{for } \tau(d_i) = 0, \\ \quad (d_{i,0} = d_i,\ d_{i,N_s} = d_{i-1}), & \end{cases} \tag{4.8}$$

subject to the conditions of a Markov partition,

$$\tau(d_i) \in \{0, 1\}, \qquad \text{for } 1 \leq i \leq N_s \tag{4.9}$$

and

$$\tau(d_{i,j}) = d_j, \qquad \text{for } 1 \leq i, j \leq N_s, \tag{4.10}$$

and the condition of the transition probabilities

$$\frac{|J_{i,j}|}{|J_i|} = p_{ij}, \qquad \text{for } 1 \leq i, j \leq N_s, \tag{4.11}$$

where $|I|$ denotes the length of an interval I. Thus the restriction of Kalman's maps τ to the interval $J_{i,j}$, denoted by $\tau_{i,j}(\omega)$, are of the form

$$\tau_{i,j}(\omega) = \begin{cases} \dfrac{|J_i|\omega + (d_{i,j}d_{j-1} - d_{i,j-1}d_j)}{|J_{i,j}|}, \\ \qquad \omega \in J_{i,j} \quad \text{for } \tau(d_i) = 1, \\ \dfrac{-|J_i|\omega + (d_{i,j-1}d_j - d_{i,j}d_{j-1})}{|J_{i,j}|}, \\ \qquad \omega \in J_{i,j} \quad \text{for } \tau(d_i) = 0. \end{cases} \tag{4.12}$$

Condition (4.9) says that Kalman's procedure admits us to design 2^{N_s} different PM onto maps. However, each of these maps is equivalent to one another in terms of statistical properties of sequences of N_s symbols.

Condition (4.11) tells us that an absolutely continuous invariant (or briefly ACI) measure in each interval J_i is required to be uniform (this will be guaranteed as below). But we know that the existence of an ACI measure for the nonlinear mappings has to be first guaranteed in using chaotic dynamics. Ulam [38] posed the problem of the existence of ACI measure for the mapping and defined the transition probability from a subinterval I_i to a subinterval J_j for the mapping by

$$t_{i,j} = \frac{m[I_i \cap \tau^{-1}(I_j)]}{m(I_i)}, \ 1 \leq i, j \leq M \tag{4.13}$$

to approximate the Perron-Frobenius (or briefly P-F) operator of the mapping (for its definition [39] see (4.57) and (4.59)), where $m(\cdot)$ denotes the Lebesgue measure and

$$J = \bigcup_{i=1}^{M} I_i. \tag{4.14}$$

But by Lasota-Yorke's theorem [40] (which gave an answer for the Ulam's problem) we can observe that Kalman's map always has an ACI measure. Introduce a one-to-one mapping $\sigma(\cdot)$ for M subintervals J_{k_i} of J satisfying $J = \bigcup_{i=1}^{M} J_{k_i}$, defined as

$$\sigma(J_{k_i}) = I_i, \ 1 \leq i, k_i \leq M, \tag{4.15}$$

and denote it simply as

$$\sigma = \begin{pmatrix} J_{k_1} & J_{k_2} & \cdots & J_{k_M} \\ I_1 & I_2 & \cdots & I_M \end{pmatrix}. \tag{4.16}$$

We introduce the $M \times M$ dimensional matrix $T(\sigma) = \{t_{ij}\}_{i,j=1}^{M}$ and call it an Ulam's transition matrix associated with σ. Then taking

$$\sigma = \begin{pmatrix} J_{1,1} & J_{1,2} & \cdots & J_{N_s,1} & \cdots & J_{N_s,N_s} \\ I_1 & I_2 & \cdots & I_{N_s^2-N_s+1} & \cdots & I_{N_s^2} \end{pmatrix} \tag{4.17}$$

for N_s^2 subintervals $J_{i,j}$ $(1 \leq i, j \leq N_s)$ of Kalman's map, we see that

$$T(\sigma) = \begin{bmatrix} P_1 & P_2 & \cdots & P_{N_s} \\ P_1 & P_2 & \cdots & P_{N_s} \\ \vdots & \vdots & \vdots & \vdots \\ P_1 & P_2 & \cdots & P_{N_s} \end{bmatrix}, \tag{4.18}$$

where

$$P_i = \begin{bmatrix} \mathbf{0} \\ \vdots \\ \mathbf{0} \\ \mathbf{p}_i \\ \mathbf{0} \\ \vdots \\ \mathbf{0} \end{bmatrix}, \quad \mathbf{p}_i = (p_{i1}, p_{i2}, \cdots, p_{iN_s}), \quad \mathbf{0} = \overbrace{(0, \cdots, 0)}^{N_s}. \tag{4.19}$$

We now observe that

$$\Lambda(T(\sigma)) = \Lambda(P) \cup \{0\}^{N_s^2 - N_s}, \tag{4.20}$$

where $\Lambda(C)$ denotes a set of all eigenvalues of a matrix C. Let $Z_0, Z_1, \cdots$ be a sequence of random variables taking values in S associated with a real-valued sequence $\{\omega_n\}_{n=0}^{\infty}$ such that if $\omega_n \in J_i$, then $Z_n = i$. Since a Markov chain $Z_0, Z_1, \cdots$ is governed by Kalman's map $\tau(\cdot)$ or equivalently $T(\sigma)$, relation (4.20) says that "*a Markov chain is embedded into chaotic dynamics of the map $\tau(\omega)$ or $T(\sigma)$.*"

On the other hand, Kalman derived transition probabilities from a subinterval J_i to a subinterval J_j, denoted by t'_{ij}, defined as

$$t'_{i,j} = \frac{|J_i \cap \tau^{-1}(J_j)|}{|J_i|}, \ 1 \leq i, j \leq N_s \tag{4.21}$$

and gave an $N_s \times N_s$-dimensional matrix $T' = \{t'_{ij}\}_{i,j=1}^{N_s}$. If we can prove that the invariant density keeps constant in each subinterval J_i for $i = 1, 2, \cdots, N_s$, then we know that T' is justified and is equal to transition matrix P or equivalently the Ulam's transition matrix $T(\sigma')$ associated with

$$\sigma' = \begin{pmatrix} J_1 & J_2 & \cdots & J_{N_s} \\ I_1 & I_2 & \cdots & I_{N_s} \end{pmatrix}. \tag{4.22}$$

We briefly show that the above mapping σ' needs the coarse-graining process and the coarse-grained matrix T' is justified as follows. Equations (4.21) and (4.22) imply the transition probability from the subinterval $J_{k,i}$ to the subinterval $J_{l,j}$ in the coarse-grained subinterval J_l $(1 \leq i, j, k, l \leq N_s)$. Consequently, Kalman deduced an N_s-state Markov chain from an N_s^2-subinterval dynamics.

From Boyarsky's results [41], we observe that the Kalman's PLM onto map admits the unique invariant density such as

$$f^*(\omega) = \sum_{i,j=1}^{N_s} f^*_{i,j} I_{J_{i,j}}(\omega), \tag{4.23}$$

where $I_S(\omega)$ is an indicator with an interval $S \subset [d, e]$, defined by

$$I_S(\omega) = \begin{cases} 1 & \omega \in S \\ 0 & \omega \notin S \end{cases}. \tag{4.24}$$

Consider the left eigenvector of P with the eigenvalue 1, i.e., the stationary distribution, and the one of $T(\sigma)$ denoted, respectively, by $\mathbf{u}^t_{N_s}$ and $\mathbf{u}^t_{N_s^2}$. If the eigenvector $\mathbf{u}^t_{N_s}$ is described by

$$\mathbf{u}^t_{N_s} = (u_1, u_2, \cdots, u_{N_s}), \tag{4.25}$$

then

$$\mathbf{u}_{N_s^2}^t = (u_1 p_{11}, u_1 p_{12}, \cdots, u_1 p_{1,N_s}, \cdots, u_{N_s} p_{N_s,1}, \cdots, u_{N_s} p_{N_s,N_s}). \tag{4.26}$$

Dividing each element $u_i p_{i,j}$ of stationary distribution $\mathbf{u}_{N_s^2}^t$ by the length of the corresponding subinterval $|J_{i,j}|$, and using relation (4.11), we get an N_s^2-dimensional vector of the invariant density,

$$\mathbf{f}^* = (\overbrace{f_1^*, \cdots, f_1^*}^{N_s}, \overbrace{f_2^*, \cdots, f_2^*}^{N_s}, \cdots, \overbrace{f_{N_s}^*, \cdots, f_{N_s}^*}^{N_s}), \quad f_i^* = \frac{u_i}{|J_i|}, \tag{4.27}$$

which completes the justification of T'.

4.2.2.1 Minimizing the Number of Subintervals of Kalman's Maps

Using the condition that each row sum of transition matrix is equal to unity, together with Kalman's assumption (4.3), we can observe that P has

$$\operatorname{rank} P \cdot (N_s - 1) \tag{4.28}$$

independent variables, where $\operatorname{rank} C$ denotes the rank of a matrix C. For simplicity, we assume

$$\operatorname{rank} P = N_s. \tag{4.29}$$

Thus P has $(N_s^2 - N_s)$ independent variables.

As discussed above, Kalman embedded an N_s-state Markov chain with P into its chaotic dynamics of the PLM onto maps with $N_s^2 - 1$ partition points, while P has $(N_s^2 - N_s)$ independent variables. Kalman used the initial probabilities as the quantities to be previously prescribed as well as the Markov transition probabilities p_{ij}.[4] However, ergodic maps have a stationary distribution irrespective of the initial probabilities. This implies that Kalman's procedure includes $(N_s - 1)$ redundant parameters. In fact, there are several classes of chaotic dynamics of the PLM onto maps with $(N_s^2 - N_s)$ partition points into which a prescribed transition matrix P can be embedded, as follows.

For simplicity, let us assume (4.29) and give a simple method of embedding P into chaotic dynamics of the PLM onto maps. First, choose at random the k-th row vector of P among N_s linearly-independent row vectors $\{\mathbf{p}_i\}_{i=1}^{N_s}$ of P, say $\mathbf{p}_k$, and divide the interval $J = [0, 1]$ into N_s subintervals such as

$$J_\ell = (d_{\ell-1}, d_\ell], \quad 1 \le \ell \le N_s, \quad d_0 = 0, \quad d_{N_s} = 1, \tag{4.30}$$

subject to the condition of probabilities

$$|J_\ell| = p_{k\ell}. \tag{4.31}$$

[4]Recently, Kalman insisted on "the separation of *randomness* from *probability*" [42].

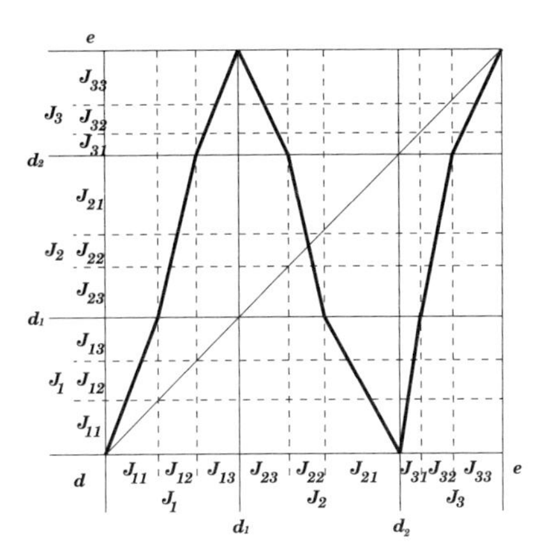

Figure 4.2: An example of a 3-state Markov chain embedding chaotic dynamics of a PLM onto map with 9 subintervals.

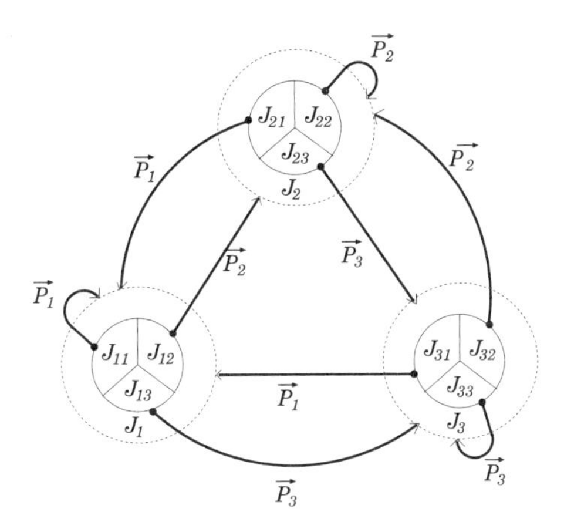

Figure 4.3: The state diagram for the 3-state Markov chain.

The restriction of τ to the interval J_k is given by

$$\tau_k(\omega) = \begin{cases} \dfrac{\omega - d_{k-1}}{|J_k|} & \omega \in J_k \quad \text{for } \tau(d_k) = 1, \\ \dfrac{-\omega + d_k}{|J_k|} & \omega \in J_k \quad \text{for } \tau(d_k) = 0. \end{cases} \tag{4.32}$$

Next divide the remaining $(N_s - 1)$ subintervals J_i $(i \neq k)$ into N_s subintervals $J_{i,j}$ $(1 \leq j \leq N_s)$ defined by (4.7) and (4.8). Each subinterval J_i is divided into N_s subintervals $J_{i,j}$ so as to satisfy (4.11). The restriction of τ to the interval $J_{i,j}$ $(i \neq k)$ is given by (4.12). Thus we get a PLM onto map with $(N_s^2 - N_s + 1)$

subintervals. Ulam's transition matrix associated with a one-to-one mapping,

$$\sigma'' = \begin{pmatrix} J_{1,1} & J_{1,2} & \cdots & J_{k-1,N_s} & J_k & J_{k+1,1} & \cdots & J_{N_s,N_s} \\ I_1 & I_2 & \cdots & I_{kN_s} & I_{kN_s+1} & I_{kN_s+2} & \cdots & I_{N_s^2-N_s+1} \end{pmatrix}, \tag{4.33}$$

is given by $T(\sigma'')$, whose set of all eigenvalues is described by

$$\Lambda(T(\sigma'')) = \Lambda(P) \cup \{0\}^{N_s^2-2N_s+1}. \tag{4.34}$$

The left eigenvector $\mathbf{u}^t_{(N_s^2-N_s+1)}$ of $T(\sigma'')$ with the eigenvalue 1 is given by

$$\mathbf{u}^t_{(N_s^2-N_s+1)} = (u_1p_{11}, u_1p_{12}, \cdots, u_1p_{1,N_s}, \cdots, u_{k-1}p_{k-1,1}, \cdots, u_{k-1}p_{k-1,N_s}, u_k,$$
$$u_{k+1}p_{k+1,1}, \cdots, u_{k+1}p_{k+1,N_s}, \cdots, u_{N_s}p_{N_s,1}, \cdots, u_{N_s}p_{N_s,N_s}). \tag{4.35}$$

Figure 4.2 shows an example of a 3-state Markov chain embedding chaotic dynamics of a Kalman map with 9 subintervals. Figure 4.3 shows its state diagram, where $\mathbf{p}_i$ ($i = 1, 2, 3$) are respectively given by (4.19) with $N_s = 3$. Figure 4.4 shows an example of a 3-state Markov chain with chaotic dynamics of a PLM onto map with the minimum number, i.e., 7 of subintervals. Its Ulam's transition matrix $T(\sigma'')$ is given by

$$T(\sigma'') = \begin{bmatrix} p_{11} & p_{12} & p_{13} & 0 & 0 & 0 & 0 \\ 0 & 0 & 0 & 1 & 0 & 0 & 0 \\ 0 & 0 & 0 & 0 & p_{31} & p_{32} & p_{33} \\ p_{21}p_{11} & p_{21}p_{12} & p_{21}p_{13} & p_{22} & p_{23}p_{31} & p_{23}p_{32} & p_{23}p_{33} \\ p_{11} & p_{12} & p_{13} & 0 & 0 & 0 & 0 \\ 0 & 0 & 0 & 1 & 0 & 0 & 0 \\ 0 & 0 & 0 & 0 & p_{31} & p_{32} & p_{33} \end{bmatrix}. \tag{4.36}$$

Its state diagram is shown in Figure 4.5. If we coarse-grain the subintervals $\{J_{11}, J_{12}, J_{13}\}$ and $\{J_{31}, J_{32}, J_{33}\}$ into J_1 and J_3, respectively, we see that a 3-state Markov chain is embedded into chaotic dynamics of the PLM onto map with 7 subintervals. Using the P-F operator associated with the map, we remark the power spectra of real-valued iterates generated by Kalman's map and the improved Kalman's map are equivalent (see reference [69]).

Lastly, note that the PLM onto maps with $(\mathrm{rank}P \cdot (N_s - 1) + 1)$ subintervals allow us to embed a prescribed transition matrix P satisfying (4.3) into their chaotic dynamics. We also remark here that a Markov chain described by P with $rankP = 1$, i.e., $\Lambda(P) = \{1\} \cup \{0\}^{N_s-1}$, referred to as the Bernoulli shift $B(p_1, p_2, \cdots, p_{N_s})$ with $|J_i| = p_i$ $(1 \leq i \leq N_s)$, $\sum_{i=1}^{N_s} p_i = 1$ is embedded into chaotic dynamics of the PLM onto map with N_s subintervals.

4.3 How to Generate Sequences of Random Variables

4.3.1 Bernoulli Shift and Rademacher Function

It is well known that a sequence of independent and identically distributed (i.i.d.) binary random variables is fundamental as a model for an information source in

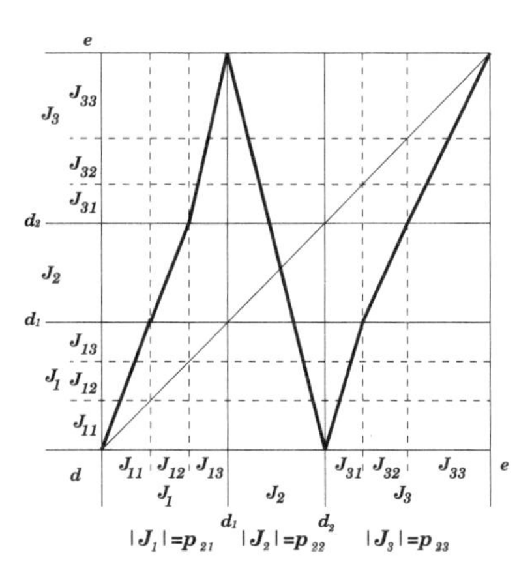

Figure 4.4: An example of a 3-state Markov chain embedding chaotic dynamics of a PLM onto map with 7 subintervals.

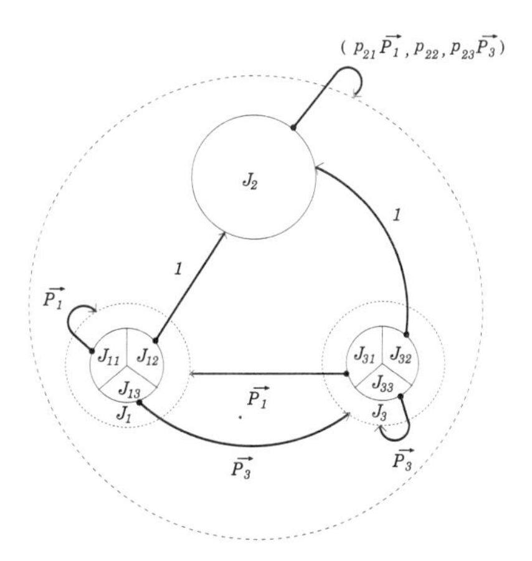

Figure 4.5: The state diagram for the 3-state Markov chain.

communication theory. Let us start with reviewing fundamental subjects from the well-known textbooks of elementary probability theory [3–5].

Let ω be a point drawn at random from the unit interval $(0, 1)$. With each ω associate its nonterminating binary expansion

$$\omega = \sum_{k=1}^{\infty} \frac{d_k(\omega)}{2^k} = 0.d_1(\omega)d_2(\omega)\cdots, \text{ where } d_i(\omega) \in \{0, 1\}. \tag{4.37}$$

Imagine now a coin with faces labeled 1 and 0 instead of the usual heads (H) and tails (T). If ω is drawn at random, $\{d_k(\omega)\}_{k=1}^{\infty}$ behaves as if it resulted from an infinite sequence of tosses of a fair coin. To see this, consider first the event that the first n tosses give the outcomes $u_1, u_2, \cdots, u_n$ in sequence. Then the set

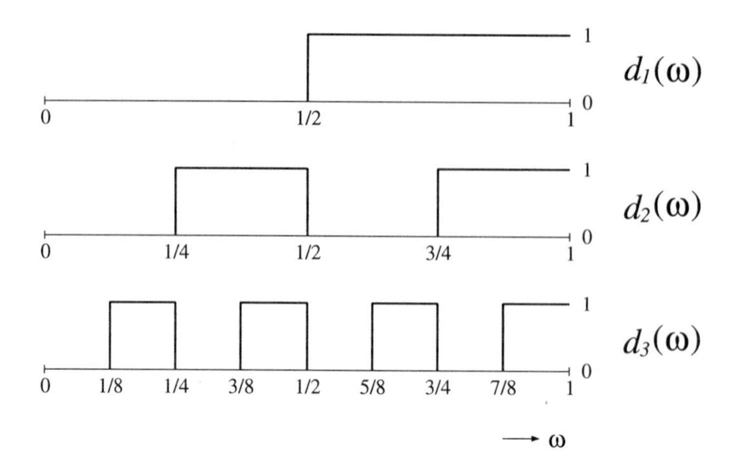

Figure 4.6: Binary functions $d_k(\omega)$ defined by Rademacher functions $r_k(\omega)$..

of ω for which $d_i(\omega) = u_i, i = 1, 2, \cdots, n$ is represented by

$$[\omega : d_i(\omega) = u_i, 1 \leq i \leq n] = \left(\sum_{i=1}^{n} \frac{u_i}{2^i}, \sum_{i=1}^{n} \frac{u_i}{2^i} + \sum_{i=n+1}^{\infty} \frac{1}{2^i}\right] \tag{4.38}$$

which is referred to as the dyadic interval.

On the other hand, define the map $\tau_B(\cdot)$, referred to as the *Bernoulli shift* or *dyadic map* by

$$\tau_B(\omega) = 2\omega, \quad \text{mod } 1 \tag{4.39}$$

which gives

$$\tau_B(\omega) = 0.d_2(\omega)d_3(\omega)\cdots. \tag{4.40}$$

This implies that $\tau_B(\cdot)$ shifts the digits one place to the left, namely,

$$d_k(\tau_B(\omega)) = d_{k+1}(\omega), \quad \text{for } k \geq 1. \tag{4.41}$$

The functions $d_k(\omega), k = 1, 2, \cdots$ furnish us with a *model* of independent tosses of a "fair" coin as follows. The argument becomes simpler if the $d_k(\omega)$ are replaced by the *Rademacher functions* [3, 5],

$$r_k(\omega) = 1 - 2d_k(\omega), \quad k = 1, 2, \cdots. \tag{4.42}$$

For any c_k, the function $\sum_{k=1}^{n} c_k r_k(\omega)$ satisfies the following equation (see [3])

$$\int_0^1 \exp[i \sum_{k=1}^{n} c_k r_k(\omega)] d\omega = \prod_{k=1}^{n} \int_0^1 \exp[i c_k r_k(\omega)] d\omega \tag{4.43}$$

which implies that $r_k(\omega)$ (and hence $d_k(\omega)$), $k = 1, 2, \cdots$ are "independent random variables." Figure 4.6 shows the binary functions $d_k(\omega)$ defined by Rademacher functions $r_k(\omega)$.

4.3.2 EDP and CSP

In the application of chaotic dynamics to cryptosystem, the dyadic map is impracticable with the help of a computer with its limited accuracy in the sense that the period of its chaotic orbit is very short. Of crucial importance is the question whether any maps other than the dyadic map can realize coin tossing or not.

To discuss this problem, let us start by giving three simple methods [56, 57] for obtaining binary sequences from chaotic real-valued sequences $\{\omega_n\}_{n=0}^{\infty}$, generated by an ergodic map $\tau(\cdot) : J = [d, e] \to J$

$$\omega_{n+1} = \tau(\omega_n),\ \omega_n = \tau^n(\omega_0) \in J,\ n = 1, 2, \cdots . \tag{4.44}$$

Method-1: We define a threshold function $\Theta_t(\omega)$ as

$$\Theta_t(\omega) = \begin{cases} 0 & \text{for } \omega < t \\ 1 & \text{for } \omega \geq t \end{cases} \tag{4.45}$$

and define its complementary function

$$\overline{\Theta}_t(\omega) = 1 - \Theta_t(\omega). \tag{4.46}$$

Using these functions, we can obtain a binary sequence $\{\Theta_t(\tau^n(\omega))\}_{n=0}^{\infty}$, which is referred to as a *chaotic threshold sequence.*

Method-2: We write the value of ω ($|\omega| \leq 1$) in a binary representation:

$$|\omega| = 0.A_1(\omega)A_2(\omega)\cdots A_i(\omega)\cdots,\ \text{where } A_i(\omega) \in \{0, 1\}. \tag{4.47}$$

The i-th bit $A_i(\omega)$ can be expressed as

$$A_i(\omega) = \sum_{r=1}^{2^i}(-1)^r \left\{\Theta_{\frac{r}{2^i}}(\omega) + \overline{\Theta}_{-\frac{r}{2^i}}(\omega)\right\}. \tag{4.48}$$

Thus we can obtain a binary sequence $\{A_i(\omega_n)\}_{n=0}^{\infty}$ which we call a *chaotic bit sequence.* Since $\Theta_t(\omega)$ can be regarded as a Boolean function whose variable, seed ω, is not binary but real-valued, $A_i(\omega)$ can be rewritten by

$$A_i(\omega) = \bigoplus_{r=1}^{2^i} \left\{\Theta_{-\frac{r}{2^i}}(\omega) \oplus \Theta_{\frac{r}{2^i}}(\omega)\right\} \tag{4.49}$$

where $\oplus$ denotes modulo 2 addition.

Furthermore, for any map $\tau(\cdot)$ defined on the interval $[d, e]$, we can give a generalized version of the second method [64] as follows.

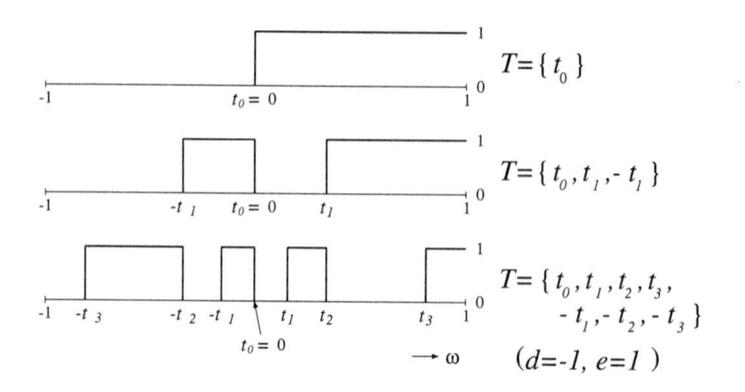

Figure 4.7: *Independent* pulse functions defined by $C_T(\omega)$.

Method-3: We define a binary function with the set of thresholds, $T = \{t_r\}_{r=0}^N$, defined by

$$C_T(\omega) = \bigoplus_{r=0}^{N} \Theta_{t_r}(\omega) = \sum_{r=0}^{N} (-1)^r \Theta_{t_r}(\omega). \tag{4.50}$$

This function $C_T(\omega)$ gives the following two simple examples.

Example 1.1 of $C_T(\omega)$: When $N = 0$, $C_T(\omega) = \Theta_{t_0}(\omega)$.

Example 1.2 of $C_T(\omega)$: We write the value of $\dfrac{\omega - d}{e - d} \in [0, 1]$ in a binary representation:

$$\frac{\omega - d}{e - d} = 0.B_1(\omega)B_2(\omega) \cdots B_i(\omega) \cdots, \text{ where } B_i(\omega) \in \{0, 1\} \tag{4.51}$$

which implies that

$$C_T(\omega) = B_i(\omega), \text{ when } N = 2^i, t_r = (e - d)\frac{r}{2^i} + d. \tag{4.52}$$

If the interval $J = [0, 1]$, then $A_i(\omega) = B_i(\omega)$.

Thus each of $\{A_i(\tau^n(\omega))\}_{n=0}^\infty$ and $\{B_i(\tau^n(\omega))\}_{n=0}^\infty$ is referred to as a *chaotic bit sequence*. Furthermore, it is reasonable that the binary function $C_T(\omega)$ is referred to as the binary pulse function. Figure 4.7 shows examples of these pulse functions defined by $C_T(\omega)$. Note that $B_i(\omega) = d_i(\omega)$ when $\tau(\cdot) = \tau_B(\cdot)$.

Now let us move on to conditions for such *a pulse function* to generate a sequence of i.i.d. binary random variables according to a real-valued orbit $\{\tau^n(\omega)\}_{n=0}^\infty$.

We consider a piecewise monotonic and ergodic map τ (4.44) that satisfies the following properties:

(i) There is a partition $d = d_0 < d_1 < \cdots < d_{N_\tau} = e$ of $[d, e]$ such that for each integer $i = 1, \cdots, N_\tau$ ($N_\tau \geq 2$) the restriction of τ to the interval $[d_{i-1}, d_i)$, denoted by τ_i ($1 \leq i \leq N_\tau$), is a C^2 function; as well as

(ii) $\tau((d_{i-1}, d_i)) = (d, e)$, that is, τ_i is onto;

(iii) $\tau(\omega)$ has a unique absolutely continuous invariant (ACI) measure, denoted by $f^*(\omega)d\omega$ and is mixing on J with respect to this measure.

It is known from the Birchoff individual ergodic theorem [39] that the time average of any L_1 function $F(\cdot)$ along a chaotic real-valued trajectory $\{\omega_n\}_{n=0}^{\infty}$ generated by the map (4.44), denoted by T-average$\{F\}$, defined by

$$\text{T-average}\{F\} = \lim_{T\to\infty} \frac{1}{T} \sum_{n=0}^{T-1} F(\omega_n) \tag{4.53}$$

is equal to almost everywhere to the ensemble average of $F(\cdot)$ over the interval J, denoted by $\langle F \rangle$, defined by

$$\langle F \rangle = \int_J F(\omega) f^*(\omega)\, d\omega. \tag{4.54}$$

We, however, use the expectation notation $\mathbf{E}[F]$ instead of $\langle F \rangle$ because we concentrate our interest on the stationary distribution $f^*(\omega)d\omega$ only, that is,

$$\mathbf{E}[F] = \int_J F(\omega) f^*(\omega)\, d\omega. \tag{4.55}$$

It should be noted that expectation $\mathbf{E}[\cdot]$ is used here in a sense applicable to deterministic sequences of the form (4.44), and differs from the standard one for stochastic sequences which will involve a joint probability density function.

Also note that the function $f^*(\omega)$ is the eigenfunction with the eigenvalue 1 of the following P-F equation

$$P_\tau f^*(\omega) = f^*(\omega). \tag{4.56}$$

where P_τ is the P-F operator of the map $\tau(\cdot)$ with the interval $J = [d, e]$ [39], which is defined by

$$P_\tau H(\omega) = \frac{d}{d\omega} \int_{\tau^{-1}([d,\omega])} H(y) dy. \tag{4.57}$$

Let $G(\omega)$ and $H(\omega)$ be any two L_1 functions of bounded variation. The P-F operator P_τ has the important property:

$$\int_J G(\omega) P_\tau\{H(\omega)\} d\omega = \int_J G(\tau(\omega)) H(\omega) d\omega. \tag{4.58}$$

For the above class of maps, we have [39]

$$P_\tau H(\omega) = \sum_{i=1}^{N_\tau} |g_i'(\omega)| H(g_i(\omega)), \tag{4.59}$$

where $g_i(\omega)$ is the i–th preimage of $\tau(\cdot)$, namely,

$$g_i(\omega) = \tau_i^{-1}(\omega). \tag{4.60}$$

Several famous ergodic maps and their ACI measures are listed as follows [43].

(1) R-adic map

$$\tau(\omega) = R\omega \bmod 1,\ R = 2, 3, 4, \cdots, \omega \in [0,1], \quad f^*(\omega)d\omega = d\omega \tag{4.61}$$

(2) tent map

$$\tau(\omega) = \begin{cases} 2\omega, & \omega \in [0, \frac{1}{2}) \\ 2(1-\omega) & \omega \in [\frac{1}{2}, 1], \end{cases} \quad \omega \in [0,1], \quad f^*(\omega)d\omega = d\omega \tag{4.62}$$

(3) logistic map [44]

$$\tau(\omega) = 4\omega(1-\omega), \omega \in [0,1], \quad f^*(\omega)d\omega = \frac{d\omega}{\pi\sqrt{\omega(1-\omega)}} \tag{4.63}$$

(4) Chebyshev map of degree k [43, 45, 46]

$$\tau(\omega) = \cos(k \cos^{-1} \omega),\ k = 2, 3, 4, \cdots, \omega \in [-1,1], \quad f^*(\omega)d\omega = \frac{d\omega}{\pi\sqrt{1-\omega^2}}. \tag{4.64}$$

Of major importance to the investigation of statistical properties of two sequences $\{G(\tau^n(\omega))\}_{n=0}^{\infty}$ and $\{H(\tau^n(\omega))\}_{n=0}^{\infty}$ is their 2nd-order cross-covariance function, which is defined by

$$\begin{aligned} \widetilde{\rho}^{(2)}(\ell; G, H) &= \int_J (G(\omega) - \mathbf{E}[G])(H(\tau^\ell(\omega)) - \mathbf{E}[H]) f^*(\omega) d\omega, \\ & \quad \text{where,} \quad \ell = 0, 1, 2, \cdots. \end{aligned} \tag{4.65}$$

Note that when $G = H$, these denote the auto-covariance function.

Using Equation (4.58), we get

$$\widetilde{\rho}^{(2)}(\ell; G, H) = \int_J P_\tau^\ell \{G(\omega) f^*(\omega)\} H(\omega) d\omega - \mathbf{E}[G]\mathbf{E}[H]. \tag{4.66}$$

For several maps, such as the tent map (4.62), the logistic map (4.63), and the Chebyshev maps (4.64), the auto-covariance functions of real-valued sequences were already evaluated [43, 47].

We now define an interesting class of maps as follows [64, 68, 70].

Definition 1.1: If the above piecewise monotonic maps with properties (i), (ii), and (iii) satisfy

$$|g_i'(\omega)| f^*(g_i(\omega)) = \frac{1}{N_\tau} f^*(\omega),\ 1 \leq i \leq N_\tau, \tag{4.67}$$

then the map is referred to as the *equidistributivity property* (or briefly *EDP*).

Note that this class contains the well-known maps, such as the R-adic map (4.61), the tent map (4.62), the logistic map (4.63), and the Chebyshev map of degree k (4.64), where $N_\tau = R, 2, 2, k$, respectively. Thus we give the following interesting lemma [64] which is very useful in evaluating covariance functions of chaotic threshold and bit sequences.

Lemma 1.1: For the piecewise monotonic maps satisfying Equation (4.67), we can get

$$P_\tau\{(\Theta_t(\omega) - \mathbf{E}[\Theta_t]) f^*(\omega)\} = \frac{s(\tau'(t))}{N_\tau}(\Theta_{\tau(t)}(\omega) - \mathbf{E}[\Theta_{\tau(t)}]) f^*(\omega) \tag{4.68}$$

where $s(\omega)$ is the signum function defined by

$$s(\omega) = \begin{cases} -1 & \text{for } \omega < 0 \\ 1 & \text{for } \omega \geq 0. \end{cases} \tag{4.69}$$

Lemma 1.1 gives

$$P_\tau\{\Theta_{d_i}(\omega) f^*(\omega)\} = \mathbf{E}[\Theta_{d_i}] f^*(\omega). \tag{4.70}$$

Now, if $T = \{t_r\}_{r=0}^{2M}$ is the set of *symmetric thresholds*, defined by

$$t_r + t_{2M-r} = d + e, \quad r = 0, 1, \cdots, 2M, \tag{4.71}$$

then $C_T(\omega)$ is referred to as a *symmetric binary function.* [64]

Next let us restrict our attention to the map satisfying

$$f^*(d + e - \omega) = f^*(\omega), \quad \omega \in [d, e], \tag{4.72}$$

which is referred to as a *symmetric property of the invariant measure.* Note that such a class of maps contains the well-known maps, such as the R-adic map, the tent map, the logistic map, and the Chebyshev map.

Remark 1.1: For the maps with the symmetric property of the invariant measure (4.72), we get

$$\mathbf{E}[C_T] = \frac{1}{2}. \tag{4.73}$$

Furthermore, we consider a somewhat restricted class of piecewise monotonic maps satisfying Equation (4.67) which also satisfies the *symmetric property of the map*

$$\tau(d+e-\omega)=\tau(\omega), \quad \omega\in[d,e]. \tag{4.74}$$

Such a class includes the tent map, the logistic map, and the Chebyshev map of even degree k. The fact that τ is monotonic and onto gives

$$\tau\left(\frac{d+e}{2}\right)=d \text{ or } e. \tag{4.75}$$

The following lemma [64] plays an important role in estimating the covariance functions of symmetric binary sequences as shown in the corollary that follows.

Lemma 1.2: For the piecewise monotonic maps satisfying both Equation (4.67) and Equation (4.72), and their symmetric binary functions, we can get

$$P_\tau\{C_T(\omega)f^*(\omega)\}=\mathbf{E}[C_T]f^*(\omega). \tag{4.76}$$

which is a generalized version of Equation (4.70).

We now define the symmetry property of the any function $H(\cdot)$ associated with a class of maps $\tau(\cdot)$ satisfying the EDP as follows

Definition 1.2: For a class of maps with the *EDP* and any L_1 function $H(\cdot)$, if we get

$$\frac{1}{N_\tau}\sum_{i=1}^{N_\tau}H(g_i(\omega))=\mathbf{E}[H], \tag{4.77}$$

then the function $H(\cdot)$ is said to satisfy the *constant summation property (CSP)*. [68, 70]

Note that the r.h.s. of Equation (4.77) is independent on the seed ω.

Remark 1.2: It is obvious from Equation (4.59) that the CSP of the symmetric binary function $C_T(\omega)$, namely,

$$\frac{1}{N_\tau}\sum_{i=1}^{N_\tau}C_T(g_i(\omega))=\mathbf{E}[C_T], \tag{4.78}$$

implies Equation (4.76) automatically holds.

Corollary 1.1: Consider the piecewise monotonic maps with both Equation (4.67) and Equation (4.74). Denote two different sets of symmetric thresholds by $T=\{t_r\}_{r=0}^{2M}$ and $T'=\{t'_r\}_{r=0}^{2M'}$, where

$$d=t_0<t_1<\cdots<t_{2M}=e, \text{ with } t_r+t_{2M-r}=d+e,\ 0\le r\le 2M, \tag{4.79}$$

$$d=t'_0<t'_1<\cdots<t'_{2M'}=e, \text{ with } t'_r+t'_{2M'-r}=d+e,\ 0\le r\le 2M'. \tag{4.80}$$

Then we can obtain

$$\widetilde{\rho}^{(2)}(\ell; C_T, C_{T'}) = \begin{cases} Q^C_{TT'} & \text{for } \ell = 0 \\ 0 & \text{for } \ell \geq 1 \end{cases} \tag{4.81}$$

where

$$Q^C_{TT} = \mathbf{E}[C_T] = \frac{1}{2}, \quad Q^C_{TT'} = \mathbf{E}[C_T C_{T'}] = \int_{J^C_{TT'}} d\omega, \tag{4.82}$$

$$J^C_{TT'} = \left(\bigcup_{r=1}^{M} J^C_T(r)\right) \bigcap \left(\bigcup_{s=1}^{M'} J^C_{T'}(s)\right), \text{ where } J^C_T(r) = [\mathbf{E}[\Theta_{t_{2r}}], \mathbf{E}[\Theta_{t_{2r-1}}]]\,. \tag{4.83}$$

Note that the symmetric binary function $C_T(\omega)$ is a generalized version of the Rademacher function for the dyadic map.

Remark 1.3: Let $C_T = B_i$ and $C_{T'} = B_j$. Then $\rho^{(2)}(0; B_i, B_j) = Q_{ij}$. We get $Q_{ij} = \frac{1}{4}$ for the maps with the uniform invariant density. On the other hand, for the maps with the nonuniform invariant densities, such as the logistic and the Chebyshev map, we can get [64]

$$\lim_{\substack{i\to\infty \\ \text{or } j\to\infty}} Q_{ij} = \frac{1}{4} \quad \text{for } i \neq j. \tag{4.84}$$

Next, let $\mathbf{U} = U_0 U_1 \cdots U_{m-1}$ be an arbitrary string of m binary digits where $U_n \in \{0, 1\}$ $(0 \leq n \leq m-1)$. Then there are 2^m possible strings. Let $\mathbf{u}^{(r)} = u^{(r)}_0 u^{(r)}_1 \cdots u^{(r)}_{m-1}$ be the r-th string. Introducing a binary random variable

$$\Gamma_n(\omega; G, \mathbf{u}^{(r)}) = G(\omega) u^{(r)}_n + \overline{G}(\omega) \overline{u}^{(r)}_n \tag{4.85}$$

for any binary function $G(\omega)$. Then the probability of the event $\mathbf{u}^{(r)}$ in an infinite binary sequence $\{G(\tau^n(\omega))\}^{\infty}_{n=0}$ is given by

$$\begin{aligned} \text{Prob}\{\mathbf{u}^{(r)}; G\} &= \int_J \left\{ \prod_{n=0}^{m-1} \Gamma_n(\tau^n(\omega); G, \mathbf{u}^{(r)}) \right\} f^*(\omega) d\omega \\ &= \rho^{(m)}(\underbrace{1, 1, \cdots, 1}_{m-1}; \Gamma_0(G, \mathbf{u}^{(r)}), \Gamma_1(G, \mathbf{u}^{(r)}), \cdots, \Gamma_{m-1}(G, \mathbf{u}^{(r)})). \end{aligned} \tag{4.86}$$

where $\rho^{(m)}(\cdot, \cdot)$ is the ensemble-average of the higher-order (m-th order) correlation function defined by

$$\begin{aligned} &\rho^{(m)}(\ell_{m-1}, \ell_{m-2}, \cdots, \ell_1; H_m, H_{m-1}, \cdots, H_1) \\ &= \int_J H_m(\omega) H_{m-1}(\tau^{\ell_{m-1}}(\omega)) H_{m-2}(\tau^{\ell_{m-1}+\ell_{m-2}}(\omega)) \cdot \\ &\quad \cdots H_1(\tau^{\ell_{m-1}+\ell_{m-2}+\cdots+\ell_1}(\omega)) f^*(\omega) d\omega \quad \text{for all integers } \ell_i \geq 0, \end{aligned} \tag{4.87}$$

where each of $H_i(\omega)$ denotes an L_1 real-valued function ($i = 1, 2, \cdots, m$). It is, in general, difficult to evaluate such higher-order correlation functions explicitly. However, it is simplified if the CSP (or Equation (4.77)) is satisfied. In fact, we can get [64] the following theorem for any binary function $H(\omega)$ with the CSP (or Equation (4.77)) [64, 68, 70].

Theorem: Any binary function $H(\omega)$ with the CSP gives

$$\text{Prob}\{\mathbf{u}^{(r)}; H\} = (\mathbf{E}[H])^s(1 - \mathbf{E}[H])^{m-s}, \tag{4.88}$$

where s is the number of 1's in $\{u_n^{(r)}\}_{n=0}^{m-1}$.

This implies that $H(\tau^n(\omega))\}_{n=0}^{\infty}$ is a sequence of i.i.d. binary random variables in the sense that it can realize a Bernoulli sequence with probability $\mathbf{E}[H]$. Note that we can get a fair Bernoulli sequence when $\mathbf{E}[H] = \frac{1}{2}$, that is, an m-distributed binary random sequence.

4.3.3 Design of Sequences of p-Ary Random Variables

We turn now to the next problem. We observe that the CSP is of cruicial importance in designing of sequences of i.i.d. p-ary random variables [65] which are a generalized version of the binary case in the previous subsection (see Equation (4.76)) as follows. To do this, we introduce several definitions.

Definition 1.3: For PL onto maps and PM onto maps with the EDP, the partition $\{J_i\}_{i=1}^{N_\tau}$ is referred to as *trivial partition.*

Remark 1.4: Each of N_τ indicators $I_{J_i}(\omega)$ for the *trivial partition* $\{J_i\}_{i=1}^{N_\tau}$ satisfies

$$P_\tau\{I_{J_i}(\omega)f^*(\omega)\} = \mathbf{E}[I_{J_i}]f^*(\omega), \; J_i = [d_{i-1}, d_i), \; i = 1, 2, \cdots, N_\tau \tag{4.89}$$

that is, the random variable $I_{J_i}(\omega)$ satisfies the CSP.

Definition 1.4: Define a random variable taking a value in a set $\Gamma = \{\gamma_1, \cdots, \gamma_p\}$ by the linear combination of $I_{A_i}(\omega)$ given by

$$X(\omega) = \sum_{i=1}^{p} \gamma_i I_{A_i}(\omega) \tag{4.90}$$

which is called *an elementary random variable.* [4, 5]

The probability of a random variable X taking the value γ_i, denoted by $\text{Prob}\{X = \gamma_i\}$ is given by

$$\text{Prob}\{X = \gamma_i\} = \int_{A_i} f^*(\omega)d\omega = m(A_i). \tag{4.91}$$

Thus, the expectation $\mathbf{E}[X]$ is given by

$$\mathbf{E}[X] = \sum_{i=1}^{p} \gamma_i m(A_i). \tag{4.92}$$

Suppose that $p = N_\tau$ and $A_i = J_i (i = 1, 2, \cdots, N_\tau)$. Then, from Equation (4.89), we can easily obtain

$$P_\tau\{X(\omega) f^*(\omega)\} = \mathbf{E}[X] f^*(\omega) \tag{4.93}$$

which is a generalized version of Equation (4.76). Furthermore, it can be proven that $\{X(\tau^n(\omega))\}_{n=0}^{\infty}$ is a sequence of i.i.d. N_τ-ary random variables which is m-distributed.

Remark 1.5: For $p \leq N_\tau$, nonlinear ergodic maps $\tau(\cdot)$ can generate a sequence of i.i.d. p-ary random variables by arbitrarily constructing new indicators such as $I_{J_i \cup J_j}(\omega)$. If $\text{Prob}(X = \gamma_i) = \frac{1}{p}$ for all $i = 1, 2, \cdots, p$, then $\{X(\tau^n(\omega))\}_{n=0}^{\infty}$ is a sequence of *m-distributed* p-ary random variables.

It is an interesting issue from the theoretical and practical points of view to get a lot of sequences of i.i.d. p-ary random variables from a single chaotic real-valued orbit. However, the above trivial partition $\{J_i\}_{i=1}^{N_\tau}$ alone is incapable of producing a lot of elementary random variables from a single orbit. Thus we consider *nontrivial* partitions which can produce many kinds of elementary random variables.

Definition 1.5: Let $\{\widetilde{J}_i\}_{i=1}^{N_\tau}$ be a nontrivial partition of J satisfying

$$\left.\begin{array}{l} \widetilde{J}_i \bigcap \widetilde{J}_j = \phi, \ (i \neq j) \\ \bigcup_{i=1}^{N_\tau} \widetilde{J}_i = J. \end{array}\right\} \tag{4.94}$$

If the above nontrivial partition $\{\widetilde{J}_i\}_{i=1}^{N_\tau}$ satisfies

$$P_\tau\{I_{\widetilde{J}_i}(\omega) f^*(\omega)\} = \mathbf{E}[I_{\widetilde{J}_i}] f^*(\omega), \ i = 1, 2, \cdots, N_\tau, \tag{4.95}$$

then, it is obvious that an elementary random variable

$$X(\omega) = \sum_{i=1}^{N_\tau} \gamma_i I_{\widetilde{J}_i}(\omega) \tag{4.96}$$

produces a sequence of i.i.d. N_τ-ary random variables. Hence we concentrate our attention on designing many kinds of nontrivial partitions satisfying Equation (4.95). Owing to limited space, details of design of such nontrivial partitions are omitted here. But, we give a simple example of such design.

Example 1.3: Figure 4.8 shows trivial and nontrivial partitions which can generates sequences of i.i.d. N_τ-ary random variables for the PL map with the uniform measure $f^*(\omega)d\omega = d\omega$, where (a) the PL onto map with $N_\tau = 6$, (b) a trivial partition $\{J_i\}_{i=1}^6$, being shaded, and (c) a nontrivial partition $\{\widetilde{J}_i\}_{i=1}^6$, where $\widetilde{J}_i$ is composed of three shaded subsubintervals. Note that we divide the subinterval J_i into disjoint subsubintervals $\{J_{i,r}\}_{r=1}^N$ to get nontrivial partitions and the subsubintervals $\{J_{i,r}\}_{i=1}^{N_\tau}$, being shaded with the same pattern, are pairwise-interval interchangeable.

Next, we evaluate the cross-correlation function between two sequences of p-ary random variables. Let $X(\omega)$ and $Y(\omega)$ be two elementary N_τ-ary random variables obtained by two sets of indicators $\{I_{J_i^X}(\omega)\}_{i=1}^{N_\tau}$ and $\{I_{J_j^Y}(\omega)\}_{j=1}^{N_\tau}$ satisfying Equation (4.95), respectively given by

$$X(\omega) = \sum_{i=1}^{N_\tau} \gamma_i I_{J_i^X}(\omega), \tag{4.97}$$

$$Y(\omega) = \sum_{i=1}^{N_\tau} \gamma_i I_{J_i^Y}(\omega), \tag{4.98}$$

which can produce sequences of i.i.d. N_τ-ary random variables $\{X(\tau^n(\omega))\}_{n=0}^\infty$ and $\{Y(\tau^n(\omega))\}_{n=0}^\infty$.

Lemma 1.3: The covariance function between two sequences of i.i.d. N_τ-ary random variables $\{X(\tau^n(\omega))\}_{n=0}^\infty$ and $\{Y(\tau^n(\omega))\}_{n=0}^\infty$ from a seed ω is given by

$$\begin{aligned}
\widetilde{\rho}^{(2)}(\ell; X, Y) &= \int_J (X(\omega) - \mathbf{E}[X])(Y(\tau^\ell(\omega)) - \mathbf{E}[Y]) f^*(\omega)d\omega, \\
&= \sum_{i=1}^{N_\tau}\sum_{j=1}^{N_\tau} \gamma_i\gamma_j \int_J I_{J_i^X}(\omega) I_{J_j^Y}(\tau^\ell(\omega)) f^*(\omega)d\omega - \mathbf{E}[X]\mathbf{E}[Y] \\
&= \begin{cases} \displaystyle\sum_{i=1}^{N_\tau}\sum_{j=1}^{N_\tau} \gamma_i\gamma_j \int_{J_i^X \cap J_j^Y} f^*(\omega)d\omega - \mathbf{E}[X]\mathbf{E}[Y] & \text{for } \ell = 0, \\ 0 & \text{for } \ell \geq 1. \end{cases}
\end{aligned} \tag{4.99}$$

Note that the above covariance function is equal to 0 for an arbitrary set $\{\gamma_i\}_{i=1}^{N_\tau}$ when $\ell \geq 1$. It should also be noted that if two N_τ-ary sequences are *completely independent* of each other, the covariance function should have zero value for all ℓ and any set $\{\gamma_i\}_{i=1}^{N_\tau}$. As in Lemma 1.3, the covariance function may have nonzero value when $\ell = 0$ as far as we are intending to get sequences of i.i.d. N_τ-ary random variables generated by a single chaotic real-valued orbit. We can, however, appropriately design elementary random variables in order that the covariance function has zero value for all ℓ and an arbitrary set $\{\gamma_i\}_{i=1}^{N_\tau}$. The

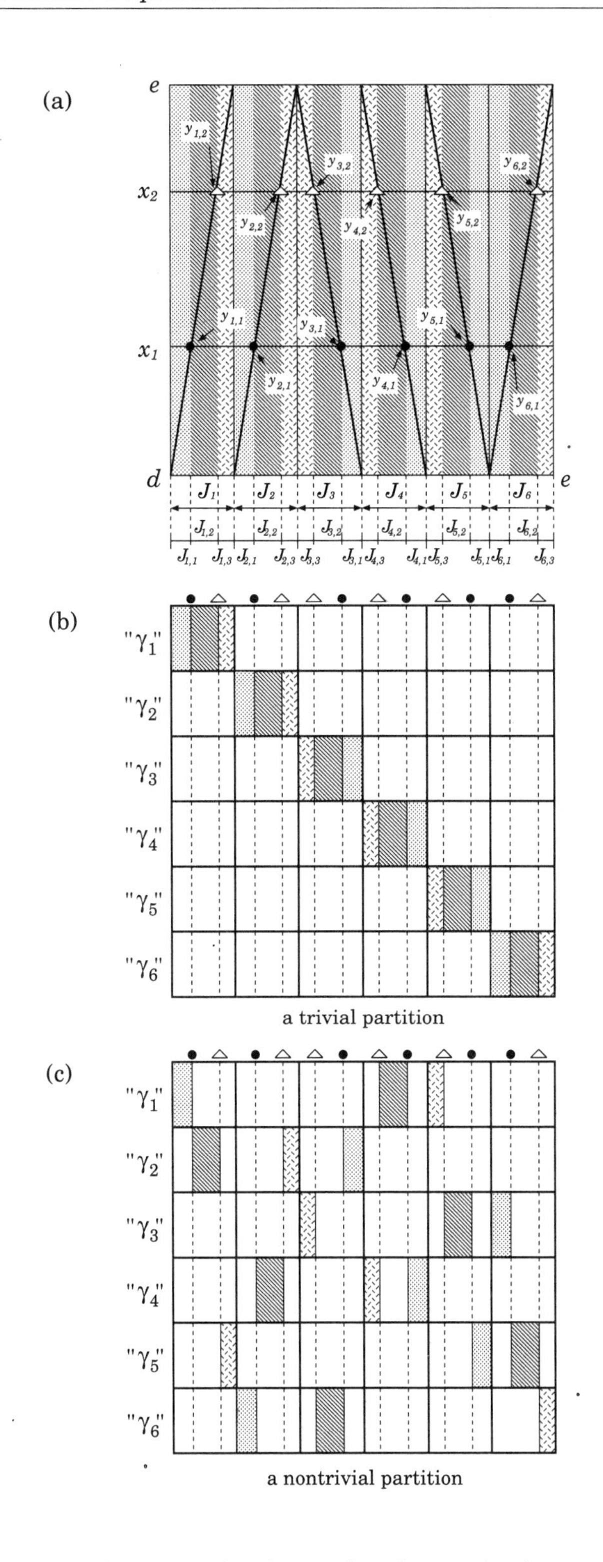

Figure 4.8: An example of trivial and nontrivial partitions.

details of such design are also omitted here. But we give an example as follows.

Example 1.4: Figure 4.9 shows an example of partitions and appropriate mappings of values $\gamma_i (i = 1, 2, 3)$ for the 3-adic map which produce sequences of i.i.d. ternary random variables of depth s, $\widehat{X}^{(s)}(\omega) = \sum_{i=1}^{3} \gamma_i I_{\widehat{J}_i^{(s)}}(\omega)$, being completely uncorrelated with each other, (a) $s = 1$, (b) $s = 2$, (c) $s = 3$, (d) $s = 4$, and (e) $s = 5$. Note that the interval $J_i^{(s)}$ is composed of three shaded subsubintervals. Notice that for the case (e), the subsubintervals are enlarged.

By designing many different sets of nontrivial partitions, we can get a lot of sequences of i.i.d. p-ary random variables from a single chaotic real-valued orbit. It has been shown that we can design many sets of *nontrivial partitions* based on trivial partitions. Such an indicator with nontrivial partitions is proven to be of crucial importance in giving a sufficient condition for the map to produce sequences of i.i.d. p-ary random variables because it is a generalized version of the Rademacher function for the dyadic map. This condition is a weakened version of the previous sufficient condition for the binary case.

4.3.4 Correlational Properties of Sequences of Real-Valued Random Variables

Let us now move on to a real-valued sequence $\{\tau^n(\omega_0)\}_{n=0}^{\infty}$ generated by the map (4.44). We show that the CSP leads us to define two *nonlinear correlational properties* of such a real-valued sequence.

In order to investigate randomness of such a chaotic sequence, we usually evaluate linear dependency with correlation coefficients, defined by

$$\mathrm{Corr}(X_n, X_{n+\ell}) = \frac{\mathbf{E}[X_n X_{n+\ell}] - \mathbf{E}[X]^2}{\mathbf{E}[X^2] - \mathbf{E}[X]^2}, \quad \ell = 1, 2, \cdots, \tag{4.100}$$

where $X_n = \tau^n(\omega), n = 0, 1, 2, \cdots$.

On the other hand, so as to evaluate independence of such a chaotic sequence, we must consider correlation coefficients between the N_1-th power sequence $\{X_n^{N_1}\}_{n=0}^{\infty}$ and the N_2-th power sequence $\{X_n^{N_2}\}_{n=0}^{\infty}$, defined by

$$\mathrm{Corr}(X_n^{N_1}, X_{n+\ell}^{N_2}) = \frac{\mathbf{E}[X_n^{N_1} X_{n+\ell}^{N_2}] - \mathbf{E}[X^{N_1}]\mathbf{E}[X^{N_2}]}{\mathbf{E}[X^{N_1+N_2}] - \mathbf{E}[X^{N_1}]\mathbf{E}[X^{N_2}]}, \quad \ell = 1, 2, \cdots, \ N_1, N_2 \geq 0, \tag{4.101}$$

which is a nonlinear type of correlation of the $\{X_n\}_{n=0}^{\infty}$ sequence. [48]

The notion of independence is of central importance in probability theory. For m stochastic random variables, $X_0, X_1, \cdots, X_{m-1}$, we know Kac's theorem in the following.

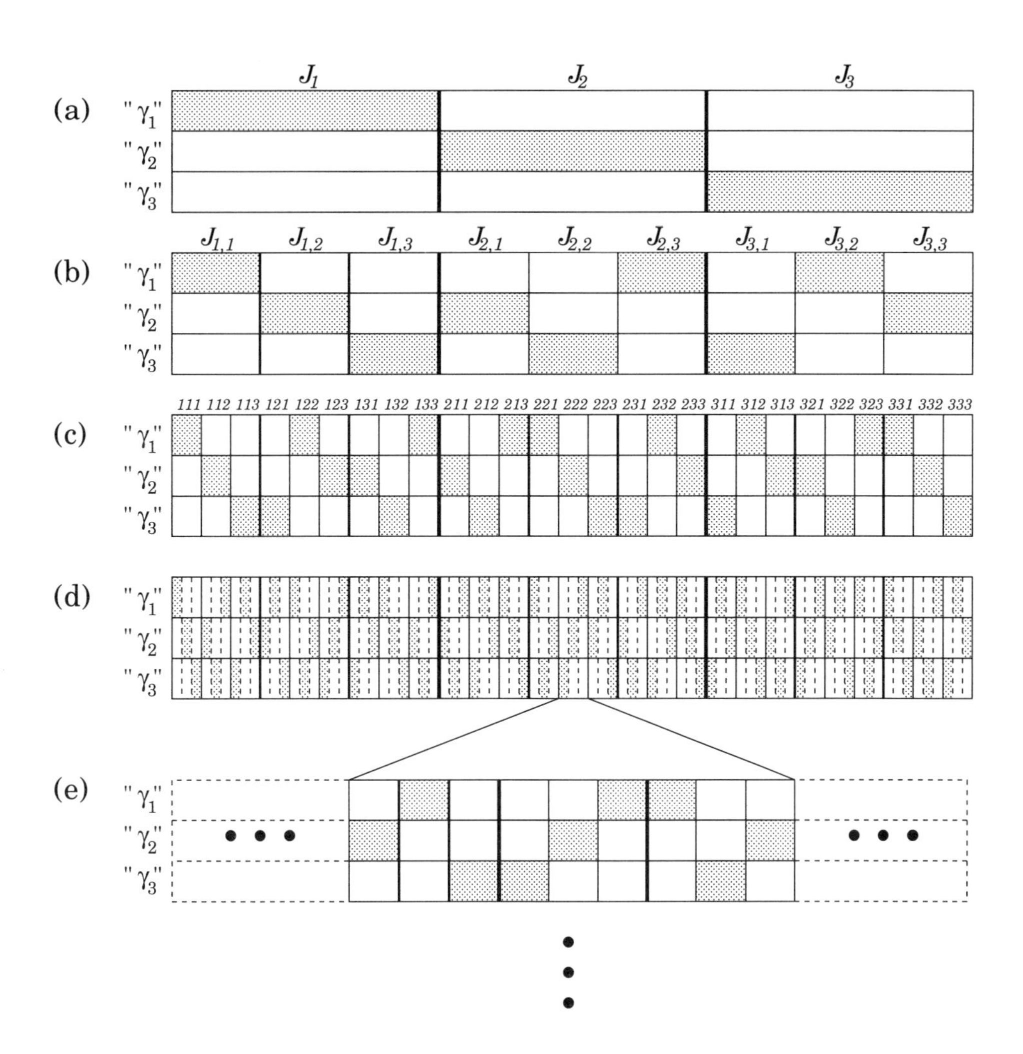

Figure 4.9: An example of partitions and appropriate mappings of values $\gamma_i (i = 1, 2, 3)$ for the 3-adic map which produce sequences of i.i.d. ternary random variables.

Definition 1.6: [4] m random variables, $X_0, X_1, \cdots, X_{m-1}$ are *independent* if and only if

$$\mathbf{E}[\prod_{n=0}^{m-1} e^{jt_n X_n}] = \prod_{n=0}^{m-1} \mathbf{E}[e^{jt_n X_n}] \quad \text{for } \forall t_i \in \Re, \quad i = 0, 1, 2, \cdots, m-1, \tag{4.102}$$

where $\Re$ denotes the set of real numbers. Note that (4.102) holds for sequences of i.i.d. binary random variables, $B_0(\omega), B_1(\omega), \cdots, B_{m-1}(\omega), \cdots$, where $B_n(\omega) = B(\tau^n(\omega))$ $n = 0, 1, \cdots$ is generated by a class of maps which is discussed in the previous section [64].

Remark 1.6: m random variables generated by a nonlinear ergodic map (4.44), $X_0, X_1, \cdots, X_{m-1}$ are independent if and only if

$$\text{Corr}(X_n^{N_1}, X_{n+\ell}^{N_2}) = 0, \quad \text{for } \forall N_1, N_2, \ell \in \mathcal{N}, \tag{4.103}$$

where $\mathcal{N}$ denotes the set of natural numbers.

Definition 1.7: For a given k ($k = 2, 3, \cdots$), a sequence is referred to as being *k-th-order correlated* if $X_n^{N_1}$ and $X_{n+\ell}^{N_2}$ are uncorrelated whenever $N_1 < k$ irrespective of $N_2, \ell \in \mathcal{N}$.

Remark 1.7: If a stationary sequence $\{X_n\}_{n=0}^{\infty}$ is *2nd-order correlated*, then the sequence is linearly uncorrelated, i.e.,

$$\mathbf{E}[X_n X_{n+\ell_1} X_{n+\ell_2} \cdots X_{n+\ell_{m-1}}] = \mathbf{E}[X]^m, \tag{4.104}$$

for $0 < \ell_1 < \cdots < \ell_{m-1}$, $\ell_i \in \mathcal{N}$, $i = 1, 2, \cdots, m-1$.

Furthermore, we here introduce the notion of a second correlated property in the following.

Definition 1.8: For a given k ($k = 2, 3, \cdots$) and some $N_1 \in \mathcal{N}$, a sequence is referred to as being *d-delay correlated* if $X_n^{N_1}$ and $X_{n+\ell}^{N_2}$ are uncorrelated whenever $\ell > d$ irrespective of $N_2 \in \mathcal{N}$.

We principally consider the Chebyshev map of degree k (Equation (4.64)). As a starting point for this work it is known [43, 47] that the linear correlations (4.100) are all zero ($\ell \geq 1$).

Let us turn now to evaluations of higher-order statistics of a real-valued sequence. To do this, our main task is to evaluate the joint moment function

$$\mathbf{E}[X_n^{N_1} X_{n+\ell}^{N_2}] = \int_J \omega^{N_1} \{\tau^\ell(\omega)\}^{N_2} f^*(\omega) d\omega. \tag{4.105}$$

Using the definition of the P-F operator (Equation (4.59)), we get

$$P_\tau\{\omega^N f^*(\omega)\} = \frac{1}{k} f^*(\omega) \sum_{i=0}^{k-1} \{g_i(\omega)\}^N, \quad N = 0, 1, 2, \cdots. \tag{4.106}$$

By virtue of (4.58) and (4.106), we easily get

$$\mathbf{E}[X_n^{N_1} X_{n+\ell}^{N_2}] = \frac{1}{k}\int_J \{\tau^{\ell-1}(\omega)\}^{N_2} f^*(\omega) \left[\sum_{i=0}^{k-1}\{g_i(\omega)\}^{N_1}\right] d\omega, \quad \ell = 1, 2, \cdots. \tag{4.107}$$

Hence, we have to evaluate the summation term in (4.107) as the first step in this specific case.

For the Chebyshev map we know from [56] that

$$g_i(\omega) = \cos\left(\frac{i\pi + \cos^{-1}\{(-1)^i\omega\}}{k}\right), \quad i = 0, 1, \cdots, k-1. \tag{4.108}$$

Using familiar trigonometric formulae, we get

$$\sum_{i=0}^{k-1}\{g_i(\omega)\}^N = \begin{cases} \begin{cases} \dfrac{k}{2^{N-1}} \displaystyle\sum_{n=1}^{[\frac{N}{k}]} \binom{N}{\frac{1}{2}(N-nk)} T_n(\omega) & (k\text{: odd}) \\ 0 & (k\text{: even}) \end{cases}, & \text{for } N = 2M+1, \\ \begin{cases} \dfrac{k}{2^{N-1}} \displaystyle\sum_{n:\,\text{even}}^{[\frac{N}{k}]} \binom{N}{\frac{1}{2}(N-nk)} T_n(\omega) + \dfrac{k}{2^N}\binom{N}{\frac{1}{2}N} & (k\text{: odd}) \\ \dfrac{k}{2^{N-1}} \displaystyle\sum_{n=1}^{[\frac{N}{k}]} \binom{N}{\frac{1}{2}(N-nk)} T_n(\omega) + \dfrac{k}{2^N}\binom{N}{\frac{1}{2}N} & (k\text{: even}) \end{cases} & \text{for } N = 2M. \end{cases} \tag{4.109}$$

where M denotes an integer and $[x]$ denotes the maximum integer not exceeding x. If an integer a is divisible by another integer $b(\neq 0)$, we denote it by $b \mid a$. Unless a is divisible by b, we denote it by $b \not| \, a$ (a, $b \in \mathcal{Z}$, $\mathcal{Z}$ denotes the set of integers).

We are now ready for the proof of one of the main results.

Lemma 1.4: For a given natural number N, the Chebyshev map of degree k greater than N, $T_k(\omega)$ satisfies

$$P_{T_k}\{\omega^N f^*(\omega)\} = \mathbf{E}[X^N] f^*(\omega). \tag{4.110}$$

Proof: From the stationarity of deterministic process, we get

$$\mathbf{E}[X^N] = \int_J \omega^N f^*(\omega) d\omega = \int_J P_{T_k}\{\omega^N f^*(\omega)\} d\omega. \tag{4.111}$$

Using the orthogonal relation of $T_n(\omega)$, in conjunction with (4.106) and (4.109), we have

$$\mathbf{E}[X^N] = \begin{cases} 0 & \text{for } N = 2M+1 \\ \dfrac{1}{2^N}\begin{pmatrix} N \\ \frac{1}{2}N \end{pmatrix} & \text{for } N = 2M \end{cases}, \quad M = 0, 1, 2, \cdots . \tag{4.112}$$

and hence by (4.109),

$$\sum_{i=0}^{k-1} \{g_i(\omega)\}^N = k\mathbf{E}[X^N], \quad N = 0, 1, 2, \cdots, k-1, \tag{4.113}$$

because $[N/k] = 0$ ($N = 0,\ 1,\ 2, \cdots$) for any $k > N$. This, in conjunction with (4.106), completes the proof.

Lemma 1.4 implies that, for the Chebyshev map of degree k greater than N, the Nth power of ω, i.e., $H(\omega) = \omega^N$ satisfies the *CSP*. Thus, we immediately obtain

Corollary 1.2: For a given N_1, the Chebyshev map of degree k greater than N_1 satisfies

$$\mathbf{E}[X_n^{N_1} X_{n+\ell}^{N_2}] = \mathbf{E}[X^{N_1}]\mathbf{E}[X^{N_2}], \quad N_2, \ell \in \mathcal{N}, \tag{4.114}$$

so that the correlations (4.101) are zero.

This implies that a chaotic real-valued sequence generated by the Chebyshev map of degree k is *k-th-order correlated.* Furthermore, Corollary 1.2 is generalized as

Remark 1.8: For given $N_i, i = 1, 2, \cdots, m$ the Chebyshev map of degree k greater than $\max[N_1, N_2, \cdots, N_{m-1}]$ satisfies

$$\mathbf{E}[X_n^{N_1} X_{n+\ell_1}^{N_2} X_{n+\ell_2}^{N_3} \cdots X_{n+\ell_{m-1}}^{N_m}] = \mathbf{E}[X^{N_1}]\mathbf{E}[X^{N_2}]\mathbf{E}[X^{N_3}] \cdots \mathbf{E}[X^{N_m}] \tag{4.115}$$

for $0 < \ell_1 < \cdots < \ell_{m-1}$, $N_i \geq 0, i = 1, 2, \cdots, m$ and $m \geq 1$. Note that (4.115) holds for any N_m.

It is obvious that P_τ is a linear operator and $P_\tau^\ell = P_{\tau^\ell}$, where

$$P_\tau^\ell = \underbrace{P_\tau \circ P_\tau \circ \cdots \circ P_\tau}_{\ell} .$$

Thus, the P-F operator for the Chebyshev map of degree k, P_{T_k} has the following property,

$$P_{T_k}^\ell = P_{T_{k^\ell}}, \tag{4.116}$$

because the Chebyshev polynomials has the semi-group property, i.e.,

$$T_m(T_n(\omega)) = T_{mn}(\omega). \tag{4.117}$$

From elementary facts about the theory of numbers, we know that

$$N = k^{e_k(N)} + r, \quad 0 \le r < k, \tag{4.118}$$

where $N \in \mathcal{N}$, $e_k(N), r \in \mathcal{N} + \{0\}$, $k = 2, 3, \cdots$. Note that, for any N, there exists k and $e_k(N)$ which satisfies (4.118).

By virtue of (4.116) and (4.118), we can easily prove the following lemma.

Lemma 1.5: For any integer N and $\ell > e_k(N)$, the Chebyshev map of degree k satisfies

$$P^{\ell}_{T_k}\{\omega^N f^*(\omega)\} = \mathbf{E}[X^N] f^*(\omega). \tag{4.119}$$

Proof: Combining (4.116) and (4.106), we get

$$P^{\ell}_{T_k}\{\omega^N f^*(\omega)\} = \frac{1}{k^{\ell}} f^*(\omega) \sum_{i=0}^{k^{\ell}-1} \{h_i(\omega)\}^N, \quad N = 0, 1, 2, \cdots, \tag{4.120}$$

where $h_i(\omega)$ is the i-th preimage of ω in the map $T_{k^{\ell}}(\omega)$. When $\ell > e_k(N)$, $[N/k^{\ell}] = 0$, which implies, by using (4.109),

$$\sum_{i=0}^{k^{\ell}-1} \{h_i(\omega)\}^N = k^{\ell} \mathbf{E}[X^N], \quad N = 1, 2, \cdots. \tag{4.121}$$

Hence, in conjunction with (4.120), the result follows.

Thus if $\ell > e_k(N)$, for the Chebyshev map of degree k, the Nth power of ω satisfies *CSP*.

Corollary 1.3: For given N_1 and k, there is an integer $e_k(N_1)$ such that

$$N_1 = k^{e_k(N_1)} + r, \quad 0 \le r < k \tag{4.122}$$

If $\ell > e_k(N_1)$, the Chebyshev map of degree k satisfies

$$\mathbf{E}[X_n^{N_1} X_{n+\ell}^{N_2}] = \mathbf{E}[X^{N_1}]\mathbf{E}[X^{N_2}], \text{ for } \forall N_2 \in \mathcal{N}, \tag{4.123}$$

so that the correlations (4.101) are zero, which implies that a chaotic real-valued sequence generated by the Chebyshev map of degree k is $e_k(N_1)$-*delay correlated.*

Corollary 1.3 is generalized as

Remark 1.9: For given k and N_i $(i = 1, 2, \cdots, m-1)$, there is an integer $e_k(N_i)$ such that

$$N_i = k^{e_k(N_i)} + r, \quad 0 \le r < k, \tag{4.124}$$

If $\ell_i > e_k(N_i)$, the Chebyshev map of degree k satisfies

$$\mathbf{E}[X_n^{N_1} X_{n+\ell_1}^{N_2} X_{n+\ell_1+\ell_2}^{N_3} \cdots X_{n+\ell_1+\ell_2+\cdots+\ell_{m-1}}^{N_m}] = \mathbf{E}[X^{N_1}]\mathbf{E}[X^{N_2}]\mathbf{E}[X^{N_3}] \cdots \mathbf{E}[X^{N_m}],$$
$$\text{for} \,\forall\, N_m \in \mathcal{N}, \quad N_i \ge 0, \ (1 \le i \le m), \ m \ge 1. \tag{4.125}$$

Note that (4.125) holds for any N_m. Other simple ergodic maps with the EDP are discussed in reference [70].

In order to evaluate dependence of a real-valued trajectory generated by the Chebyshev map, we evaluate the characteristic function as follows. Note that analytic expressions of characteristic functions are already derived in [47] by using the orthogonal property of the Chebyshev polynomials. We give a simple proof [70] of Geisel and Fairen's expressions by using the following lemma:

Lemma 1.6: For a given Chebyshev polynomial of degree n, the Chebyshev map of degree k satisfies

$$P_{T_k}\{T_n(\omega)f^*(\omega)\} = \begin{cases} T_{\frac{n}{k}}(\omega)f^*(\omega) & \text{for } k \mid n \\ 0 & \text{for } k \not| \, n \end{cases} \tag{4.126}$$

Proof: From (4.108), we get

$$\sum_{i=0}^{k-1} T_n(g_i(\omega)) = \begin{cases} k\, T_{\frac{n}{k}}(\omega) & \text{for } k \mid n \\ 0 & \text{for } k \not| \, n. \end{cases} \tag{4.127}$$

which, in conjunction with (4.106), completes the proof.

Let $J_\nu(z)$ be the familiar Bessel function of degree ν defined by [49]

$$J_\nu(z) = \left(\frac{z}{2}\right)^\nu \sum_{r=0}^{\infty} \frac{(-z^2/4)^r}{r!(\nu+r)!}. \tag{4.128}$$

Then the fundamental properties of Chebyshev polynomials

$$e^{jt_0\omega} = \sum_{m=-\infty}^{\infty} j^m J_m(t_0)T_m(\omega), \quad \mathbf{E}[e^{jt_0X}] = J_0(t_0) \tag{4.129}$$

lead us to get [70]

Corollary 1.4: [47] The Chebyshev map of degree k satisfies

$$\begin{aligned} \mathbf{E}[e^{jt_0X_n}e^{jt_1X_{n+\ell}}] &= \mathbf{E}[e^{jt_0X}]\mathbf{E}[e^{jt_1X}] + \sum_{\substack{m=-\infty \\ m\neq 0}}^{\infty} j^{k^\ell m} J_{k^\ell m}(t_0)\, j^m J_m(t_1) \\ & \text{for } \forall t_0, t_1 \in \Re, \quad \ell \in \mathcal{N}. \end{aligned} \tag{4.130}$$

4.4 Applications to Communication Systems

4.4.1 Stream Cipher System

The central problem in stream cipher cryptography is the difficulty of efficiently generating long running-key sequences from a short and random key [8]. It is

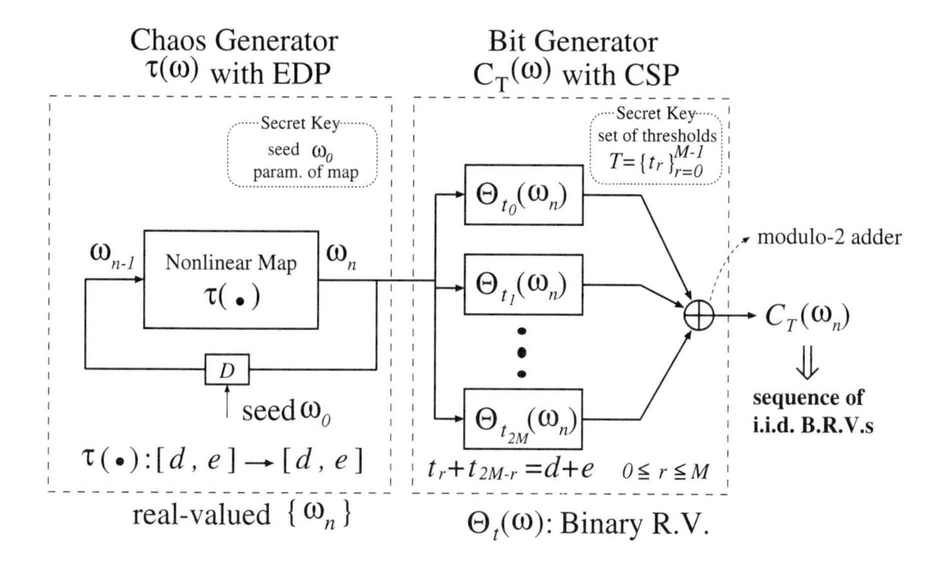

Figure 4.10: Running-key sequence generator for a stream cipher system based on chaotic dynamics, e.g., *symmetric binary sequence* $\{C_T(\omega_n)\}_{n=0}^{\infty}$.

obvious that sequences of i.i.d. binary random variables are good examples of running-key sequences.

Such a situation motivated us to propose a stream cipher system whose running-key sequence are sequences of *i.i.d.* binary random variables based on chaotic dynamics, e.g., symmetric binary sequence $\{C_T(\omega_n)\}_{n=0}^{\infty}$ with the symmetric threshold set $T = \{t_r\}_{r=0}^{2M}$ as shown in Figure 4.10 [55, 61–63]

$$C_T(\omega) = \Theta_{t_0}(\omega) \oplus \Theta_{t_2}(\omega) \oplus \cdots \oplus \Theta_{t_{2M}}(\omega), \tag{4.131}$$

$$= \sum_{r=0}^{2M} (-1)^{r-1} \Theta_{t_r}(\omega), \tag{4.132}$$

$$\text{where} \quad t_r + t_{2M-r} = d + e, \quad r = 0, 1, \cdots, 2M. \tag{4.133}$$

The time-averaged crosscorrelation function between sequences $\{G(\omega_n)\}_{n=0}^{\infty}$ with an initial seed ω_0 and $\{H(\omega_n{}')\}_{n=0}^{\infty}$ with an initial seed $\omega_0{}'$ is defined by

$$r_N(\ell, \omega_0, \omega_0{}'; G, H) = \frac{1}{N} \sum_{n=0}^{N-1} G(\omega_n) H(\omega'_{n+\ell}) \tag{4.134}$$

where the subscripts are taken mod N. Note that $r_N(\ell, \omega_0, \omega_0{}'; \Theta_t, \Theta_{t'})$ has a large value only at $\ell = 0$ because of its no-correlation property. The fluctuated and smooth curves in Figure 4.11 indicate respectively the time-average $r_{64}(\ell, \omega_0, \omega_0; G, H)$ of cross-correlation between $\Theta_t(\cdot)$ and $\Theta'_t(\cdot)$ versus t', and the ensemble-averaged one $\rho(\ell; \Theta_t, \Theta_{t'})$, where $\rho(\ell; G, H)$ is defined by

$$\rho(\ell; G, H) = \int_J G(\omega) H(\tau^\ell(\omega)) f^*(\omega)\, d\omega. \tag{4.135}$$

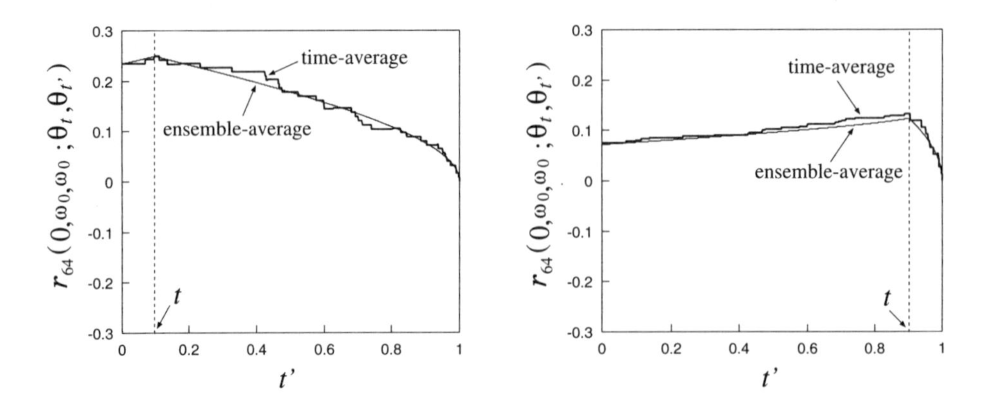

Figure 4.11: $r_{64}(0, \omega_0, \omega_0; \Theta_t, \Theta_{t'})$ versus t' to search t.

This figure implies the estimation of the parameter $T = \{t_r\}_{r=0}^{2M}$ is easy if the seed ω_0 and parameters of the map are previously known.

Figure 4.12 shows an illustration of cryptanalysis using the statistic $r_{64}(0, \omega_0, {\omega_0}'; C_T, C_T)$ to search a 8-digit seed ω_0, a partial secret key of a bit sequence $\{C_T(\omega_n)\}_{n=0}^{\infty}$ provided that all of other parameters are known within the limited accuracy. In this search all possible 2^8 ω_0''s are scanned. We can find that we need exhaustive searches of ω_0' even if the parameters of the map and the set of thresholds are previously known because $r_{64}(0, \omega_0, \omega_0'; C_T, C_T)$ has a peak only when ω_0' is equal to ω_0 completely in a given precision [55, 61–63]. This implies that this strategy is computationally infeasible because of the large key space of ω_0. It is noteworthy that most of the existing chaos cryptosystems have not been capable of using the *sensitive dependence on initial conditions property* primarily because they are based on analog circuits.

4.4.2 Image Transmission Using SS Techniques

In image transmission systems through CDMA channels using spread spectrum techniques [50], the main problem is how to efficiently transmit an image whose information is huge. Since information of a color image is about three times as large as a monochrome one, an efficient transmission is required. Spreading sequences of the same period have been usually used [12]. This implies that each of the erroneously transmitted bits occurs with approximately with the same probability. To reduce a much wider bandwidth for the transmission of images, we consider a CDMA system in which spreading sequences of a longer period are assigned to more significant bits rather than to less ones [60, 66, 67]. This technique is analogous to the Shannon-Fano encoding [26, 27]. Sequences of i.i.d. binary random variables are the best choice for such sequences of variable-period. The use of such sequences inevitably makes a CDMA system asynchronous. But we can get reconstructed images of better quality [60, 66, 67] as follows.

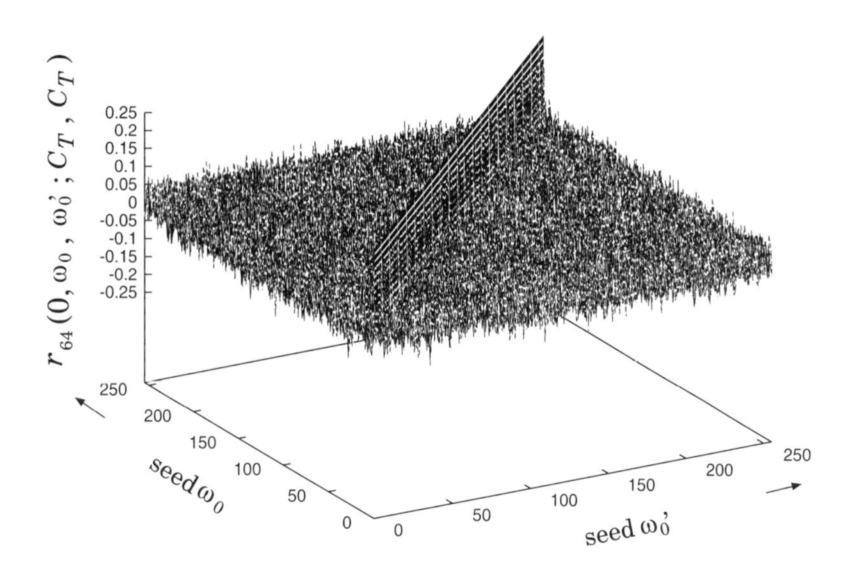

Figure 4.12: Cryptanalysis using $r_{64}(0, \omega_0, \omega_0'; C_T, C_T)$.

A basic image communication system using spread spectrum technique is illustrated in Figure 4.13 [60, 66, 67]. First, images are partitioned into small blocks (8 × 8 pixels). In using YIQ signals, RGB signals are transformed into YIQ signals. Signal Y is more significant than the others. The discrete cosine transform (DCT) of two-dimensional (2-D) signal in each block is computed. Next the 2-D DCT coefficients are quantized and appropriate numbers of bits are assigned to them. Note that encoding and decoding are implemented block by block. We assign more bits to low frequency coefficients than to high ones, and to signal Y than to the others. The bit allocation map we use is shown in Figure 4.14.

We transmit only the first 15 DCT coefficients of signal Y, 6 DCT coefficients of signal I, and 3 DCT coefficients of signal Q, namely 84bits/block, when we use YIQ signals. In using RGB signals, only the first 15 DCT coefficients of each signal, namely 162 bits/block. Furthermore, the n-th significant bit of the m-th coefficient, for example y_m, is denoted by $y_{m\text{-}n}$, $y_{0\text{-}1}$ is the most significant bit (MSB) of the DCT coefficient y_0 and $y_{0\text{-}8}$ is the least significant bit (LSB).

For an image, we assign each bit to CDMA channels (*e.g.*, 6 channels) appropriately. Let $t(x_{m\text{-}n})$ or $t_{m\text{-}n}$, where $x = y$, i, and q, be the period of the spreading sequence to be assigned to $y_{m\text{-}n}$, $i_{m\text{-}n}$, and $q_{m\text{-}n}$, respectively. Assume that the total number of chips $t(x_{m\text{-}n})$ in the i-th channel, denoted by T_i, called

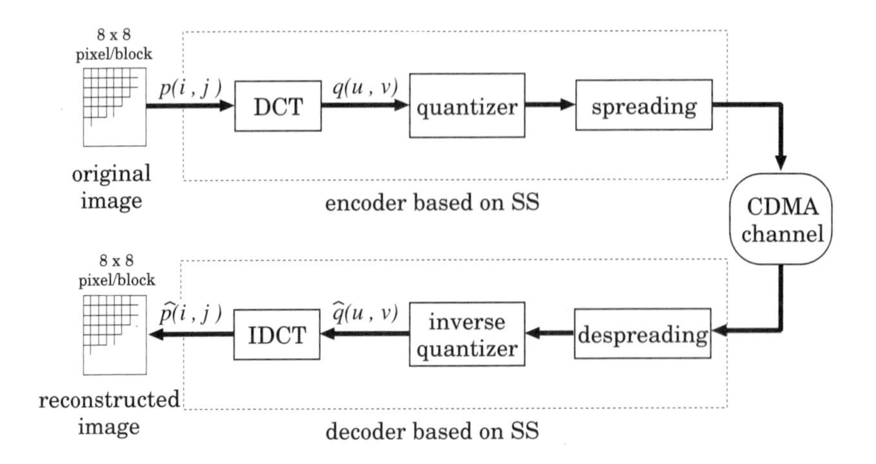

Figure 4.13: Image communication system through CDMA channels using SS techniques.

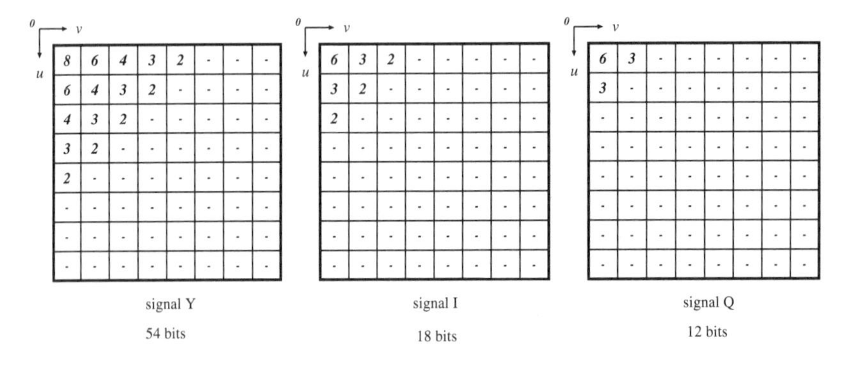

Figure 4.14: Bit allocation map.

the "block-period of the channel," defined by

$$T_i = \sum_{y_{m-n}, i_{m-n}, q_{m-n} \in \text{ the } i\text{-th channel}} t_{m-n}, \quad 1 \leq i \leq 6. \tag{4.136}$$

is equal to others, i.e., $T_i = T$, $1 \leq i \leq 6$. We call T the "block-period."

There are various kinds of models according to assignments of each bit to 6 channels. First, consider the simplest model as shown in Figure 4.15, called "YIQ Model-1," where spreading sequences of fixed period are used.

Now define the mean color distance error (MCDE) by

$$\begin{aligned} MCDE &= \frac{1}{N \times 8 \times 8} \sum_{k=1}^{N} \sum_{i=0}^{7} \sum_{j=0}^{7} (|R_k(i,j) - \widehat{R}_k(i,j)| \\ &+ |G_k(i,j) - \widehat{G}_k(i,j)| + |B_k(i,j) - \widehat{B}_k(i,j)|) \end{aligned}$$

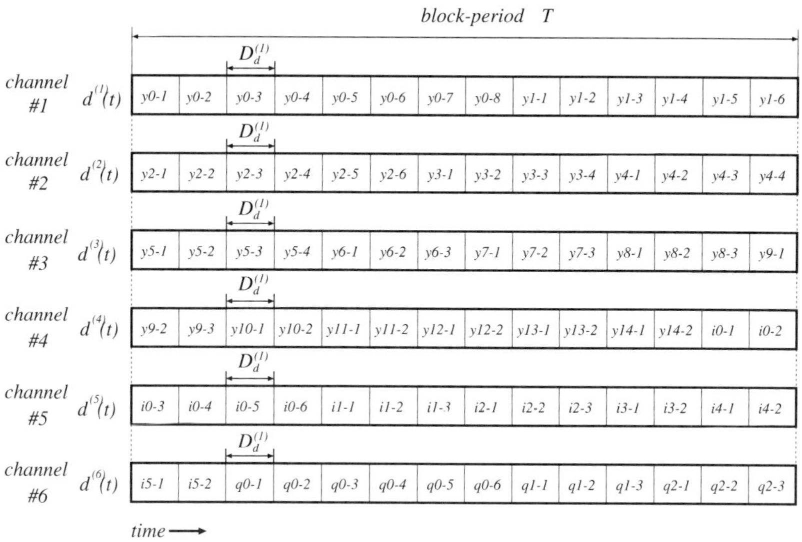

block-period T

channel #1 $d^{(1)}(t)$	y0-1	y0-2	y0-3	y0-4	y0-5	y0-6	y0-7	y0-8	y1-1	y1-2	y1-3	y1-4	y1-5	y1-6
channel #2 $d^{(2)}(t)$	y2-1	y2-2	y2-3	y2-4	y2-5	y2-6	y3-1	y3-2	y3-3	y3-4	y4-1	y4-2	y4-3	y4-4
channel #3 $d^{(3)}(t)$	y5-1	y5-2	y5-3	y5-4	y6-1	y6-2	y6-3	y7-1	y7-2	y7-3	y8-1	y8-2	y8-3	y9-1
channel #4 $d^{(4)}(t)$	y9-2	y9-3	y10-1	y10-2	y11-1	y11-2	y12-1	y12-2	y13-1	y13-2	y14-1	y14-2	i0-1	i0-2
channel #5 $d^{(5)}(t)$	i0-3	i0-4	i0-5	i0-6	i1-1	i1-2	i1-3	i2-1	i2-2	i2-3	i3-1	i3-2	i4-1	i4-2
channel #6 $d^{(6)}(t)$	i5-1	i5-2	q0-1	q0-2	q0-3	q0-4	q0-5	q0-6	q1-1	q1-2	q1-3	q2-1	q2-2	q2-3

$D_d^{(1)}$ time →

Figure 4.15: YIQ Model-1 with spreading sequences of fixed period.

where $R_k(i,j)$ and $\widehat{R}_k(i,j)$ denote intensity values of the (i,j) element of the k-th block in the original image and in the reconstructed image, respectively. In order to investigate the significance of each bit, we compute the values of MCDE when each bit is transmitted erroneously [66, 67]. The MCDE enables us to calculate the insignificance of each bit defined by (MCDE - lowerbound of MCDE)$^{-1}$ which means the number of the erroneously transmitted bit to cause a constant MCDE.

YIQ Model-1 is not efficient because each bit is equiprobably transmitted in error. As is well known, for a constant number of channels, spreading sequences of longer periods can reduce bit error rates more than ones of shorter periods. This motivates us to assign spreading sequences of an appropriate period to the bit according to its significance like the Shannon-Fano encoding. To do this, we should investigate the bit error rates for various periods of spreading sequences and for various numbers of channels. As spreading sequences of variable-period, we have used a sequence of i.i.d. binary random variables, i.e., *chaotic bit sequence*, generated by chaotic dynamics [56]. Such sequences are quite different from LFSR sequences such as M sequences, Kasami sequences, and Gold sequences [12]. Interferences from other channels of sequences of i.i.d. binary random variables form Gaussian distribution. Thus, we can estimate bit error probability of CDMA system theoretically. On the other hand, that of Gold sequences is not Gaussian distribution. However, we can see that there is scarcely any difference between bit error probability of chaotic bit sequences and that of Gold sequences according to increasing number of channels. Using the results of the calculated bit error rates for chaotic bit sequences, we can construct "YIQ Model-2" where spreading sequences of variable-period are used as shown

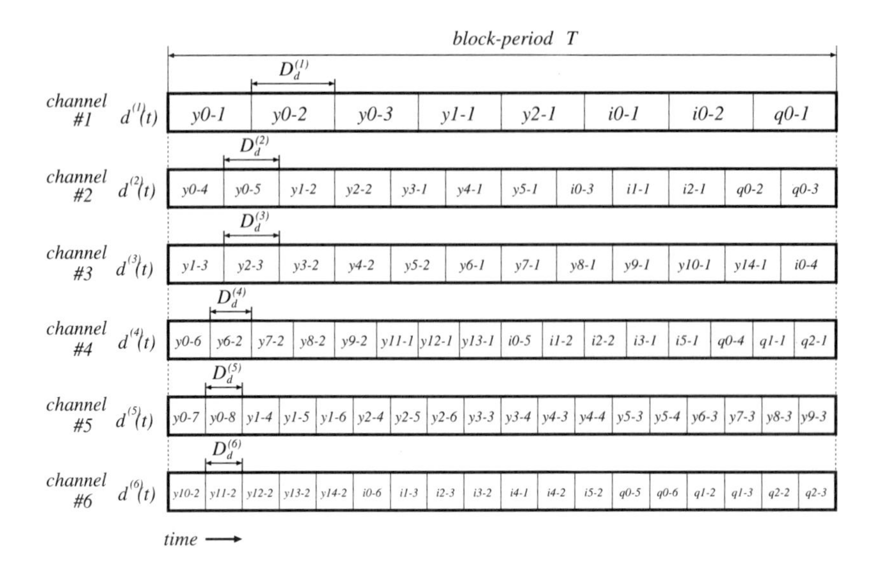

Figure 4.16: YIQ Model-2 with spreading sequences of variable-period.

in Figure 4.16. Similarly, we can construct "RGB Model-1" and "RGB Model-2."

As a criterion to evaluate the quality of reconstructed images, we use the mean square error (MSE) defined by

$$\begin{aligned} MSE &= \frac{1}{N \times 8 \times 8} \sum_{k=1}^{N} \sum_{i=0}^{7} \sum_{j=0}^{7} ([R_k(i,j) - \widehat{R}_k(i,j)]^2 \\ &+ [G_k(i,j) - \widehat{G}_k(i,j)]^2 + [B_k(i,j) - \widehat{B}_k(i,j)]^2) \end{aligned}$$

Figure 4.17 shows bit error rates versus block-periods using YIQ and RGB signals. Furthermore, MSEs versus block-periods are shown in Figure 4.18. We can find that the bit error rates in the four models are similar to each other. On the other hand, the MSE performance of Model-2 has shown to be much better than that of Model-1. Furthermore, that of YIQ Model-2 is better than that of RGB Model-2. This implies that the quality of reconstructed images in YIQ Model-2 is drastically improved.

4.4.3 Interference Properties

In quasi-synchronous DS/CDMA systems, the i-th correlation receiver outputs

$$Z_p^{(i)} = d_p^{(i)} N + \eta^{(i)} + I_{J,p}^{(i)} \tag{4.137}$$

where $\{d_p^{(i)}\}_{p=-\infty}^{\infty}$ is a binary sequence of information symbols of the i-th channel, $\eta^{(i)}$ is noise component, and $I_{J,p}^{(i)}$ denotes multiple-access interference (MAI) [51,

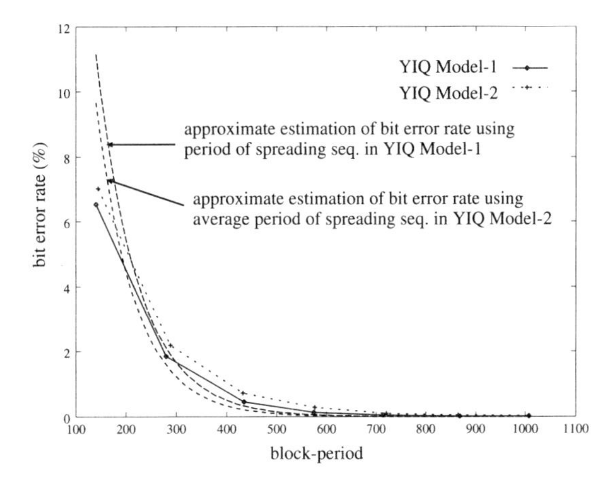

Figure 4.17: Bit error rates versus block-periods using YIQ signals.

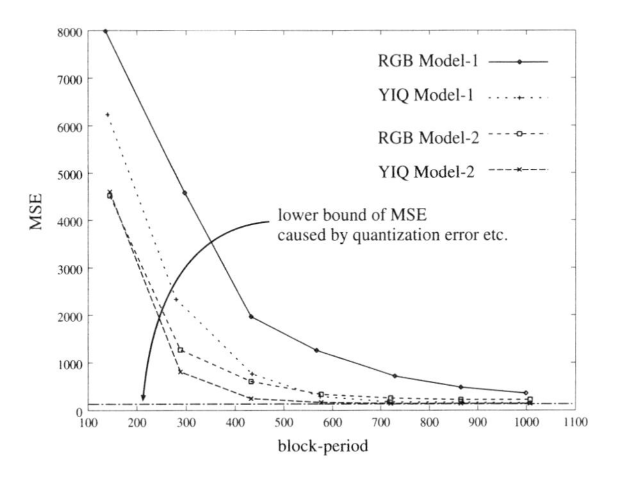

Figure 4.18: MSEs versus block-periods.

52] from other $J-1$ channels given by

$$I_{J,p}^{(i)} = \sum_{\substack{j=1 \\ j \neq i}}^{J} \left\{ \frac{d_p^{(j)} + d_{p+1}^{(j)}}{2} R^E(\ell_{ij}; X^{(i)}, X^{(j)}) + \frac{d_p^{(j)} - d_{p+1}^{(j)}}{2} R^O(\ell_{ij}; X^{(i)}, X^{(j)}) \right\} \tag{4.138}$$

where $R^E(\ell; X, Y)$ and $R^O(\ell; X, Y)$ denote the even and the odd cross-correlation functions between two spreading sequences $X = \{X_n\}_{n=0}^{N-1}$ and $Y = \{Y_n\}_{n=0}^{N-1}$, respectively $(0 \leq \ell \leq N-1)$. They are defined by [12]

$$R_N^E(\ell; X, Y) = R_N^A(\ell; X, Y) + R_N^A(N-\ell; Y, X) \tag{4.139}$$

$$R_N^O(\ell; X, Y) = R_N^A(\ell; X, Y) - R_N^A(N-\ell; Y, X) \tag{4.140}$$

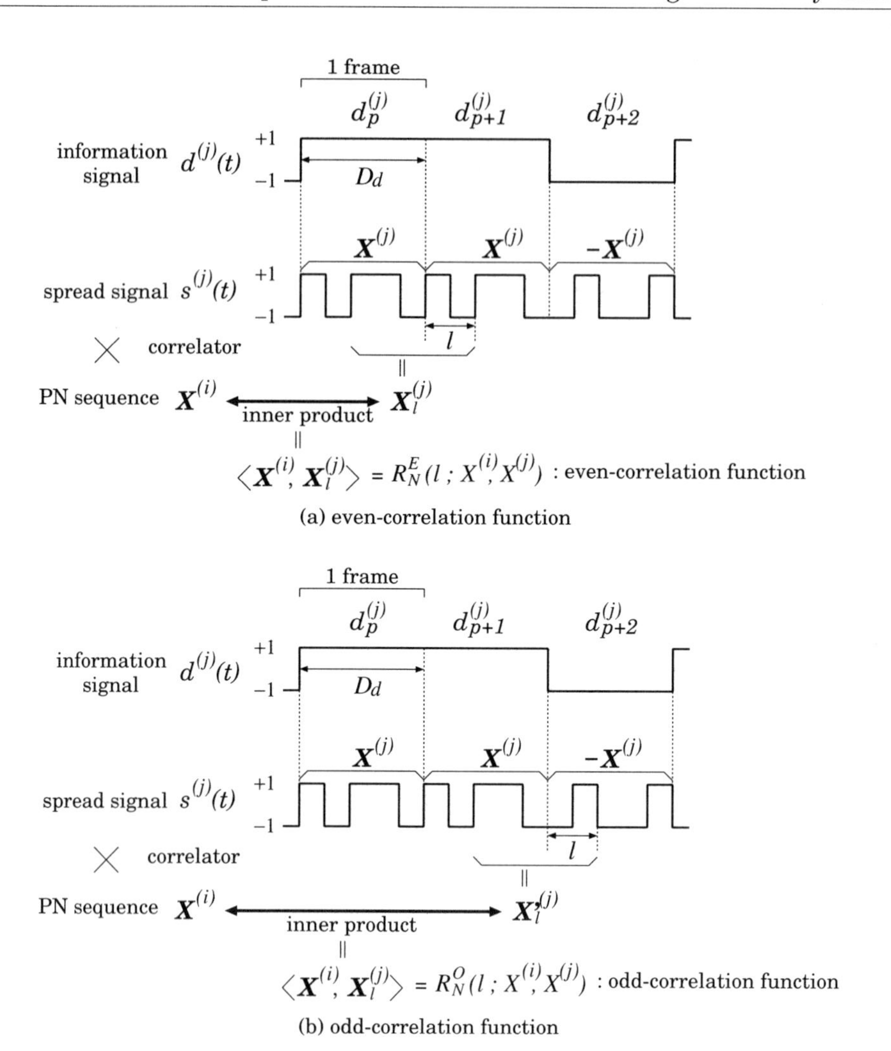

Figure 4.19: Two types of correlation functions.

where $R_N^A(\ell; X, Y)$ is called the *aperiodic cross-correlation function* or the *partial correlation function* between two spreading sequences X and Y, defined by [12]

$$R_N^A(\ell; X, Y) = \sum_{n=0}^{N-1-\ell} X_n Y_{n+\ell}. \tag{4.141}$$

These even- and odd-correlation functions are illustrated in Figure 4.19.

The bit error rate (or briefly BER) is defined by

$$\mathrm{BER} = \Pr\{d_p^{(i)} = -1 \text{ and } Z_p^{(i)} > 0\} + \Pr\{d_p^{(i)} = 1 \text{ and } Z_p^{(i)} < 0\} \tag{4.142}$$

which depends on such MAI. As shown in the previous subsection, it is important to investigate bit error rates for various periods of spreading sequences and for

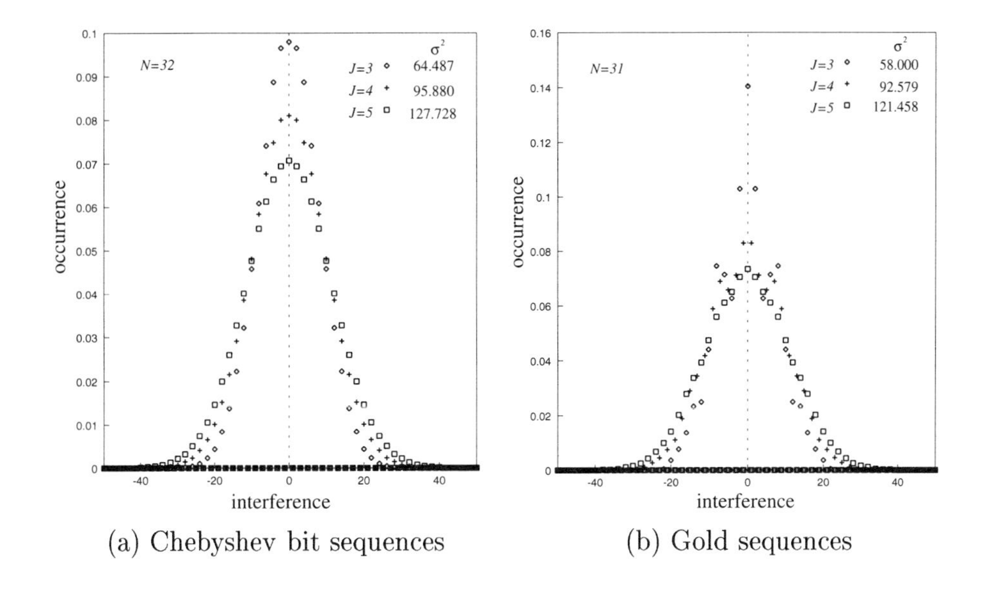

(a) Chebyshev bit sequences (b) Gold sequences

Figure 4.20: Multiple-access interferences from other channels in quasi-synchronous CDMA systems.

various numbers of channels. For evaluations of such bit error rates, empirical and analytical methods have been proposed [51–54].

It follows from (4.138) that if information data of each channel are independent and $\Pr\{d_p^{(j)} = 1\} = \Pr\{d_p^{(j)} = -1\} = \frac{1}{2}$ for all j, each even and odd cross-correlation value causes the MAI equally. Thus, in order to reduce the MAI, the absolute values of such even and odd cross-correlation functions, which depend on the set of spreading sequences, are desired to be small.

By numerical simulations, we have investigated interferences from other channels when we use chaotic bit sequences and Gold sequences as spreading sequences [56–59], as shown in Figure 4.20. Interferences of chaotic bit sequences form the Gaussian distribution, which implies that we can estimate bit error probability of CDMA system theoretically. On the other hand, that of Gold sequences is not the Gaussian distribution. In Figure 4.21, we can see that there is scarcely any difference between bit error probability of Chebyshev bit sequences and that of Gold sequences according to increasing number of channels.

In conventional analysis [51], the noise component $\eta^{(i)}$ and the MAI $I_{J,p}^{(i)}$ are considered to be a common noise, such as the additive white Gaussian noise. However, Figure 4.20 implies that we can estimate the MAI itself theoretically. This motivates us to evaluate distributions of correlation values of spreading sequences of i.i.d. binary random variables based on the Central Limit Theorem because such distribution dominates the bit error rates.

Now we introduce the following Central Limit Theorem (CLT) in order to discuss distributions of correlation values of spreading sequences.

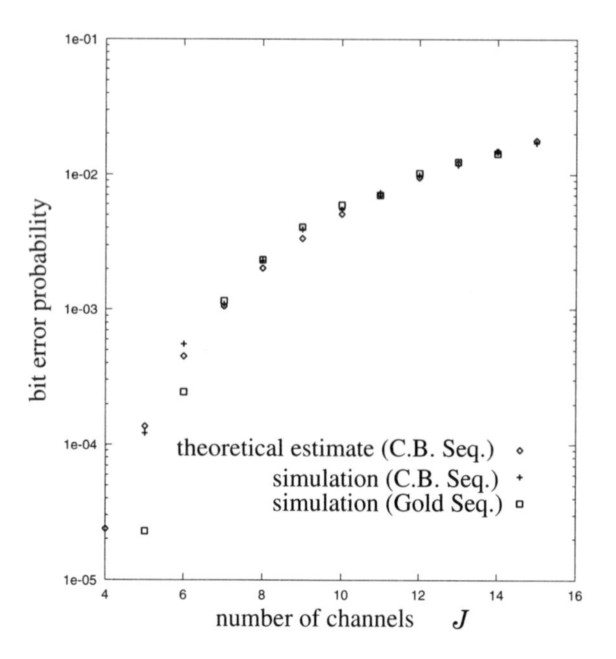

Figure 4.21: Bit error rates in quasi-synchronous CDMA systems.

Central Limit Theorem: Let $V_0, V_1, \cdots$ be i.i.d. random variables. Consider $S_N = V_0 + V_1 + \cdots + V_{N-1}$. Then $S_N/\sqrt{N}$ tends to the Gaussian distribution with mean 0 and variance σ^2, where the variance σ^2 is given by

$$\sigma^2 = \lim_{N\to\infty} \frac{1}{N} E[S_N^2]. \tag{4.143}$$

Let $\{X_n\}_{n=0}^{\infty}$ and $\{Y_n\}_{n=0}^{\infty}$ be balanced sequences of i.i.d. $\{-1,1\}$-valued random variables and mutually independent, satisfying

$$E[X_n] = E[Y_n] = 0, \tag{4.144}$$

$$\left.\begin{array}{lll} E[X_nX_{n+\ell}] & = E[X_n]E[X_n] & = 0 \\ E[Y_nY_{n+\ell}] & = E[Y_n]E[Y_n] & = 0 \end{array}\right\} \quad \text{for } \ell \geq 1, \tag{4.145}$$

$$E[X_nY_{n+\ell}] = E[X_n]E[Y_n] = 0 \quad \text{for } \ell \geq 0. \tag{4.146}$$

Then, we have

$$E[R_N^A(\ell; X, Y)] = 0,\ E[R_N^E(\ell; X, Y)] = E[R_N^O(\ell; X, Y)] = 0 \quad \text{for } \ell \geq 0, \tag{4.147}$$

which leads us to get

Lemma 1.7: Let $\{X_n\}_{n=0}^{\infty}$ and $\{Y_n\}_{n=0}^{\infty}$ be balanced sequences of i.i.d. $\{-1,1\}$-valued random variables and mutually independent, satisfying (4.144), (4.145), and (4.146). Then their correlation function $R_N(\omega, \ell; X, Y)/\sqrt{N}$ $(\ell \geq 0)$ tends to the Gaussian distribution with mean 0 and variance $\sigma^2 = 1$.

Proof: Define

$$V_{n,\ell} = X_n Y_{n+\ell} \quad \text{for } \ell \geq 0 \tag{4.148}$$

Then we can write

$$R_N(\ell; X, Y) = \sum_{n=0}^{N-1} V_{n,\ell} \tag{4.149}$$

which corresponds to the correlation function. Since $\{V_{n,\ell}\}_{n=0}^{\infty}$ is a sequence of i.i.d. random variables, the Central Limit Theorem [5] holds for the correlation function $R_N(\ell; X, Y)/\sqrt{N}$ automatically. From independence of X_n and Y_n, it follows that

$$E[X_n Y_{n+\ell} X_m Y_{m+\ell}] = \delta_{n,m} \tag{4.150}$$

where

$$\delta_{n,m} = \begin{cases} 1 & n = m \\ 0 & n \neq m. \end{cases} \tag{4.151}$$

This leads us to get

$$\sigma^2 = \lim_{N\to\infty} \frac{1}{N} E[R_N(\ell; X, Y)^2] = 1 \tag{4.152}$$

which completes the proof.

Corollary 1.5: From Lemma 1.7, the CLT also holds for even and odd correlation functions. Their variances are, respectively, given by

$$(\sigma^E)^2(\ell; X, Y) = (\sigma^O)^2(\ell; X, Y) = 1. \tag{4.153}$$

Note that these variances are independent of ℓ.

As discussed before, we can easily get sequences of i.i.d. binary random variables using chaotic dynamics. Let $\{g(\tau^n(\omega))\}_{n=0}^{\infty}$ and $\{h(\tau^n(\omega))\}_{n=0}^{\infty}$ be sequences of i.i.d. $\{0, 1\}$-valued random variables and put $X_n(\omega) = 2g(\tau^n(\omega)) - 1$, $Y_n(\omega') = 2h(\tau^n(\omega')) - 1$, where $2g(\cdot) - 1$ and $2h(\cdot) - 1$ imply the transformation $\{0, 1\} \to \{-1, 1\}$, and where ω and ω' are chosen *statistically independently*. Thus $\{X_n(\omega)\}_{n=0}^{\infty}$ and $\{Y_n(\omega')\}_{n=0}^{\infty}$ are mutually independent sequences of i.i.d. binary random variables satisfying (4.144), (4.145), and (4.146). Hence, as discussed in the above, the CLT for their correlation functions also holds automatically, where the mean and the variance are 0 and 1, respectively.

It should be noted that $\{X_n(\omega)\}_{n=0}^{\infty}$ and $\{Y_n(\omega)\}_{n=0}^{\infty}$, which are generated from a common seed ω, are not completely independent of each other. In order to discuss such a case, we introduce the following definition and theorem.

Definition 1.9: If $(V_1, \cdots, V_k)$ and $(V_{k+n}, \cdots, V_{k+n+\ell})$ are independent whenever $n > m$, then the sequence $\{V_n\}_{n=0}^{\infty}$ is called *m-dependent* [5]. Note that an independent sequence is 0-dependent.

Central Limit Theorem (m-dependent version) [5]**:** Suppose that V_0, V_1, $\cdots$ is stationary and m-dependent and that $E[V_n] = 0$ and $E[V_n^{12}] < \infty$. The Central Limit Theorem (CLT) also holds for such random variables.

We can show that $V_{n,\ell} = X_n(\omega)Y_{n+\ell}(\omega)$ is ℓ-dependent random variable. Hence, the above CLT holds for such two sequences generated from a common seed ω when $\ell \geq 1$. For $\ell = 0$, we should discuss it more carefully, but it is omitted because such discussions are somewhat complicated.

4.5 Concluding Remarks

We have reviewed how Kalman embedded a prescribed Markov information source into chaotic dynamics and we gave its improved embedding algorithm. We have given a simple sufficient condition for a class of ergodic maps and their associated $p \geq 2-$ary functions to generate a sequence of i.i.d. $p-$ary random variables and shown that its condition is of crucial importance in evaluating independence of a real-valued sequence. We have given two applications of sequences of i.i.d. binary random variables; (1) to a running-key sequence in a stream-cipher system and (2) to a spreading spectrum sequence in quasi-asynchronous DS/CDMA systems. Furthermore, we have evaluated distributions of even- and odd-correlation values of spreading sequences of i.i.d. binary random variables based on the central limit theorem.

References

[1] von Neumann, J., Summary written by Forsythe, G. E., "Various techniques used in connection with random digit," *National Bureau of Standards, Applied Math. Series*, vol. 12, 36, 1951. Reprinted in Collected Works of von Neumann, vol. 5, 768, 1963.

[2] Knuth, D., *The Art of Computer Programming 2, Seminumerical Algorithms*, 2nd ed., Addison-Wesley, Reading, MA, 1981.

[3] Kac, M., *Statistical Independence in Probability Analysis and Number Theory*, The Mathematical Association of America, 1959.

[4] Loève, M., *Probability Theory I*, Graduate Texts in Mathematics 45, Springer-Verlag, New York, 1977.

[5] Billingsley, P., *Probability and Measure.* John Wiley & Sons, New York, 1995.

[6] Simon, M. K., Omura, J. K., Scholtz, R. A., and Levitt, B. K., *Spread Spectrum Communications Handbook*, McGraw-Hill, New York, 1994.

[7] Peterson, R. L., Ziemer, R. E., and Borth, D. E., *Introduction to Spread Spectrum Communications*, Prentice-Hall, Englewood Cliffs, NJ, 1995.

[8] Shannon, C. E., "Communication Theory of Secrecy Systems," *Bell Syst. Tech. J.*, 28, 656, 1949.

[9] Massey, J. L., "An Introduction to Contemporary Cryptology," *Proc. IEEE*, vol. 76, no. 5, 533, May 1988.

[10] Tausworthe, R. C., "Random numbers generated by linear recurrence modulo two," *Mathematics of Computation*, vol. 19, 201, 1965.

[11] Lewis, T. G. and Payne, W. H., "Generalized feedback shift register pseudorandom number algorithm," *J. ACM*, vol. 20, 45, 1973.

[12] Sarwate D. V. and Pursley, M. B., "Crosscorrelation properties of pseudorandom and related sequences," *Proc. IEEE*, vol. 68, no. 3, 593, 1980.

[13] Pecora, L. M. and Carroll, T. L., "Synchronization in chaotic systems," *Physical Review Letters*, vol. 64, 821, 1990.

[14] Cuomo, K. M. and Oppenheim, A. V., "Circuit implementation of synchronized chaos with applications to communications," *Physical Review Letters*, vol. 71, 65, 1993.

[15] Heidari-Bateni G., McGillem, C.D., and Tenorio M. F., "A novel multiple-address digital communication system using chaotic signals," *Proc. 1992 IEEE Int. Conf. on Communications (ICC '92)*, 1232, 1992.

[16] Heidari-Bateni G. and McGillem, C.D., "A chaotic direct-sequence spread-spectrum communication system," *IEEE Trans. Comm.*, vol. 42, no. 2/3/4, 1524, 1994.

[17] Parlitz, U. and Ergezinger, S., "Robust communication based on chaotic spreading sequences," *Physics Letters A*, vol. 188, 146, 1994.

[18] Mazzini, G., Setti, G., and Rovatti, R., "Chaotic complex spreading sequences for asynchronous DS-CDMA part I : system modeling and results," *IEEE Trans. Circuit Syst.*, vol. CAS-44, no. 10, 937, 1997.

[19] Götz, M., Kelber, K., and Schwarz, W. "Discrete-time chaotic encryption systems part I : statistical design approach," *IEEE Trans. Circuit Syst.*, vol. CAS-44, no. 10, 963, 1997.

[20] Barnsley, M. F., *Fractal Image Compression*, AK Peters, Ltd., 1993.

[21] Kanaya, F., "A chaos model of a stationary discrete memoryless source and arithmetic coding," *Proc. SITA95*, 361, 1995 (in Japanese).

[22] Kanaya, F., "A chaos model of finite-order discrete Markov sources and arithmetic coding," *Proc. SITA96*, 81, 1996 (in Japanese).

[23] Kanaya, F., "A chaos model of finite-order Markov sources and arithmetic coding," *Proc. 1997 IEEE Int. Symp. on Information Theory*, 247, 1997.

[24] Kalman, R. E., "Nonlinear aspects of sampled-data control systems," *Proc. Symp. Nonlinear Circuit Analysis VI*, 273, 1956.

[25] Shannon, C. E., "A Mathematical Theory of Communication," *Bell Syst. Tech. J.*, vol. 27, 379, 623, 1948.

[26] Reza, F. M., *An Introduction to Information Theory*, McGraw-Hill, New York, 1961.

[27] Ash, R. B., *Information Theory*, Dover, New York, 1965.

[28] Goldie, C. M. and Pinch, R. G. E., *Communication Theory*, London Mathematical Society Student Texts 20, Cambridge University Press, Cambridge, 1991.

[29] Cover, T. M. and Thomas, J. A., *Elements of Information Theory*, John Wiley & Sons, New York, 1991.

[30] Viterbi, A. J. and Omura, J. K., *Principles of Digital Communication and Coding*, McGraw-Hill, New York, 1979.

[31] Blahut, R. E., *Digital Transmission of Information*, Addison-Wesley, Reading, MA, 1990.

[32] Lorenz, E. N., "Deterministic nonperiodic flow," *J. Atoms. Sci.*, vol. 20, 130, 1963.

[33] May, R. M., "Simple mathematical models with very complicated dynamics," *Nature*, vol. 261, 459, 1976.

[34] Jackson, E. Atlee, *Perspective Nonlinear Dynamics*, Cambridge University Press, Cambridge, 1989.

[35] Han, T. S. and Hoshi, M., "Interval algorithm for random number generation," *IEEE Trans. Information Theory*, vol. 43, no. 2, 599, 1997.

[36] Visweswariah, K., Kulkarni, S. R. and Verdù, S., "Source codes as random number generators," *IEEE Trans. Information Theory*, vol. 44, no. 2, 462, 1997.

[37] Kalman, R. E., "Looking back 45 years – conversation with von Neumann and Ulam – and also looking forward to 21st century," *IEICE Trans. Fundamentals*, vol. E82-A, no. 9, p. 1686, 1999.

[38] Ulam, S. M., *A Collection of Mathematical Problems*, Interscience Publishers, Inc., 1960.

[39] Lasota, A. and Mackey, M. C., *Chaos, Fractals, and Noise*, Springer-Verlag, New York, 1994.

[40] Lasota, A. and Yorke, J. A., "On the existence of invariant measures for piecewise monotonic transformations (1)," *Trans. Amer. Math. Soc.*, vol. 186, 481, 1973.

[41] Boyarsky, A. and Scarowsky, M., "On a class of transformations which have unique absolutely continuous invariant measures," *Trans. Am. Math. Soc.*, vol. 255, 243, 1979.

[42] Kalman, R. E., "Randomness and probability," *Math. Japonica*, vol. 41, no. 1, 41, 1995.

[43] Grossmann, S. and Thomae, S., "Invariant distributions and stationary correlation functions of one-dimensional discrete processes," *Z. Naturforsch.*, vol. 32a, 1353, 1977.

[44] Ulam, S. L. and von Neumann, J., "On combination of stochastic and deterministic processes," *Bull. Math. Soc.*, vol. 53, 1120, 1947.

[45] Adler, R. L. and Rivlin, T. J., "Ergodic and mixing properties of Chebyshev polynomials," *Proc. Amer. Math. Soc.*, vol. 15, 794, 1964.

[46] Rivlin, T. J., *Chebyshev Polynomials*, John Wiley & Sons, Inc., New York, 1990.

[47] Geisel, T. and Fairen, V., "Statistical properties of chaos in Chebyshev maps," *Physics Letters*, vol. 105A, no. 6, 263, 1984.

[48] Lawrance, A. J. and Spencer, N. M., "Curved chaotic map time series models and their stochastic reversals," *Scandinavian J. Statistics*, vol. 25, 371, 1998.

[49] Abramowitz, M. and Stegun, I. A., *Handbook of Mathematical Functions with Formulas, Graphs, and Mathematical Tables*, Dover, 1972.

[50] W. F. Schreiber, "Spread-spectrum television broadcasting," *SMPTE Journal*, 538, August, 1992.

[51] Pursley, M. B., "Performance evaluation for phase-coded spread-spectrum multiple-access communication — part I: system analysis," *IEEE Trans. Commun.*, vol. COM-25, no. 8, 795, 1977.

[52] Yao, K., "Error probability of asynchronous spread spectrum multiple access communication systems," *IEEE Trans. Commun.*, vol. COM-25, no. 8, 803, 1977.

[53] Holtzman, J. M., "A simple, accurate method to calculate spread-spectrum multiple-access error probabilities," *IEEE Trans. Commun.*, vol. 40, no. 3, 461, 1992.

[54] Letaief, K. B., "Efficient evaluation of the error probabilities of spread-spectrum multiple-access communications," *IEEE Trans. Commun.*, vol. 45, no. 2, 239, 1997.

[55] Kohda, T., *Discrete Dynamics and Chaos*, Corona Publishing, Tokyo, 1998 (in Japanese).

[56] Kohda, T., Tsuneda, A., and Sakae, T., "Chaotic Binary Sequences by Chebychev Maps and Their Correlation Properties," *Proc. of the IEEE Second Int. Symp. on Spread Spectrum Techniques and Applications (ISSSTA)*, 63, 1992.

[57] Kohda, T. and Tsuneda, A., "Pseudonoise sequences by chaotic nonlinear maps and their correlation properties," *IEICE Trans. Communications*, vol. E76-B, no. 8, 855, 1993.

[58] Kohda, T. and Tsuneda, A., "Explicit evaluations of correlation functions of Chebyshev binary and bit sequences based on Perron-Frobenius operator," *IEICE., Trans.*, vol. E77-A, no. 11, 1794, 1994.

[59] Kohda T. and Tsuneda, A., "Even- and odd-correlation functions of chaotic Chebyshev bit sequences for CDMA," *Proc. of the IEEE Third International Symposium on Spread Spectrum Techniques and Applications*, 391, 1994.

[60] Kohda, T., Tsuneda, A., Osiumi, A., and Ishii, K., "A study on pseudonoise-coded image communications," *Proc. of SPIE's Visual Communications and Image Processing '94*, 874, 1994.

[61] Kohda, T. and Tsuneda, A., "Enciphering/deciphering apparatus and method incorporating random variable and keystream generation," U.S. Patent Application Serial No. 08/734919, 1999.

[62] Kohda, T. and Tsuneda, A., "Chaotic bit sequences for stream cipher cryptography and their correlation functions," *Proc. SPIE's Int. Symp. on Information, Communications and Computer Technology, Applications and Systems*, vol. 2612, 86, 1995.

[63] Kohda, T. and Tsuneda, A., "Stream cipher systems based on chaotic binary sequences," *Proc. SCIS'96*, SCIS96-11C, 1996.

[64] Kohda, T. and Tsuneda, A., "Statistics of chaotic binary sequences," *IEEE Trans. Information Theory*, vol. 43, no. 1, 104, 1997.

[65] Kohda, T. and Tsuneda, A., "Design of sequences of p-ary random variables," *Proc. of 1997 IEEE Int. Symp. on Information Theory*, 76, 1997.

[66] Kohda, T., Ishii. K., and Tsuneda, A., "Image transmission systems through CDMA channels using spreading sequences of variable-period," *Proc. of the IEEE Fourth Int. Symp. on Spread Spectrum Techniques and Applications*, 781, Sept. 1996.

[67] Kohda, T., Ookubo, Y., and Ishii, K., "A color image communication using YIQ signals by spreading spectrum techniques," *Proc. of the IEEE Fifth Int. Symp. on Spread Spectrum Techniques and Applications*, 743, 1998.

[68] Kohda, T., "Sequences of i.i.d. binary random variables using chaotic dynamics," *Sequences and Their Applications – Proceedings of SETA '98*, Ding, C., Helleseth, T. and Niederreiter, H., eds., 297, Springer-Verlag, New York, 1999.

[69] Kohda, T. and Fujisaki, H., "Kalman's recognition of chaotic dynamics in designing Markov information sources," *IEICE Trans. Fundamentals*, vol. E82-A, no. 9, 1747, 1999.

[70] Kohda, T., Tsuneda, A., and Lawrance, A. J., "Correlational properties of Chebyshev chaotic sequences," to appear in *J. Time Series Analysis*.

Part II

CHAOS AT SIGNAL LEVEL

Chapter 5

Overview of Digital Communications

Géza Kolumbán
Department of Measurement and Information Systems
Technical University of Budapest
H-1521 Budapest, Hungary
kolumban@mit.bme.hu

Michael Peter Kennedy
Department of Microelectronic Engineering
University College Cork
Cork, Ireland
Peter.Kennedy@ucc.ie

5.1 Introduction

In digital communications, the digital information has to be transmitted over a band-pass analog channel. Because digital information cannot be directly transmitted over an analog channel, first it has to be mapped to analog sample functions having finite length. Based on the noisy and distorted received sample functions the receiver must recognize the digital information transmitted.

This basic structure of digital telecommunications can be recognized in both conventional and chaotic modulation schemes. On the other hand the application of chaotic signals as information carriers requires a few brand new solutions.

To elaborate a unified framework for the study, analysis, and comparison of conventional telecommunications systems, the basic ideas of conventional digital telecommunications have to be surveyed first.

In this chapter, we will describe the major components of a conventional digital telecommunications system and recall that the primary source of errors is the analog channel. We will explain why a realistic channel model must include at least Additive White Gaussian Noise (AWGN) and band-limiting. We will review

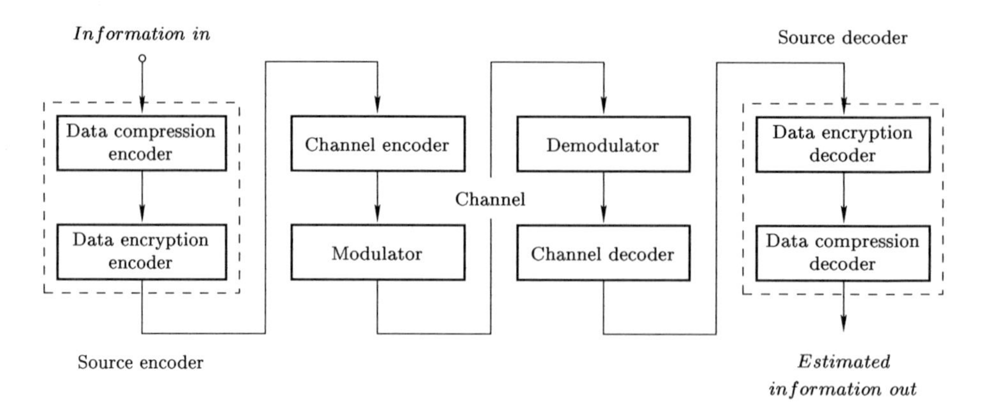

Figure 5.1: Digital communications system showing source and channel coding, modulation, and channel.

the notion of Bit Error Rate (BER) as a way of comparing digital modulation schemes.

In Section 5.3, we will show how a signal set may be constructed from a limited set of orthonormal basis functions and explain the advantages of this choice.

In Section 5.4, we will show that the primary motivation for carrier synchronization is to permit coherent detection, the benefits of which are improved noise performance and bandwidth efficiency.

We will illustrate these concepts in the case of a classical Binary Phase Shift Keying (BPSK) with coherent detection in Section 5.5.

Under poor propagation conditions, where carrier synchronization cannot be maintained, the advantages of coherent detection are lost. In such circumstances, a noncoherent receiver offers a more robust and less complex solution, as will be shown in Section 5.6.

5.2 Digital Communications System: Structure

Communications system theory is concerned with the transmission of information from a source to a receiver through a channel [1, 2].

The goal of a digital communications system, shown schematically in Figure 5.1, is to convey information from a digital information source (such as a computer, digitized speech or video, etc.) to a receiver as effectively as possible. This is accomplished by mapping the digital information to a sequence of symbols which vary some properties of an analog electromagnetic wave called the carrier. This process is called *modulation*. Modulation is always necessary because all practical telecommunications channels are band-pass analog channels which cannot transmit digital signals directly.

At the receiver, the signal to be received is selected by a channel filter, demodulated, interpreted, and the information is recovered.

Conversion of the digital information stream to an analog signal for transmission may be accompanied by encryption and coding to add end-to-end security, data compression, and error-correction capability.

Built-in error-correction is often required because real channels distort analog signals by a variety of linear and nonlinear mechanisms: attenuation, dispersion, intersymbol interference, intermodulation, PM/AM and AM/PM conversions, noise, interference, multipath effects, etc.

A *channel encoder* introduces algorithmic redundancy into the transmitted symbol sequence that can be used to reduce the probability of incorrect decisions at the receiver.

Modulation is the process by which a symbol is transformed into an analog waveform that is suitable for transmission. Common digital modulation schemes include Amplitude Shift Keying (ASK), Phase Shift Keying (PSK), Frequency Shift Keying (FSK), Continuous Phase Modulation (CPM), and Amplitude Phase Keying (APK), where a one-to-one correspondence is established between the symbols and the amplitudes, phases, frequencies, phase and phase transitions, and amplitudes and phases, respectively, of a sinusoidal carrier.

The *channel* is the physical medium through which the information-carrying analog waveform passes as it travels from the transmitter to the receiver.

The transmitted signal is invariably corrupted in the channel. Hence, the receiver never receives exactly what was transmitted. The role of the *demodulator* at the receiver is to produce from the received corrupted analog signal an estimate of the transmitted symbol sequence. The role of the *channel decoder* is to reconstruct the original bit stream, i.e., the information, from the estimated symbol sequence. Because of disturbances in real communications channels, error-free transmission is never possible.

Nonlinear dynamics has potential applications in several of the building blocks of a digital communications system: data compression, encryption, and modulation [3]. Data compression and encryption are potentially reversible, error-free digital processes. By contrast, the transmission of an analog signal through a channel and its subsequent interpretation as a stream of digital data is inherently error-prone.

In Chapters 6 and 7, we will focus on the application of chaotic signals as basis functions for digital modulation. In order to compare the use of a chaotic carrier with that of a conventional sinusoidal carrier, we must consider a realistic channel model and quantify the performance of each chaotic modulation scheme using this channel.

In this section, we will introduce the minimum requirements for a realistic channel model and the performance measures by which we will compare conventional and chaotic modulation schemes.

5.2.1 Minimum Requirements for a Channel Model

The definition of the telecommunications channel depends on the goal of the analysis performed. In the strict sense, the channel is the physical medium that carries the signal from the transmitter to the receiver. If the performance of a modulation scheme has to be evaluated, then the channel model should contain everything from the modulator output to the demodulator input. Even if the physical medium can be modeled by a constant attenuation, the following effects have to be taken into account:

- in order to get maximum power transfer, the input and output impedances of the circuits of a telecommunications system are matched. This is why thermal noise modeled as Additive White Gaussian Noise[1] is *always present* at the input to a radio-frequency (RF) receiver, and
- the bandwidth of the channel has to be limited by a so-called channel (selection) filter in order to suppress the unwanted input signals that are always present at the input of a radio receiver and that cause interference due to the nonlinearities of the receiver.

The simplest channel model that can be justified when evaluating the performance of a modulation scheme is shown in Figure 5.2. Note that the channel filter is used only to select the desired transmission frequency band at the receiver and not to model any frequency dependence of the physical transmission medium. If, in addition to noise and attenuation, other nonindealities of the physical transmission medium (such as frequency dependence, selective fading, interferences, etc.) are to be taken into account, then these should be included in the first block in Figure 5.2.

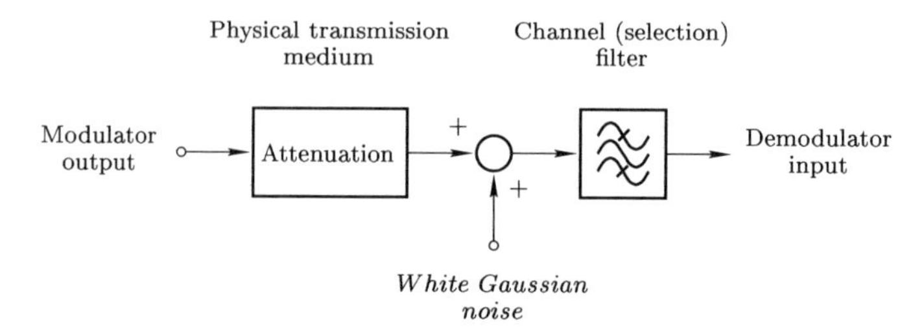

Figure 5.2: Model of an Additive White Gaussian Noise channel including the frequency selectivity of the receiver.

In the model shown in Figure 5.2, we have assumed that the received signal is corrupted by AWGN. In a real telecommunications system, the noise may not be exactly white or Gaussian. The reasons for assuming AWGN are that

- it makes calculations tractable,
- thermal noise, which is of this form, is dominant in many practical communications systems, and

[1]The definition for the Gaussian process is given in [1, 4]. The autocorrelation of white noise is a Dirac delta function multiplied by $N_0/2$ and located at $\tau = 0$, where N_0 is the power spectral density of the noise.

- experience has shown that the relative performance of different modulation schemes determined using the AWGN channel model remains valid under real channel conditions, i.e., a scheme showing better results than another for the AWGN model also performs better under real conditions [1, 2].

5.2.2 Performance Measures

The primary source of errors in a digital communications system is the analog channel. The fundamental problem of digital communications is to maximize the effectiveness of transmission through this channel, i.e., to minimize the energy required to transmit one bit of information.

The performance of a digital communications system is measured in terms of the Bit Error Rate, which gives the probability of bit errors in the received bit stream. In general, this depends on the coding scheme, the type of modulation scheme used, transmitter power, channel characteristics, and the demodulation scheme. The conventional graphical representation of performance in a linear channel with AWGN, depicted in Figure 5.3, shows BER versus E_b/N_0, where E_b is the energy per bit and N_0 is the power spectral density of the noise introduced in the channel.

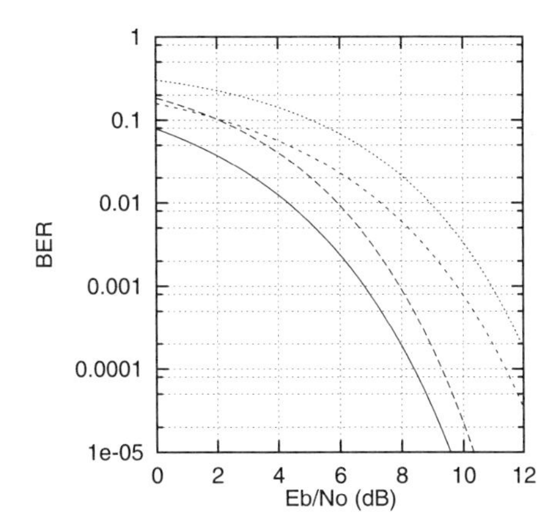

Figure 5.3: Comparison of the noise performances of digital modulation schemes. From left to right: coherent Binary Phase Shift Keying (solid), Differential Phase Shift Keying (long dash), coherent (short dash) and noncoherent (dot) binary orthogonal Frequency Shift Keying.

For a given background noise level, the BER may be reduced by increasing the energy associated with each bit, either by transmitting with higher power or for a longer period per bit. The challenge in digital communications is to achieve a specified BER with *minimum energy per bit*. A further consideration is *bandwidth efficiency*, defined as the ratio of data rate to channel bandwidth [1].

5.2.3 Factors Affecting the Choice of Modulation Scheme

For a given BER and background noise level, the main goal in the design of a digital communications system is to minimize the energy required for the transmission of each bit. The second goal is the efficient utilization of channel bandwidth. There are special applications, for example indoor radio, where other effects, such as multipath propagation, limit the overall system performance. These design requirements affect the choice of the modulation scheme to be used.

While modulation is a relatively straightforward process of mapping symbols to analog waveforms in a deterministic manner, demodulation, which is concerned with mapping corrupted stochastic analog signals back to symbols, is a more difficult and error-prone task. The theoretical background of modulation and demodulation will be discussed in the following two sections.

5.3 Modulation and Demodulation: The Basis Function Approach

Modulation is the process of mapping symbols to analog waveforms [the elements of the so-called "signal set" $s_m(t)$] in a deterministic manner. The signal $s_m(t)$ is transmitted through an analog channel where it is possibly distorted and noise is added. Demodulation is the process by which the received signal $r_m(t)$ (a corrupted stochastic analog signal) is mapped back to a sequence of symbols.

5.3.1 Orthonormal Basis Functions

Let $s_m(t), m = 1, 2, \ldots, M$ denote the elements of the signal set. Our goal is to minimize the number of special signals, called *basis functions*, that have to be known at the receiver to perform demodulation. Let us introduce N real-valued orthonormal basis functions

$$g_j(t), \quad j = 1, 2, \ldots, N,$$

where

$$\int_0^T g_l(t) g_j(t)\, dt = \begin{cases} 1, & \text{if } l = j \\ 0, & \text{elsewhere} \end{cases}$$

and T denotes the symbol duration. Then each element of the signal set can be represented as a linear combination of N basis functions

$$s_m(t) = \sum_{j=1}^{N} s_{mj} g_j(t), \quad \begin{cases} 0 \leq t \leq T \\ m = 1, 2, \ldots, M, \end{cases} \tag{5.1}$$

where $N \leq M$. In conventional digital telecommunications systems, sinusoidal basis functions are used; the most common situation involves a quadrature pair of sinusoids.

5.3.2 Signal Set Generation

The coefficient s_{mj} in Equation (5.1) may be thought of as the jth element of an N-dimensional *signal vector* $\mathbf{s}_m$. The incoming bit stream is first transformed into a symbol sequence; the elements of the signal vector are then determined from the symbols. The signals $s_m(t)$ to be transmitted are generated as a weighted sum of basis functions as given by (5.1).

5.3.2.1 Example

In the case of two basis functions $g_1(t)$ and $g_2(t)$ and two symbols, the signal to be transmitted is generated as a weighted sum of basis functions, as shown in Figure 5.4. Symbol 1 is transmitted by setting $s_{m1} = 1$ and $s_{m2} = 0$; symbol 2 is transmitted by setting $s_{m1} = 0$ and $s_{m2} = 1$.

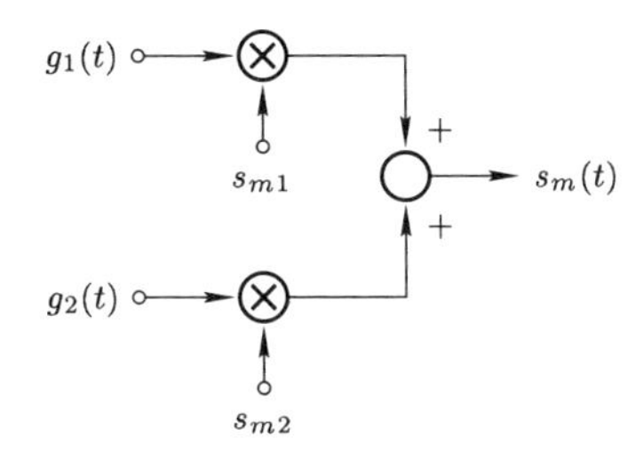

Figure 5.4: Generation of the elements of a signal set with two basis functions.

5.3.3 Recovery of the Signal Vector by Correlation

Because the basis functions are orthonormal, the elements of the signal vector can be recovered from the elements of the signal set, i.e., from the received signal, if every basis function is known at the receiver. In particular,

$$s_{mj} = \int_0^T s_m(t) g_j(t)\, dt, \quad \begin{cases} m = 1, 2, \ldots, M \\ j = 1, 2, \ldots, N \end{cases}. \tag{5.2}$$

Thus, a demodulator can be thought of as a bank of N correlators, each of which recovers the weight s_{mj} of basis function $g_j(t)$. Since there exists a one-to-one mapping between signal vectors and symbols, the transmitted symbols can be recovered by post-processing the outputs of the correlators, and the demodulated bit stream can thus be regenerated.

5.3.3.1 Example: Recovery of a binary signal vector by correlation

In the case of a binary signal set with two basis functions, the weights s_{ij} can be recovered as shown schematically in Figure 5.5:

$$s_{mj} = \int_0^T s_m(t) g_j(t)\, dt, \quad \begin{cases} m = 1, 2 \\ j = 1, 2 \end{cases}. \tag{5.3}$$

In this case, the demodulator consists of two correlators, which recover the weights s_{m1} and s_{m2} of basis functions $g_1(t)$ and $g_2(t)$, respectively.

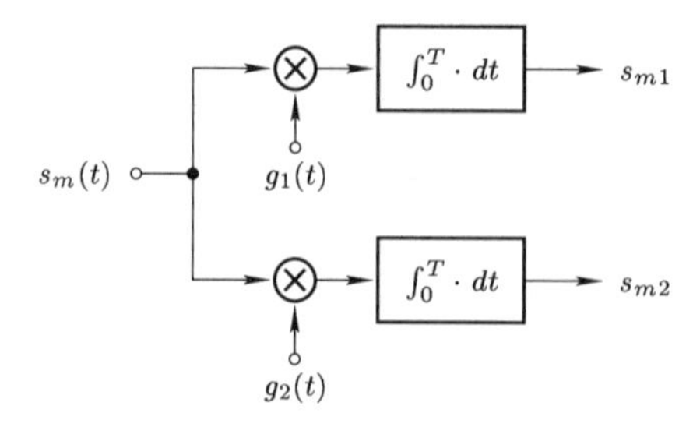

Figure 5.5: Determination of the signal vector, i.e., the set of coefficients $\{s_{mj}\}$ for two basis functions.

5.3.4 Orthonormal Basis Functions for Bandwidth Efficiency

High bandwidth efficiency requires a large signal set. The main advantage of using orthonormal basis functions is that a huge signal set can be generated from a small number of basis functions. Typically, a pair of quadrature sinusoidal signals (a cosine and a sine) is used as the set of basis functions. Since quadrature sinusoidal signals can be generated using a simple phase shifter, it is sufficient to know (or recover) only one sinusoidal signal at the receiver.

An example showing modulator and demodulator circuits for binary PSK (BPSK) will be given in Section 5.5.

5.4 Detection of a Single Symbol in Noise: Basic Receiver Configurations

The receiver must recognize the symbols sent via the channel in order to recover the information which has been transmitted. For the sake of simplicity, only the detection of a single isolated symbol is considered in this section, the effect of Intersymbol Interference (ISI), i.e., the interference between successive symbols is neglected [5].

Our goal is to minimize the average probability of symbol errors, i.e., to develop an optimum receiver configuration. For an AWGN channel and for the case when all symbols to be transmitted are equally likely, Maximum Likelihood (ML) detection has to be used in order to get an optimum receiver [6]. The ML detection method can be implemented by either correlation or matched filter receivers [1].

In this section, we will demonstrate the connection between correlation and matched filter receivers, and consider the relative merits of coherent and noncoherent detection.

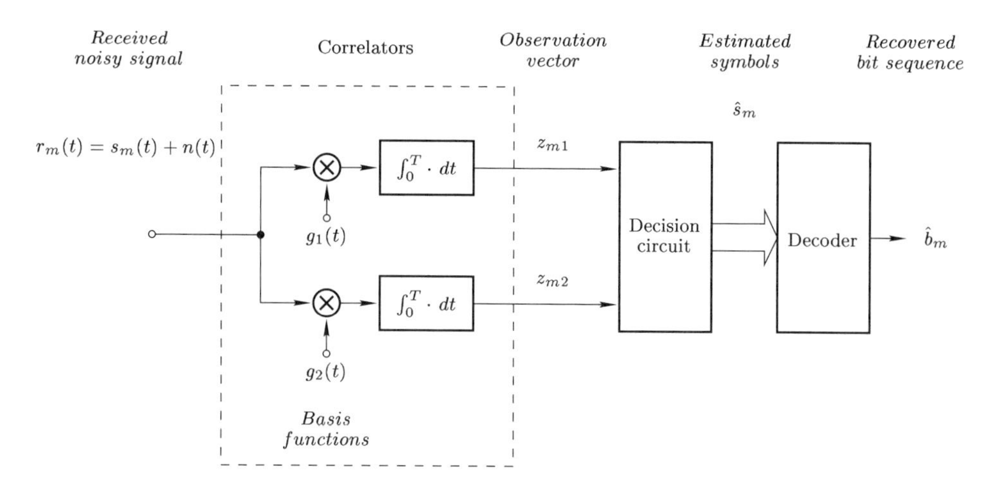

Figure 5.6: Block diagram of correlation receiver for $N = 2$.

5.4.1 Correlation and Matched Filter Receivers

5.4.1.1 Correlation receiver

Equation (5.2) shows how the signal vector can be recovered from a received signal by correlators if the N basis functions $g_j(t)$ are orthonormal and are known at the receiver. Note that, in addition to the basis functions, both the symbol duration T and the initial time instant of symbol transmission have to be known at the receiver. The latter data are called *timing information.* The idea suggested by (5.3) for $N = 2$ is exploited in the correlation receiver shown in Figure 5.6.

In any practical telecommunications system, the received signal is corrupted by noise, i.e., the input to each of the correlators is the sum of the transmitted signal $s_i(t)$ and a sample function $n(t)$ of a zero-mean, stationary, white, Gaussian noise process. The elements of the signal vector can still be estimated using correlators, although the estimates may differ from their nominal values, due to corruption in the channel.

The outputs of the correlators, called the *observation vector*, are the inputs of a decision circuit. The decision circuit applies the ML detection method, i.e., it chooses the signal vector from among all the possibilities that is the closest to the observation vector. Estimates of the symbols are determined from the signal vector and finally the demodulated bit sequence is recovered from the estimated symbol.

Note that in a correlation receiver *all the basis functions* and *the timing information* are required; these must be recovered from the (noisy) received signal.

5.4.1.2 Matched filter receiver

The observation vector can also be generated by a set of *matched filters* [1]. In a matched filter receiver a bank of matched filters replaces for the correlators in

Figure 5.6. In this case the basis functions are stored locally as the impulse responses of the matched filters, i.e., only the timing information must be recovered from the received signal.

5.4.2 Coherent and Noncoherent Receivers

5.4.2.1 Coherent receivers

Receivers in which exact copies of all the basis functions are known are called *coherent receivers.* In practice, coherent correlation receivers are used almost exclusively to demodulate ASK, PSK, and their special cases such as Quadrature Phase Shift Keying (QPSK), M-ary PSK (MPSK) and M-ary Quadrature Amplitude Modulation (MQAM) signals.

The required impulse response of a matched filter can at best be approximated by a physically-realizable analog filter. Any deviation from the ideal impulse response results in a large degradation of performance. Therefore, coherent matched filter receivers are not used in radio communications where the elements of the signal set are analog signals.

5.4.2.2 Noncoherent receivers

In applications where the propagation conditions are poor, the basis functions $g_j(t)$ cannot be recovered from the received signal. In these cases, the conventional solution is to use M-ary FSK (MFSK, $M \geq 2$) modulation and a noncoherent receiver.

The basis functions $g_j(t)$ or the elements $s_m(t)$ of the signal set are not known in a noncoherent receiver, but one or more robust characteristics of $s_m(t)$, $m = 1, 2, \ldots, M$ can be determined. Demodulation is performed by evaluating one or more selected characteristics of the received signal.

For example, M different *signaling frequencies* are used in MFSK. In a noncoherent FSK receiver, a bank of band-pass filters is applied to recognize the different signaling frequencies. The observation vector is generated by envelope detectors and the decision circuit simply selects the "largest" element of the observation vector [1].

5.4.2.3 Relative merits of coherent and noncoherent receivers

It is often claimed that the main advantage of coherent receivers over noncoherent ones is that their performance in the presence of additive channel noise is better than that of their noncoherent counterparts. Let us estimate the size of this advantage for the selected application domain: digital communications.

In a practical digital communications system, communication is not possible if the BER becomes worse than 10^{-3} or 10^{-2}, so we only consider operation below this range. In fact, the average value of "raw" BER for terrestrial microwave radio systems varies from 10^{-7} to 10^{-6}; with error correction, this can be reduced to below 10^{-9} [5].

The noise performance of coherent and noncoherent binary FSK receivers is shown in Figure 5.3. At $E_b/N_0 = 10^{-2}$, the E_b/N_0 required by the noncoherent FSK receiver is only 1.6 dB greater than the corresponding value for the coherent one. Moreover, at high values of E_b/N_0, noncoherent FSK receivers perform almost as well as coherent ones for the same E_b/N_0.

The real advantage of the coherent technique is not better noise performance but the fact that huge signal sets can be generated by means of very few orthonormal basis functions. For example, in terrestrial digital microwave radio systems, 256 signals are typically generated using a pair of quadrature sinusoidal signals. This huge signal set results in excellent *bandwidth efficiency*. Moreover, the receiver must recover *just one* sinusoidal signal from the incoming signal.

For their part, noncoherent techniques offer two advantages over coherent detection:

- when *propagation conditions are poor*, the basis functions cannot be recovered from the received signal because $r_i(t)$ differs too much from $s_i(t)$. In this case, a noncoherent receiver is the only possible solution.
- noncoherent receivers can, in principle, be implemented with *very simple circuitry*, because the basis functions do not need to be recovered.

5.5 Example: BPSK with Coherent Detection

The block diagram of a coherent BPSK transceiver can be developed from (5.1) and Figure 5.6. As shown in Figure 5.7, the most important operations performed at the receiver are:

- recovery of the basis function $\hat{g}_1(t)$ and timing information,
- determination of the observation signal z_{i1}, and
- decision-making.

In the previous sections we have assumed that a single isolated symbol is transmitted via the channel. In a real applications, a sequence of symbols has to be transmitted. This effect is considered in Figure 5.7, where a switch controlled by the timing recovery circuit assign the beginning of a new symbol. In Figure 5.7, indices i and k denote the ith transmitted bit and kth transmitted symbol, respectively.

In the case of BPSK, the bit and symbol streams are identical and two symbols are used to transmit the bit stream b_k. Thus, the signal set contains two sinusoidal signals $s_1(t)$ and $s_2(t)$. The binary symbols "1" and "0" are mapped to the signals

$$s_1(t) = \sqrt{\frac{2E_b}{T}}\cos(\omega_c t) \quad \text{and} \quad s_2(t) = -\sqrt{\frac{2E_b}{T}}\cos(\omega_c t),$$

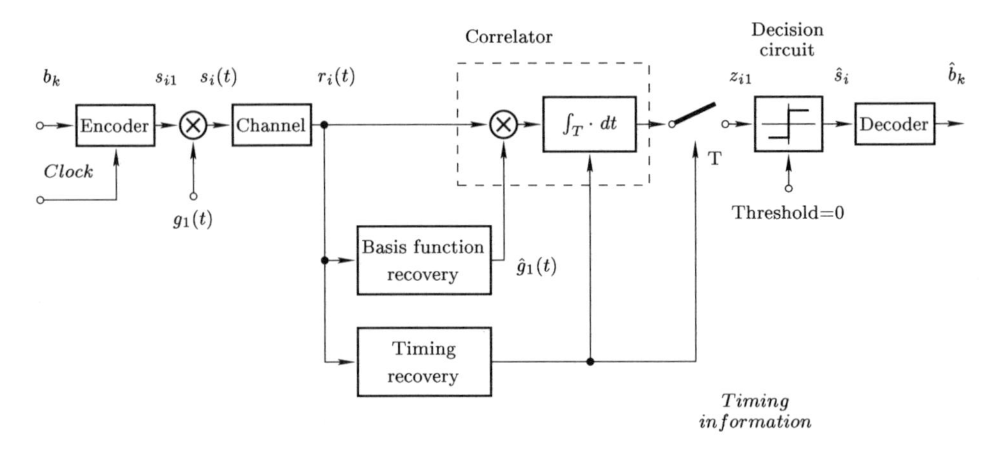

Figure 5.7: Block diagram of a coherent BPSK receiver.

respectively, where $0 \leq t \leq T$, ω_c is the carrier frequency and E_b is the transmitted energy per bit.

To simplify the recovery of the basis function, each transmitted symbol is designed to contain an integral number of cycles of the sinusoidal carrier wave. Given that there is just one basis function of unit energy

$$g_1(t) = \sqrt{\frac{2}{T}} \cos(\omega_c t), \qquad 0 \leq t < T,$$

we recover the elements of the signal vector from Equation (5.3) as

$$s_{11} = +\sqrt{E_b} \quad \text{and} \quad s_{21} = -\sqrt{E_b}.$$

Channel effects

In any practical communications system, the signal $r_i(t)$ which is present at the input to the demodulator differs from that which was transmitted, due to the effects of the channel.

Each weight s_{ij} may be estimated by correlating the received signal $r_i(t)$ with a reference signal $\hat{g}_j(t)$. In the case of BPSK, the estimate z_{i1} of s_{i1}, shown in Figure 5.7, is given by

$$\begin{aligned}
z_{i1} &= \int_T r_i(t)\hat{g}_1(t)\, dt \\
&= \int_T \Big(s_i(t) + n(t) \Big) \hat{g}_1(t)\, dt \\
&= \int_T \Big(\pm \sqrt{E_b} g_1(t) + n(t) \Big) \hat{g}_1(t)\, dt \\
&= \pm\sqrt{E_b} \int_T g_1(t)\hat{g}_1(t)\, dt + \int_T n(t)\hat{g}_1(t)\, dt.
\end{aligned}$$

Demodulation of BPSK when $\hat{g}_1(t) = g_1(t)$

If the basis function $g_1(t)$ is known exactly at the receiver, then $\hat{g}_1(t) = g_1(t)$ and

$$\begin{aligned} z_{i1} &= \pm\sqrt{E_b}\int_T g_1^2(t)\,dt + \int_T n(t)g_1(t)\,dt \\ &= \pm\sqrt{E_b} + \int_T n(t)g_1(t)\,dt \end{aligned} \tag{5.4}$$

because

$$\int_T g_1^2(t)\,dt = 1$$

by the definition of an orthonormal basis function.

In the absence of noise,

$$z_{i1} = \pm\sqrt{E_b}.$$

If one plots a histogram of the estimates of the observation signal z_{i1} for a large number of transmitted symbols, the samples are clustered at $-\sqrt{E_b}$ and $+\sqrt{E_b}$, as shown in Figure 5.8(a).

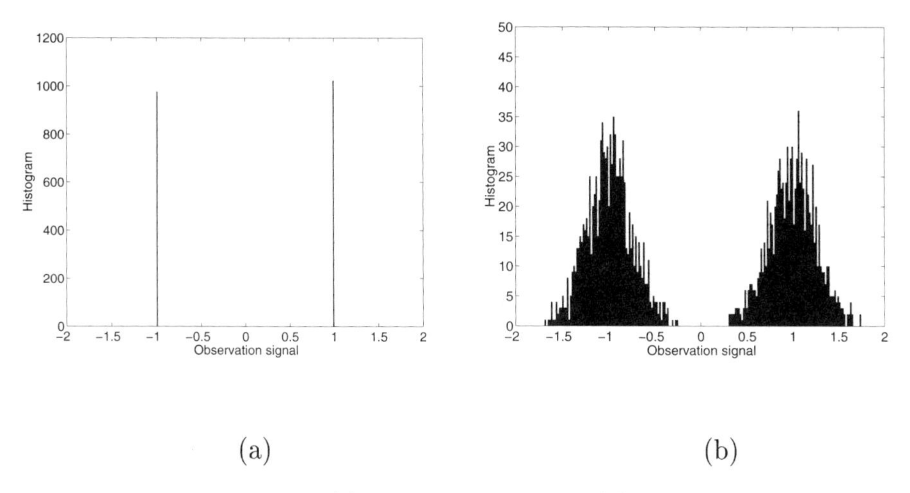

(a) (b)

Figure 5.8: Histograms of (a) the noise-free and (b) noisy observation signals in a coherent BPSK correlation receiver.

When the second term in Equation (5.4) (due to additive Gaussian noise) is non-zero, the histogram of the estimates z_{i1} for a large number of transmitted symbols consists of two bell-shaped clusters, as shown in Figure 5.8(b).

The separation of the peaks is determined by the bit energy E_b and the variance by the noise spectral density N_0. Errors occur when the clusters overlap.

To minimize the probability of error, i.e., to get an optimum receiver, Maximum Likelihood detection is used. The decision rule is simply to make the decision in favor of symbol "1" if the received signal is "closer" to $s_1(t)$, i.e., if the output of the correlator is greater than zero at the decision time instant.

If the observation signal z_{i1} is less than zero, then the receiver decides that a symbol "0" has been transmitted. The decision circuit is simply a level comparator with zero threshold.

Note that the BER (equivalently, the overlap between the clusters) is zero in the noise-free case. This is an important property of a conventional modulation scheme based on orthonormal basis functions.

5.6 Synchronization in Digital Communications

In this section, we consider two fundamental synchronization problems: *timing recovery*, which is an essential part of digital communications, and *carrier recovery,* which is necessary only for coherent detection.

5.6.1 Carrier Recovery and Timing Recovery

5.6.1.1 Carrier recovery and the need for synchronization in coherent detection

In general, coherent reception requires knowledge of the basis functions at the receiver. Because matched filter receivers cannot be implemented in the analog signal domain without a considerable implementation loss, only correlation receivers can be used for coherent detection, and synchronization must be used to recover the basis functions.

In the special case of sinusoidal basis functions, knowledge of both the frequency and phase of a carrier is required. The basis functions are typically recovered from the received noisy signal by means of a suppressed carrier Phase-Locked Loop (PLL) [7]. In conventional systems, estimation of the frequency and phase of the carrier is called *carrier recovery* [5].

5.6.1.2 Timing recovery and the need for symbol synchronization in digital communications

In any practical system, not only an isolated single symbol, but a sequence of symbols, has to be transmitted. To perform demodulation, the receiver has to know precisely the time instants at which the modulation can change its state. That is, it has to know the start and stop time instants of the individual symbols in order to assign the decision time instants and to determine the time instants when the initial conditions of the correlators have to be reset to zero. Determination of these time instants is called *timing recovery* or *symbol synchronization.*

In contrast with carrier recovery, which is an optional step that is required only by coherent receivers, timing recovery is a *necessary* operation in digital communications. The decision times at the receiver must be aligned in time (synchronized) with those corresponding to the ends of symbol intervals T at the transmitter. Symbol synchronization must be achieved as soon as possible after transmission begins, and must be maintained throughout the transmission.

This chapter is aimed at providing a clear exposition of the important issues in both conventional and chaotic modulation/demodulation techniques. Although symbol synchronization has to be solved in every digital communications system, it belongs to the decision circuit and not to the demodulation process. Therefore, the details of the timing recovery problem are not discussed in this book. The interested reader can find excellent expositions of timing recovery in [6] and [8], for example.

In the next subsection, we will discuss the advantages and disadvantages of synchronization for basis function recovery in coherent receivers.

5.6.2 Advantages and Disadvantages of Synchronization

The main advantage of synchronization is that it makes the implementation of coherent receivers possible. As mentioned in Section 5.4.2.3, the most significant feature of a coherent receiver when used with a sinusoidal carrier is that by recovering just one signal—the carrier—and regenerating the quadrature basis functions (by means of a simple phase-shifter), a pair of orthonormal basis functions can be generated and therefore a huge signal set can be used (256-QAM, for example). In addition, a coherent receiver has marginally better noise performance than its noncoherent counterpart.

However, there are significant costs associated with synchronization, in terms of synchronization time, circuit complexity, and severe penalties associated with loss of synchronization. In this subsection, we discuss these issues.

5.6.2.1 Costs associated with achieving synchronization in a coherent receiver

In conventional digital communications systems, various types of PLLs are used to perform synchronization [7].

Two basic operation modes have to be distinguished for a PLL. Under normal operating conditions, the phase-locked condition has been achieved and is maintained. The PLL simply follows the phase of an incoming signal; this is called the *tracking mode*.

Before the phase-locked condition is achieved, the PLL operates in a highly nonlinear *capture mode*. The time required for the PLL to achieve the phase-locked condition is called the *pull-in time*. The *transient time*, which is associated with the tracking mode, is always significantly shorter than the pull-in time.

In a digital communications system where synchronization is lost at the beginning of every new symbol, the received symbol can be estimated from the noisy received signal only after the basis functions have been recovered, i.e., only after the phase-locked condition has been achieved. In this case, the total detection time associated with each symbol is the sum of the pull-in time and the estimation time. In particular, a long pull-in time results in a very low symbol rate. This is why *synchronization is always maintained* in the carrier recovery circuits of conventional digital communications systems.

5.6.2.2 Penalties for failing to achieve synchronization in a coherent receiver

The block diagram of a coherent correlation receiver is shown in Figure 5.6. The received signal is always a stochastic process due to additive channel noise. The observation vector, i.e., the output of the correlators, is a random variable, where the mean value of estimation depends on the bit energy and the "goodness" of recovery of the basis functions. In conventional receivers, the variance of this estimation is determined by the noise spectral density N_0 [1]. The probability of error, i.e., the probability of making wrong decisions, depends on the mean value and variance of estimation. The main disadvantage of synchronization follows from the sensitivity of noise performance to the "goodness" of recovery of the basis functions.

The most serious problem is caused by the cycle slips in PLLs used for recovering a suppressed carrier. As a consequence of noise, the phase error in the PLL is a random process. If the variance of the phase error is large, cycle slips appear with high probability due to the periodic characteristic of the phase detector [9]. This means that the VCO phase, i.e., the phase of the recovered carrier, can slip by one or more cycles with respect to the reference phase. Every cycle slip results in a symbol error. For high Signal-to-Noise Ratio (SNR) the probability of cycle slips is low, but it increases steeply with increasing noise power. This phenomenon results in a large degradation in noise performance.

Even if the physical transmission medium can be characterized by a pure attenuation, a small error in the phase of the recovered carrier may be present due to non-ideal properties of the synchronization circuit. This error generally causes a large degradation in the noise performance of a coherent receiver [2,10].

A real telecommunications channel always causes some transformation of the basis functions. This problem is especially hard to overcome in coherent receivers if the transmission medium is time-varying. The time-dependent channel transformation requires adaptive control of the recovered basis functions. This can be accomplished only by means of a wide-band PLL circuit. However, in this case the recovered basis functions are corrupted by the channel noise passed by the PLL transfer function. This also results in a large degradation in the system's noise performance.

Disturbances and interferences passed by the channel filter may cause a transient in the carrier recovery circuit. The result of this transient is that the recovered basis functions deviate from their ideal values for a while. This also results in performance degradation.

Synchronization can also be lost from time to time due to deep and/or selective fading [11]. Loss of synchronization automatically initiates a pull-in process which means that all symbols received during the pull-in time are lost. Once again, the result is a degradation in BER.

The conclusion is that coherent receivers exploiting synchronization offer the best system performance *if synchronization can be maintained.* However, they do not offer optimum performance if the SNR is low, the propagation conditions are poor, the properties of the channel are time-varying, or if the probability of deep fading is relatively high. In these cases, a more robust modulation scheme such as FSK with a noncoherent receiver must be used.

A further disadvantage of a coherent receiver from an implementation point of view is that it generally requires more complicated circuitry than its noncoherent counterpart.

5.7 Summary

Although conventional and chaotic digital telecommunications systems seem to be very different at first glance, a systematic analysis shows that a unified framework can be developed for both of them by using the basis function approach (also known as the Gram-Schmidt orthogonalization procedure [1]).

In order to understand the basic idea of digital telecommunications and to develop the chaotic telecommunications system in a systematic way, we have surveyed the theoretical background of conventional digital communications systems in this chapter. We have shown that the digital information to be transmitted has to be mapped into analog sample functions since only analog signals can be transmitted through analog channels. These analog sample functions form the signal set.

At the transmitter, the incoming bit stream is mapped to symbols and each digital symbol is mapped, in a one-to-one deterministic manner, to an element of the analog signal set. To minimize the number of basis functions required at the receiver to perform the demodulation, each element of the signal set is represented as a linear combination of the basis functions. The weighting factors of this linear combination are the elements of the signal vector.

In coherent receivers, the basis functions have to be known either as recovered signals or as the impulse responses of matched filters. Then the elements of the signal set can be recovered by correlators or matched filters.

In a real telecommunications system, the received signal is corrupted by noise, suffers distortion, interferences, and so on. This is why the observation signals, i.e., estimations of the elements of the signal vector, are stochastic variables which can be characterized by their mean value and variance. Overlap in the histogram of the observation signals results in wrong decisions.

We have shown in this chapter that the performance of a digital telecommunications scheme can be quantified by plotting BER as a function of E_b/N_0. To illustrate the theoretical investigation, the operation of a conventional BPSK telecommunications system has been discussed in detail.

In many applications, if a very simple receiver is required or if, due to bad propagation conditions, the basis functions cannot be recovered at the receiver, a noncoherent system configuration has to be used for reception. In this case, the digital information to be transmitted is mapped to a selected characteristic of the transmitted signal, e.g., to its frequency, and that characteristic is determined during demodulation to recover the transmitted information. However, although there is no need for carrier recovery in noncoherent receivers, the timing recovery problem has to be solved both in coherent and noncoherent digital telecommunication systems.

Finally, the role of synchronization in digital communications has been discussed in this chapter. We have shown that coherent reception, which requires carrier synchronization, offers larger signal sets and better noise performance than noncoherent reception, where the basis functions are not known at the receiver. But there are strict penalties if there is a synchronization error in a coherent receiver or if synchronization is lost. We have concluded that if the propagation conditions are so bad that synchronization cannot be maintained continuously and the synchronization error cannot be kept small enough, then the overall system performance of noncoherent receivers exceeds that of their coherent counterparts.

References

[1] S. Haykin. *Communication Systems.* John Wiley & Sons, New York, 3rd edition, 1994.

[2] J. G. Proakis. *Digital Communications.* McGraw-Hill, Singapore, 1983.

[3] M. P. Kennedy. Communicating with chaos: State of the art and engineering challenges. In *Proc. 4th Int. Workshop on Nonlinear Dynamics of Electronic Systems*, page 1, Seville, 27–28 June 1996.

[4] J. S. Bendat and A. G. Piersol. *Measurement and Analysis of Random Data.* John Wiley & Sons, New York, 2nd edition, 1966.

[5] I. Frigyes, Z. Szabó, and P. Ványai. *Digital Microwave Transmission.* Elsevier Science Publishers, Amsterdam, 1989.

[6] E. A. Lee and D. G. Messerschmitt. *Digital Communications.* Kluwer Academic Publishers, Boston, 2nd edition, 1993.

[7] F. M. Gardner. *Phaselock Techniques.* John Wiley & Sons, New York, 2nd edition, 1979.

[8] M. K. Simon, S. M. Hinedi, and W. C. Lindsey. *Digital Communication Techniques: Signal Design and Detection.* Prentice-Hall, Englewood Cliffs, 1995.

[9] G. Ascheid and H. Meyr. Cycle slips in phase-locked loops: a tutorial survey. *IEEE Trans. Communications*, COM-30:2228, Oct. 1982.

[10] R. W. Lucky, J. Salz, and E. J. Weldon. *Principles of Data Communication.* McGraw-Hill, New York, 1968.

[11] R. C. Dixon. *Spread Spectrum Systems with Commercial Applications.* John Wiley & Sons, New York, 3rd edition, 1994.

Chapter 6

Chaotic Modulation Schemes

Michael Peter Kennedy
Department of Microelectronic Engineering
University College Cork
Cork, Ireland
Peter.Kennedy@ucc.ie

Géza Kolumbán and Zoltán Jákó
Department of Measurement and Information Systems
Technical University of Budapest
H-1521 Budapest, Hungary
kolumban@mit.bme.hu, jako@mit.bme.hu

6.1 Introduction

The aim of this chapter is to describe the state of the art in digital modulation schemes which use chaotic rather than periodic basis functions. Problems and opportunities arising from the non-periodicity of chaotic signals are highlighted, solutions are proposed, potential application domains are identified, and current research directions are discussed.

The generation of the analog signal set to be transmitted, mapping of the digital bit stream to symbols, and mapping of the symbols to the elements of signal set have been discussed in Chapter 5 for conventional digital communications systems. To compare conventional and chaotic communications schemes, we will use the same terminology for the discussion of chaotic modulation schemes in this chapter. Except where otherwise stated, we will consider only the case when a single isolated symbol is transmitted.

In conventional communications, the modulated signal consists of segments of periodic waveforms corresponding to the individual symbols. When sinusoidal basis functions are used without spread spectrum techniques, the transmitted signal is a narrowband signal. Consequently, multipath propagation (due to the reception of multiple copies of the transmitted signal traveling along different paths) can cause high attenuation or even dropout (in the case of catastrophic destructive interference) of the received narrowband signal.

A chaotic signal generator automatically produces a wideband noiselike signal with robust and reproducible statistical properties [1, 2]. Due to its wideband nature, a modulation scheme using chaotic basis functions is potentially more resistant to multipath propagation than one based on sinusoids.

One factor which limits the performance of all telecommunications systems is interference. In conventional systems based on periodic carrier signals, where the Spread Spectrum Code Division Multiple Access (SS-CDMA) technique is not used, signals can be made orthogonal by putting them in different frequency bands [Frequency Division Multiple Access (FDMA)] or time slots [Time Division Multiple Access (TDMA)], by ensuring that the basis functions are orthogonal to each other (using sine and cosine basis functions, for example), or by using orthogonal electromagnetic polarization, for example. If these requirements are not met, interference occurs.

In contrast with periodic signals, chaotic signals decorrelate rapidly with themselves and chaotic signals generated by different chaotic circuits are almost orthogonal. This means that the correlation, equivalently the interference, between two chaotic signals generated by unsynchronized chaotic circuits started from different initial conditions and/or having different circuit parameters is low.

6.2 Digital Communications Using Chaos

In a digital communications system, the information symbol to be transmitted is mapped by the modulator to an analog sample function which then passes through an analog channel. The analog signal in the channel is subject to a number of disturbing influences including attenuation, bandpass filtering, and additive noise. The role of the demodulator is to decide, on the basis of the received corrupted sample function, which symbol was transmitted.

In a conventional communications system, the analog sample function of duration T which represents a symbol is a linear combination of sinusoidal basis functions and the symbol duration T is an integer multiple of the period of the basis function. In a chaotic digital communications system, shown schematically in Figure 6.1, the analog sample function of duration T which represents a symbol is a *chaotic* basis function. Figure 6.1 helps to answer a frequently asked question: whether chaotic communications schemes can be considered as Spread Spectrum (SS) systems. According to Dixon [3], "a spread spectrum system ... must meet two criteria: (1) the transmitted bandwidth is much greater than the bandwidth or rate of the information being sent, and (2) some function other than the information being sent is employed to determine the resulting modulated RF bandwidth." Following this definition, chaotic communication systems form a special class of SS systems where a chaotic signal is the spreading signal.

The decision as to which symbol was transmitted is made by estimating some property of the received sample function [4, 5]. That property might be

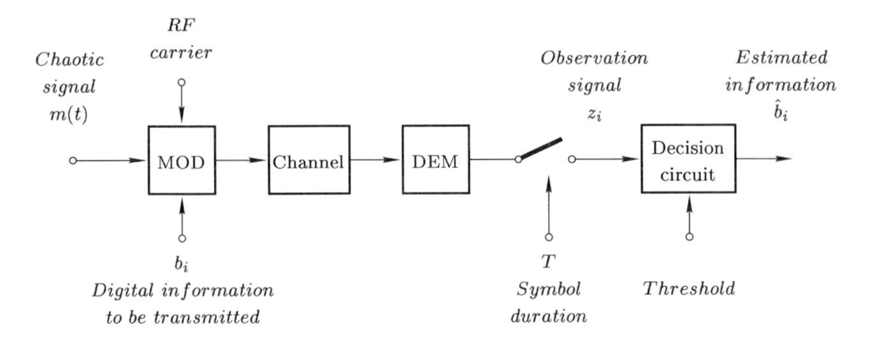

Figure 6.1: Block diagram of a chaotic communications scheme.

the energy of the chaotic sample function or the correlation measured between different parts of the transmitted signal, for example.

Since chaotic waveforms are not periodic, each sample function of duration T is different. This has the advantage that each transmitted symbol is represented by a unique analog sample function, and the correlation between chaotic sample functions is extremely low. However, it also produces a problem associated with estimating long-term statistics of a chaotic process from sample functions of finite duration. We will discuss this so-called *estimation problem* [6] in more detail in Section 6.5.4.

6.3 Chaotic Modulation

Chaotic digital modulation is concerned with mapping symbols to analog chaotic waveforms. In Chaos Shift Keying, information is carried in the weights of a combination of basis functions. Differential Chaos Shift Keying is a variant of Chaos Shift Keying where information is also carried in the correlation between parts of the basis functions.

In the following two sections, we describe in detail the operation of the Chaos Shift Keying and Differential Chaos Shift Keying modulation schemes. We will concentrate on the reception of a single isolated symbol. Problems arising from the reception of symbol streams are not treated here.

6.4 Chaos Shift Keying

Chaos Shift Keying (CSK) [7, 8] is a digital modulation scheme where chaotic signals generated by different attractors or chaotic signals generated by the same attractor but emerging from different initial conditions are used as basis functions. The number of attractors or initial conditions is equal to the number of basis functions. The attractors may be produced by the same dynamical system for different values of a bifurcation parameter or by completely different dynamical systems.

Note that the shape of each basis function is not fixed in chaotic communication. This is why the elements of the signal set which are transmitted through the channel have a different shape during every symbol period T, even if the same symbol is transmitted. As a result, the transmitted signal is never periodic.

Using the notation introduced in Chapter 5, the elements of the signal set are defined by

$$s_m(t) = \sum_{j=1}^{N} s_{mj} g_j(t), \quad j = 1, 2, \ldots, N$$

where the basis functions $g_j(t)$ are chaotic waveforms. The signals $s_m(t)$ may be produced conceptually as shown in Figure 6.2.

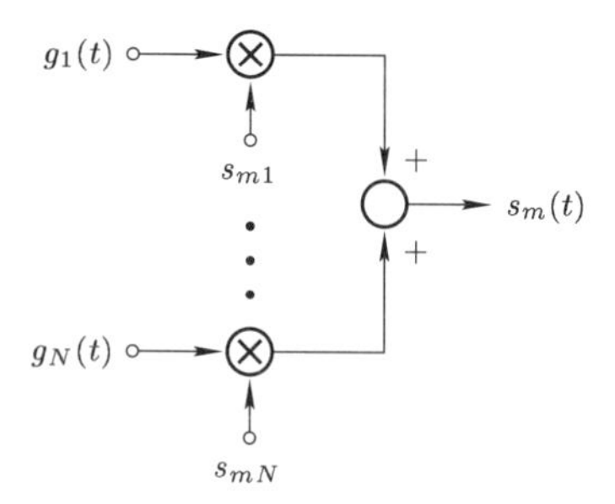

Figure 6.2: Generation of the elements of the signal set.

Chaotic basis functions are orthonormal in the mean, i.e.,

$$E\left[\int_0^T g_l(t) g_j(t) dt\right] = \begin{cases} 1 & \text{if } l = j \\ 0 & \text{otherwise} \end{cases} \tag{6.1}$$

where $E[\cdot]$ denotes the expectation operator. Equation (6.1) identifies another important characteristic of chaotic modulation schemes. The basis functions are not periodic but are chaotic signals which can be modeled only as stochastic processes [9]. Consequently, the cross-correlation and autocorrelation of basis functions evaluated for the bit duration become random numbers which can be characterized by their mean value and variance. The details of this problem, called the estimation problem, will be discussed later in Section 6.5.4.

The weights s_{mj} of the signal vector can be recovered by correlating the received signal with locally generated copies of the basis functions $g_j(t)$, as shown in Figure 6.3.

The element z_{mj} of the observation vector at the output of the jth correlator, when signal $s_m(t)$ is transmitted, is given by:

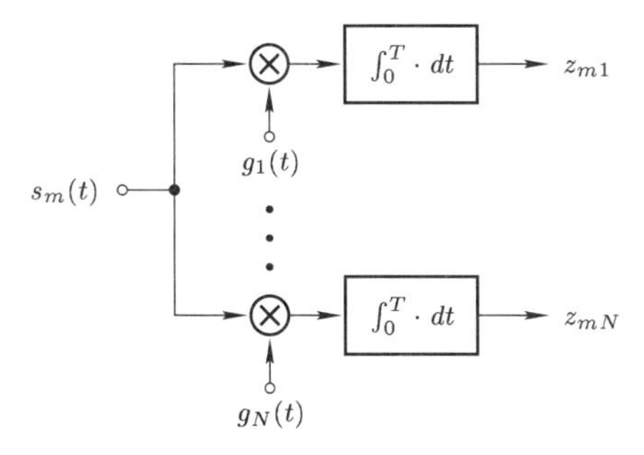

Figure 6.3: Determination of the observation signals.

$$
\begin{aligned}
z_{mj} &= \int_0^T s_m(t) g_j(t) dt \\
&= \int_0^T \left[\sum_{j=1}^{N} s_{mj} g_j(t) \right] g_j(t) dt \\
&= s_{mj} \int_0^T g_j(t) g_j(t) dt \\
&\approx s_{mj}
\end{aligned}
\tag{6.2}
$$

where $E\left[\int_0^T g_j(t)g_j(t)dt\right] = 1$.

Equation (6.2) shows that, in the case of a distortion- and noise-free channel, if the bit duration T is sufficiently long, then the observation and signal vectors are equal to each other. In real applications, the elements z_{mj} of the observation vector are random numbers because of the estimation problem and additive channel noise. This is why the observation vector can be considered only as an *estimation* of the signal vector.

6.4.1 CSK Modulation with One Basis Function

In the simplest case of binary chaos shift keying, a single chaotic basis function $g_1(t)$ is used, i.e.,

$$s_m(t) = s_{m1} g_1(t).$$

A block diagram of CSK modulator is shown in Figure 6.4.

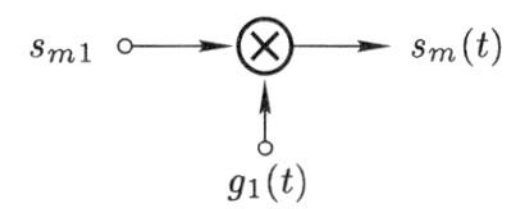

Figure 6.4: CSK modulator with one basis function.

At least three types of CSK based on a single basis function can be imagined:

Chaotic On-Off Keying (COOK): In COOK [10], symbol "1" is represented by $s_1(t) = \sqrt{2E_b}g_1(t)$ and symbol "0" is given by $s_2(t) = 0$. Equivalently,

$$s_{11} = \sqrt{2E_b}; \qquad s_{21} = 0,$$

where E_b denotes the average energy per bit and we have assumed that the probabilities of symbols "1" and "0" are equal.

Unipodal CSK: In unipodal CSK, symbols "1" and "0" are distinguished by transmitting bit energies E_{b1} and $E_{b2} = kE_{b1}$, respectively, where $0 < k < 1$. Assume that the probabilities of symbols "0" and "1" are equal and let the average bit energy be E_b. Then symbol "1" is represented by $s_1(t) = s_{11}g_1(t)$ and symbol "0" is given by $s_2(t) = s_{21}g_1(t)$, where $s_{11} = \sqrt{\frac{2E_b}{1+k}}$ and $s_{21} = \sqrt{\frac{2kE_b}{1+k}}$.

Antipodal CSK: In antipodal CSK, symbol "1" is represented by $s_1(t) = s_{11}g_1(t)$ and symbol "0" is given by $s_2(t) = s_{21}g_1(t)$, where $s_{11} = \sqrt{E_b}$ and $s_{21} = -\sqrt{E_b}$.

The upper limit on the noise performance of every modulation scheme is determined by the separation of the message points in the signal space diagram. The greater the separation, the better the noise performance. Figures 6.5(a–c) show the signal-space diagrams for these three CSK modulation schemes implemented with one basis function. The best noise performance can be achieved by the antipodal modulation scheme but this signal can be demodulated only by means of a coherent receiver.

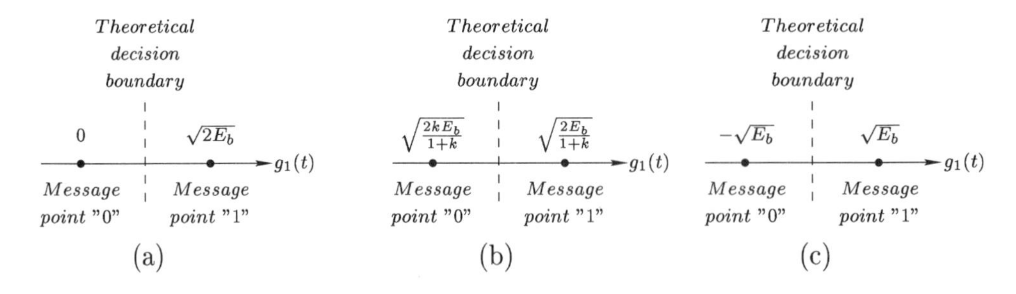

Figure 6.5: Signal-space diagrams for coherent binary (a) COOK, (b) unipodal, and (c) antipodal CSK modulation schemes.

6.4.2 Demodulation of CSK with One Basis Function

At the CSK receiver, the coefficients s_{11} and s_{21} can be recovered using coherent or noncoherent techniques.

6.4.2.1 Coherent matched filter receiver

Matched filters can be used only if the waveforms corresponding to each symbol are known in advance and pre-programmed as the impulse responses of filters. In the case of CSK modulation, the symbols are mapped to chaotic waveforms and a different sample function is generated each time a symbol is transmitted. Therefore, coherent matched filter receivers simply cannot be used in chaotic communications.

6.4.2.2 Coherent correlation receivers

In a coherent correlation receiver, shown schematically in Figure 6.6, the observation signal z_{m1} is given by:

$$\begin{aligned} z_{m1} &= \int_0^T s_m(t) g_1(t) dt \\ &= s_{m1} \int_0^T g_1^2(t) dt \\ &\approx s_{m1} \end{aligned}$$

because, by definition, $E\left[\int_0^T g_1^2(t)dt\right] = 1$.

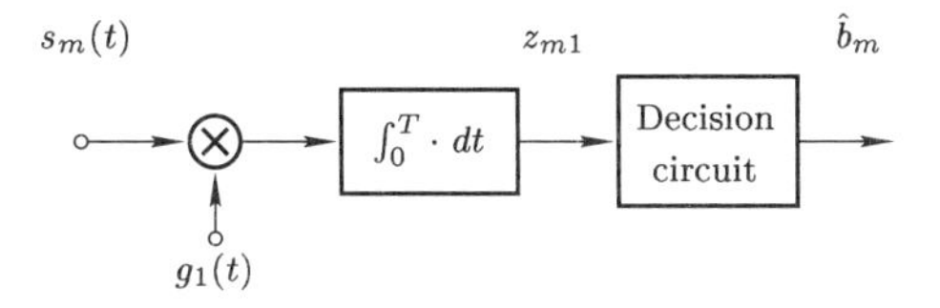

Figure 6.6: Coherent receiver for a CSK signal with one basis function.

The decision circuit decides in favor of symbol "1" if $z_{m1} > z_T$ and symbol "0" if $z_{m1} < z_T$, where z_T is the appropriate threshold for the modulation scheme.

This receiver structure can be used to demodulate COOK, unipodal CSK, and antipodal CSK, provided that the basis function $g_1(t)$ can be recovered from the received signal $s_m(t)$.

A minimum requirement of the circuitry at the receiver which generates a local copy of $g_1(t)$ is that it must do so whether $s_{11}g_1(t)$ or $s_{21}g_1(t)$ is received. Although several strategies for recovering the basis function $g_1(t)$ have been proposed in the literature, under the title "chaotic synchronization," we are not aware of any chaotic synchronization technique which can recover the basis function in this way, *independently of the modulation.*

6.4.2.3 Noncoherent receivers

The bit energies of the two symbols are different in both the COOK and unipodal CSK modulation schemes. The noncoherent demodulator shown schematically in Figure 6.7 determines the bit energy of the received signal. Note that the demodulator contains a correlator but in this case the received signal $s_m(t)$ is correlated with itself and not with the recovered basis function $g_1(t)$.

The observation signal is given by

$$\begin{aligned} z_{m1} &= \sqrt{\int_0^T s_m^2(t)dt} \\ &= \sqrt{s_{m1}^2 \int_0^T g_1^2(t)dt} \\ &= |s_{m1}|\sqrt{\int_0^T g_1^2(t)dt} \\ &\approx |s_{m1}| \end{aligned}$$

since $E\left[\int_0^T g_1^2(t)dt\right] = 1$.

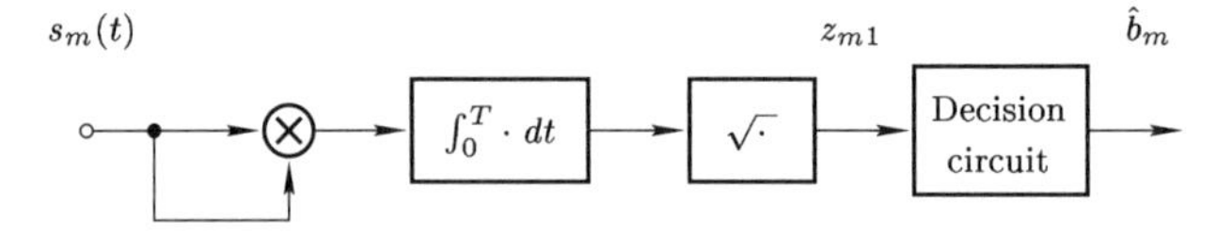

Figure 6.7: Noncoherent receiver for CSK signal with one basis function.

Histograms of samples of the observation signal z_{m1} are shown in Figure 6.8 for COOK and unipodal CSK. The decision circuit decides in favor of symbol "1" if $z_{m1} > z_T$ and symbol "0" if $z_{m1} < z_T$, where z_T is the appropriate threshold for the modulation scheme.

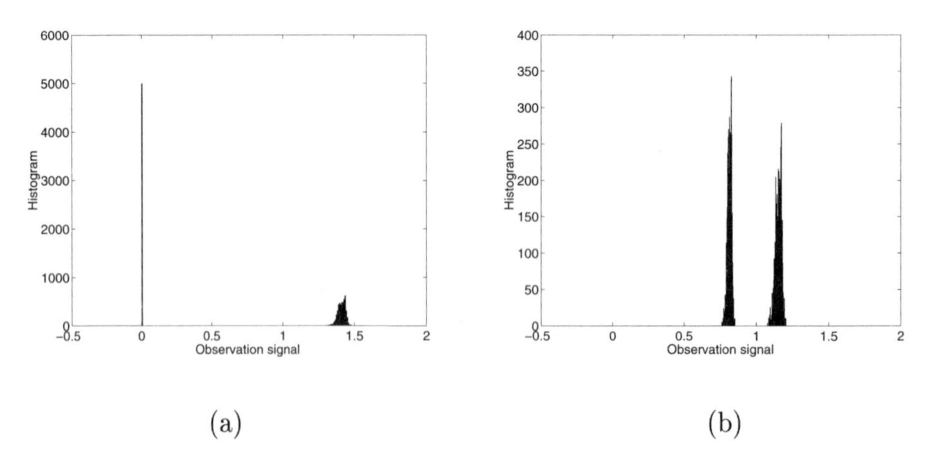

Figure 6.8: Histograms of samples of the observation signal z_{m1} for (a) COOK and (b) unipodal CSK.

The effect of the estimation problem can be seen in Figure 6.8, where the histograms are plotted for the noise-free case. Even in this case, samples of the observation signal do not coincide with E_b but are clustered about them. Due to the nonperiodic property of a chaotic basis function, only its average energy per bit is equal to unity. The evaluation of the integral

$$\int_0^T g_1^2(t)dt$$

can be considered as an *estimation* of the energy per bit of the basis function.

This receiver structure can be used to demodulate both COOK and unipodal CSK. However, because of the $|\cdot|$ function, it cannot be used to recover s_{m1}'s of opposite sign and is therefore unsuitable for demodulating antipodal CSK.

6.4.3 CSK Modulation with Two Basis Functions

Because of the difficulty in recovering chaotic basis functions independently of the modulation, as described in Section 6.4.2.2, an alternative coherent binary CSK communications scheme has been proposed which exploits two basis functions.

In this case, the two elements of the signal set are given by

$$s_m(t) = s_{m1}g_1(t) + s_{m2}(t)g_2(t).$$

A block diagram of a CSK modulator of this type is shown in Figure 6.9. The basis functions $g_1(t)$ and $g_2(t)$ could be derived from two different chaotic sources, or from the same source with different parameters.

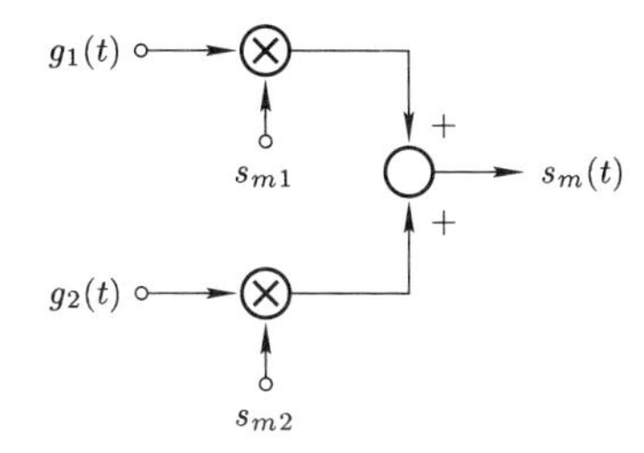

Figure 6.9: CSK modulator with two basis functions.

In a special case of CSK, also called "chaotic switching," the elements of the signal set are weighted basis functions. The transmitted sample functions $s_1(t) = s_{11}g_1(t)$ and $s_2(t) = s_{22}g_2(t)$ (representing symbols "1" and "0," respectively) are the outputs of two free-running chaotic signal generators which produce basis functions $g_1(t)$ and $g_2(t)$.

The signal vectors are $(s_{11}\ s_{12}) = \left(\sqrt{E_b}\ 0\right)$ and $(s_{21}\ s_{22}) = \left(0\ \sqrt{E_b}\right)$, where E_b denotes the average bit energy.

The signal-space diagram for coherent chaotic switching is shown in Figure 6.10. Note that the Euclidean distance between the two message points is $\sqrt{2E_b}$, which is less than that of antipodal CSK with one basis function. This implies that, under the same conditions, the noise performance of coherent chaotic switching is 3 dB worse than that of the antipodal modulation scheme described in Section 6.4.2.2.

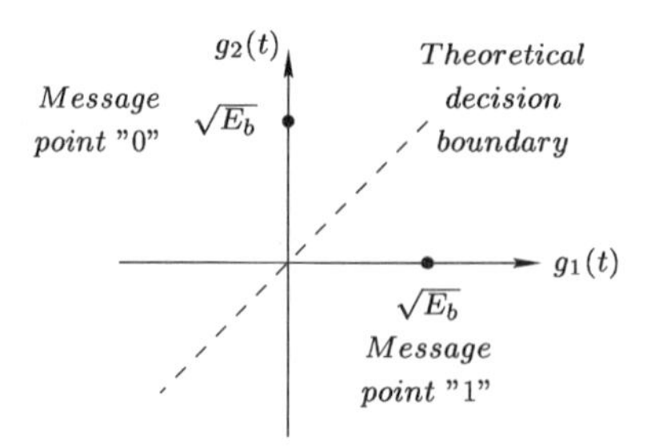

Figure 6.10: Signal-space diagram of coherent chaotic switching.

6.4.4 Demodulation of CSK with Two Basis Functions

A correlation receiver, as shown in Figure 6.11, may be used to recover the elements s_{mj} of the signal vector. In this case,

$$\begin{aligned} z_{m1} &= \int_0^T s_m(t) g_1(t) dt \\ &= \int_0^T [s_{m1} g_1(t) + s_{m2} g_2(t)] g_1(t) dt \\ &= s_{m1} \int_0^T g_1^2(t) dt + s_{m2} \int_0^T g_1(t) g_2(t) dt \end{aligned} \tag{6.3}$$

$$\begin{aligned} z_{m2} &= \int_0^T s_m(t) g_2(t) dt \\ &= \int_0^T [s_{m1} g_2(t) + s_{m2} g_2(t)] g_2(t) dt \\ &= s_{m1} \int_0^T g_1(t) g_2(t) dt + s_{m2} \int_0^T g_2^2(t) dt \end{aligned} \tag{6.4}$$

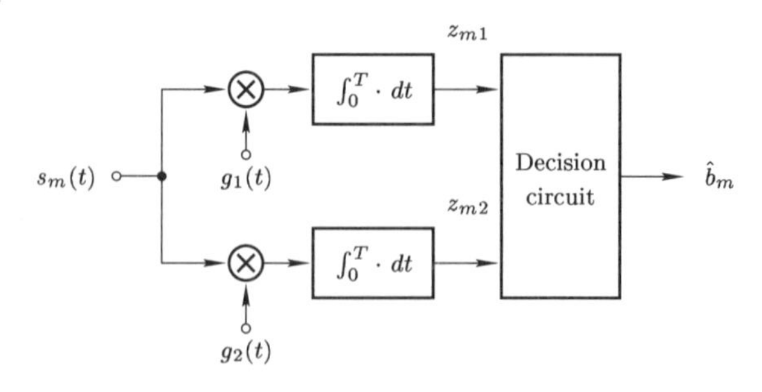

Figure 6.11: Coherent receiver for CSK signals with two basis functions.

Since (6.3) and (6.4) contain both the cross-correlations and autocorrelations of the basis functions, the estimation problem also appears in the coherent demodulation of CSK using two basis functions. Taking account of the orthonormality in the mean of the basis functions $g_1(t)$ and $g_2(t)$, the components s_{mj} of the signal vector can in principle be recovered directly, as follows:

$$z_{m1} \approx s_{m1}$$
$$z_{m2} \approx s_{m2}$$

Demodulation is simplified considerably in the case of chaotic switching, where the outputs of the correlators become

$$z_{11} = s_{11} \int_0^T g_1^2(t)dt \approx \sqrt{E_b}, \qquad z_{12} = s_{12} \int_0^T g_1(t)g_2(t)dt \approx 0$$

when symbol "1" is transmitted, and

$$z_{21} = s_{21} \int_0^T g_1(t)g_2(t)dt \approx 0, \qquad z_{22} = s_{22} \int_0^T g_2^2(t)dt \approx \sqrt{E_b}$$

when symbol "0" is transmitted, assuming once again that the basis functions $g_1(t)$ and $g_2(t)$ are approximately orthonormal in the interval $[0, T]$.

If the autocorrelation of each $g_j(t)$ with itself in each symbol interval T is larger than the cross-correlation with the other basis function, then the correlation receiver structure may be used to identify the element of the signal set which is most likely to have produced the received signal [11]. In particular, if $z_{m1} > z_{m2}$ then the decision circuit decides in favor of symbol "1"; if $z_{m1} < z_{m2}$ then the decision circuit decides in favor of symbol "0".

6.4.5 The Role of Chaotic Synchronization in Coherent Correlation Receivers

As in the case of a conventional correlation receiver based on synchronization, a local synchronized copy of each basis function $g_j(t)$ *has to be produced* in the receiver using appropriate synchronization circuitry.[1] In the case of chaotic basis functions, this topic is called *chaotic synchronization*.

Synchronizable counterparts of the circuits which produce the basis functions $g_j(t)$ at the transmitter are used to recover the basis functions in a coherent correlation receiver, as shown in Figure 6.12. Here, the received signal $r_m(t)$ tries simultaneously to synchronize all of the "synchronizable chaotic circuits" at the receiver.

For example, assume that the signal $s_m(t) = s_{m1}g_1(t)$ is transmitted. During the synchronization time T_S, which is analogous to the pull-in time in a PLL, the output $\hat{g}_1(t)$ converges to $g_1(t)$. By contrast, $\hat{g}_2(t)$ fails to synchronize with $g_1(t)$. The decision as to which symbol was transmitted is made on the basis of the "goodness" of synchronization. In the ideal case, $\hat{g}_1(t)$ is more strongly

[1] Recall that the weights s_{mj} are recovered by computing $\hat{s}_{mj} = \int_0^T r_m(t)g_j(t)dt$ for $j = 1, 2, \ldots, N$ [11].

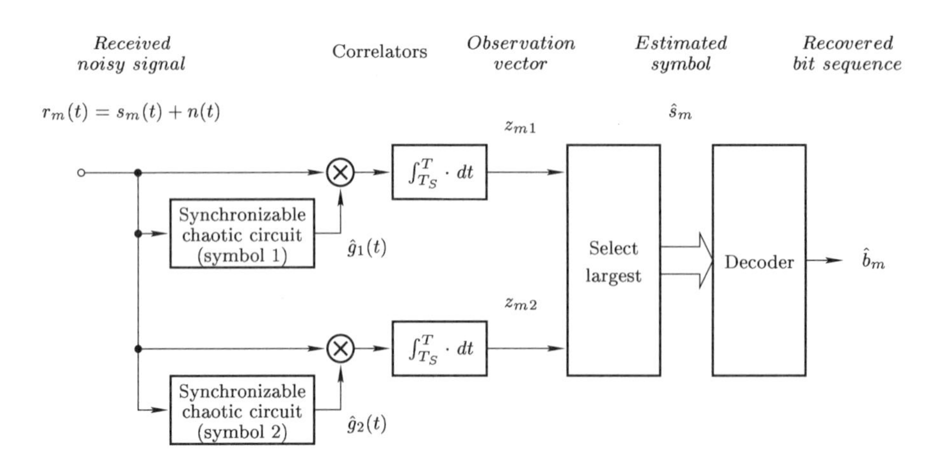

Figure 6.12: Block diagram of coherent correlation CSK receiver.

correlated with $r_m(t)$ than $\hat{g}_2(t)$ during the interval $[T_S, T]$. Hence, $z_{m1} > z_{m2}$ and the decision circuit decides that symbol "1" was transmitted.

In any realistic situation, the received signal is always corrupted by noise $n(t)$. Even in the case of perfect synchronization, the instantaneous value of the received signal may differ considerably from the recovered chaotic signal. This is why correlators *must* be used for the detection, i.e., to determine the "goodness" of synchronization. Because of the time-averaging involved in correlation, the use of correlators also tolerates loss of synchronization for short periods of time. However, any loss in synchronization results in a large degradation in noise performance.

6.4.5.1 Data rate of a coherent correlation receiver with chaotic synchronization for CSK

The disadvantage of the coherent correlation CSK receiver shown in Figure 6.12 is that synchronization is lost and recovered every time the transmitted symbol is changed [8]. The symbol duration is therefore equal to the sum of the synchronization time T_S *plus* the estimation time of the observation vector. The synchronization time puts an upper bound on the symbol rate and thus the data rate.

The other disadvantage of this technique is that the energy transmitted to recover the synchronization is lost from the demodulation point of view. The lost part of the energy per bit results in a degradation in noise performance.

To maximize the data rate in conventional digital systems, synchronization is always maintained. If the transmitted signal does not contain a signal that can be used as a reference for synchronization (in the case of suppressed carrier modulation schemes, for example) then a nonlinear operation is used to regenerate the reference signal at the receiver [12, 13]. This idea could also be exploited in chaotic communications if a synchronization technique could be found which was sufficiently insensitive to some parameter of the chaotic basis functions. In that case, a selected parameter could be varied according to the modulation and synchronization could be maintained continuously. The symbols to be transmitted would then be mapped to the selected parameter of the chaotic sample function and only one attractor would be necessary; this is analogous to a conventional modulation technique where the attractor is a periodic trajectory whose amplitude, frequency, or phase might be controlled by the modulation.

If synchronization of a chaotic circuit could be maintained in the presence of other chaotic signals, then it would be possible to increase the size of the signal set by generating $s_m(t)$ as a weighted sum of basis functions with more than one nonzero weight. To our knowledge, none of the chaotic synchronization techniques which exist in the literature is sufficiently robust to permit augmentation of the signal set in this way [14].

6.5 Differential Chaos Shift Keying

Differential Chaos Shift Keying (DCSK) is a variant of CSK with two basis functions, the important feature of this scheme being that the basis functions consist of repeated segments of a chaotic waveform. This property means that, in addition to a coherent correlation receiver, a simple differentially coherent technique can be used for demodulation.

6.5.1 DCSK Modulation

In binary DCSK, as in the simplest form of CSK with two basis functions described in Section 6.4.3, the two elements of the signal set are given by

$$s_m(t) = s_{m1}g_1(t) + s_{m2}g_2(t) \tag{6.5}$$

where

$$(s_{11} \;\; s_{12}) = \left(\sqrt{E_b} \;\; 0\right) \quad \text{and} \quad (s_{21} \;\; s_{22}) = \left(0 \;\; \sqrt{E_b}\right). \tag{6.6}$$

In the case of DCSK, the basis functions have the special form

$$\begin{aligned} g_1(t) &= \begin{cases} +\frac{1}{\sqrt{E_b}}c(t), & 0 \le t < T/2 \\ +\frac{1}{\sqrt{E_b}}c(t-T/2), & T/2 \le t < T \end{cases} \\ g_2(t) &= \begin{cases} +\frac{1}{\sqrt{E_b}}c(t), & 0 \le t < T/2 \\ -\frac{1}{\sqrt{E_b}}c(t-T/2), & T/2 \le t < T \end{cases} \end{aligned} \tag{6.7}$$

where $c(t)$ is a chaotic waveform. The first half of the basis function is called the reference chip, while the second one is the information-bearing chip. This is shown schematically in Figure 6.13.

In binary DCSK, bit "1" is sent by transmitting $s_1(t) = \sqrt{E_b}g_1(t)$, while for bit "0," $s_2(t) = \sqrt{E_b}g_2(t)$.

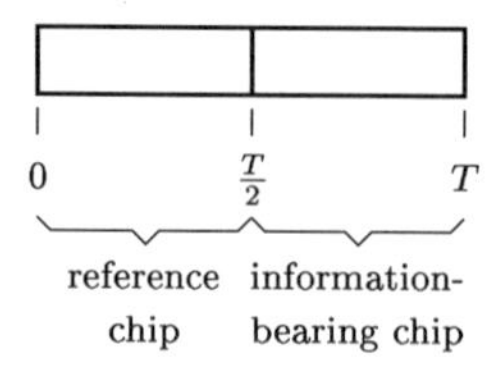

Figure 6.13: Signal $g_j(t)$ consists of two segments, the first of which we call the reference chip and the second the information-bearing chip.

Figure 6.14 shows a block diagram of a DCSK modulator. The modulation driver, delay circuit and switch are used to generate the appropriate basis functions according to the modulation input b_m. A typical DCSK signal corresponding to the binary sequence 1010 is shown in Figure 6.15. In this example, the chaotic signal $c(t)$ was produced by a simulated discrete-time Bernoulli shift generator [15,16] with a clock rate of 40 MHz and with 16-bit amplitude resolution. The bit duration was set to $T = 2$ μs.

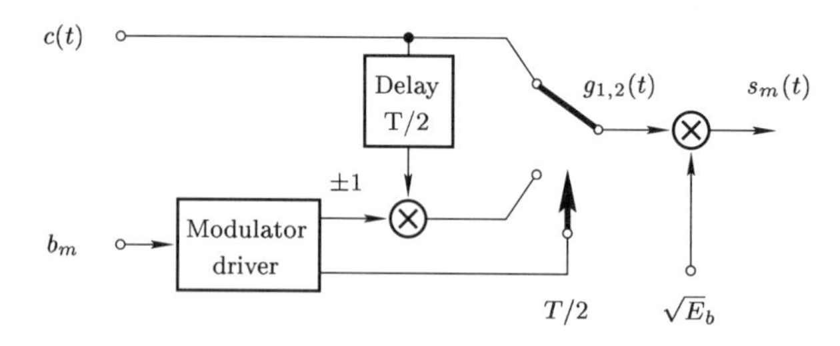

Figure 6.14: Block diagram of a DCSK modulator.

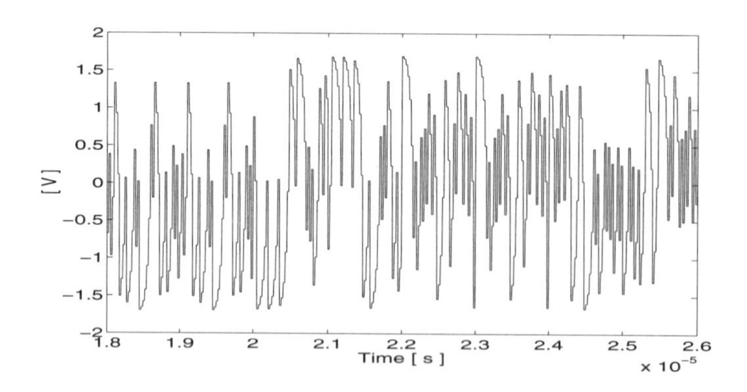

Figure 6.15: DCSK signal corresponding to binary sequence 1010.

6.5.2 DCSK Demodulation

Like any other modulated signals, a DCSK signal can be demodulated using a coherent correlation receiver. The unique property of a DCSK signal that the information is also mapped into the correlation between the reference and information-bearing chips makes it possible, in addition, to demodulate the signal using a differential coherent receiver.

6.5.2.1 Coherent demodulation of DCSK

Since the DCSK modulation scheme is a variant of CSK with two basis functions, it can be demodulated by the coherent receiver shown in Figure 6.11. For bits "1" and "0" the observation vectors become

$$(z_{11} \ z_{12}) \approx \left(\sqrt{E_b} \ 0\right) \quad \text{and} \quad (z_{21} \ z_{22}) \approx \left(0 \ \sqrt{E_b}\right), \tag{6.8}$$

respectively. The decision is a level comparator which compares z_{m1} to z_{m2}.

Equation (6.8) shows that the signal-space diagram plotted in Figure 6.10 for coherent chaotic switching is also valid for coherent DCSK. Note that the Euclidean distance between the two message points is $\sqrt{2E_b}$, which is the same as in conventional coherent FSK.

6.5.2.2 Differentially coherent demodulation of DCSK

We have already explained that recovery of the basis functions by chaotic synchronization in a chaotic correlation receiver is difficult. However, the structure of a DCSK basis function—it consists of a piece of chaotic waveform followed by a non-inverted or inverted copy of itself—makes it possible to perform the demodulation by evaluating the correlation between the reference and information-bearing chips.

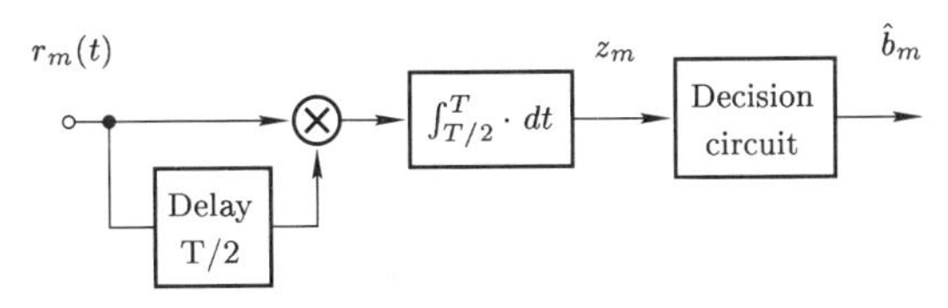

Figure 6.16: Block diagram of a differentially coherent DCSK receiver.

The block diagram of a differentially coherent DCSK receiver is shown in Figure 6.16, where the received signal is delayed by half of the bit duration $(T/2)$ and the correlation between the received signal and the delayed copy of itself is determined. In this case,

$$\begin{aligned} z_m &= \int_{T/2}^{T} s_m(t) s_m(t - T/2) dt \\ &= \int_{T/2}^{T} E_b g_m(t) g_m(t - T/2) dt \end{aligned}$$

Recognizing that $E[\int_{T/2}^{T} g_m^2(t) dt] = 1/2$, we have that

$$z_1 \approx +E_b/2 \quad \text{and} \quad z_2 \approx -E_b/2.$$

The decision as to which symbol was transmitted can be made by a simple level comparator with its threshold set to zero [17].

The signal-space diagram for differentially coherent DCSK is shown in Figure 6.17. Note that the Euclidean distance between the two message points is E_b. This signal-space diagram cannot be compared directly with those we have shown for coherent demodulators because in the differentially coherent DCSK receiver a *noisy* reference signal is correlated with a noisy information-bearing signal, while a noise-free reference is available, in principle, in the coherent case.

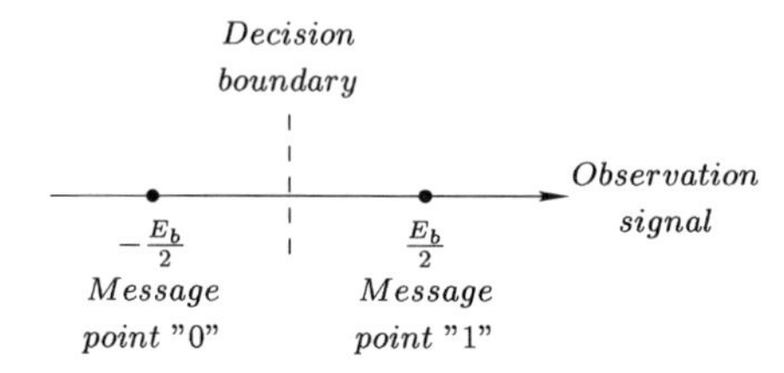

Figure 6.17: Signal-space diagram of differentially coherent DCSK.

Figure 6.18 shows a plot of the correlator output for the transmitted signal shown in Figure 6.15. The observation variable z_m is positive at the decision time instants $10T$ and $12T$, and is negative at $11T$ and $13T$, allowing the transmitted bit sequence (1010) to be recovered.

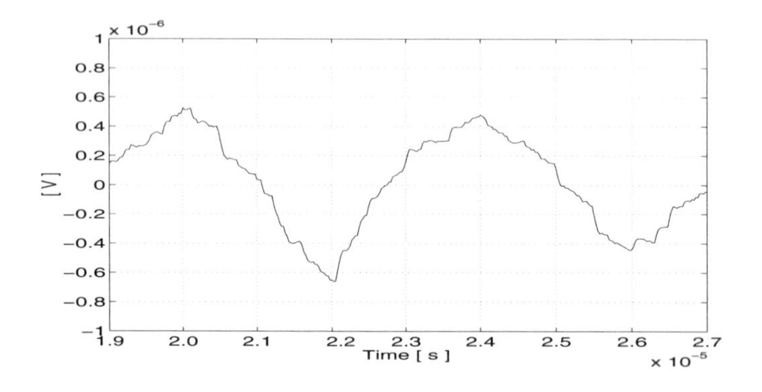

Figure 6.18: Correlator output corresponding to sequence 1010.

The histogram of samples of the observation signal z_m in the absence of channel noise is shown in Figure 6.19. Note that the samples are clustered around $+E_b/2$ and $-E_b/2$, as expected.

6.5.3 Qualitative Advantages of DCSK Over CSK

An advantage of DCSK results from the fact that the information is carried by the correlation between the reference and information-bearing chips. These sample functions pass through the same channel, thereby rendering the modulation scheme insensitive to channel distortion. DCSK can also operate over a

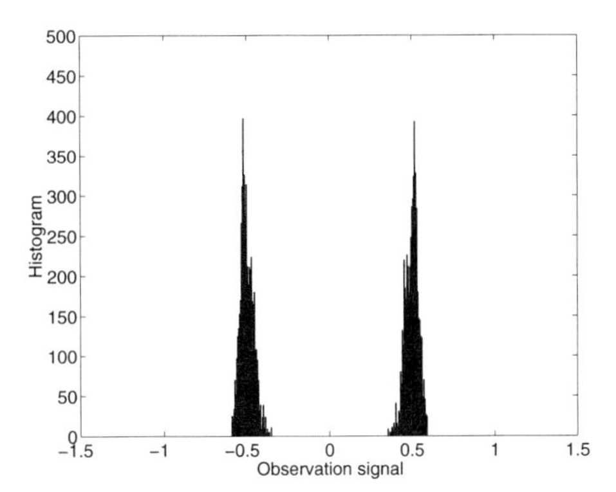

Figure 6.19: Histogram of samples of the observation signal in a noise-free DCSK system.

time-varying channel if the parameters of the channel remain constant for the bit duration T.

Conventional coherent receivers cannot receive pure "1" or "0" bit sequences [12]; an extra circuit called a scrambler has to be used. The differentially coherent reception of DCSK does not suffer from this problem; it can demodulate pure "1" and "0" bit sequences.

Furthermore, the performance of a differentially coherent DCSK system in a noisy channel is insensitive to the precise shape of the transmitted sample functions because only the correlation between the reference and information-bearing chips is used to demodulate the signal. However, a special problem arises in chaotic modulation schemes due to variations in the energy per bit from one symbol to the next. We refer to this as the *estimation problem* [6].

6.5.4 The Estimation Problem

In conventional modulation schemes using periodic basis functions, $g_1(t)$ is periodic and the bit duration T is an integer multiple of the period of the basis function; hence, $\int_T g_1^2(t)dt$ is constant.

By contrast, chaotic signals are inherently nonperiodic and $g_1(t)$ is different in every interval of length T. Consequently, $\int_T g_1^2(t)dt$ is different for every symbol.

Figures 6.20 and 6.21 show histograms of samples of $\int_T g_1^2(t)dt$ for periodic and chaotic waveforms $g_1(\cdot)$, respectively. In the periodic case, all samples lie at $E[\int_T g_1^2(t)dt]$ with zero variance. By contrast, the samples in the chaotic case are centered at $E[\int_T g_1^2(t)dt]$ with non-zero variance.

Let the equivalent statistical bandwidth[2] [18] of the chaotic signal be defined

[2]In a chaotic stochastic process, the ensemble of sample functions is generated by the same chaotic attractor starting from all possible initial conditions [6].

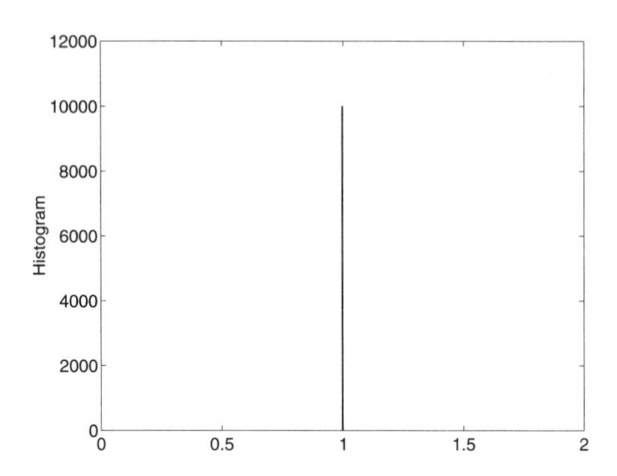

Figure 6.20: Samples of $\int_T g_1^2(t)dt$ for periodic basis function $g_1(t)$.

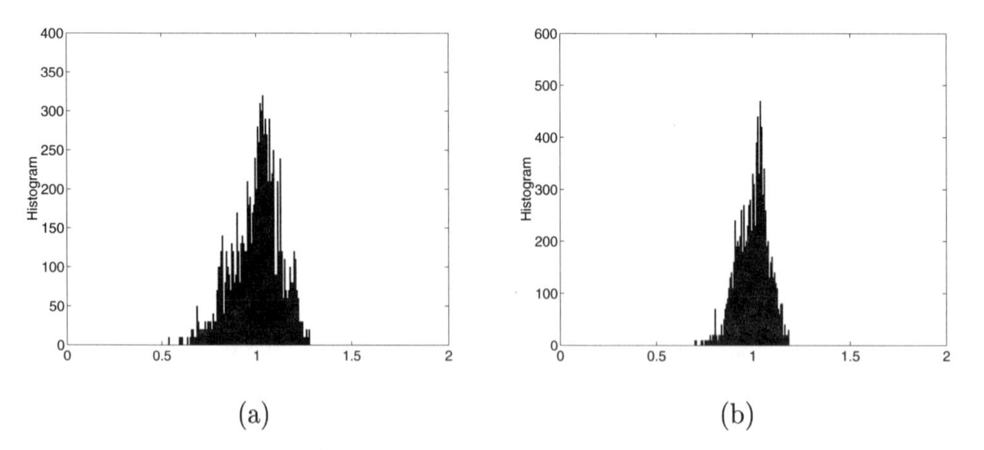

Figure 6.21: Samples of $\int_T g_1^2(t)dt$ for chaotic basis function $g_1(t)$ with (a) short and (b) long bit duration.

by

$$BW_{eq} = \frac{1}{S_C(0)} \int_{-\infty}^{\infty} S_C(f)df$$

where $S_C(f)$ is the power spectral density associated with the stationary chaotic stochastic process [6]. Then the standard deviation of samples of $\int_T g_1^2(t)dt$ scales approximately as $1/(TBW_{eq})$, as shown in Figure 6.22 [6]. Note that the variance of estimation can be reduced by increasing the statistical bandwidth of the transmitted chaotic signal or by increasing the bit duration T [18].

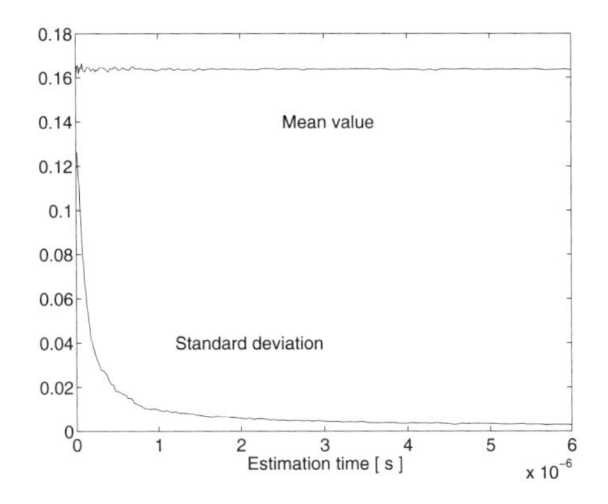

Figure 6.22: Mean and standard deviation of the estimation of $\int_T g_1^2(t)dt$ versus the estimation time.

Since the bandwidth of the signal is restricted to an allocated region of the radio frequency spectrum, the only way to meet a given specification for BER and E_b/N_0 is to increase the bit duration, thereby placing an unnecessarily restrictive upper bound on the data rate.

One way to improve the data rate is to use a multilevel modulation scheme, examples of which are described in [19]. Alternatively, one may solve the estimation problem directly by modifying the modulation scheme such that the transmitted energy for each symbol is kept constant. FM-DCSK [20], which will be discussed in detail in the next section, is an example of the latter approach.

6.5.5 FM-DCSK

The objective of FM-DCSK is to generate a wide-band DCSK signal with constant bit energy E_b. The instantaneous power of an FM signal does not depend on the modulation, provided that the latter is slowly-varying compared to the carrier. Let the chaotic signal be the input of an FM modulator. If the wide-band output of the FM modulator is varied using the DCSK technique, then the correlator output at the receiver has zero variance in the noise-free case and the estimation problem is solved.

Note that, as in the DCSK technique, every information bit is transmitted in two pieces; the first sample function serves as a reference, while the second one carries the information. The operation of the modulator is the same as in DCSK, the only difference being that not the chaotic signal itself, but the FM modulated signal is the input to the DCSK modulator (see Figure 6.23).

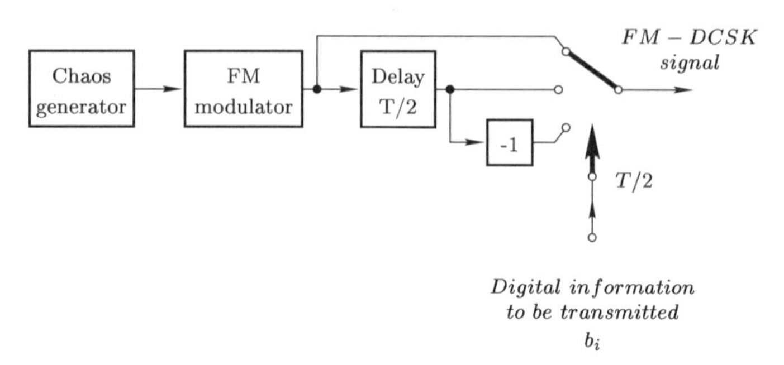

Figure 6.23: Block diagram of an FM-DCSK modulator.

The input of the FM modulator is a chaotic signal, which can be generated by any chaotic circuit.

Note that the only difference between the DCSK and FM-DCSK modulation schemes is that FM modulated chaotic signals, rather than the chaotic signals themselves, are used as basis functions in FM-DCSK. The advantage of this is that, like conventional modulation schemes, the energy of the basis function(s) is constant, i.e., it is always unity. Consequently, the estimation problem does not appear. The same idea can be applied to the CSK modulation schemes in which only one basis function is used. If this is done, the estimation problem can be eliminated in both the COOK and unipodal CSK modulation schemes.

The simplest FM-DCSK receiver can be built by using the differentially coherent demodulator shown in Figure 6.16. Note that in FM-DCSK the correlation between the reference and information-bearing chips of the FM-modulated signal carries the information; hence there is no need for an FM demodulator at the receiver. The input of the differentially coherent DCSK demodulator is the FM-DCSK signal in this case. For more details on the implementation of an FM-DCSK system and its detailed performance evaluation, see Chapter 7.

The use of an FM chaotic signal as the basis function has a very important effect on the implementation of a chaotic communications system. Chaotic signals are almost exclusively low-pass signals. However, only a band-pass RF signal can be used in radio communications. This is why in the case of most chaotic modulation schemes the output of the modulator is a low-pass signal and an extra conventional RF modulator is necessary to generate the RF band-pass signal to be transmitted.

However, if an FM modulated chaotic signal is used as the carrier, as in the case of FM-DCSK, the output of the chaotic modulator is an RF band-pass signal and there is no need for an extra RF modulator and demodulator at the transmitter and receiver, respectively.

6.6 Chaotic Communication Architectures: A Qualitative Comparison

Noise performance is the most important characteristic of a modulation scheme and receiver configuration. Since all of the chaotic modulation techniques discussed in this chapter can be considered under the unifying umbrella of the basis function approach, we also consider their noise performance in this context.

In the previous sections, we saw that chaotic transmissions (CSK and DCSK) can be demodulated using correlation in one of three ways:

- a coherent correlation receiver, where the basis functions are recovered by synchronization,
- a noncoherent correlation receiver, where the basis functions are derived directly from the received signal, and
- a differentially coherent receiver, where the information is carried by the correlation between the reference and information-bearing chips.

The difference between these schemes is primarily in the manner in which the reference signals are generated for the correlators. Therefore, we will analyze all three receiver configurations using the common block diagram shown in Figure 6.24. Because the channel (selection) filter plays an important role in the DCSK receiver, it is also included explicitly in this figure.

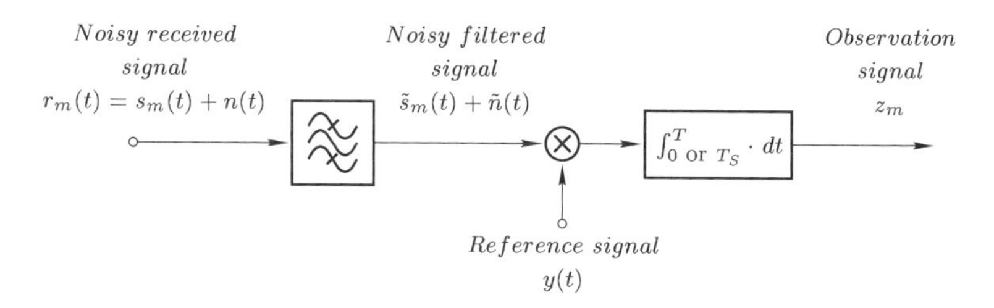

Figure 6.24: General block diagram of a digital chaotic communications receiver.

Comparing the coherent receivers developed for one (see Figure 6.6) and two (see Figure 6.11) basis functions, we conclude that the general block diagram of Figure 6.24 is valid for both cases. The only difference is that in the case of two basis functions two correlators have to be used.

For simplicity in the following, we consider only a binary transmission system with a single basis function $g_1(t)$ and we assume that the bit duration is sufficiently long that the effect of the estimation problem can be neglected. The elements of the signal set are given by $s_m(t) = s_{m1}g_1(t)$. We denote by $y(t)$ and $\tilde{r}_m(t)$, respectively, the reference signal and the filtered version of the noisy received signal which emerges from the channel filter.

In this case, we have

$$\begin{aligned}\tilde{r}_m(t) &= \tilde{s}_m(t) + \tilde{n}(t),\\ &= s_{m1}\tilde{g}_1(t) + \tilde{n}(t)\end{aligned}$$

where $\tilde{g}_1(t)$ is the filtered version of the basis function at the receiver.

The decision is performed based on the observation signal z_m. The probability of wrong decisions, and therefore the BER, depends on the mean value and variance of the observation signal [13].

6.6.1 Noncoherent Correlation Receiver

In a noncoherent correlation receiver, the reference signal $y(t)$ is equal to the noisy filtered signal $\tilde{s}_m(t) + \tilde{n}(t)$, and the observation signal can be expressed as

$$\begin{aligned}z_m &= \int_0^T [\tilde{s}_m(t) + \tilde{n}(t)]^2 dt\\ &= \int_0^T \tilde{s}_m^2(t)dt + 2\int_0^T \tilde{s}_m(t)\tilde{n}(t)dt + \int_0^T \tilde{n}^2(t)dt \qquad (6.9)\\ &= s_{m1}^2 \int_0^T \tilde{g}_1^2(t)dt + 2s_{m1}\int_0^T \tilde{g}_1(t)\tilde{n}(t)dt + \int_0^T \tilde{n}^2(t)dt.\end{aligned}$$

In the noise-free case, if the signal $s_m(t)$ emerges unchanged from the channel then the observation signal is equal to the bit energy E_b of the transmitted symbol, i.e.,

$$z_m = s_{m1}^2 \int_0^T \tilde{g}_1^2(t)dt \approx E_b.$$

Filtering and distortion of the signal in the channel [such that $\tilde{g}_1(t) \neq g_1(t)$] and the presence of noise in the channel cause z_m to differ from E_b. In particular, z_m is a *random variable* which depends on the bit energy, filtering, distortion, and noise in the channel. The mean value of z_m depends on both the bit energy of the chaotic signal and the filtered noise [see the first and third terms in (6.9), respectively]. Consequently, the receiver is a *biased estimator*; the threshold level of the comparator used as a decision circuit depends explicitly on the noise level. In addition, for low and moderate SNR the variance of estimation is large due to the third term in (6.9).

The histogram of observed values of z_m is plotted in Figure 6.25 for a large number of transmitted symbols. Because binary modulation is used, the histogram has two distinct peaks. For a given noise level, channel filter, and chaotic signal, the best noise performance can be achieved if the distance between the two peaks is a maximum. The separation of the peaks is determined by the distance between the elements of the signal set. Note, in particular, that the first term in (6.9) is *always positive,* so the mean values of both peaks lie on the same side of zero.

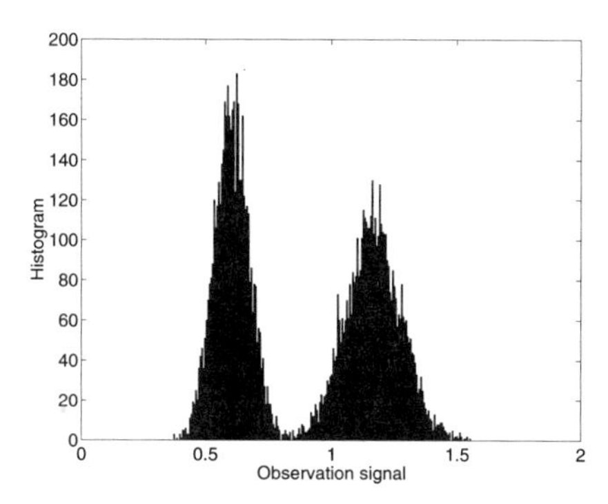

Figure 6.25: Histogram of samples of z_m in a noncoherent correlation receiver.

The maximum distance between the mean values of the observation signals occurs for COOK. In this case, the distance between the elements of the signal set is equal to twice the mean value of the energy per bit.

A noncoherent correlation receiver has a serious problem: it is a biased estimator, which makes the implementation of the decision circuit very difficult.

Is there a way to produce an unbiased estimator which ensures the maximum distance between the elements of a binary signal set, while maintaining similar bit energies? The answer is yes; coherent CSK and DCSK offer potential solutions.

6.6.2 Coherent Correlation Receiver with Chaotic Synchronization

In a coherent correlation receiver, the basis functions $g_j(t)$ are recovered by synchronization from the noisy filtered received signal. The chaotic synchronization techniques which have been published to date are sensitive to both noise and distortion in the channel. In particular, the basis functions $g_j(t)$ cannot be recovered exactly when $r_m(t) \neq s_m(t)$.

Therefore, let $\hat{g}_1(t)$ denote the recovered chaotic basis function, where $\hat{g}_1(t) \approx \tilde{g}_1(t)$ if $t > T_S$.[3] This corresponds to our reference signal $y(t)$ in Figure 6.24.

As explained in Section 6.4.5, we assume that synchronization is lost and recovered at the beginning of every new symbol. Since the synchronization transient cannot be used to transmit information, the observation variable must be estimated during the interval $T_S < t \leq T$. Let $s_m(t), m = 1, 2$, denote the elements of the signal set for binary CSK modulation with a single basis function $g_1(t)$.

[3]Recall that T_S is the synchronization time.

Then the observation signal is given by:

$$\begin{aligned} z_m &= \int_{T_S}^{T} [\tilde{s}_m(t) + \tilde{n}(t)]\hat{g}_1(t)dt \\ &= \int_{T_S}^{T} [s_{m1}\tilde{g}_1(t) + \tilde{n}(t)]\hat{g}_1(t)dt \\ &= s_{m1}\int_{T_S}^{T} \tilde{g}_1(t)\hat{g}_1(t)dt + \int_{T_S}^{T} \tilde{n}(t)\hat{g}_1(t)dt. \end{aligned} \tag{6.10}$$

Note that z_m is again a *random variable*, whose mean value depends on the bit energy of the chaotic signal and the "goodness" of synchronization [see the first term in (6.10)].

The energy of the transmitted signal $s_m(t)$ corresponding to one bit is defined by

$$E_b = \int_0^T s_1^2(t)dt = s_{m1}^2 \int_0^T g_1^2(t)dt$$

In a noise-free channel with exact recovery of the basis function and permanent synchronization, $\tilde{g}_1(t) = g_1(t)$, $\hat{g}_1(t) = g_1(t)$, and $T_S = 0$. The observation variable in this case,

$$z_m = s_{m1}\int_0^T g_1^2(t)dt \approx s_{m1},$$

is proportional to the square root of the bit energy.

In general, z_m is an estimator of $s_{m1}\int_0^T g_1^2(t)dt$. The variance of the estimation is determined by the chaotic signal and the filtered noise. Note that the noise has no *direct* influence on the variance of estimation. As shown by the second term of (6.10), the variance of estimation is influenced only by the cross-correlation of the noise and the recovered chaotic signal.

For a given chaotic signal and bandwidth of the channel filter, the variance of estimation is inversely proportional to the observation time $(T - T_S)$. The mean value of estimation does not depend on the noise; thus, the receiver is an *unbiased estimator*. In particular, this means that the threshold level required by the level comparator does not depend on the channel noise.

As in the case of a conventional receiver and periodic basis functions, the noise performance of a coherent correlation receiver using chaotic basis functions is theoretically excellent. However, the BER also depends on the "goodness" of synchronization [equivalently, the closeness of the reference signal $\hat{g}_1(t)$ to the desired chaotic basis function $g_1(t)$]. Any synchronization error, especially loss of synchronization, results in a large degradation in the noise performance of a correlation receiver. Recall that a digital communications link is automatically severed at the system level if the BER increases above a pre-determined threshold. Therefore, synchronization-based receivers are not suitable for noisy propagation environments.

The synchronization time is another factor which degrades the noise performance of this system. During the synchronization transient, information cannot be transmitted; therefore this component of the bit energy is completely lost and results in reduced noise performance.

Although a coherent CSK correlation receiver is superior to a noncoherent receiver in the sense that it provides an unbiased estimator, its performance depends critically on the ability to regenerate the basis functions at the receiver. Existing chaos synchronization techniques are not sufficiently robust for a practical communications system, nor are we aware of a synchronization scheme which can permit antipodal modulation with a single basis function. By contrast, the differentially coherent detection of DCSK and FM-DCSK provides an unbiased estimator which allows us to use a simple zero-threshold comparator as a decision circuit. In this case, the distance between the two peaks of histogram of observation signal is as large as E_b.

6.6.3 Differentially Coherent DCSK and FM-DCSK Receivers

In a binary differentially coherent DCSK or FM-DCSK receiver, the reference signal $y(t)$ is a delayed version of the filtered noisy signal. Note that different sample functions of filtered noise corrupt the inputs of the correlator. If the time-varying channel varies slowly compared to the symbol rate, then the observation signal is

$$z_m = \int_{T/2}^{T} [\tilde{s}_m(t) + \tilde{n}(t)][\tilde{s}_m(t - T/2) + \tilde{n}(t - T/2)]dt. \tag{6.11}$$

Substituting (6.6) and (6.7) into (6.5) the filtered DCSK or FM-DCSK signal is obtained as

$$\tilde{s}_m(t) = \begin{cases} \tilde{c}(t), & 0 \leq t < T/2 \\ (-1)^{m+1}\tilde{c}(t), & T/2 \leq t < T. \end{cases} \tag{6.12}$$

To get the observation signal, (6.12) has to be substituted into (6.11)

$$\begin{aligned} z_m =&(-1)^{m+1} \int_{T/2}^{T} \tilde{c}^2(t)dt + \int_{T/2}^{T} \tilde{n}(t)\tilde{c}(t - T/2)dt \\ &+ (-1)^{m+1} \int_{T/2}^{T} \tilde{c}(t)\tilde{n}(t - T/2)dt + \int_{T/2}^{T} \tilde{n}(t)\tilde{n}(t - T/2)dt, \end{aligned} \tag{6.13}$$

where the fact that $\tilde{c}(t) = \tilde{c}(t - T/2)$ in the interval $[T/2, T]$ has been exploited. The signals $\tilde{n}(t)$ and $\tilde{n}(t - T/2)$ denote the sample functions of filtered noise that corrupt the reference and information-bearing parts of the received signal, respectively.

For DCSK and FM-DCSK, respectively, the mean value of first term and first term, respectively, are equal to $\pm E_b/2$ in (6.13). The last three terms containing the filtered channel noise are zero mean. This shows that the receiver is an *unbiased estimator* in this case, i.e., the threshold level of the decision circuit is zero and is independent of the noise level.

Consider the DCSK modulation scheme first. In this case the mean value of the estimation depends on the bit energy of the chaotic signal; the variance of estimation is determined by the chaotic signal and the filtered noise. This is why the estimation problem also appears in DCSK.

For FM-DCSK the first term in (6.13) is equal to $\pm E_b/2$ and the estimation problem does not appear. This is why the noise performance of FM-DCSK is superior among the chaotic modulation schemes.

6.6.4 Coherent CSK versus DCSK and FM-DCSK

The main advantage of DCSK and FM-DCSK over CSK is that both the reference and information-bearing components of the transmitted signal pass through the *same* channel so they undergo the same transformation. This transformation does not change the correlation that carries the information, provided that the time-varying channel remains almost constant during the symbol period.

Because there is no need for synchronization, differentially coherent DCSK and FM-DCSK receivers can be used even under poor propagation conditions. However, the symbol rate is halved compared with a synchronization-based receiver in which synchronization is always maintained.

Recall, however, that the synchronization time T_S of a coherent receiver is wasted—no information can be carried during this interval. If each symbol must be synchronized independently and the synchronization time T_S is comparable to the correlation time $(T - T_S)$, then DCSK and FM-DCSK systems can in principle operate at the *same symbol rate* as a synchronization-based coherent receiver, with the added advantage of superior performance under poor propagation conditions.

Thus, synchronization-based recovery of chaotic basis functions from a noisy received signal offers superior performance to DCSK and FM-DCSK, in terms of data rate, only if synchronization can be maintained. This advantage is lost if the modulation technique requires the loss and recovery of synchronization at the beginning of every new symbol or if poor propagation conditions make it impossible to maintain synchronization.

6.7 Noise Performance

In this chapter so far, we have described existing chaotic modulation schemes, have provided a qualitative comparison of their noise performance, and have discussed the estimation problem which results from the nonperiodic nature of chaotic signals. Next, we plot noise performance curves for the most important chaotic modulation schemes and show how the estimation problem degrades the noise performance of CSK, COOK, and DCSK systems.

We assume in this section that the low-pass outputs of the CSK, COOK and DCSK modulators are converted into the RF domain by a multiplier, i.e., double sideband-suppressed carrier (DSB-SC) modulation is the secondary modulation scheme. Every noise performance curve shown here was determined by computer simulation. A detailed description of the simulator can be found in Chapter 7.

As expected, the worst noise performance is obtained using noncoherent unipodal CSK+AM/DSB-SC modulation (for the block diagram of the demodulator, see Figure 6.7). The noise performance is shown in Figure 6.26, where one basis function with weighting factors of $s_{11} = 0.8165$ and $s_{21} = 1.1547$ was used. The ratio of the two transmitted bit energies is 2 in this case. The bit duration, 2 μs, is sufficiently long that the estimation problem does not manifest itself in the plotted region. Figure 6.26 shows that the E_b/N_0 has to be about 24 dB at the demodulator input to obtain a BER of 10^{-3}. This value is unacceptably high for practical telecommunications applications.

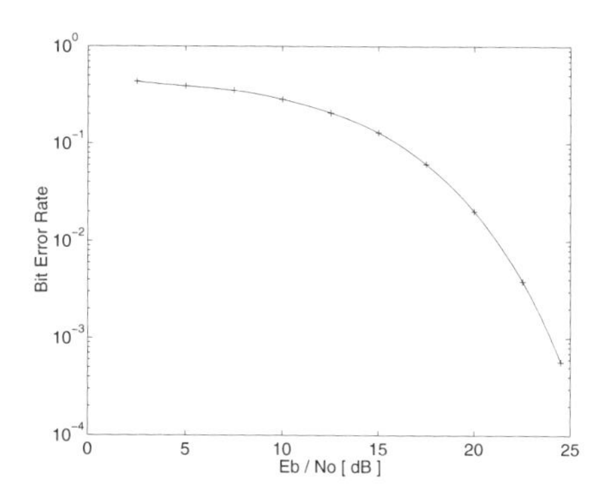

Figure 6.26: Noise performance of the noncoherent unipodal CSK+AM/DSB-SC modulation scheme for $s_{11} = 0.8165$ and $s_{21} = 1.1547$.

COOK can be considered as a special case of CSK where the separation between the message points on the signal-space diagram is maximized. The noise performance of the COOK+AM/DSB-SC modulation scheme for 0.25 μs and 2 μs bit durations is shown in Figure 6.27. Note that, if the bit duration is too short, the noise performance of COOK is degraded significantly due to the estimation problem.

The noise performance of COOK is acceptable: an E_b/N_0 ratio of about 13.5 dB is required to achieve a BER of 10^{-3}. However, this noise performance can be achieved only if the threshold of the decision device is varied adaptively as a function of E_b/N_0. The optimum threshold is plotted in Figure 6.28 as a function of E_b/N_0.

Proper control of the decision threshold requires a continuous determination of E_b/N_0 at the receiver which, in turn, increases the complexity of the demodulator circuitry. If the probabilities of symbols "0" and "1" are both 0.5 then, according to Bayes' rule, the decision threshold has to be set to the crossing point of the two histograms.

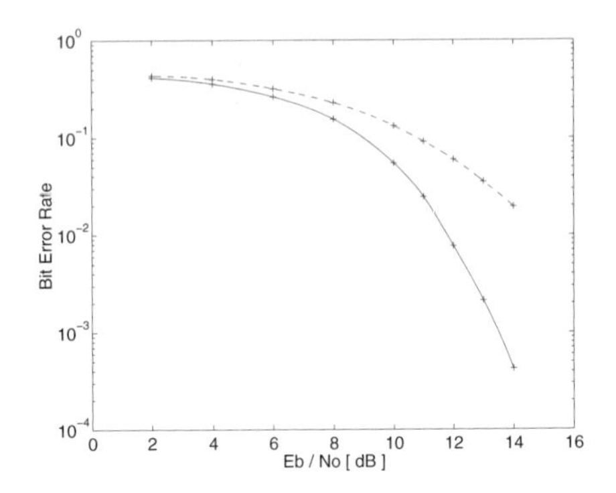

Figure 6.27: Noise performance of the COOK+AM/DSB-SC modulation scheme for bit durations of 0.25 μs (dashed) and 2 μs (solid).

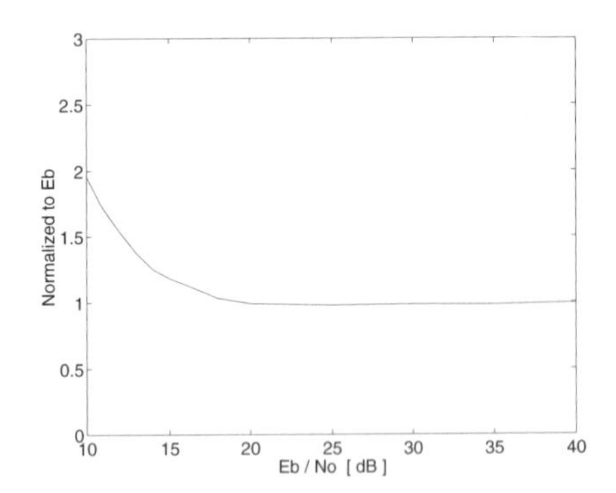

Figure 6.28: Optimum decision threshold as a function of E_b/N_0.

For example, the histogram of the observation signal is plotted in Figure 6.29 for the case $E_b/N_0 = 12$ dB. If $E_b = 1$ then the optimum decision threshold in this case is 1.54.

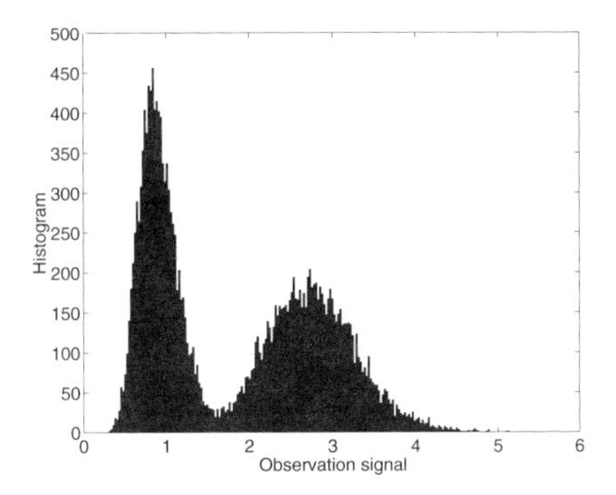

Figure 6.29: Determination of the optimum decision threshold for E_b/N_0=12 dB.

There is no need for adaptive threshold control in the DCSK modulation scheme. The noise performance of a DCSK+AM/DSB-SC system for bit durations of 0.25 μs and 2 μs is shown in Figure 6.30.

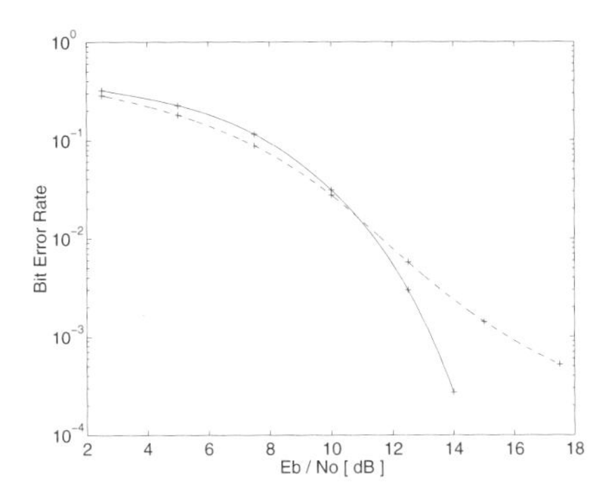

Figure 6.30: Noise performance of the DCSK+AM/DSB-SC modulation scheme for bit durations of 0.25 μs (dashed) and 2 μs (solid).

The noise performance of DCSK+AM/DSB-SC is equal to that of COOK+AM/DSB-SC, i.e., if E_b/N_0 is equal to 13.5 dB, then the BER is about 10^{-3}. Note that the estimation problem also has a very strong influence on the noise performance in this case and limits the attainable data rate of the telecommunications system.

The output of the FM-DCSK modulator is a band-pass RF signal. Hence, there is no need for an extra RF modulator at the transmitter nor for an extra RF demodulator at the receiver. The other advantage of FM-DCSK is that the estimation problem does not appear.

The noise performance of FM-DCSK is plotted in Figure 6.31 for bit durations of 0.25 μs and 2 μs. To achieve a BER of 10^{-3}, the E_b/N_0 has to be about 13 dB at the input of the demodulator.

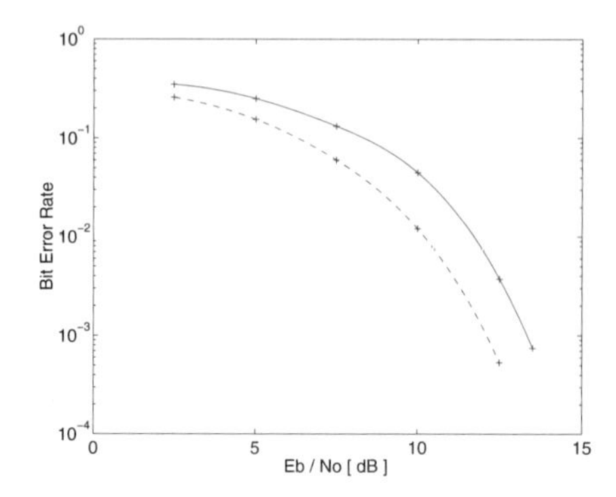

Figure 6.31: Noise performance of the FM-DCSK modulation scheme for bit durations of 0.25 μs (dashed) and 2 μs (solid).

Figures 6.30 and 6.31 show that the noise performance of these modulation schemes depends on the bit duration. This effect is discussed in detail and further data on the noise and multipath performance of the FM-DCSK scheme and the effects of the main system parameters on the overall system performance are presented in Chapter 7.

6.8 Summary

In this chapter, we have described digital chaotic modulation schemes using the basis functions approach.

Each element of the signal set in CSK is a weighted sum of one or more chaotic basis functions. The optimum CSK modulation scheme using a single basis function is antipodal CSK.

We have noted the following properties of CSK receivers:

- A matched filter receiver structure cannot be used because of the fact that the reference sample function differs from one symbol to the next.

- A noncoherent correlation receiver can be used to demodulate unipodal CSK and COOK, but not the more efficient antipodal CSK.

- A coherent correlation receiver can be used to demodulate COOK, unipodal CSK and antipodal CSK, provided that the basis function(s) can be recovered at the receiver independently of the modulation. A robust coherent CSK receiver structure is not known; all known coherent CSK receivers perform worse than the best noncoherent correlation receivers.

- A CSK communications scheme using two basis functions has been demonstrated but it achieves its optimum performance only when the basis functions are orthogonal and synchronization is maintained throughout the transmission. Neither of these problems has been resolved satisfactorily in the literature.

In DCSK, each element of the signal set is represented by a chaotic sample function followed by an inverted or noninverted copy of that function. Differentially coherent DCSK communications systems have the following properties:

- Synchronization between the transmitter and receiver is not required.

- The threshold of the decision circuit is always zero, independently of the SNR at the input of the demodulator.

The nonperiodic nature of chaotic signals causes a degradation in the noise performance and imposes an upper bound on the data rate of every digital chaotic communications system unless the energy per bit is kept constant for each bit; this is called the estimation problem.

By maintaining a constant energy per symbol, FM-DCSK can achieve the noise performance of DCSK, without an artificial upper bound on the data rate. FM-DCSK has the best noise performance of all known digital chaotic modulation schemes.

Enhanced versions of DCSK have been proposed to improve its performance. In [19], a multilevel DCSK modulation scheme has been proposed. By sending more than one information-bearing chip after each reference chip and applying averaging, the noise performance and data rate of DCSK can be improved [21]. A limited code division multiple access capability can also be achieved [19]

References

[1] M. P. Kennedy. Three steps to chaos—Part I: Evolution. *IEEE Trans. Circuits and Systems—Part I: Fundamental Theory and Applications, Special Issue on Chaos in Nonlinear Electronic Circuits—Part A: Tutorial and Reviews*, 40(10):640, Oct. 1993.

[2] M. P. Kennedy. Three steps to chaos—Part II: A Chua's circuit primer. *IEEE Trans. Circuits and Systems—Part I: Fundamental Theory and Applications, Special Issue on Chaos in Nonlinear Electronic Circuits—Part A: Tutorial and Reviews*, 40(10):657, Oct. 1993.

[3] R. C. Dixon. *Spread Spectrum Systems with Commercial Applications*. John Wiley & Sons, New York, 3rd edition, 1994.

[4] H. Papadopoulos and G. W. Wornell. Maximum likelihood estimation of a class of chaotic signals. *IEEE Trans. Information Theory*, 41(1), Jan. 1995.

[5] S. H. Isabelle and G. W. Wornell. Statistical analysis and spectral estimation techniques for one-dimensional chaotic signals. *IEEE Trans. Signal Processing*, 45(6), June 1997.

[6] G. Kolumbán, M. P. Kennedy, and G. Kis. Determination of symbol duration in chaos-based communications. In *Proc. Int. Specialist Workshop on Nonlinear Dynamics of Electronic Systems*, page 217, Moscow, June 26–27 1997.

[7] M. P. Kennedy and H. Dedieu. Experimental demonstration of binary chaos shift keying using self-synchronizing Chua's circuits. In A. C. Davies and W. Schwarz, editors, *Proc. Int. Specialist Workshop on Nonlinear Dynamics of Electronic Systems*, page 67, Dresden, 23–24 July 1993.

[8] H. Dedieu, M. P. Kennedy, and M. Hasler. Chaos shift keying: Modulation and demodulation of a chaotic carrier using self-synchronizing Chua's circuits. *IEEE Trans. Circuits and Systems—Part II: Analog and Digital Signal Processing, Special Issue on Chaos in Nonlinear Electronic Circuits—Part C: Applications*, 40(10):634, Oct. 1993.

[9] G. Kis. Required bandwidth of chaotic signals used in chaotic modulation schemes. In *Proc. Int. Specialist Workshop on Nonlinear Dynamics of Electronic Systems*, page 191, Budapest, Hungary, 1998.

[10] G. Kolumbán, M. P. Kennedy, and G. Kis. Performance improvement of chaotic communications systems. In *Proc. European Conf. on Circuit Theory and Design*, pages 284–289, Budapest, Hungary, 30 Aug.–3 Sept. 1997.

[11] G. Kolumbán, M. P. Kennedy, and L. O. Chua. The role of synchronization in digital communications using chaos – Part I: Fundamentals of digital communications. *IEEE Trans. Circuits and Systems–Part I: Fundamental Theory and Applications*, 44(10):927, Oct. 1997.

[12] I. Frigyes, Z. Szabó, and P. Ványai. *Digital Microwave Transmission*. Elsevier Science Publishers, Amsterdam, 1989.

[13] S. Haykin. *Communication Systems*. John Wiley & Sons, New York, 3rd edition, 1994.

[14] G. Kolumbán, H. Dedieu, J. Schweizer, J. Ennitis, and B. Vizvári. Performance evaluation and comparison of chaos communication systems. In *Proc. Int. Specialist Workshop on Nonlinear Dynamics of Electronic Systems*, page 105, Seville, June 27–28 1996.

[15] M. Delgado-Restituto, F. Medeiro, and A. Rodríguez-Vázquez. Nonlinear switched-current CMOS IC for random signal generation. *Electron. Lett.*, 29(25), Dec. 1993.

[16] E. Ott. *Chaos in Dynamical Systems.* Cambridge University Press, Cambridge, 1993.

[17] G. Kolumbán, B. Vizvári, W. Schwarz, and A. Abel. Differential chaos shift keying: A robust coding for chaotic communication. In *Proc. Int. Specialist Workshop on Nonlinear Dynamics of Electronic Systems*, page 87, Seville, June 27–28 1996.

[18] J. S. Bendat and A. G. Piersol. *Measurement and Analysis of Random Data.* John Wiley & Sons, New York, 2nd edition, 1966.

[19] G. Kolumbán, M. P. Kennedy, and G. Kis. Multilevel Differential Chaos Shift Keying. In *Proc. Int. Specialist Workshop on Nonlinear Dynamics of Electronic Systems*, page 191, Moscow, Russia, 1997.

[20] G. Kolumbán, G. Kis, Z. Jákó, and M. P. Kennedy. FM-DCSK: a robust modulation scheme for chaotic communications. *IEICE Trans. on Fundamentals of Electronics, Communications and Computer Sciences*, E81-A(9):1798, Sep. 1998.

[21] G. Kolumbán, Z. Jákó, and M. P. Kennedy. Enhanced versions of DCSK and FM-DCSK data transmission systems. In *Proc. ISCAS'99*, volume IV, page 475, Orlando, USA, June 1999.

Chapter 7

Performance Evaluation of FM-DCSK

Géza Kolumbán
Department of Measurement and Information Systems
Budapest University of Technology and Economics
H-1521 Budapest, Hungary
kolumban@mit.bme.hu

Michael Peter Kennedy
Department of Microelectronic Engineering
University College Cork
Cork, Ireland
Peter.Kennedy@ucc.ie

Gábor Kis
Department of Measurement and Information Systems
Budapest University of Technology and Economics
H-1521 Budapest, Hungary
kisg@mit.bme.hu

7.1 Introduction

Local Area Networks (LANs) are used everywhere from production lines to offices to connect computers and data processing units together. In many applications such as production lines, warehouses, hospitals, reconfigurable office buildings, etc., mobility is the top priority, so the LAN has to be established via a radio connection [1].

In other applications such as shopping centers or universities, a wireless implementation of a LAN provides convenient, reconfigurable access to the data available on computer networks.

In Wireless Local Area Networks (WLANs) a radio connection is used to implement the LAN. In this application multipath propagation limits the overall system performance and the power spectral density of the transmitted WLAN signals is bounded in order to minimize interference with other users in the same frequency band. Spread spectrum systems [2] offer the classical solution to this problem.

Spread Spectrum (SS) technology has been used for many decades in military applications. The U.S. Federal Communications Commission (FCC) has made available three frequency bands, 902–928 MHz, 2400–2483.5 MHz, and 5728–5750 MHz to satisfy the demand for using this technology in commercial applications. In these frequency bands, called Industrial Scientific and Medical (ISM) bands, license and service provider fees are not required and new radio communications techniques for which standards have not yet been approved can be tested.

To provide interoperability among the WLAN systems offered by the different vendors, the IEEE 802.11 WLAN standard has been developed. This standard covers both Direct Sequence (DS) and Frequency Hopping (FH) SS techniques for WLAN radio communications. IEEE 802.11-compliant WLAN systems operate in the 2.4 GHz ISM band. The standard provides rules for both the Media Access Control (MAC) and Physical (PHY) layers of the network.

A shared media protocol called Carrier Sense, Multiple Access, Collision Avoidance (CSMA/CA) is used by the IEEE 802.11 standard. In this CA protocol, collision is avoided by sending a short "ready to send" message first which tells the other nodes the duration and destination of the message to be sent and prevents them from transmitting during this period. The desired receiving station then issues a "clear to send" message. Finally, error-free reception of the message frame is acknowledged by the receiver.

Two radio-frequency (RF) SS techniques and one infrared technology are described in the PHY layer specification of the IEEE 802.11 standard. Both RF SS systems operate in the 2.4 GHz ISM band having an RF bandwidth of 83 MHz. The typical data rate is 1 Mb/s.

In the FH-SS, seventy-five frequency hopping channels with a separation of 1 MHz have been assigned. The hopping rate must be at least 2.5 hop/s. The IEEE 802.11 standard defines three sets of twenty-two hopping patterns and each frequency hopping channel must be used at least once every 30 seconds. Gaussian Frequency Shift Keying (GFSK) modulation is typically used and an entire data frame is sent before the system hops to another channel.

In the DS-SS version, the data stream is spread over a 22 MHz wide frequency band by means of an 11-chip Baker sequence which is the only valid spreading code. Binary Phase Shift Keying (BPSK) and Quadrature Phase Shift Keying (QPSK) modulation are used for 1 Mb/s and 2 Mb/s data rates, respectively. The 83 MHz wide RF band is divided into eleven channels with a channel spacing of 5 MHz.

In chaotic communications systems, the digital information to be transmitted is mapped to inherently wide-band signals. In this sense these systems offer a novel solution for spread spectrum communications.

Of the chaotic modulation schemes published to date, FM-DCSK offers the best noise performance and the best robustness against multipath and channel imperfections. The objective of this chapter is to provide a detailed performance evaluation of the FM-DCSK system.

First the spectral properties of an FM-DCSK signal will be discussed and the properties of fast and slow FM-DCSK systems will be identified.

The performance evaluation of every digital communications system is a very time-consuming task. By developing a discrete-time low-pass model of the FM-DCSK system, we will show how the simulation time can be reduced by a factor which varies from 100 to 1000, depending on the simulation task performed.

Next, the noise performance of FM-DCSK will be evaluated and the effects of the main system parameters on the overall system performance will be determined.

One of the most important potential applications of FM-DCSK is in data communications over multipath channels. In the last section, a tapped delay line model of multipath channels will be introduced and the qualitative and quantitative behavior of FM-DCSK in multipath channels will be determined.

The data of the multipath channel used in the simulations are typical for a WLAN application.

7.2 Spectrum of Transmitted FM-DCSK Signal

The objectives of spreading the transmitted signal in a WLAN application are twofold:

- to overcome the multipath propagation problem and
- to reduce the transmitted power spectral density in order to minimize interference with other radio communications in the same frequency band.

Depending on the parameters of the FM-DCSK modulator we can implement a fast or a slow FM-DCSK system.

The fast and slow FM-DCSK techniques both spread the RF signal to be transmitted and reduce its average power spectral density. These two techniques have the same noise performance, but their influence on other radio links and their multipath performance are qualitatively different.

7.2.1 Fast and Slow Spreading Techniques

In an FM-DCSK modulator, a chaotic signal is fed into an FM modulator to generate a wide-band RF band-pass signal. DCSK modulation is then applied to this wide-band signal.

Let the chaotic spreading signal be generated by a discrete-time chaotic circuit. The output of the chaotic signal generator is converted into the continuous-time domain by a zero-order hold circuit, as shown in Figure 7.1. The hold time is determined by the *chip rate*. Depending on the chip rate T_{chip}, we can implement fast or slow spreading systems. The shape of FM-DCSK spectrum depends on the parameters of chaotic signal, chip rate, gain of FM modulator and bit duration.

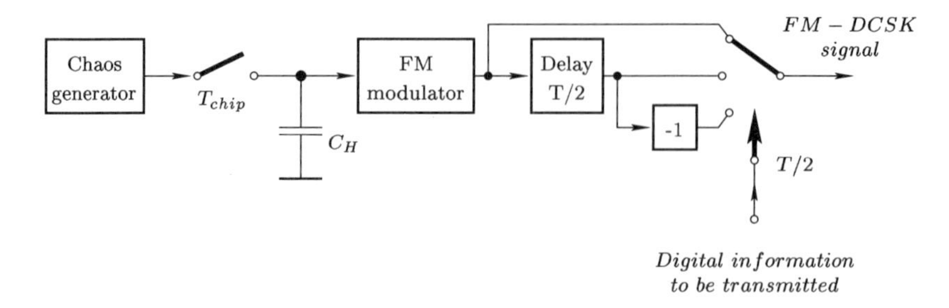

Figure 7.1: Block diagram of FM-DCSK transmitter including the zero-order hold circuit.

To obtain the best multipath and interference performance, the spectra of the FM-DCSK signals must be free from periodic components. In our simulations the chip rates are 20 MHz and 10 MHz, respectively, for the fast and slow FM-DCSK systems, respectively, and the digital information to be transmitted is modeled by a random binary sequence. The center frequency of the FM-DCSK signal is 36 MHz and its RF bandwidth is 17 MHz, corresponding to the IEEE 802.11 WLAN standard. To minimize interference with other radio channels, the power generated outside the 17 MHz RF bandwidth has to be kept as low as possible.

Simulated spectra of the chaotic FM modulator output and the FM-DCSK signal are plotted in Figure 7.2 for the fast FM-DCSK system. To illustrate the relationship between the bit duration and the skirt of the FM-DCSK signal, Figure 7.2(b) shows spectra for $T = 2$ μs and $T = 16$ μs.

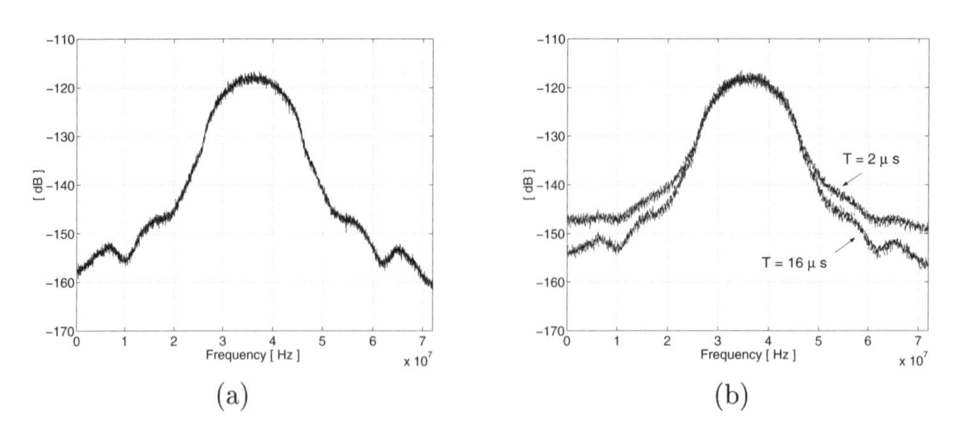

Figure 7.2: Output spectra of the (a) FM and (b) DCSK modulators in a fast FM-DCSK system. T is set to 2 μs and 16 μs to illustrate the effect of bit duration on the height of the skirts.

Inspecting Figure 7.2(b), we can conclude that by choosing appropriate values for the chip rate and peak frequency deviation of the FM modulator, DCSK modulation does not cause periodic components in the spectrum, but it does increase the skirt considerably. The skirt can be lowered by increasing the bit duration, but increasing T reduces the attainable data rate. Alternatively, the unwanted skirt can be suppressed, without reducing the data rate, by means of a band-pass filter at the transmitter.

Simulated spectra of the slow FM-DCSK system are shown in Figure 7.3. Note that the in-band shapes of the spectra of the fast and slow FM-DCSK signals differ considerably from each other. The fast system has a strong curvature in band, while the slow system is significantly flatter. Note again that the effect of DCSK modulation is to raise the skirt of FM-DCSK spectrum.

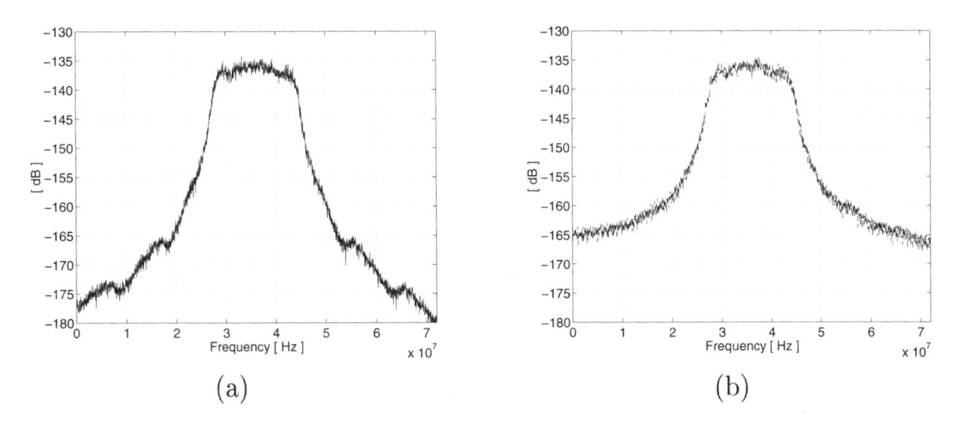

Figure 7.3: Spectra of a slow FM-DCSK system (a) at the output of the FM modulator and (b) at the output of the DCSK modulator.

7.2.2 Differences Between Fast and Slow FM-DCSK Systems

The average power spectral density is reduced in both the fast and slow FM-DCSK, but the ways it is done are very different. This difference has serious implications for other radio communications systems operating in or near the same frequency band.

In the fast FM-DCSK system, the instantaneous frequency of the transmitted signal is constant only for a very short time interval (the chip time). Thus, the transmitted energy is never concentrated about a certain frequency; rather, it is spread continuously over the entire RF channel bandwidth. If this signal interferes with another radio channel in the same band then it simply increases the background "noise level," and so reduces the quality of that channel, but it does not interrupt it. If the power spectral density of the fast FM-DCSK system remains under the noise floor of the other users then its presence may not even be noticed.

In the slow FM-DCSK system, the chip rate is relatively low and the instantaneous frequency of the transmitted signal remains constant for a relatively long time interval. Consequently, the slow FM-DCSK signal appears like a relatively narrow-band signal which changes its center frequency. A narrow-band radio in the same band which is interfered with by this signal will experience a pop or burst of noise when the slow FM-DCSK signal hits its channel. In this case, spreading shares the pain of channel interruption over all narrow-band radio systems using the same frequency band.

Note that in this sense the slow FM-DCSK is qualitatively very similar to the slow Frequency Hopping Spread Spectrum systems [2]. However, there is a fundamental difference: in slow FH-SS, a couple of bits are transmitted at every frequency, while slow FM-DCSK visits many different frequencies during one bit duration. Thus, in the case of slow FH-SS, a couple of bits can be lost due to multipath, but this problem does not appear in slow FM-DCSK.

Figures 7.2(b) and 7.3(b) show that the in-band shapes of the radiated spectra are different for the fast and slow FM-DCSK systems. We will show later that the fast FM-DCSK system performs better than the slow one in the WLAN application because of the difference in the shapes of their spectra.

7.3 Tools for System Performance Evaluation

Chaotic communications systems present several challenges for the system developer. Most importantly, theoretical results for noise performance of FM-DCSK are not yet available. Even if this problem were solved, the performance of an FM-DCSK system could be determined analytically only in the simplest cases. The determination of the system performance under various propagation conditions, the selection of the main system parameters such as bandwidth, bit duration, etc., for different applications, and the support of circuit and subsystem development all require a comparative performance evaluation by computer simulation.

If the computer simulation of a band-pass RF system is performed directly in the RF domain, then the sampling frequency (equivalently, the time-step) of the simulation depends on both the carrier frequency and the bandwidth of the transmitted signal. A high carrier frequency results in a high sampling frequency and consequently a long simulation time. However, the performance of a band-pass system does not depend on the value of the carrier so it should be possible in principle to remove the carrier frequency from the analysis.

Indeed, it is well known that a low-pass equivalent model can be developed for every band-pass system [3]. By developing such an equivalent, the carrier frequency can be removed from the model of an RF communications system and the sampling frequency is then determined exclusively by half the bandwidth of the RF band-pass signal. This reduces significantly the computational effort required to characterize the performance of a chaotic communications system.

In this section, we first summarize the theoretical basis of the analytic signal approach [3] in order to establish the relationship between the original RF band-pass system and its low-pass equivalent. Then the development of a low-pass equivalent model for the RF FM-DCSK system is illustrated. Simulation of a discrete-time equivalent of the corresponding low-pass analog model in a Matlab environment is also explained. Finally, the results of simulations performed on the RF band-pass model and its discrete-time low-pass equivalent are compared to verify the equivalence.

7.3.1 Determination of Low-Pass Equivalent Model

In an FM-DCSK system, shown schematically in Figure 7.4, the low-pass chaotic signal $m(t)$ is fed into an FM modulator to obtain a wide-band RF band-pass signal $y(t)$. The binary information b_i drives the DCSK modulator and the modulator output $s(t)$ passes through the telecommunications channel. The received signal $r(t)$ is applied to the input of the FM-DCSK demodulator. Note that band-pass FM signals, rather than low-pass chaotic signals, are correlated directly in the FM-DCSK demodulator. The output $z(t)$ of the FM-DCSK demodulator is sampled at the decision time instants $iT, i = 1, 2, \ldots$ to produce the observation signal z_i. The estimated bit stream $\hat{b}_i$ is recovered from the observation signal by a level comparator.

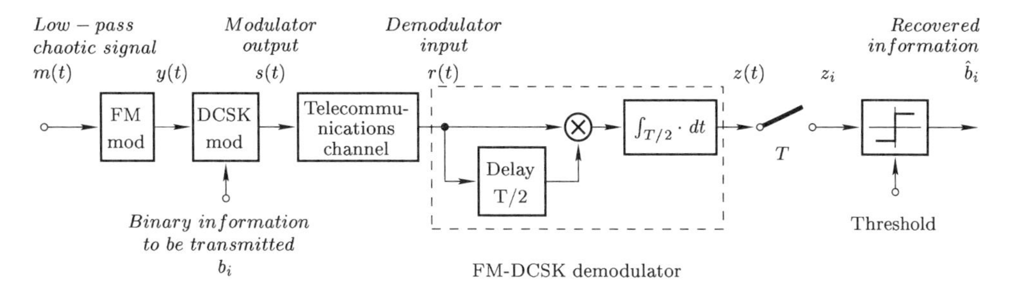

Figure 7.4: Block diagram of an FM-DCSK RF system.

Note that the system can be separated into an analog part and a decision circuit. The input and output of the analog part are the chaotic signal $m(t)$ and the demodulator output signal $z(t)$, respectively. Both the chaotic and observation signals are low-pass signals. During the development of the low-pass equivalent model, the relationship between $z(t)$ and $m(t)$ must be established. Additional system-level problems such as timing recovery and the decision circuit are not considered in this section.

7.3.1.1 Representation of band-pass signals

A signal $x(t)$ is referred to as a *band-pass signal* if its energy is non-negligible only in a frequency band of total extent $2BW$ centered about a carrier frequency f_c. Every band-pass signal can be expressed as a product of a slowly-varying signal $\tilde{x}(t)$ and the carrier

$$x(t) = Re\left[\tilde{x}(t)e^{j\omega_c t}\right], \tag{7.1}$$

where $\tilde{x}(t)$ is called the complex envelope. In general, $\tilde{x}(t)$ is a complex-valued quantity; it can be expressed in terms of its *in-phase* and *quadrature* components as

$$\tilde{x}(t) = x_I(t) + jx_Q(t).$$

Both $x_I(t)$ and $x_Q(t)$ are low-pass signals, limited to the frequency band $-BW \leq f \leq BW$.

Note that the low-pass complex envelope $\tilde{x}(t)$ carries all of the information, except the carrier frequency, of the original band-pass signal $x(t)$. This means that if the complex envelope of a signal is given, then that signal is completely characterized. Knowing the carrier frequency too means that the original band-pass signal can be reconstructed perfectly.

The complex envelope can also be expressed in polar form

$$\tilde{x}(t) = a(t)e^{j\phi(t)}, \tag{7.2}$$

where $a(t)$ and $\phi(t)$ are both slowly-varying real-valued low-pass functions.

The in-phase and quadrature components, $x_I(t)$ and $x_Q(t)$, respectively, of the complex envelope can be generated from the band-pass signal $x(t)$ using the scheme shown in Figure 7.5, where both ideal low-pass filters are identical, each having a bandwidth BW.

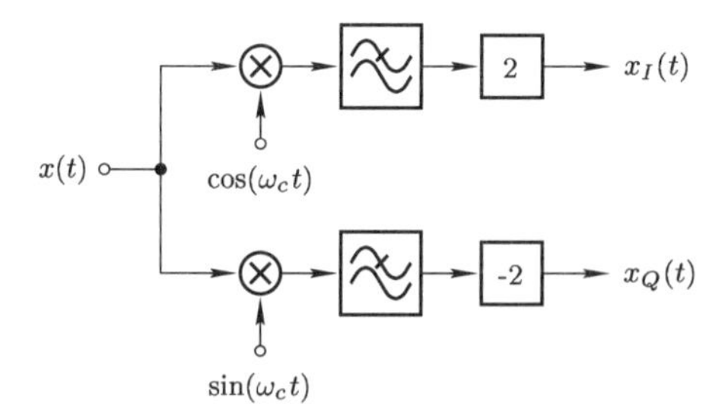

Figure 7.5: Generation of the in-phase and quadrature components of a band-pass signal $x(t)$.

The original band-pass signal $x(t)$ can be reconstructed from the in-phase and quadrature components of $\tilde{x}(t)$ as shown in Figure 7.6.

7.3.1.2 *Representation of band-pass systems*

Let the band-pass input signal $x(t)$ be applied to a linear time-invariant band-pass system with impulse response $h(t)$, and let the bandwidth of the band-pass system be equal to $2B$ and centered about the carrier frequency f_c. Then, by

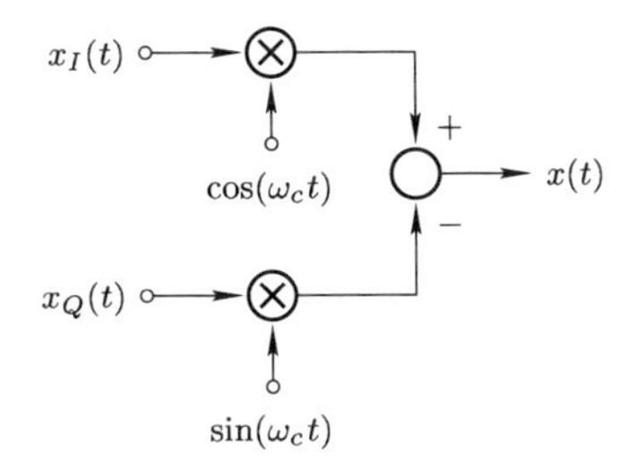

Figure 7.6: Reconstruction of the original band-pass signal from its in-phase and quadrature components.

analogy with the representation of band-pass signals, the impulse response of the band-pass system can be expressed as the product of a slowly-varying complex impulse response $\tilde{h}(t)$ and the carrier

$$h(t) = Re\left[\tilde{h}(t)e^{j\omega_c t}\right].$$

In general, the complex impulse response is a complex-valued quantity which may be expressed in terms of its in-phase and quadrature components

$$\tilde{h}(t) = h_I(t) + jh_Q(t).$$

Note that $\tilde{h}(t)$, $h_I(t)$ and $h_Q(t)$ are all low-pass functions limited to the frequency band $-B \leq f \leq B$.

Let $x(t)$ and $\sigma(t)$ be the input and output of a band-pass system, respectively. Then the complex envelope of the output can be expressed as

$$\tilde{\sigma}(t) = \frac{1}{2}\int_{-\infty}^{\infty} \tilde{h}(\tau)\tilde{x}(t-\tau)\, d\tau. \tag{7.3}$$

Equation (7.3) shows that, up to a scaling factor of 2, the complex envelope $\sigma(t)$ of the output is obtained by convolving the complex impulse response of the band-pass system with the complex envelope $\tilde{x}(t)$ of the input. However, note that in general both $\tilde{h}(t)$ and $\tilde{x}(t)$ are complex-valued quantities. Consequently, four real-valued convolutions must be evaluated in order to solve Equation (7.3).

7.3.1.3 Representation of band-pass Gaussian noise

In the low-pass equivalent model, every signal must be replaced by its complex envelope. The channel noise $n(t)$ corrupting the received signal is always band-pass noise in any practical implementation of a communications system.

Let $n(t)$ be a band-pass Gaussian random process. Because of the band-pass property, $n(t)$ always has zero mean. Let the spectrum of $n(t)$ be symmetric about the carrier frequency f_c.

Then the band-pass RF channel noise can also be expressed in the form of (7.1), i.e., the channel noise can be represented by its slowly-varying low-pass complex envelope

$$\tilde{n}(t) = n_I(t) + jn_Q(t).$$

During our computer simulations, only the low-pass in-phase and quadrature components $n_I(t)$ and $n_Q(t)$, respectively, of the channel noise are generated. This is possible because of the following properties [3]:

- both $n_I(t)$ and $n_Q(t)$ are Gaussian random processes,
- both $n_I(t)$ and $n_Q(t)$ have zero mean values,
- the in-phase and quadrature components have the same variance as the channel noise $n(t)$,
- the correlation functions of $n_I(t)$ and $n_Q(t)$ can be expressed as

  $$\begin{aligned} E[n_I(t)n_Q(t+\tau)] =& E[-n_Q(t)n_I(t+\tau)] = 0 \\ E[n_I(t)n_I(t+\tau)] =& E[n_Q(t)n_Q(t+\tau)], \end{aligned}$$

  where $E[\cdot]$ denotes the time averaging operation. The first equation shows that $n_I(t)$ and $n_Q(t)$ are independent, while the second one means that the autocorrelation functions of $n_I(t)$ and $n_Q(t)$ are equal to each other.

- both the in-phase and quadrature components of the noise have the same power spectral density which is related to the power spectral density $S_N(f)$ of $n(t)$ as follows

  $$S_I(f) = S_Q(f) = \begin{cases} \frac{1}{2}\Big[S_N(f - f_c) + S_N(f + f_c)\Big], & -BW \le f \le BW \\ 0, & \text{elsewhere,} \end{cases}$$

  where $S_N(f)$ occupies the frequency band $f_c - BW \le |f| \le f_c + BW$ and $f_c > BW$.

The most important conclusion of these relationships is that the equivalent RF channel noise can be generated directly in the low-frequency domain.

7.3.1.4 Low-pass equivalent of an FM-DCSK system

The block diagram of the FM-DCSK RF system for which the low-pass equivalent has to be developed is shown in Figure 7.7. The attenuation of the telecommunications channel is assumed to be $20\log(K)$. The attenuated signal $c(t)$ is corrupted by the channel noise $n(t)$, which is modeled by a Gaussian stochastic process. From the received signal $w(t)$, the channel filter, characterized by its impulse response $h(t)$, selects the desired signal $r(t)$. To derive the low-pass equivalent model, the correlator output $z(t)$ has to be obtained as a function of $m(t)$ and b_i.

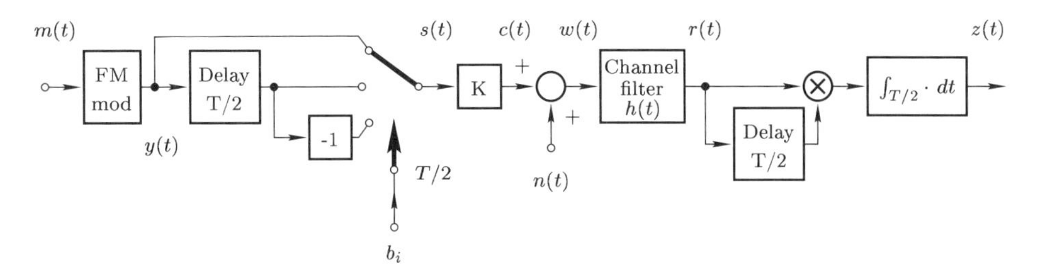

Figure 7.7: Block diagram of the FM-DCSK RF system.

First we replace the channel filter by its low-pass equivalent. Using the method shown in Figure 7.5, the in-phase and quadrature components $w_I(t)$ and $w_Q(t)$, respectively, of $w(t)$ are generated. The in-phase and quadrature components $r_I(t)$ and $r_Q(t)$ of the channel filter output are calculated according to the relationship given in (7.3). Then the channel filter output $r(t)$ is reconstructed in the RF domain by means of the method shown in Figure 7.6. The transmitter and receiver parts of the redrawn block diagram are given in Figures 7.8 and 7.9, respectively.

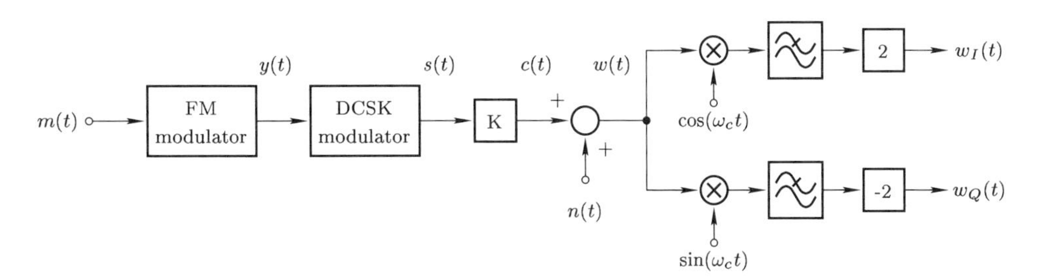

Figure 7.8: Transmitter part of the FM-DCSK system after substitution of the low-pass equivalent of the channel filter.

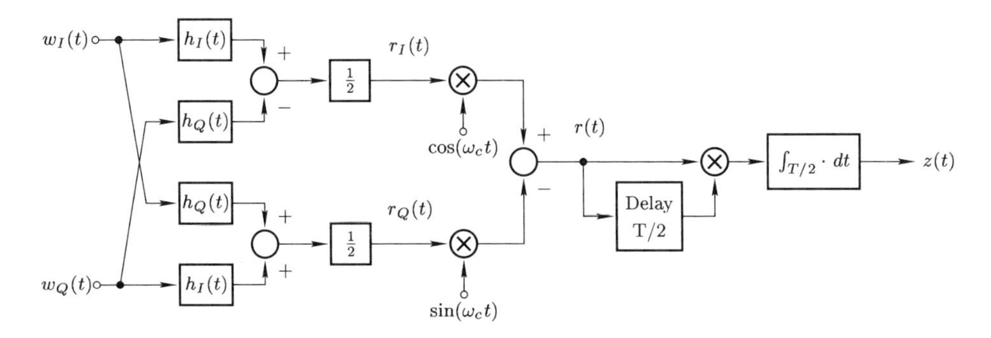

Figure 7.9: Receiver part of the FM-DCSK system after substitution of the low-pass equivalent of the channel filter.

Next we have to simplify both the transmitter and receiver parts of the equivalent system. Transformation of Figure 7.8 requires that the DCSK modulator, channel noise and attenuation be replaced by their low-pass equivalents.

Using (7.1), the input of the channel filter can be obtained as

$$\begin{aligned} w(t) = c(t) + n(t) &= Re\big[\tilde{c}(t)e^{j\omega_c t}\big] + Re\big[\tilde{n}(t)e^{j\omega_c t}\big] \\ &= Re\big(\left[\tilde{c}(t) + \tilde{n}(t)\right] e^{j\omega_c t}\big). \end{aligned} \tag{7.4}$$

Equation (7.4) means that the in-phase and quadrature components of the equivalent low-pass channel noise can be added directly to the in-phase and quadrature components of the input of the channel filter respectively. Thus,

$$w_I(t) = c_I(t) + n_I(t) \quad \text{and} \quad w_Q(t) = c_Q(t) + n_Q(t). \tag{7.5}$$

Because the relationship between the properties of $n(t)$ and $\tilde{n}(t)$ is known (see Section 7.3.1.3), it suffices to generate only the in-phase and quadrature components of the channel noise and to add them to $c_I(t)$ and $c_Q(t)$, respectively.

The delay caused by the telecommunications channel has no effect on the system performance. Let us assume that the telecommunications channel has a pure attenuation, i.e., K is a real number. Using (7.1) we may write

$$c(t) = Ks(t) = K\,Re\big[\tilde{s}(t)\,e^{j\omega_c t}\big] = Re\big[K\tilde{s}(t)\,e^{j\omega_c t}\big]. \tag{7.6}$$

Note that a pure attenuation can be shifted directly to the low-pass domain.

The output of the DCSK modulator is

$$s(t) = \begin{cases} y(t), & t_i \le t < t_i + T/2 \\ a_i y(t - T/2), & t_i + T/2 \le t < t_i + T, \end{cases} \tag{7.7}$$

where a_i can be ± 1 depending on the binary information b_i to be transmitted. Equation (7.7) shows that in the first time slot ($t_i \le t < t_i + T/2$) the DCSK modulator has no influence on the incoming signal. In the second time slot ($t_i + T/2 \le t < t_i + T$), the input signal is delayed and multiplied by a_i. Because a_i is a real number, using (7.1), we have that

$$\begin{aligned} s(t) = a_i y(t - T/2) &= a_i\,Re\big[\tilde{y}(t - T/2)e^{j\omega_c t}e^{-j\omega_c T/2}\big] \\ &= Re\big[a_i\tilde{y}(t - T/2)e^{j\omega_c t}\big], \end{aligned} \tag{7.8}$$

where

$$\frac{\omega_c T}{2} = 2\pi l, \qquad l = 1, 2, \ldots \tag{7.9}$$

was assumed. Equation (7.9) means that half of the bit duration T is equal to an entire multiple of the carrier period $T_c = 2\pi/\omega_c$.

Equation (7.8) shows that DCSK modulation can be applied directly to the complex envelope of $y(t)$.

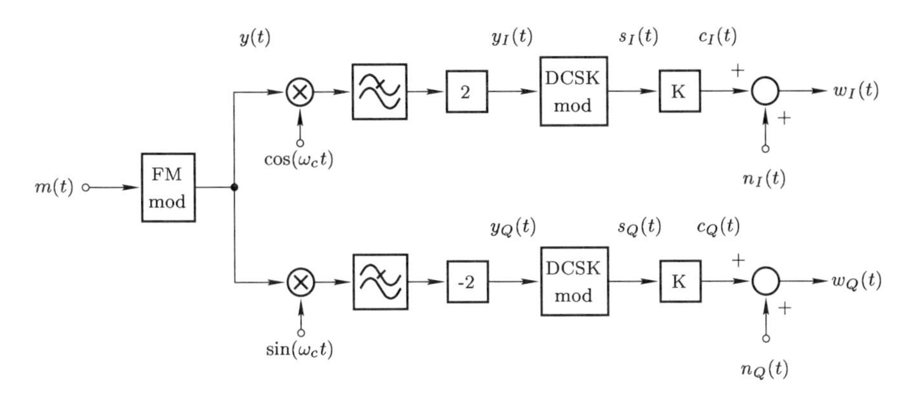

Figure 7.10: Transmitter part of the FM-DCSK system after substituting low-pass equivalents for the channel attenuation and DCSK modulator.

Using the transformations described by Equations (7.5), (7.6) and (7.8), the transmitter part of the FM-DCSK system given in Figure 7.8 can be modified further. The resulting block diagram is shown in Figure 7.10.

The goal of the next step is to remove the FM modulator from Figure 7.10. To do this, direct relationships must be found between $m(t)$ and $\tilde{y}(t) = y_I(t)+jy_Q(t)$.

An important feature of the analytic signal approach is that it gives a very simple description of every modulation scheme. Let the FM modulated signal be expressed as

$$y(t) = A_c cos\big[\omega_c t + 2\pi k_f \int_0^t m(\tau)\, d\tau\big], \tag{7.10}$$

where A_c is the amplitude of the carrier signal and k_f denotes the gain of the FM modulator.

Substituting (7.2) into (7.1), we obtain

$$y(t) = Re\big[\tilde{y}(t)e^{j\omega_c t}\big] = Re\big[a(t)e^{j\phi(t)}e^{j\omega_c t}\big] = a(t)\cos\big[\omega_c t + \phi(t)\big]. \tag{7.11}$$

By comparing (7.10) and (7.11), the following relationships can be derived

$$a(t) = A_c,$$
$$\phi(t) = 2\pi k_f \int_0^t m(\tau)\, d\tau.$$

Expressing the complex envelope in both Cartesian and polar forms, we have

$$\tilde{y}(t) = y_I(t) + jy_Q(t) = a(t)e^{j\phi(t)}$$

and the required equations are obtained as

$$\begin{aligned} y_I(t) &= A_c \cos\big(2\pi k_f \int_0^t m(\tau)\, d\tau\big), \\ y_Q(t) &= A_c \sin\big(2\pi k_f \int_0^t m(\tau)\, d\tau\big). \end{aligned} \tag{7.12}$$

These equations show that the complex envelope of the FM modulator output can be expressed directly from the low-pass chaotic signal $m(t)$. Taking account of (7.12), the low-pass equivalent of the transmitter part can be constructed from Figure 7.10, as shown on the left side of Figure 7.11.

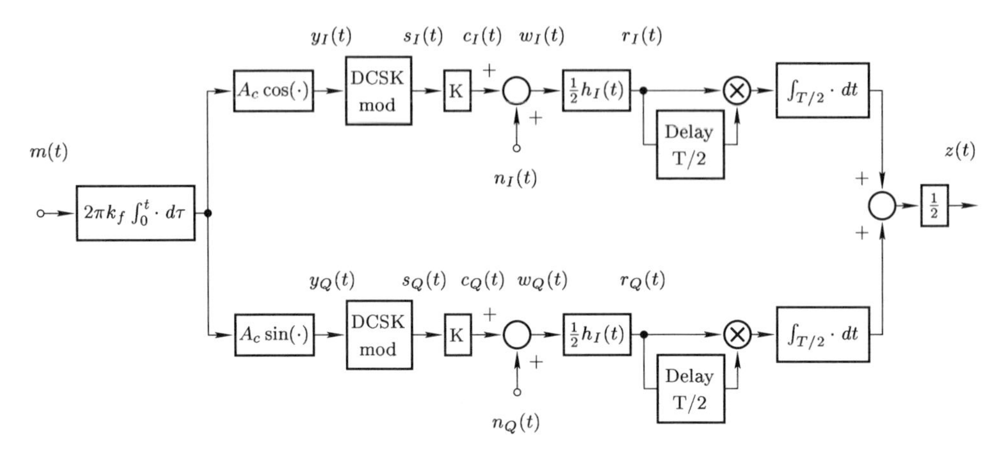

Figure 7.11: Low-pass equivalent model of the FM-DCSK chaotic communications system.

The next task is to develop the low-pass equivalent for the receiver. The computational effort required by the receiver part shown in Figure 7.9 can be reduced considerably if the complex impulse response of the channel filter is real-valued, i.e., $h_Q(t) = 0$. This can be achieved by applying a zero-phase channel filter. Because the channel filter is used only to restrict the bandwidth of the received signal, the use of a zero-phase filter does not limit the validity of our model. Although a zero-phase filter cannot be built, it is very easy to implement by computer simulation [4].

Finally, the FM-DCSK demodulator must be replaced by its low-pass equivalent. The output of the DCSK demodulator during one information bit period is

$$z(t) = \int r(t) r(t - T/2)\, dt, \tag{7.13}$$

where the integral has to be evaluated from the beginning to the end of each information bit period.

If $r(t)$ is expressed in canonical form [3].

$$r(t) = r_I(t)\cos(\omega_c t) - r_Q(t)\sin(\omega_c t) \tag{7.14}$$

and if half of the bit duration is equal to an entire multiple of the carrier period, then substituting (7.14) into (7.13) we obtain four terms:

$$\begin{aligned} z(t) =& \int r_I(t) r_I(t-T/2)\cos^2(\omega_c t)\, dt \\ &- \int r_I(t) r_Q(t-T/2)\sin(\omega_c t)\cos(\omega_c t)\, dt \\ &- \int r_Q(t) r_I(t-T/2)\sin(\omega_c t)\cos(\omega_c t)\, dt \\ &+ \int r_Q(t) r_Q(t-T/2)\sin^2(\omega_c t)\, dt, \end{aligned} \tag{7.15}$$

where $r_I(t)$ and $r_Q(t)$ are slowly-varying functions compared to the sinusoidal terms.

Using well-known trigonometric identities, and taking account of the fact that $r_I(t)$ and $r_Q(t)$ can be considered constant for half of the carrier period, many terms can be canceled in (7.15) [5]. After these simplifications, we obtain

$$z(t) = \frac{1}{2}\left[\int r_I(t) r_I(t-T/2)\, dt + \int r_Q(t) r_Q(t-T/2)\, dt\right]. \tag{7.16}$$

Equation (7.16) demonstrates that FM-DCSK demodulation can be performed without recovering the RF signal; the output $z(t)$ of the FM-DCSK demodulator can be expressed directly from the complex envelope of the channel filter output $r(t)$.

From these observations, the low-pass equivalent model of the receiver part can be developed. The low-pass equivalent model of the whole FM-DCSK system is shown in Figure 7.11. Note that all high frequency signals and the carrier frequency have been removed. Consequently, the sampling frequency required by the computer simulations is determined exclusively by the slowly-varying low-pass signals.

7.3.2 Simulation in the Discrete-Time Domain

The inputs and outputs of the low-pass equivalent models developed in Section 7.3.1 are analog signals. Since computer simulations are carried out in the discrete-time domain, the signals can be determined by using:

- numerical integration, or
- a discrete-time equivalent.

Numerical integration requires a large oversampling rate (OSR). In order to determine the noise performance of the system under test, millions of bits must be transmitted. With a large OSR, the resulting simulation time becomes extremely long.

The simulation time can be reduced significantly if a discrete-time equivalent of the low-pass model is used, where every analog signal is represented by its sampled equivalent. Uniform sampling is used and the discrete-time model is determined such that the differences between the sampled versions of each analog signal and its discrete-time equivalent signal are minimized.

There are three basic techniques (impulse invariance, backward-difference approximation and bilinear transformation) to transform a continuous-time system with a rational transfer function to a discrete-time equivalent [4]. To avoid aliasing and to minimize in-band distortion, the bilinear transformation has been selected to obtain the discrete-time equivalent of each analog block of the low-pass equivalent model. Another advantage of the bilinear transformation is that a stable continuous-time system will always map to a stable discrete-time system.

In the case of the bilinear transformation, the mapping between the s-plane and z-plane is given by

$$s = \frac{2}{T_s}\left[\frac{1 - z^{-1}}{1 + z^{-1}}\right],$$

where T_s is the sampling rate in the discrete-time equivalent.

Let the continuous-time system be characterized by its transfer function $H_{cont}(s)$. Then the transfer function of the equivalent discrete-time system is given by [4]

$$H_{disc}(z) = H_{cont}(s)\Bigg|_{s=\frac{2}{T_s}\left[\frac{1-z^{-1}}{1+z^{-1}}\right]}.$$

The unit circle in the z-plane maps onto the $j\omega$-axis in the s-plane and the relationship between the continuous-time frequency variable ω and its discrete-time counterpart Ω is given by

$$\omega = 2\tan^{-1}\left(\frac{\Omega T_s}{2}\right).$$

This equation shows that the in-band distortion of the bilinear transformation is negligible but that a nonlinear warping of the frequency axis appears. This distortion can be reduced by increasing the OSR.

7.3.3 FM-DCSK Simulation in a Matlab Environment

Matlab provides a computing environment for high-performance numeric computation and visualization. Although a Telecommunications Toolbox is available in Matlab, it is not suitable for very-fast simulation because it uses mainly RF models. The Simulink Toolbox is also very slow since the differential equations used to describe the operation of analog circuits are solved by numerical integration. These techniques are extremely inefficient when tens of millions of bits (and their corresponding nonperiodic FM-DCSK waveforms) have to be transmitted in order to determine the noise performance of the system under test.

To minimize the simulation time required, an FM-DCSK simulator can use the discrete-time low-pass equivalent model discussed above. Subroutines which have to be performed many times in loops and which contain many calls to other subroutines make Matlab extremely slow. These subroutines can be written in C, compiled, and linked into Matlab.

In a simulator based on these principles, every analog signal is represented by its discrete-time equivalent. The simulator consists of subroutines, each of which describes the operation of one circuit block. Every block is represented by its low-pass equivalent model and the bilinear transformation is used to obtain its discrete-time equivalent. The inputs and outputs of the blocks are the in-phase and quadrature components of complex envelopes, and in the case of certain blocks, the digital information signal. Uniform sampling is applied; the sampling frequency is determined by the spectra of the low-pass signals and by the oversampling requirement of bilinear transformation. The required E_b/N_0 has to be entered and then the discrete-time low-pass equivalent noise is generated by the Matlab random number generator. These subroutines appear as built-in functions and are called by a conventional m-file.

As outputs, the simulator can determine the time waveforms and frequency spectra of every signal. It can also "measure" the Signal-to-Noise Ratio (SNR) at any point in the system under test and can calculate the Bit Error Rate (BER). Because the Matlab data structure is always used, post processing and visualization of results is straightforward. Comparative simulations performed in Matlab without any modification and with the FM-DCSK simulator built in this way have demonstrated that the simulation time can be reduced by a factor which varies from 100 to 1000, depending on the simulation task performed.

7.3.4 Verification of the FM-DCSK Simulator

The results obtained by the FM-DCSK simulator cannot be verified analytically, because no theoretical expression has yet been derived to express the BER of the continuous-time FM-DCSK system in terms of E_b/N_0. In order to verify the operation of the FM-DCSK simulator, two different techniques have been developed.

The noise performance of Continuous-Phase Frequency Shift Keying (CPFSK) is known in analytical form [3]. The basic idea behind the first verification method is to build a simulator for CPFSK using the building blocks of the FM-DCSK simulator. If the results of the analytical approach correspond to those obtained from the simulation, then we can conclude that our blocks are functioning correctly.

In the second method, the simulator is verified by comparing its results with those we can obtain by a direct simulation of the FM-DCSK system in the RF domain.

7.3.4.1 Verification by means of CPFSK modulation scheme

In a binary CPFSK system, symbols "1" and "0" are mapped into two sinusoidal waves which differ in frequency. The basis functions are the two sinusoidal signals characterized by their frequencies ω_1 and ω_2.

The demodulator contains two correlators with a common input; the reference signals of these correlators are the two locally generated coherent basis functions with frequencies ω_1 and ω_2 [3]. The correlator outputs are then subtracted from each other. If the result of subtraction is greater than zero, the receiver decides in favor of symbol "1." In the opposite case, the decision is made in favor of symbol "0."

The FM-DCSK demodulator correlates the reference and information-bearing parts of the incoming signal. If every CPFSK signal is mapped into four sample functions such that the first and second basis functions of CPFSK are transmitted by the first and third sample functions, and the information-bearing CPFSK signal is inserted into the second and fourth time slots, then we can build a CPFSK simulator from the subroutines developed for the FM-DCSK simulator.

Using the FM-DCSK simulator modified to emulate a CPFSK system, its noise performance has been determined. The following system parameter values were used in the simulation: the bit duration was 1 μs, the difference between the frequencies of the basis functions was set to 2 MHz, and the RF bandwidth of the CPFSK system was set to 4 MHz. The sampling frequencies of the simulator and the correlator were set to $f_s = 40$ MHz and $f_{corr} = 20$ MHz, respectively.

The results are plotted in Figure 7.12, where the BER values obtained using the simulator are denoted by "+" marks and the theoretical result is given by the solid curve. The results plotted in this figure are identical, showing that the subroutines developed for the FM-DCSK system are correct.

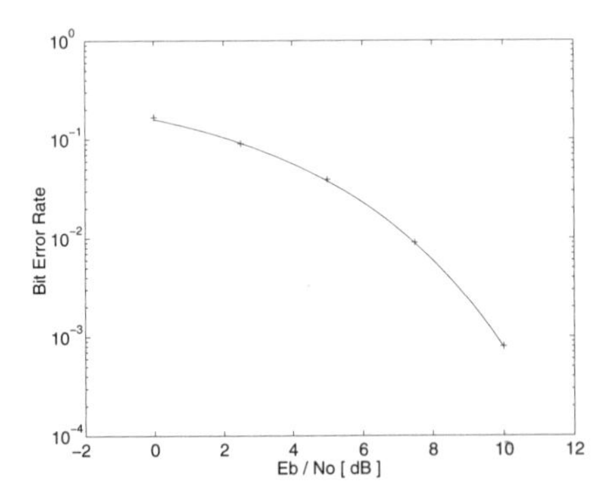

Figure 7.12: Verification of the FM-DCSK simulator: simulated ("+" marks) and theoretical (solid curve) noise performance of the CPFSK modulation scheme.

7.3.4.2 Verification by direct RF simulation

The low-pass equivalent model developed for the FM-DCSK simulator has also been verified by comparing the results of a direct simulation performed using the radio-frequency model shown in Figure 7.4 with that of the simulation of the discrete-time low-pass equivalent model. The BER curves calculated using the discrete-time low-pass equivalent and RF models are shown in Figure 7.13 as solid and dashed curves, respectively. The values of the main system parameters were as follows: bit duration was 2 μs, the RF bandwidth was set to 17 MHz, the chip rate was 20 MHz, the center frequency of the FM-DCSK modulator was $f_c = 36$ MHz, and the sampling frequencies of the simulator and the correlator were set to $f_s = 40$ MHz and $f_{corr} = 20$ MHz, respectively.

The BER values calculated using the low-pass and RF models are in almost perfect agreement, showing that the subroutines developed for the FM-DCSK system are correct. The source of the slight difference is that the number of bits sent via the RF channel was limited by the simulation time available.

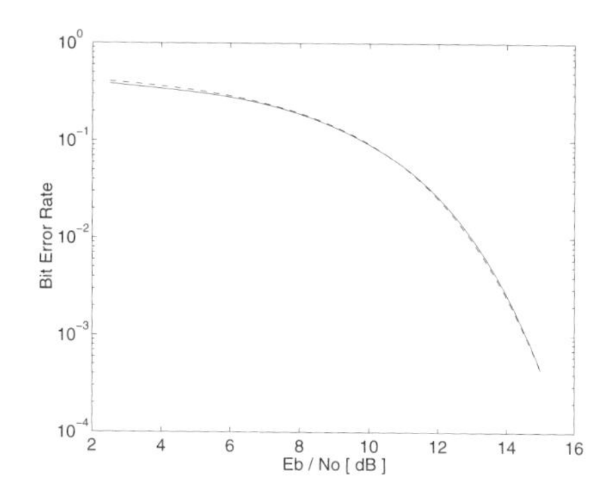

Figure 7.13: Verification of the FM-DCSK simulator: noise performance of FM-DCSK determined using the RF model (dashed curve) and the discrete-time low-pass equivalent (solid curve) model.

7.4 Noise Performance of FM-DCSK

In a digital communications system, the analog sample functions carrying the information pass through a telecommunications channel in which they are corrupted by noise and may suffer from distortion and multipath. The demodulator must decide, on the basis of the corrupted and distorted received signal, which bit was most likely transmitted. When wrong decisions are made, bit errors occur. The quality of a digital communications system is characterized by the BER which quantifies the average number of bit errors for specified channel conditions.

We have shown in Section 7.2 that either a fast or a slow FM-DCSK system can be implemented, depending on the parameters of modulator. Although the interferences caused by the two implementations and their multipath behaviors are different, the noise performance of fast and slow FM-DCSK is the same in the case of a single-path Additive White Gaussian Noise (AWGN) channel.

In this section the noise performance of FM-DCSK modulation is determined, assuming a linear band-pass channel with Additive White Gaussian Noise. The channel model, including multipath effects, is shown in Figure 7.14. In this section, only the effects of the main system parameters on noise performance are studied. The effects of multipath propagation will be discussed in Section 7.5.

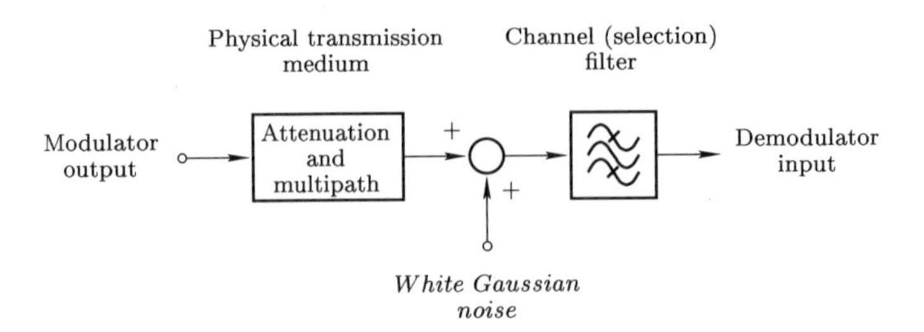

Figurc 7.14: Model of AWGN RF channel including the effects of multipath propagation.

7.4.1 Relationship between SNR and BER

To characterize the performance of a digital communication scheme, one must determine the BER as a function of the ratio of the signal energy per bit to the noise spectral density. However, in a physical system, the SNR rather than the E_b/N_0 can be measured. Knowing the total bandwidth of the RF channel selection filter B_{RF} and the bit duration T, the SNR at the input of the demodulator can be related to the E_b/N_0 as follows:

$$SNR = \frac{E_b}{N_0} - 10\log_{10}(TB_{RF}), \tag{7.17}$$

where both SNR and E_b/N_0 are expressed in dB. A digital communications system is typically expected to perform with a BER of less than 10^{-3}. In many applications such as a Wireless Local Area Network, the SNR may be as low as 0 dB in the worst case situation [1,6].

7.4.2 Noise Performance in an AWGN Channel

To the best of our knowledge, an analytical expression for the noise performance of the continuous-time FM-DCSK modulation scheme in an AWGN channel is not available in the literature. The noise performance determined by *computer simulation* using the AWGN channel model of Figure 7.14 is shown in Figure 7.15, where the BER is plotted as a function of E_b/N_0. For a given E_b/N_0, the required SNR at the demodulator input can be calculated from (7.17). The main system parameters used in the simulations were: bit duration $T = 2\ \mu$s and RF channel bandwidth $B_{RF} = 17$ MHz. These parameters have been selected to obtain an IEEE 802.11-compliant WLAN system with a data rate of 500 kb/s. For comparison, the noise performance of the noncoherent Frequency Shift Keying (FSK) [3] and GFSK [6] modulation schemes are also plotted in Figure 7.15.

Note that the noise performance of these modulation schemes is much worse than that of coherent Binary Phase Shift Keying or FSK modulation (see [3], for example). However, recall that FM-DCSK is intended for special applications such as WLAN and industrial applications, indoor radio, and mobile communications, where the synchronization requirements of coherent demodulators cannot be satisfied, where the transmitted power spectral density must be low to

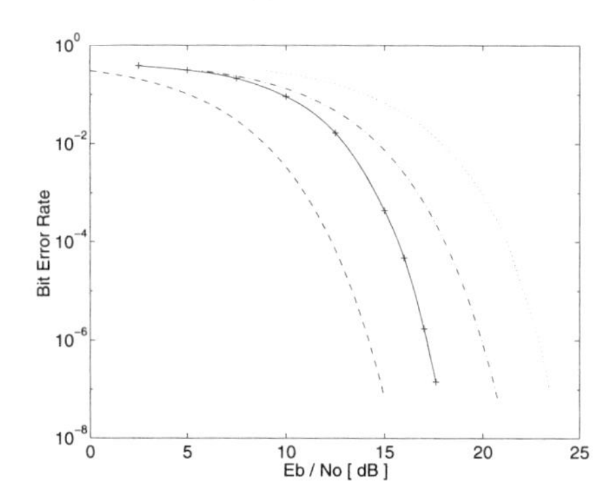

Figure 7.15: Noise performance of noncoherent FSK (dashed curve), FM-DCSK (solid curve) and GFSK modulation schemes. For GFSK the performance is given for two modulation indices: $\beta = 0.31$ (dash-dot curve) and $\beta = 0.22$ (dotted curve).

avoid interfering with other telecommunications systems, and where multipath propagation and industrial disturbances limit the performance of the telecommunications system. In these applications, the noise performance is an important, but by no means the most important, system parameter. Other properties of the modulation scheme, such as robustness to channel nonidealities, are more significant.

In these applications, FM-DCSK has many advantages:

- the demodulation is performed without synchronization,
- it is not sensitive to the particular waveform transmitted, so there is no need for complicated control circuitry in the chaotic signal generator to keep the control parameters constant in the presence of temperature variations, aging, etc.,
- because both the reference and information-bearing parts of the FM-DCSK signal pass through the same telecommunication channel, it is not sensitive to channel distortion,
- it can operate over a time-varying channel if the variations in the channel parameters are negligible over half the bit duration, and
- it can transmit pure "0" and "1" sequences, i.e., there is no need for a scrambler circuit.

7.4.3 Effect of Main System Parameters on Noise Performance

When developing a new telecommunications system, a trade-off between the main system parameters is inevitable. In our case the most important system parameters are the following:

- bandwidth of the channel selection filter,
- bit duration, and
- bandwidth of the transmitted signal.

For example, the bit duration determines the data rate, while a larger transmitted bandwidth offers enhanced multipath performance.

7.4.3.1 Effect of channel selection filter

The channel (selection) filter at the receiver is used to maximize the SNR at the demodulator input and to prevent out-of-band signals from entering the demodulator. The most important parameter of the channel filter is its RF bandwidth.

Figure 7.16 shows the noise performance of the FM-DCSK modulation scheme for three different filter bandwidths: 17, 34, and 51 MHz. Recall that the bandwidth of the transmitted RF FM-DCSK signal is 17 MHz in each case.

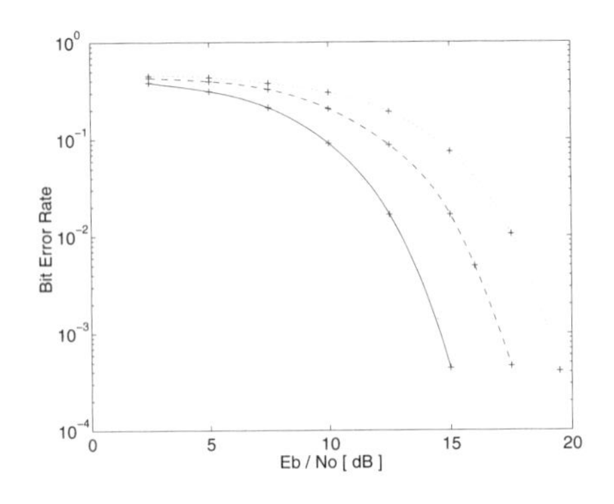

Figure 7.16: Noise performance of FM-DCSK for channel selection filter bandwidths of 17 MHz (solid curve), 34 MHz (dashed curve), and 51 MHz (dotted curve).

The lower limit on the bandwidth of the channel selection filter is determined by the fact that the filter should suppress excess noise, but it must not significantly reduce the power of the received signal. On the other hand, Figure 7.16 shows that if the bandwidth of the channel filter exceeds the bandwidth of the FM-DCSK signal, then the system suffers a considerable degradation in noise performance. In the optimum case, the bandwidths of the channel selection filter and the FM-DCSK signal should be the same, i.e., the filter is matched to the FM-DCSK signal so that it maximizes the SNR at the input of the demodulator.

7.4.3.2 Effect of bit duration

At the demodulator of a conventional telecommunications system, the received signal is compared, in some sense, to a noise-free reference. In a correlation receiver the noise-free reference is recovered by synchronization, while in a matched filter receiver it is stored as the impulse response of a matched filter. If the energy per bit is kept constant, then the noise performance of these modulation schemes is independent of the bit duration and the RF bandwidth of the transmitted signal [3].

In FM-DCSK, the reference part of the signal is transmitted via the same telecommunications channel as the information-bearing part, which means that it is also corrupted by noise. An important question arises due to this basic difference, namely, whether the bit duration has any influence on the noise performance of an FM-DCSK system.

The effect of bit duration is shown in Figure 7.17, where the RF bandwidth of the FM-DCSK signal and the bandwidth of the channel selection filter at the receiver are both 17 MHz.

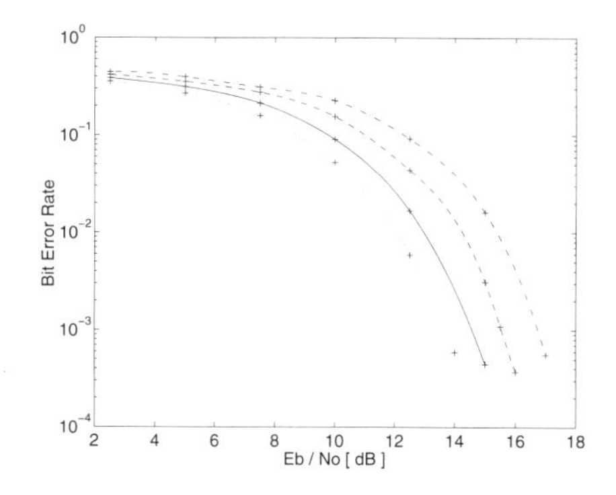

Figure 7.17: Noise performance of FM-DCSK modulation for bit durations of 1 μs (dotted curve), 2 μs (solid curve), 4 μs (dashed curve), and 8 μs (dash-dot curve).

Figure 7.17 shows that the bit duration has a strong influence on the noise performance; the latter becomes worse if the bit duration is increased. A similar result has been derived analytically for a discrete-time DCSK modulation scheme [7].

We conclude that reducing the bit duration improves the noise performance of FM-DCSK and results in a higher data rate. However, it also increases the skirt of the generated FM-DCSK signal [see Figure 7.2(b)] and makes the system more sensitive to timing recovery errors.

7.4.3.3 Effect of the RF bandwidth

The noise performance of conventional modulation schemes is independent of the bandwidth of the transmitted signal. To demonstrate the effect of RF bandwidth on the noise performance of FM-DCSK modulation, we show the results of computer simulations for three different RF bandwidths: 8 MHz, 12 MHz, and 17 MHz. The bit duration was 2 μs in every case and the bandwidth of the channel selection filter was matched to the RF bandwidth.

Figure 7.18 shows the effect of the RF bandwidth of the FM-DCSK signal on the noise performance. For example, if BER= 10^{-3} is specified, then the required values of E_b/N_0 are 13.7, 14.2, and 14.6 dB for RF bandwidths of 8, 12, and 17 MHz, respectively. Although reducing the RF bandwidth improves the noise performance, it also degrades the multipath performance, as we shall see later. Thus, the choice of RF bandwidth is a trade-off between noise performance and multipath performance.

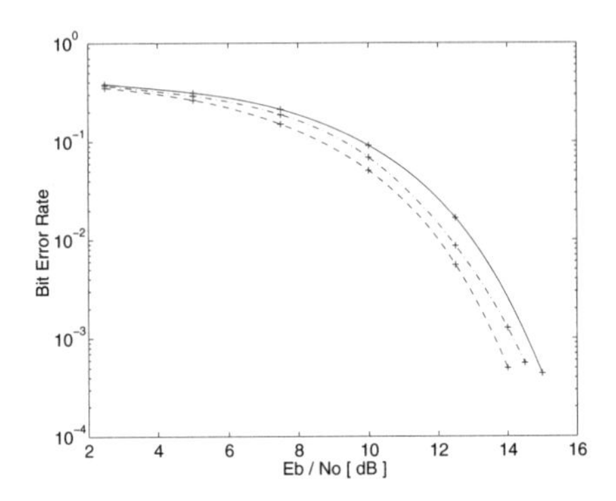

Figure 7.18: The effect of RF bandwidth on the noise performance of the FM-DCSK modulation scheme. Curves are shown for three RF bandwidths: 8 MHz (dashed curve), 12 MHz (dash-dot curve), and 17 MHz (solid curve).

7.5 Operation in a Multipath Environment

In many applications such as WLAN, mobile communications, and indoor radio, the received signal contains components which have traveled from the transmitter to the receiver via multiple propagation paths with differing delays; this is called *multipath propagation* [2,3]. The components arriving via different propagation paths may add destructively, resulting in deep frequency-selective fading. Conventional narrow-band systems fail catastrophically if a *multipath-related null*, defined below, coincides with the carrier frequency.

In the applications mentioned above, the distance between the transmitter and receiver is relatively short, i.e., the attenuation of the telecommunications channel is moderate. The effect which limits the performance of communications in such an environment is not the additive channel (thermal) noise N_0, but deep frequency-selective fading caused by multipath propagation. In these applications, the most important system parameter is the sensitivity to multipath.

Figure 7.15 shows that the noise performance of FM-DCSK in a single-path AWGN channel is better than that of GFSK, but it is much worse than that of coherent modulation schemes. However, FM-DCSK has potentially lower sensitivity to multipath, because

- the demodulation is performed without synchronization, and

- the transmitted signal is a wide-band signal which cannot be completely canceled by a multipath-related null.

In this section, the performance degradation of the FM-DCSK modulation scheme resulting from multipath propagation is determined by computer simulation.

7.5.1 Model of Multipath Channel

The tapped delay line model of a time-invariant multipath radio channel having N propagation paths is shown in Figure 7.19. The radiated power is split and travels along the N paths, each characterized by a delay T_l and gain k_l, where $l = 1, 2, \ldots, N$.

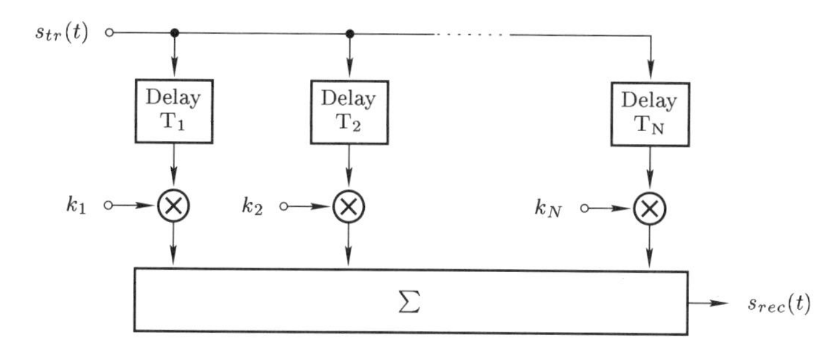

Figure 7.19: Tapped delay line model of an RF multipath radio channel.

If a narrow-band telecommunications system is considered, then in the worst case two paths exist and the two received signals cancel each other completely at the carrier frequency ω_c, i.e.,

$$\Delta\tau\omega_c = (2n+1)\pi, \quad n = 1, 2, 3, \ldots,$$

where $\Delta\tau = T_2 - T_1$ denotes the additional delay of the second path.

Let the two-ray multipath channel be characterized by its frequency response shown in Figure 7.20. Note that the multipath-related nulls, where the attenuation becomes infinitely large, appear at

$$f_{null} = \frac{2n+1}{2\Delta\tau}, \quad n = 1, 2, 3 \ldots \tag{7.18}$$

Let the bandwidth of fading be defined as the frequency range over which the attenuation of the multipath channel is greater than 10 dB. Then the bandwidth of multipath fading can be expressed as

$$\Delta f_{null} \approx \frac{0.1}{\Delta\tau}. \tag{7.19}$$

Equations (7.18) and (7.19) show that the center frequencies of the multipath-related nulls, the distances between them, and their bandwidths, are determined by $\Delta\tau$; a shorter delay accentuates the problem.

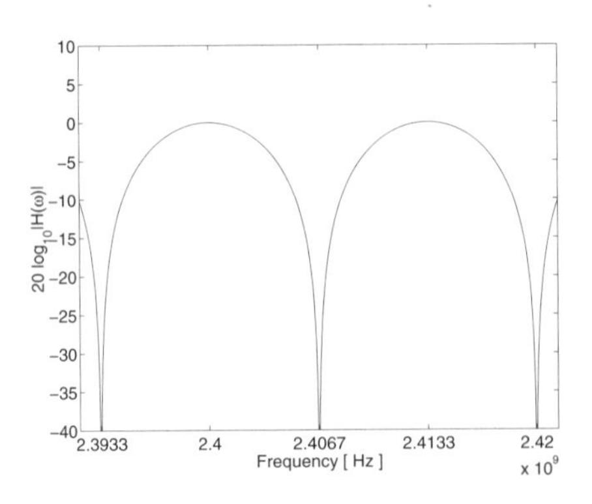

Figure 7.20: Magnitude of frequency response of a two-ray multipath channel.

In WLAN applications, the typical values of $\Delta\tau$ are 91 ns for large warehouses and 75 ns for office buildings [6]. If $\Delta\tau = 75$ ns, then the distance between two adjacent multipath-related nulls is 13.33 MHz. In the case of the three IEEE 802.11-compliant telecommunications channels in the 2.4 GHz ISM band [6], if off-the-shelf channel selection filters are used, then the RF bandwidth of the FM-DCSK signal should be 17 MHz. This means that at most two multipath-related nulls may appear in any of the three channels.

Equation (7.18) shows that the frequencies of the multipath-related nulls are determined by the additional delay of the second path $\Delta\tau$. The number of multipath-related nulls appearing in a WLAN channel and their positions relative to the FM-DCSK center frequency depend on the exact value of $\Delta\tau$. In a real application, $\Delta\tau$ may vary, thus changing the frequencies of the multipath-related nulls. To quantify this effect in the following, but using the same multipath channel for every simulation, the additional delay of the second path is kept constant but the center frequency of the FM-DCSK signal is varied.

7.5.2 Qualitative Behavior of FM-DCSK in a Two-Ray Multipath Channel

To illustrate the effect of the two-ray multipath channel on the received signal, Figure 7.21 shows the spectra of the transmitted and received signals of the fast FM-DCSK system for $T = 2$ μs and RF bandwidth $B_{RF} = 17$ MHz. The two possible extreme cases are shown in the figure; in the first case, the multipath-related null coincides with the center frequency of the FM-DCSK signal, while in the second case the two nulls appear symmetrically about the center frequency. Due to the rounded shape of the fast FM-DCSK spectrum, the loss in the received energy per bit E_b is almost the same in both cases. We expect, therefore, that the multipath performance of the fast FM-DCSK experiences low sensitivity to the relative positions of the center frequency and the multipath-related nulls.

The corresponding spectra for slow FM-DCSK are shown in Figure 7.22. Due to the low chip rate in this case, the spectrum of slow FM-DCSK becomes

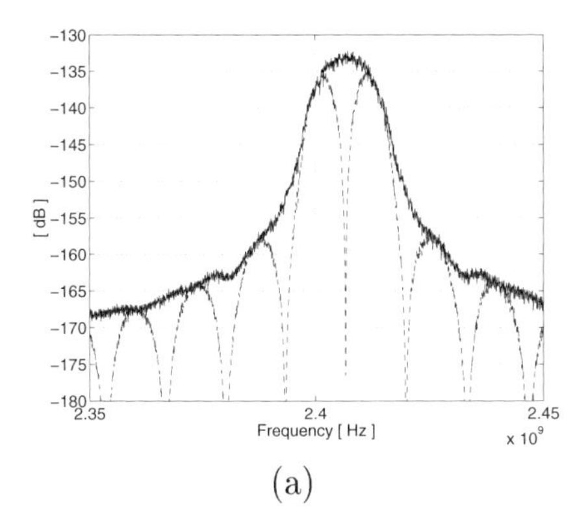

(a)

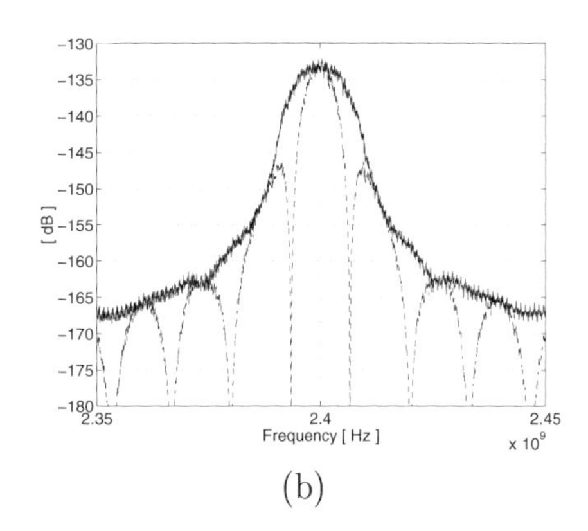

(b)

Figure 7.21: The transmitted (solid curve) and received (dashed curve) spectra of a fast FM-DCSK system (a) when the center frequency of the FM-DCSK coincides with a multipath-related null and (b) when two nulls appear symmetrically about the center frequency.

rectangular; it is similar to that of a slowly swept FM oscillator. Note that the loss in E_b is significantly different in the two extreme cases. Consequently, we expect that the multipath performance of slow FM-DCSK should depend on the relative positions of the center frequency and the multipath-related nulls.

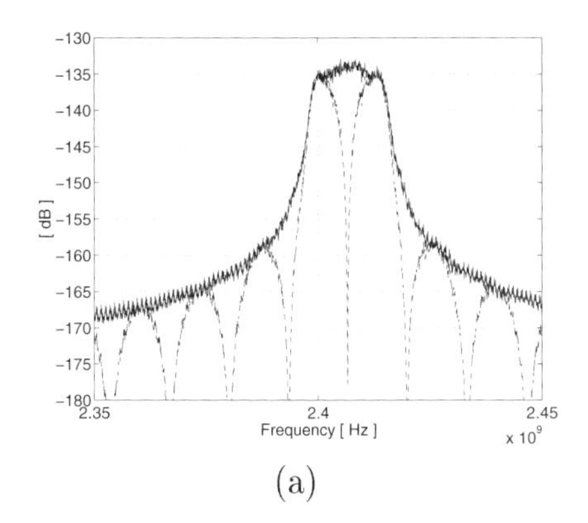

(a)

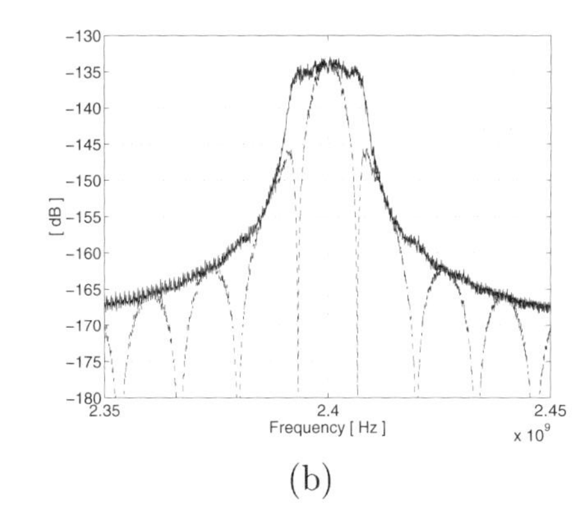

(b)

Figure 7.22: The transmitted (solid curve) and received (dashed curve) spectra of a slow FM-DCSK system (a) if the center frequency of FM-DCSK coincides with a multipath-related null and (b) if two nulls appear symmetrically about the center frequency.

It follows from the frequency response of the two-ray multipath channel shown in Figure 7.20 that the required RF bandwidth of an FM-DCSK transmission depends on the worst case additional delay $\Delta\tau$ to be considered in a given application. A shorter delay requires larger transmission bandwidth. This effect can be seen clearly in Figure 7.23, where $\Delta\tau$ has been reduced from 75 ns to 25 ns. In this case, if the multipath-related null coincides with the center frequency of the FM-DCSK system, then almost the entire bit energy E_b is lost, resulting in a very poor Bit Error Rate.

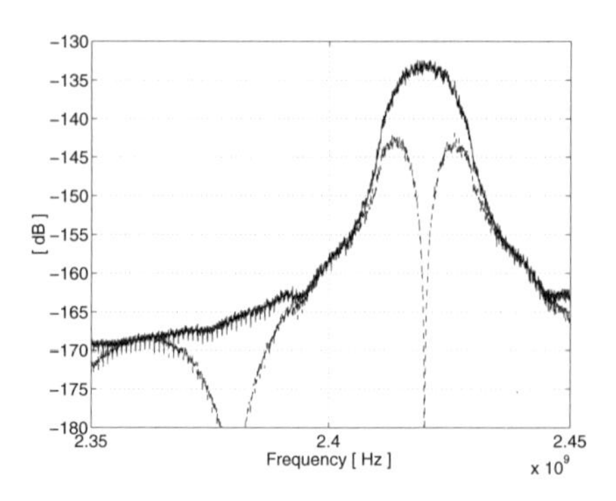

Figure 7.23: The transmitted (solid curve) and received (dashed curve) spectra of a fast FM-DCSK system if the additional delay of the second path is reduced to 25 ns.

7.5.3 Quantitative Behavior of FM-DCSK in a Two-Ray Multipath Channel

Figure 7.20 shows qualitatively why conventional narrow-band systems can fail catastrophically to operate over a multipath channel. Due to high attenuation appearing about the multipath-related nulls, the SNR becomes extremely low at the input of the receiver. Consequently, the demodulator cannot operate. The situation becomes even worse if a carrier recovery circuit is used, because such a circuit typically cannot synchronize unless the input signal level exceeds a minimum threshold.

In the FM-DCSK system, the power of the radiated signal is spread over a wide frequency range. The appearance of a multipath-related null means that part of the transmitted power is lost, but the system still operates. Of course, the lower SNR at the demodulator input results in a worse BER. The special feature of FM-DCSK that it does not use synchronization to perform the demodulation makes it even more robust against multipath.

7.5.3.1 Difference in performance degradation of fast and slow FM-DCSK

Figures 7.2 and 7.3 show that the shapes of the average power spectral densities of the fast and slow FM-DCSK signals are different. The in-band spectrum is rounded for the fast FM-DCSK system and rectangular for the slow one. In this context "average" means that very long sample functions (of length $1000T$) have been used to calculate the spectra.

The consequence of this difference is highlighted in Figure 7.24, where the two-ray multipath performance of fast and slow FM-DCSK is plotted for different center frequencies of the FM-DCSK signal. The curves plotted belong to the best and worst results for both cases. Note that the average performance degradation for the fast and slow FM-DCSK systems is the same, but the variation in performance loss as a function of the relative positions of the center frequency and the multipath-related nulls is higher in the slow FM-DCSK system. This effect is caused by the different shapes of the fast and slow FM-DCSK spectra.

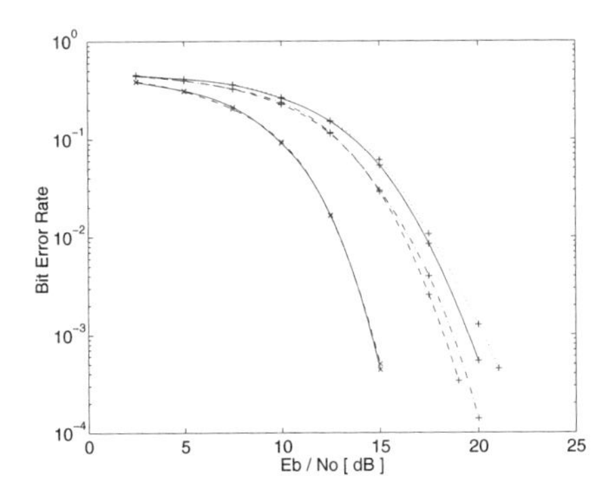

Figure 7.24: Variation in the multipath performance degradation of the fast (solid and dash-dot curves) and slow (dotted and dashed curves) FM-DCSK systems. For comparison, the noise performance of the fast (solid curve with '×' marks) and slow (dashed curve with '×' marks) systems without multipath is also shown.

In Figure 7.24, the bit duration is 2 μs, $B_{RF} = 17$ MHz, $\Delta\tau = 75$ ns and the transmitted signal propagates via two paths, the gain of each path being equal to 1/2.

Because the multipath performance of the fast FM-DCSK is better than that of the slow one, we will consider mainly the fast FM-DCSK system with a bit duration of 2 μs in the following simulations.

7.5.3.2 Degradation due to multipath with equal attenuation of both paths

The performance degradation of the fast FM-DCSK system due to multipath propagation is shown in Figure 7.25, where $\Delta\tau = 75$ ns. The solid curve marked with ×'s shows the noise performance without multipath propagation. To determine the multipath performance in this example, we assume that the transmitted signal can propagate via two paths, the gain of each path being equal to 1/2.

We noted above that the relative positions of the multipath-related nulls and the center frequency of the FM-DCSK signal might influence the multipath performance. This effect is apparent in Figure 7.25, where the model of the multipath channel was fixed as shown in Figure 7.19, but the center frequency of the FM-DCSK signal was varied from 2.4 GHz to 2.412 GHz in steps of 2 MHz.

If a Bit Error Rate of 10^{-3} is required, then the average loss due to multipath is only 4.8 dB and the variation in the loss is less than 1.2 dB for fast FM-DCSK. These results confirm our prediction in 1997 [8] and 1998 [9] that the FM-DCSK modulation scheme could outperform conventional narrow-band modulation schemes under poor propagation conditions.

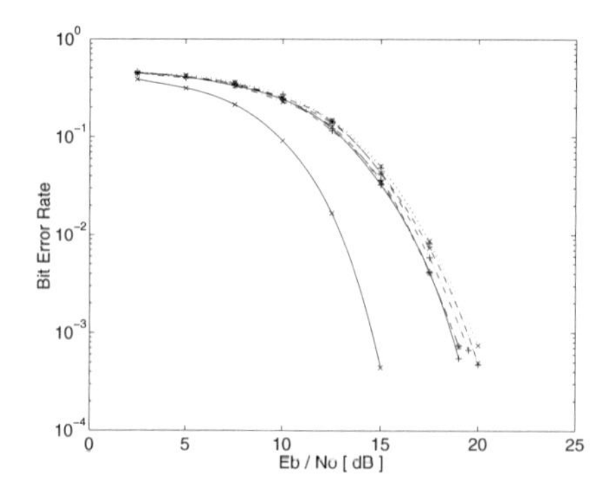

Figure 7.25: Performance degradation caused by two-ray multipath propagation in a fast FM-DCSK system. To change the relative positions of the multipath-related nulls, the center frequency of the FM-DCSK transmission was varied from 2.4 GHz to 2.412 GHz in steps of 2 MHz. For comparison, the noise performance of FM-DCSK without multipath (solid curve with '×' marks) is also shown.

7.5.3.3 Degradation due to multipath with unequal attenuation of both paths

Figure 7.25 showed the system performance achieved when two propagation paths exist and both of them suffer equal attenuation. The degradation in system performance is less if the attenuations of the two paths are different; this is illustrated in Figure 7.26, where the difference between the attenuation of the two paths is 10 dB. The performance degradation has been determined for different FM-DCSK center frequencies; the dashed and dotted curves correspond to the best and worst cases, respectively. Under these propagation conditions, the average performance loss is only 2.5 dB at BER=10^{-3} for the fast FM-DCSK system.

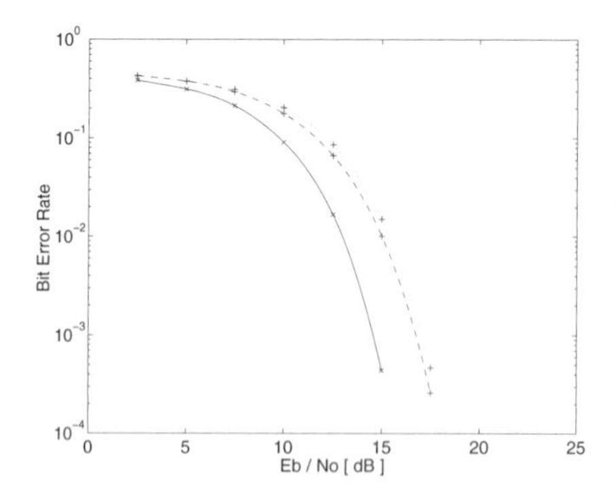

Figure 7.26: Performance degradation of fast FM-DCSK due to multipath propagation when the attenuation of one propagation path is 10 dB higher than that of the other. For comparison, the noise performance of fast FM-DCSK without multipath (solid curve with '×' marks) is also shown.

7.5.3.4 Performance degradation in terms of bandwidth

The bandwidth of the transmitted FM-DCSK signal has the strongest influence on the multipath performance of the system. This effect is illustrated in Figure 7.27, which shows the worst-case performance degradation in the fast and slow FM-DCSK systems when the RF bandwidth is reduced from 17 MHz to 8 MHz. The attenuations along the two propagation paths are the same and the multipath-related nulls coincide with the center frequency of the FM-DCSK signal. The performance degradation is unacceptably large, showing that 8 MHz bandwidth is not sufficient if data communication has to be established in a multipath environment where $\Delta\tau$ can be as short as 75 ns.

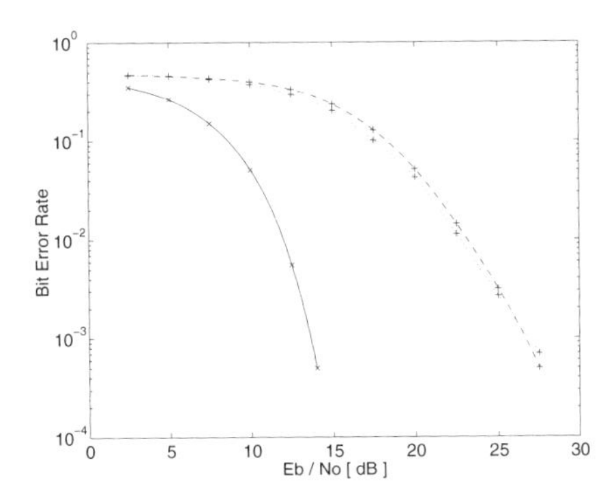

Figure 7.27: Worst-case performance degradation caused by multipath propagation in the fast (dashed curve with '+' marks) and slow (dotted curve with '+' marks) FM-DCSK systems when the RF channel bandwidth is reduced from 17 MHz to 8 MHz. For comparison, the noise performance of FM-DCSK with an 8 MHz RF channel bandwidth without multipath is also shown (solid curve with '×' marks).

7.5.4 Quantitative Behavior of the FM-DCSK in a Multi-Ray Multipath Channel

Depending on the size and interior furnishing of an office building or warehouse, more than two propagation paths may be present in a WLAN application. To check the system performance of fast FM-DCSK under these more complex propagation conditions, multipath channels having three, four, and eight paths have also been considered. In these simulations, the attenuations along the different paths have been selected at random.

7.5.4.1 Performance degradation in a three-ray multipath channel

The performance degradation of the fast FM-DCSK system due to multipath is shown in Figure 7.28, where the gains and additional delays of the three propagation paths are [0.44 0.16 0.4] and [0ns 25ns 125ns], respectively. The solid curve marked with ×'s shows the noise performance without multipath propagation. As before, the center frequency of the FM-DCSK signal was varied. The dotted and dashed curves show the best and worst results, respectively. The average performance loss is 5.7 dB at BER=10^{-3}.

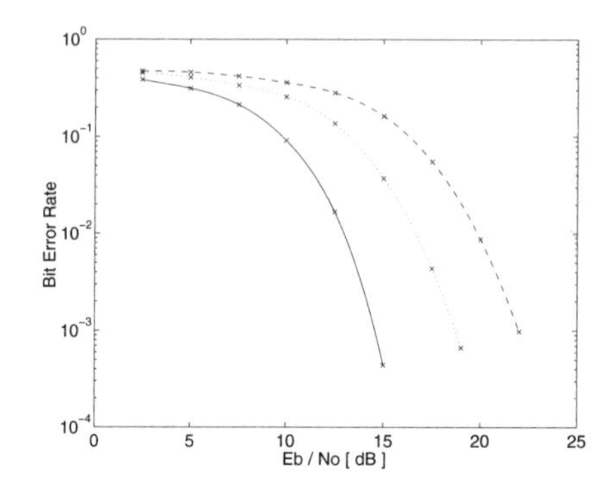

Figure 7.28: Performance degradation caused by a three-ray multipath channel in a fast FM-DCSK system. The dotted and dashed curves show the best and worst results as a function of the FM-DCSK center frequency. For comparison, the noise performance without multipath (solid curve with '×' marks) is also plotted.

7.5.4.2 Performance degradation in a four-ray multipath channel

Figure 7.29 shows the performance degradation of a fast FM-DCSK system where four propagation paths with gains [0.44 0.33 0.03 0.2] and additional delays [0ns 75ns 125ns 175ns], respectively, are considered. The dotted and dashed curves show the best and worst results, respectively, as a function of the FM-DCSK center frequency. The average loss in system performance is 6.4 dB at BER=10^{-3}.

7.5.4.3 Performance degradation in an eight-ray multipath channel

The frequency response of an eight-ray multipath channel is shown in Figure 7.30, where the gains and delays of the propagation paths are [0.36 0.15 0.12 0.12 0.10 0.08 0.04 0.03] and [0ns 80ns 110ns 160ns 190ns 220ns 230ns 240ns], respectively.

The performance degradation is plotted in Figure 7.31, where as before, the center frequency of the fast FM-DCSK signal is varied. The dotted and dashed curves show the best and worst results respectively. The average performance loss is 8.4 dB at BER=10^{-3}.

7.6 Summary

In chaotic modulation schemes the transmitted signal is a wide-band signal. The objectives of using a wide-band signal as the carrier are twofold: (a) to overcome

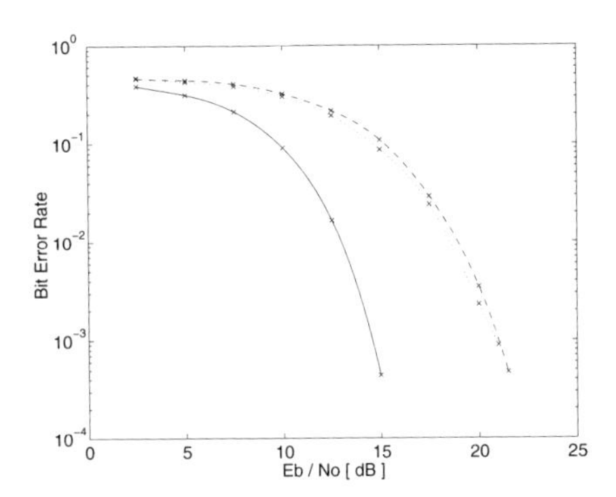

Figure 7.29: Performance degradation caused by a four-ray multipath channel in a fast FM-DCSK system. The dotted and dashed curves show the best and worst results when the FM-DCSK center frequency is varied. For comparison, the noise performance without multipath (solid curve with '×' marks) is also plotted.

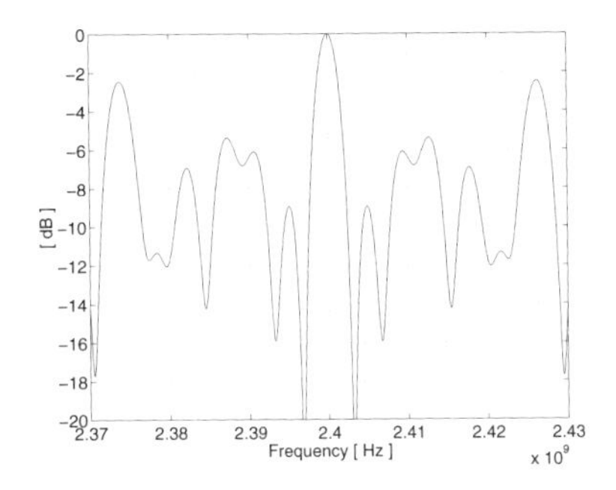

Figure 7.30: Frequency response of an eight-ray multipath channel.

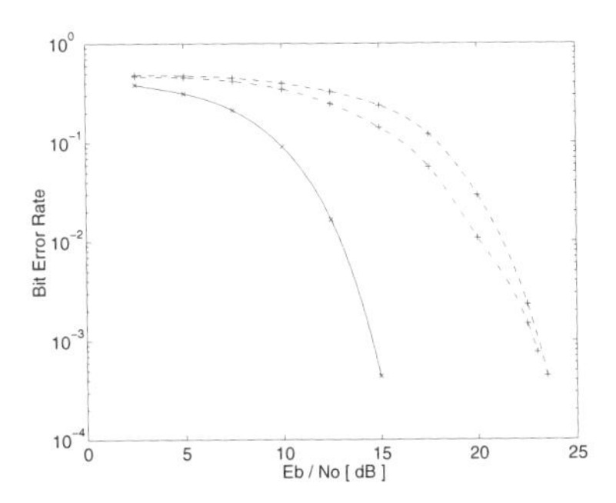

Figure 7.31: Performance degradation caused by an eight-ray multipath channel in a fast FM-DCSK system. The dotted and dashed curves show the best and worst results when the FM-DCSK center frequency is varied. For comparison, the noise performance without multipath (solid curve with '×' marks) is also plotted.

the multipath propagation problem and (b) to reduce the transmitted power spectral density to avoid interfering with other radio communications. In this sense chaotic communications systems offer a novel solution for spread spectrum communications.

Of the chaotic communications schemes published to date, FM-DCSK offers the best noise performance and the best robustness against multipath and channel imperfections. Although FM-DCSK belongs to the SS communications family, it differs from conventional DS-SS and FH-SS systems in a few basic respects:

- the narrow-band carrier is not spread by a spreading code but the carrier into which the digital information is mapped is a wide-band signal,
- because chaotic signals are not periodic, the problems which appear in conventional SS systems due to the finite length of the spreading code do not exist in FM-DCSK,
- except for timing recovery, demodulation of the transmitted bit stream is performed without synchronization in FM-DCSK, making this system much more robust against channel imperfections than modulation schemes which require carrier synchronization,
- at present, only a very limited CDMA capability can be achieved with the FM-DCSK modulation scheme, and
- in contrast to slow FH-SS, many different frequencies are visited during the transmission of one information bit, even in the slow FM-DCSK scheme.

In this chapter we have evaluated the performance of the FM-DCSK modulation scheme for different propagation conditions. Because the shape of the transmitted spectrum has the strongest influence on the multipath performance and interference caused to other radio channels, we have first analyzed in detail the shape of the spectra observed at different points in the FM-DCSK modulator. The differences between fast and slow FM-DCSK systems in terms of interference with other users have been identified, showing that fast FM-DCSK increases the background noise level in overlapping narrow-band communications systems, while slow FM-DCSK causes a pop or burst of noise which can interrupt communications in narrow-band systems which share the same frequency band.

The development of a new telecommunications system requires a fast simulator to determine the expected system performance under different channel conditions and to check the contribution of circuit and subsystem nonidealities to the overall implementation loss. Using the fact that every band-pass signal and system can be modeled by its low-pass equivalent, we have developed an ultra fast simulator where the sampling rate of simulation is determined exclusively by half of the RF bandwidth. To avoid computationally intensive numerical integration, a discrete-time equivalent of the low-pass model has been developed by means of the bilinear transformation. It is worth mentioning that the simulation technique shown here can be applied to any band-pass system having band-pass inputs.

Due to the wide-band property of chaotic basis functions, FM-DCSK offers excellent multipath performance. In the last part of this chapter we have determined by simulation the system performance of FM-DCSK in an IEEE 802.11-compliant WLAN application operating in the 2.4 GHz ISM frequency band. The results show that FM-DCSK performs extremely well over a radio channel suffering from multipath. If two propagation paths with equal attenuation are present, $\Delta\tau \geq 75$ ns, and the bandwidth of the FM-DCSK signal is 17 MHz, then the average performance loss is less than 5 dB for fast FM-DCSK. We re-emphasize that conventional narrow-band systems fail catastrophically under these conditions. In the case of three or four propagation paths, the loss in FM-DCSK system performance is less than 6.5 dB in the cases studied. The performance degradation in an eight-ray multipath channel is about 8.4 dB.

References

[1] S. Jost and C. Palmer. New standards and radio chipset solutions enable untethered information systems: $PRISM^{TM}$ 2.4GHz "Antenna-to-bits" 802.11 DSSS radio chipset solution. Technical report, Harris Corporation, http://www.semi.harris.com/prism/papers, 1998.

[2] R. C. Dixon. *Spread Spectrum Systems with Commercial Applications.* John Wiley & Sons, New York, 3rd edition, 1994.

[3] S. Haykin. *Communication Systems.* John Wiley & Sons, New York, 3rd edition, 1994.

[4] A. V. Oppenheim, A. S. Willsky, and I. T. Young. *Signals and Systems.* Prentice-Hall, Englewood Cliffs, 1983.

[5] G. Kolumbán, M. P. Kennedy, G. Kis, and Z. Jákó. FM–DCSK: A novel method for chaotic communications. In *Proc. IEEE–ISCAS'98*, volume IV, page 477, Monterey, CA, May 1998.

[6] C. Andren. A comparison of frequency hopping and direct sequence spread spectrum modulation for IEEE 802.11 applications at 2.4GHz. Technical report, Harris Corporation, http://www.semi.harris.com/prism/papers, 1997.

[7] A. Abel, M. Götz, and W. Schwarz. Statistical analysis of chaotic communication schemes. In *Proc. IEEE Int. Symposium on Circuits and Systems*, volume IV, page 465, Monterey, CA, May 31–June 3 1998.

[8] G. Kolumbán, M. P. Kennedy, and L. O. Chua. The role of synchronization in digital communications using chaos – Part I: Fundamentals of digital communications. *IEEE Trans. Circuits and Systems–Part I: Fundamental Theory and Applications*, 44(10):927, Oct. 1997.

[9] G. Kolumbán, M. P. Kennedy, and L. O. Chua. The role of synchronization in digital communication using chaos–Part II: Chaotic modulation and chaotic synchronization. *IEEE Trans. Circuits and Systems–Part I: Fundamental Theory and Applications*, 45(11):1129, Nov. 1998.

Chapter 8

Noise Filtering in Chaos-Based Communication

Thomas Schimming, Hervé Dedieu, and Martin Hasler
Dept. of Electrical Engineering, Swiss Federal Institute of Technology Lausanne DE-CIRC, CH-1015 Lausanne, Switzerland
thomas.schimming@epfl.ch, herve.dedieu@epfl.ch, martin.hasler@epfl.ch

Maciej Ogorzałek
Department of Electrical Engineering, University of Mining and Metallurgy al. Mickiewicza 30, 30-059 Kraków, Poland
maciej@zet.agh.edu.pl

8.1 Introduction

Communication schemes employing chaotic carriers (as proposed so far) adopt one of two techniques to recover the hidden information at the receiver side of the transmitting channel. The first technique requires coherent behavior of the transmitter and receiver, i.e., synchronization, while the second one employs statistical properties of the received signal (non-coherent receiver).

Chaotic chaos communication schemes typically compete with traditional spread spectrum techniques and have been pursued for a number of reasons, including the simplicity of generating non-repeating sequences and easily achieved delta-like autocorrelation properties. In fact, the use of chaotic carriers instead of random ones produces similar spectrum spreading. The benefits of this, as in standard spread-spectrum communications, include increased robustness to frequency selective fading and the fact that spectrum spreading makes the interception of the sent messages difficult or impossible using any spectral-type analysis. Typically, spread spectrum systems are used at very low signal-to-noise ratios (which is compensated by a long symbol length). Consequently, a good chaos-based scheme's performance characteristic needs to possess a good scaling behavior with respect to the symbol length, that is, allow the tradeoff between length and target signal-to-noise ratio.

For the coherent approach, a carrier signal reference is needed at the receiver. For classical schemes that are based on sinusoidal waveforms or stored pseudo-random codes this is not a problem, but due to the inherent sensitivity to perturbations, it is one for chaos-based systems. At the beginning, chaos synchronization seemed to give a way of overcoming the problem (which is a major problem for spread spectrum systems) – self-synchronizing chaotic systems find the seed values "themselves." What had been overlooked when the first chaos communication schemes were proposed is that chaos possessed very specific properties – first, synchronization is precluded by the fact that the dynamic behavior is extremely sensitive to changes in initial conditions, parameters, and noise contaminating the synchronizing signals; secondly, chaotic trajectories have regular geometric structures and this regularity and its description in terms of various measures which is used in detection also changes subject to contamination through noise signals.

From the transmission point of view usually we can assume that the parameters of the information sender and receiver are fixed with good precision and it becomes of paramount importance for minimizing the transmission error to have at the receiver side as "clean" signals as possible.

Consequently, let us consider the following problem: **Given observations of a signal which is corrupted in some way (e.g., contaminated by other sources, noise, distorted due to reflections, multipath effects, etc.) how can one develop and perform a cleaning procedure which could separate the signal of interest from other signals?**

Often, experimental measurement data are of discrete-time nature and we will assume that we have a time series $x(k)$ representing a scalar signal measured and stored from the process under consideration. With the framework we will use here, this is the case for samples of a signal transmitted through a communication channel (which represents our primary interest).

We will assume that this signal consists of the information carrying signal $u(k)$ plus a given number of contaminating signals $s_i(k)$:

$$x(k) = u(k) + s_1(k) + s_2(k) + \ldots \tag{8.1}$$

To be able to distinguish our signal $u(k)$ from the others we must find some individual characteristic of $u(k)$ which the other signals $s_i(k)$ or the measured signal $x(k)$ do not possess. In the case of conventional methods such as linear filtering, separation is done using the spectra as the criteria for distinction between different signals. There, in some cases, one can assume that the signal of interest has much lower or much higher frequency than all the contaminating signals. In other cases one can also distinguish sharp peaks in the spectrum from a flat broadband floor assumed to be produced by noise. Fourier (spectral) methods are however of little use in the case when the signal $u(k)$ is chaotic itself and has a broadband spectrum similar to the one that we assume for the noise. Filtering in such a case distorts also the signal of interest, and the coinciding spectra of noise and chaos make it impossible to distinguish the two spectrally.

What can be done if the signal we want to separate from noise is chaotic as in the case of a chaotic carrier signal? One has to use other signal characteristics such as the fact that the chaotic signal is produced by a deterministic system in contrast to the noise (which is assumed to have "only" a statistical description). Signals produced by deterministic chaotic systems possess internal structure which becomes easily visible looking at the motion in the state space. Trajectories move on attractors which are smooth submanifolds of the entire state space. In the case of "clean" dynamics it can be assumed that the evolution of the system can be described by an n-dimensional map:

$$u(n+1) = f((u(k)) \tag{8.2}$$

In practical problems one can distinguish two cases:

A. measurement noise – which we assume is also the case of distortion introduced by the transmission channel in telecommunications – when the measured signal is corrupted by an additive (multiplicative can be considered also) contaminating noise:

$$x(k) = u(k) + w(k) \tag{8.3}$$

B. dynamic noise – when contaminating signals accounting for modeling errors (including parameter mismatch) in the system evolution are considered: in the system evolution:

$$x(n+1) = f(x(k)) + v(k) \tag{8.4}$$

In the first case we consider a noise removal problem to reproduce the signal $u(k)$ while in the second case we speak of the so-called shadowing problem, e.g., reproducing the exact (noise-free) trajectory.

8.2 Chaos-Based Communication

All our considerations apply to the modulation-demodulation/detection part in the classical communication block scheme (Figure 8.1); we consider neither source nor channel coding.

For simplicity we will only consider discrete time. However, it is well known that any band limited system can be transformed to discrete time by means of sampling. In fact, (classical) communication systems are typically analyzed in an equivalent low pass discrete time representation [3], even though they are actually operated as continuous time band pass systems.

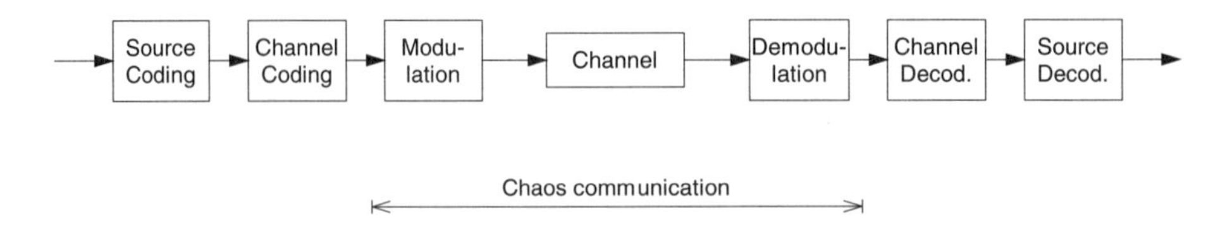

Figure 8.1: Communication system - block scheme.

8.2.1 Classes of Chaos-Based Schemes

For the communication schemes under investigation here we choose to abstract the general structure as shown in Figure 8.2 (as introduced in [17]). Essentially, for each b of M symbols to be transmitted over a given channel, a signal set V_b is chosen and a signal $\boldsymbol{x}$ thereof is sent from the transmitter to the receiver. The receiver in turn has to decide which symbol was sent, based on the received signal $\boldsymbol{y}$. This is in general accomplished by computing some generalized likelihoods (in a wide sense; or scores) λ_m (quantities that reflect the merit of detecting symbol m based on $\boldsymbol{y}$) and then choosing the detected symbol $\hat{b}$ according to the highest wide-sense likelihood (score).

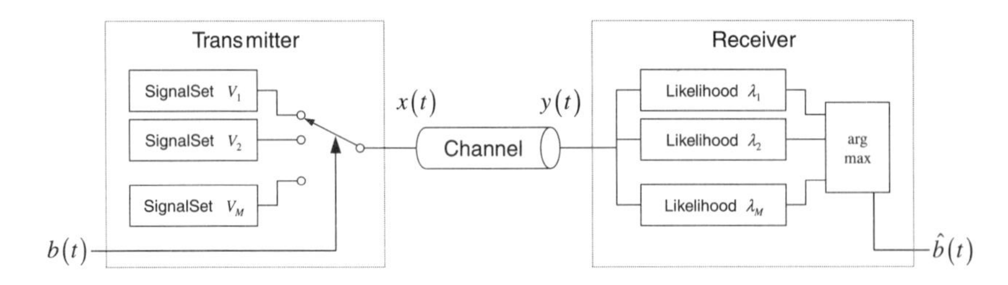

Figure 8.2: Communication system with M symbols.

The above framework is general enough to cover the chaotic schemes of interest as well as a number of classic schemes. The key difference between the schemes lies in the choice of the signal set. For efficiency reasons, the signal sets can only be of finite length N (on a given channel N has the meaning of a block length and thus is related to the speed of communication). Thus, $\boldsymbol{x} = [x_1, \ldots, x_N]$ is a vector in $\mathbb{I}^N$ (where $\mathbb{I}$ is the set on which x_k is defined, typically a subset of $\mathbb{R}$).

It is often convenient to view these signal sets as generated by dynamical systems based on a set of initial phases. Often, side conditions to assure phase continuity are in place (not directly visible in Figure 8.2).

8.2.1.1 CSK - Chaos shift keying

One of the first chaotic communication schemes proposed [20] was chaos shift keying (CSK), also called chaotic switching. Mostly, binary ($M = 2$) CSK is considered, each of the M signal sets belonging to a different chaotic system. Depending on the actual setup, a phase continuity condition (that relates the initial condition (phase) of a block to the previous block) is implied by the fact that the entire systems are not switched at the output but instead a parameter inside a single generator.

The signal sets V_b are thus defined by signals $\boldsymbol{x}_b$ originating from chaos generator $b = 1 \ldots M$. For example, assuming chaos generators based on iterated one-dimensional maps, we have

$$V_b = \left\{ \boldsymbol{x}_b \in \mathbb{I}^N : x_b(k+1) = f_b\left(x_b(k)\right), \quad x_b(1) \propto \rho_b(x_1) \right\}, \tag{8.5}$$

where the initial condition is drawn according to a given distribution $\rho_b(\cdot)$. To ensure chaos, for $\rho_b(x_1)$ it is normally preferable to use the natural invariant density [31] of the map $f_b(\cdot)$.

The classical receiver for CSK is based on the fact that chaos generators can typically synchronize an identical driven version of itself (for suitable coupling). [40] A wide-sense likelihood (score) can then be assigned in a straightforward fashion based on the synchronization error of the incoming signal $\boldsymbol{y}$ for each of the M driven systems at the receiver.

Given $\hat{\boldsymbol{x}}_m$ as the output of the m-th driven generator (using synchronization), a weighted square error likelihood (score) can now be introduced as

$$\lambda_m = -\sum_{k=1}^{N} c(k) \|\hat{x}_m(k) - y(k)\|^2, \tag{8.6}$$

where $c(k)$ is an appropriately chosen weighting of the error (typically increasing with k). [36]

Again, for the example of iterated maps, error feedback synchronization has been used mostly (there is some degree of freedom for devising synchronization schemes, see for example [35]). In this case,

$$\hat{x}_m(k+1) = f_m(\hat{x}_m(k)) + \varepsilon\left(\hat{x}_m(k) - y(k)\right) \tag{8.7}$$

where ε is the coupling constant.

In the field of classical communication schemes, CSK is conceptually related to frequency shift keying (FSK), in which the dynamical systems are sinusoidal oscillators with different frequencies. As for CSK, there is also a phase-continuous FSK variant [3]. Note that a FSK receiver is typically based on PLLs, which again involves synchronization.

8.2.1.2 CPSK - Chaotic phase shift keying

Another way of designing a chaos-based communication scheme is to always use the same dynamical system (chaos generator) for producing the signal $\boldsymbol{x}$ but changing its (initial) phase (or initial condition). Again, this results in M signal sets V_b, which now differ in the way initial conditions are chosen. For example, assuming chaos generators based on iterated one-dimensional maps as for Equation (8.5), we have

$$V_b = \left\{ \boldsymbol{x}_b \in \mathbb{I}^N : x_b(k+1) = f\left(x_b(k)\right), \quad x_b(1) \propto \rho_b(x_1) \right\}, \tag{8.8}$$

where

$$\rho_{b_1}(x_1)\rho_{b_2}(x_1) = 0 \quad \forall x_1,\ b_1 \neq b_2, \tag{8.9}$$

i.e., the support sets for the initial conditions for each symbol are disjointed. Again, for assuring chaos we might want to impose completeness,

$$\sum_{b=1}^{M} \rho_b(x_1) = \rho(x_1) \tag{8.10}$$

where $\rho(x_1)$ is again the natural invariant density. However, scenarios with some forbidden trajectories are also possible (to avoid symbols with insufficient distance, see [17]).

For the receiver, a direct estimation $\hat{x}(1)$ of the initial condition $x_b(1)$ followed by a classification can be used. Typically it is helpful to calculate estimates $\hat{x}(1)_m$ (conditioned on a hypothesis m about the received symbol $\hat{b}$) [17,45]. The wide-sense likelihoods are then the same as for CSK, shown in Equation (8.6) [46].

Alternatively, recovery of the initial condition through synchronization followed by a classification would be another possibility [35, and references therein] but in its presently known form this often has robustness problems if the noise level is high.

The conceptually related classical scheme is of course phase shift keying (PSK) with only a discrete set of phases allowed (since no chaos is used, the phase differences will not grow and thus sufficient distance is required from the beginning). Traditionally, PSK detectors are phase detectors, implemented for example by correlating with a reference signal [3]. In the non-chaotic case, it is relatively easy to synthesize a (phase-synchronized) reference signal at the receiver. Note that in the chaotic case, it is more difficult to synthesize the (phase-synchronized) reference signal due to the inherent sensitivity of chaotic systems to phase error.

8.2.1.3 DCSK - Differential chaos shift keying

The principal problem of receivers for chaotic schemes is that a phase-synchronized reference signal is typically difficult to establish. To overcome this problem, differential methods can be used that transmit the reference to be used for detection. The original DCSK proposal [48] is in fact based on very general signal sets, such that

$$V_b = \left\{ \boldsymbol{x}_b \in \mathbb{I}^N : \ x_b(k + \frac{N}{2}) = \alpha(b)\ x_b(k), \quad k = 1, \ldots, \frac{N}{2} - 1 \right\} \tag{8.11}$$

i.e., a block is divided in two parts, the first containing an (arbitrary) reference signal, the second being a modulated version of it (for binary DCSK, $\alpha(b) = \pm 1$). Using chaotic reference signals is a design choice but is not implied by the setup.

In the case that one-dimensional maps are used as in the examples shown before, the signal sets are then

$$V_b = \left\{ \boldsymbol{x}_b \in \mathbb{I}^N : \begin{cases} x_b(k+1) = f\left(x_b(k)\right) \\ x_b(k+\frac{N}{2}) = \alpha(b)\ x_b(k) \end{cases}, \quad k = 1, \ldots, \frac{N}{2} - 1 \right\}. \tag{8.12}$$

The receiver structure for this scheme is (by design) based on correlating the first and the second part of a block. This naturally induces the likelihoods (scores) defined as

$$\lambda_m = -\alpha(m) \sum_{k=1}^{N/2-1} y(k) y(k + \frac{N}{2}), \tag{8.13}$$

which is presented like this mainly for the sake of compatibility with the framework shown in Figure 8.2. Obviously, for $M = 2$ this simplifies to a decision based on the sign of the correlation.

8.2.2 Channel Noise

So far, we have introduced a number of chaos-based communication schemes that can be found in the literature. We have then presented them in the framework of signal sets (for the transmitter) and wide-sense likelihoods (for the receiver). The choice of the particular likelihood measures has been relatively ad-hoc, without explicitly taking into account the properties of the communication channel.

However, it is the communication channel with its inherent limitations that makes the communication problem challenging. The most abstract description, in terms of information theory [9], of the channel's features is its capacity. It gives a notion of the maximal amount of information in Shannon's sense that can be transmitted without errors over a given channel in a given time, summarizing all effects of interference and other signal corruption.

Unfortunately, for real channels it is a challenging problem to calculate their capacity exactly and then to design a communication scheme which can actually achieve this capacity, that is, transmit at the rate given by the channel capacity. As a consequence, simple channel models have been introduced, where the consideration of these problems is easier.

8.2.2.1 Gaussian noise

Resulting from the fact that signal energy is a central issuc for implementation, the additive Gaussian noise channel has been a very popular and perhaps the most basic channel model, since Gaussian noise is the stochastic process with the highest entropy of all stochastic processes given a certain energy (power if N is infinite). Thus, Gaussian noise can be considered a worst case noise situation if the energy level is the reference of interest. In a way it reflects the absence of any knowledge about the noise except its average power.

In a real communication system, where transmitter and receiver are far apart, channel features other than just the noise are important as well, such as the propagation delay from transmitter to receiver and a possible signal attenuation on the channel.

For the Additive White Gaussian Noise (AWGN) channel model we assume that we can deal with propagation delay and signal attenuation in a transparent way and do not include it into our channel model. Additionally, as we are considering only discrete time here, we do not have the problem of defining band limited white Gaussian noise (continuous time), all we need is to consider is an additive random noise vector $\boldsymbol{w}$ with i.i.d. Gaussian elements of zero mean and variance σ^2

$$\boldsymbol{y} = \boldsymbol{x} + \boldsymbol{w}, \qquad w_k \propto \mathcal{N}_\sigma. \tag{8.14}$$

It should be stressed that AWGN is only one possible (very simple) channel model, appropriate for some radio channels, wire-based channels, channels with thermal noise sources, etc. On other channels, for example fiberoptic channels, the main source of signal corruption is distortion, not (Gaussian) noise, thus this model is no more valid.

With the noise defined like this, it is straightforward to set up a probability space in which $\boldsymbol{X}$ is the vector process of the channel input (comprised of realizations $\boldsymbol{x}$), and $\boldsymbol{Y}$ is the vector process of the channel output. $\boldsymbol{W}$ is the noise process. The noise is now completely characterized by $p(\boldsymbol{y}|\boldsymbol{x})$, which describes the action of the channel. Assuming independence of $\boldsymbol{X}$ and $\boldsymbol{W}$ (as usual), the AWGN model implies

$$p(\boldsymbol{y}|\boldsymbol{x}) = \frac{1}{(2\pi)^N \sigma^{2N}} e^{-\frac{\|\boldsymbol{y}-\boldsymbol{x}\|^2}{2\sigma^2}} \tag{8.15}$$

Methods that imply Gaussian noise use this relation either explicitly or implicitly, sometimes in its logarithmic equivalent

$$\log p(\boldsymbol{y}|\boldsymbol{x}) = Z - \frac{1}{2\sigma^2}\|\boldsymbol{y} - \boldsymbol{x}\|^2 \tag{8.16}$$

$$= Z + \frac{1}{2\sigma^2}\left(2\langle \boldsymbol{x}, \boldsymbol{y}\rangle - \|\boldsymbol{x}\|^2 - \|\boldsymbol{y}\|^2\right) \tag{8.17}$$

Qualitatively it can be said that non-stochastic approaches that rely on the Euclidean distance $\|\boldsymbol{y} - \boldsymbol{x}\|$ often implicitly assume Gaussian noise by involving likelihoods derived from Equation (8.16), while approaches relying on correlations $\langle \boldsymbol{x}, \boldsymbol{y}\rangle$ additionally assume the signal energies to be constant. Here, we will prefer the stochastic view of the problem, rather than a geometric view which often arises from the relation with the Euclidean distance.

8.2.2.2 Other (more complicated) noise

While the AWGN model introduced before is relatively simple and allows us to analyze the performance of (in particular linear) schemes in a compact and convenient way, it does not always reflect the reality of a given communication channel to a satisfactory extent.

Most realistic channels have in fact a non-trivial impulse response (filtering channels), that can in the worst case be highly dependent on the position of the transmitter and the receiver, as in the context of mobile communication with multiple propagation paths. Also, for multi-user systems, the main interference in the channel is caused by other users, and it cannot always be assumed that this interference can be adequately modeled as Gaussian noise (in particular for the case of a low number of strong interferers).

As for the Gaussian noise, a (more complicated) noise model again results in a probability space which allows us to describe $p(\boldsymbol{y}|\boldsymbol{x})$ in a more or less compact way. Here we will limit ourselves to considerations about AWGN, but by replacing $p(\boldsymbol{y}|\boldsymbol{x})$, any noise model can be considered instead.

8.2.3 The Role of Noise Reduction

The design of a receiver for a given scheme can be based on a number of different ideas. The key issue is always implementation simplicity. Consequently, a simple detector with a simple preprocessing is preferred. Of course, there is always the possibility to devise instead a single block optimized receiver setup featuring a complicated classification rule. In the past, however, communication engineers have preferred easier to analyze sub-components.

8.2.3.1 Noise pre-filtering

The most obvious way to enhance the performance of a system that works well at low noise levels is to add a preprocessing step in which noise is suppressed. In that way, the applicability of the system is extended to higher noise levels. Thus, the (usually simple) detector of the system is retained.

For traditional communication schemes based on sinusoidal waveforms (or equivalently, narrow band spectra), it is easy to distinguish noise and information content using a matched filter which in this case will be a relatively sharp (linear) band pass filter. For chaos-based communication schemes, things are more complicated, here the simple spectral separation of signal and noise no longer applies, thus noise can only be removed using a noise reduction method that is equivalent to a nonlinear matched filter.

However, it is not immediately clear in which sense the noise reduction has to be performed by the pre-filter. In particular, it is in general not obvious that a reduction of the noise's energy by the pre-filter provides the desired performance increase at the detector.

8.2.3.2 Direct detection

The other extreme is an approach that directly classifies the incoming signal $\boldsymbol{y}$ based on the statistical information that is available about the system (probability space), as well as the information available about the dynamics in the transmitter. In general, this allows us to derive the optimal classifier, but usually the receiver obtained is complex and is not directly suitable for implementation [36]. In fact, the associated computational effort typically grows exponentially with the length N of $\boldsymbol{y}$. The optimal direct detection approach is typically useful to analyze the performance bounds of a given system and to serve as a base for suboptimal detection schemes with reduced complexity. Here we will concentrate on noise reduction techniques. For direct detection, please refer to [36, 46].

8.2.3.3 Detection on internals of the noise pre-filter

As will be shown below, a noise pre-filter often already internally computes some likelihood figures related to its input $\boldsymbol{y}$. Since the receiver that we consider in our framework (see Figure 8.2) is based on likelihoods as well, it seems intuitive to use the likelihoods present in the noise reduction pre-filter to do the detection. In this way we can eliminate the inherent problem of pre-filtering that was already stated above, namely to judge the applicability of a particular noise reduction target such as minimizing the energy of the remaining noise.

8.3 Noise Reduction

As outlined in the previous section, noise reduction is a key element of any communication scheme, chaos-based or not. In the context of sinusoidal waveforms it can simply be accomplished by linear filtering. However, when chaos is used, this requires more involved procedures which will be introduced formally in the following sections.

8.3.1 Problem Definition

In order to precisely define our notion of noise reduction, we must first establish the notion of noise and in which sense it is to be reduced. Here we will, for simplicity, only consider additive white Gaussian noise as introduced in Section 8.2.2.1. Consequently, following Equation 8.15, we introduce the probability space defined by the joint probability

$$p(\boldsymbol{x}, \boldsymbol{y}) = p(\boldsymbol{x})p(\boldsymbol{y}|\boldsymbol{x}) \tag{8.18}$$

$$= p(\boldsymbol{x})\frac{1}{(2\pi)^N \sigma^{2N}} e^{-\frac{\|\boldsymbol{y}-\boldsymbol{x}\|^2}{2\sigma^2}} \tag{8.19}$$

where for dynamical systems, there is no stochasticity for $x_2, \ldots, x_N$ once x_1 is known. (For simplicity, we derive the result for one-dimensional systems. An extension to more-dimensional systems is straightforward.) Thus, for one-dimensional maps $f(\cdot)$ such that $x_{n+1} = f(x_n)$ with natural invariant density

$\rho(x_1)$ in AWGN,

$$p(\boldsymbol{x}, \boldsymbol{y}) = \rho(x_1)\frac{1}{(2\pi)^N \sigma^{2N}} e^{-\frac{\sum_{n=1}^{N} \left\| y_n - f^{(n-1)}(x_1) \right\|^2}{2\sigma^2}} \tag{8.20}$$

Given the probability space, all statistical quantities of interest can now be computed based on appropriate marginalization and expectations.

Note that by replacing $p(\boldsymbol{y}|\boldsymbol{x})$ in the probability space model, the resulting framework can be applied to any stochastically described channel model. However, it might not be possible to express some of the results shown in the following in a compact form any more as the channel model gets more sophisticated.

Due to the relation of Gaussian noise to its energy (Section 8.2.2.1), we define (as usually done [10]) the sense in which noise reduction is optimal in terms of energies, namely as maximizing the processing gain

$$P_G = \frac{E(\|\boldsymbol{Y} - \boldsymbol{X}\|^2)}{E(\|\hat{\boldsymbol{X}} - \boldsymbol{X}\|^2)}. \tag{8.21}$$

Then, the **noise reduction problem** is:
Find an optimal estimator ϕ that maximizes the processing gain P_G, taking into account all the dynamical and statistical information we possess.

As presented here, it appears that the solution is necessarily a parallel estimator that will find the complete optimal estimate $\hat{\boldsymbol{x}} = \phi(\boldsymbol{y})$ simultaneously. However, since the dynamical system we use here induces a Markov property (that is, X_n only depends on X_{n-1}), an iterative solution $\hat{x}_n = \psi(\hat{x}_{n-1}, \boldsymbol{y})$ also exists in general [12].

8.3.2 Parallel Solution of the Noise Reduction Problem

The estimator maximizing P_G (or equivalently, minimizing the residual noise variance, since $E(\|\boldsymbol{Y} - \boldsymbol{X}\|^2)$ in Equation (8.21) can be considered fixed) is the (conditional mean) Bayesian estimator [12], such that

$$\hat{\boldsymbol{x}} = \phi(\boldsymbol{y}) = E(\boldsymbol{X}|\boldsymbol{Y} = \boldsymbol{y}) \tag{8.22}$$

or more explicitly, integrating over the support set $\mathbb{I}$ of the state space,

$$\hat{x}_n = \int_{\mathbb{I}} f^{(n-1)}(x_1) p(\boldsymbol{x}|\boldsymbol{y}) dx_1 \tag{8.23}$$

that contains the posterior probability $p(\boldsymbol{x}|\boldsymbol{y})$, which can be expressed in terms of $p(\boldsymbol{y}|\boldsymbol{x})$ using Bayes' formula yielding

$$\hat{x}_n = \phi_n(\boldsymbol{y}) = \frac{\int_{\mathbb{I}} f^{(n-1)}(x_1)\rho(x_1)p(\boldsymbol{y}|x_1)dx_1}{\int_{\mathbb{I}} \rho(x_1)p(\boldsymbol{y}|x_1)dx_1}. \tag{8.24}$$

The associated expectation of the residual noise variance $E||\hat{\boldsymbol{X}} - \boldsymbol{X}||^2$ is then

$$E||\,(\boldsymbol{X}|\boldsymbol{Y}) - \boldsymbol{X}||^2 = \int_{\mathbb{I}} \rho(x_1) \int_{-\infty^N}^{+\infty^N} p(\boldsymbol{y}|x_1) \sum_{n=1}^{N} \left\| \phi_n(\boldsymbol{y}) - f^{n-1}(x_1) \right\|^2 d\boldsymbol{y} dx_1 \tag{8.25}$$

Note that, in general, solving the integral (8.24) necessary to calculate the optimal estimate $\hat{\boldsymbol{x}}$ results in a nonlinear problem that has no explicit solution. In the following, we give an overview of solution methods that exploit the particular form of the prior probability $\rho(x_1)$ and the likelihood $p(\boldsymbol{y}|x_1)$.

8.3.2.1 Explicit solution

In the simplest case, the prior probability is discrete or uniform and the map $f(\cdot)$ is linear, as it is the case for PSK. In this case and a handful of other lucky exceptions, Equation (8.24) can be calculated explicitly.

However, in the cases of interest to us, namely where $f(\cdot)$ is nonlinear, there is no straightforward closed form solution to the integrals in Equation (8.24).

8.3.2.2 Discretization

Sometimes, due to the nonlinearity of $f(\cdot)$, there might not be a single explicit solution to the integrals in Equation (8.24), but by discretizing the domain of definition $\mathbb{I}$ into several $\mathbb{I}_k$, the integrals can be solved. This applies in particular to the case where $f(\cdot)$ is piecewise linear, as then essentially, for every *piece*, an explicit solution can be found, which are then combined to obtain $\phi(\boldsymbol{y})$.

$$\hat{x}_n = \phi_n(\boldsymbol{y}) = \frac{\sum_{k=1}^{K} \int_{\mathbb{I}_k} f_k^{(n-1)}(x_1)\rho_k(x_1)p(\boldsymbol{y}|x_1)dx_1}{\sum_{k=1}^{K} \int_{\mathbb{I}_k} \rho_k(x_1)p(\boldsymbol{y}|x_1)dx_1}. \tag{8.26}$$

For piecewise linear Markov maps, there exist $\mathbb{I}_k$ such that $f^{(n-1)}(\cdot)$ is linear and $\rho_k(\cdot)$ is constant. Assigning symbols to each of the intervals of definition of the piecewise linear map defines a symbolic sequence $\boldsymbol{s} = s_1, \ldots, s_N$ corresponding to $\boldsymbol{x}$. In the piecewise linear case, the $\mathbb{I}_k$ correspond to the possible symbolic sequences that the map $f(\cdot)$ admits [37]. Thus, the problem discretizes into 2^{N-1} linear subproblems (for two symbol symbolic dynamics where all sequences exist). For a detailed calculation, refer to [37].

8.3.2.3 Optimization

In the case where calculating the estimator explicitly is not desired or not possible, the estimate can be found by solving the minimization problem directly using some optimization (that is, local gradient based) technique. Two interpretations of this technique are possible, a stochastic one and a deterministic one. Here we will take the stochastic point of view.

Instead of minimizing the expectation of the error from Equation (8.25) (which is not straightforward in this case), one can hope to achieve the same by minimizing at each estimation the individual estimation error. This corresponds to a maximum a posteriori (MAP) estimation, which, in contrast to the above mentioned conditional mean Bayesian estimation, returns the most probable estimate according to the posterior probability, and not the conditional mean [12]. Note that for Gaussian distributions, the MAP estimate and the conditional mean coincide. However, most chaotic systems of interest do not have a Gaussian invariant probability distribution.

Thus, for obtaining a MAP estimate, we compute

$$\hat{\boldsymbol{x}} = \arg\max_{\boldsymbol{x}} p(\boldsymbol{x}|\boldsymbol{y}) \tag{8.27}$$

$$= \arg\max_{\boldsymbol{x}} \frac{p(\boldsymbol{y}|\boldsymbol{x})p(\boldsymbol{x})}{p(\boldsymbol{y})} \tag{8.28}$$

or, in logarithmic terms,

$$\hat{\boldsymbol{x}} = \arg\min_{\boldsymbol{x}} \big(-\log p(\boldsymbol{x}) - \log p(\boldsymbol{y}|\boldsymbol{x}) + \log p(\boldsymbol{y})\big) \tag{8.29}$$

$$= \arg\min_{\boldsymbol{x}} \big(C(\boldsymbol{x}, \boldsymbol{y}) + C_0\big) \tag{8.30}$$

where $\log p(\boldsymbol{y})$ and any other constant with respect to the minimization can be disregarded. The MAP problem now leads to a directly implementable estimator based on optimization.

Numerous noise reduction methods based on optimization can be found in the literature [15,43]. They are often motivated by a Lagrangian argument that can be found in optimal control theory [11]. The cost function there is not typically motivated with a statistical argument, but instead is chosen in a relatively ad-hoc manner.

Here, we present the cost function most often used with the (statistical) MAP approach and interpret the different terms it contains in this framework. The ad-hoc cost functions are typically chosen [43] as

$$C(\boldsymbol{x}, \boldsymbol{y}) = \underbrace{\sum_{n=2}^{N} \|x_n - f(x_{n-1})\|^2}_{\text{dynamical cost } C_1(\boldsymbol{x})} + \Gamma \underbrace{\sum_{n=1}^{N} \|x_n - y_n\|^2}_{\text{similarity cost } C_2(\boldsymbol{x}, \boldsymbol{y})}, \tag{8.31}$$

composed of a *dynamical cost* $C_1(\boldsymbol{x})$ (to penalize dynamically "wrong" solutions) and a *similarity cost* $C_2(\boldsymbol{x}, \boldsymbol{y})$ (to penalize solutions far away from the data). A weight factor (Lagrangian multiplier) Γ combines the two parts. Typically, both cost terms are Euclidean distances. For the *similarity cost*, correlation is also used [22,23,43].

Relating Equation (8.31) to (8.29), the method can be discussed with respect to its statistical implications. As outlined in Section 8.2.2.1, Equation (8.16), choosing Euclidean distance for the *similarity cost* corresponds to an additive

Gaussian noise. Using correlation instead of distance, $\|\boldsymbol{x}\| = const$ is implied. Motivating the *dynamical cost* is less straightforward. In the deterministic interpretation, it provides a constraint on the dynamics. In the stochastic interpretation, when relating it to Equation (8.29), a Gaussian dynamical noise in the feedback loop is implied as physical interpretation (cf. Figure 8.3).

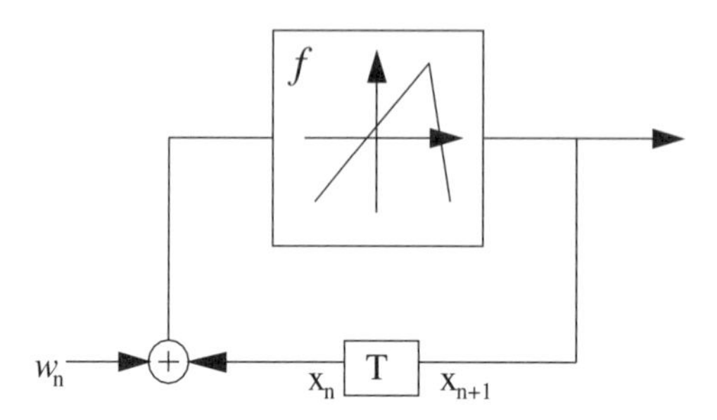

Figure 8.3: Noise in the feedback loop.

Then, the weight factor Γ, as can be verified easily, is defined by the ratio of the variances of the channel noise and the dynamic loop noise. Its choice in the literature [22, 23, 43] has been mostly empirical, particularly since the Gaussian loop noise model is not very realistic and thus it is difficult to find a corresponding variance that is realistic.

Note that this loop noise $\boldsymbol{w}$ in Figure 8.3 summarizes the modeling error (parameter uncertainty and calculation imprecisions) of $C_1(\boldsymbol{x})$ with respect to the real chaotic system that produced $\boldsymbol{x}$. In the literature, this approach using optimization is sometimes termed "approximately constrained" [15]. Assuming exact knowledge of the model (system), a variation of it, considering an exactly constrained framework ($\Gamma \gg 1$) has been considered as well [15].

The most important problem with the local gradient based techniques used for the optimization-based approach is the inherent sensitivity to local minima in $C(\boldsymbol{x}, \boldsymbol{y})$ whose presence are due to the nonlinearity of $f(\cdot)$. Additionally, the absence of knowledge about the speed of convergence makes it difficult to judge the computational effort actually necessary.

As outlined before, being a MAP technique, additional error is induced if the posterior probability is such that the mean value does not correspond to (or is close to) its maximum. Again, this is difficult to judge in the generality in which this method is usually applied.

8.3.3 Iterative Solution of the Noise Reduction Problem

While the parallel solution presented above is intuitive and in the spirit of (traditional) Bayesian parameter estimation [12], it does not necessarily provide an efficient way to calculate actual estimates.

Iterative solutions emerge from the inherent Markov property in the case that the signal $\boldsymbol{x}$ to be estimated is generated by a dynamical system. Considering the initial condition of the dynamical system to be a stochastic variable distributed according to the natural invariant density of the dynamical system [13], it can be viewed as a stochastic dynamical system, i.e.,

$$X_{n+1} = f(X_n), \quad X_1 \propto \rho(x_1) \tag{8.32}$$

$$Y_n = X_n + W_n, \tag{8.33}$$

its statistical properties are completely described in the density space by the Frobenius-Perron operator (FPO) $\mathcal{P}$ and the operator $\mathcal{O}$ [13,19], such that

$$p(x_n|\boldsymbol{y}_n) = \mathcal{O}\big(p(x_n, \boldsymbol{y}_{n-1}), y_n\big) \tag{8.34}$$

$$p(x_{n+1}|\boldsymbol{y}_n) = \mathcal{P}\big(p(x_n, \boldsymbol{y}_n)\big) \tag{8.35}$$

where $\boldsymbol{y}_n = [y_1, \dots, y_n]^T$.

Thus, to solve Equation (8.24) iteratively, an iteration over two functionals, $p(x_n|\boldsymbol{y}_n)$ and $p(x_{n+1}|\boldsymbol{y}_n)$ allows to find the densities of interest at any given time n. Note that, since the iteration is performed over functionals, this results in a problem of infinite dimension. In a few cases, it can be solved explicitly, otherwise an approximation that reduces the problem through quantization or parameterization is necessary to obtain a solution [16].

Finally, with the knowledge of the functional $p(x_n|\boldsymbol{y}_n)$ for all $n = 1, \dots, N$, we can calculate the Bayesian minimum variance estimate $E(\boldsymbol{X}|\boldsymbol{Y} = \boldsymbol{y})$ as introduced in Equation (8.23).

8.3.3.1 Explicit solution

The well known explicit solution to the problem is the Kalman filter [16], which exists in the case that $X_{n+1} = f(X_n)$ is a linear dynamical system. The Kalman filter proves to be optimal for Gaussian distribution of the involved random variables [11], however, it relies on the linear (possibly time-variant) nature of the underlying dynamical system.

Since chaotic systems (the point of our interest) are inherently nonlinear, the explicit solution to the recursive Kalman filter does not directly apply (except for a class of Lure systems).

8.3.3.2 Quantization

The main point of quantization is the reduction of the infinite-dimensional problem in the density space to a finite dimensional problem, where $\mathcal{O}$ and $\mathcal{P}$ are now transition probability matrices.

In some cases, a natural discretization (according to the symbolic dynamics of the map) is available (cf. Section 8.3.2.2); it is not available in others or exists in too complicated a manner. In the latter case, an uniform quantization of $\mathbb{I}$ into $\mathbb{I}_k$ can be used to approximate discrete transition probabilities [15,45]. Here we will concentrate on the uniform quantization.

Without loss of generality we assume that the state space $\mathbb{I}$ of the signal x_k can be normalized on an interval of length 1, i.e., $[-\frac{1}{2}, +\frac{1}{2}]$. Suppose that we describe the state space $[\frac{-1}{2}, \frac{+1}{2}]$ using K non overlapping intervals $\mathbb{I}_k$, such that

$$\mathbb{I}_k = \left((k-1)\,\Delta_B - \frac{1}{2}, k\Delta_B - \frac{1}{2} \right] \qquad \Delta_B = \frac{1}{K} \qquad k = 1 \dots K \tag{8.36}$$

the center of which is defined as $I_c(k)$, i.e.,

$$I_c\,(k) = \frac{2k-1}{2}\Delta_B - \frac{1}{2} \tag{8.37}$$

We construct a resulting discrete system such that at time n it is in state $s_n = k$ if $x_n \in \mathbb{I}_k$. The symbolic sequence $\boldsymbol{s} = s_1, \dots, s_N$ thus corresponds to the intervals $\mathbb{I}_k$ that $\boldsymbol{x}$ visits as $n = 1 \dots N$.

If K is chosen large enough we could see our noise cleaning problem as the following problem; given N consecutive observations $y_1 \dots y_N$, find the most likely state sequence $\hat{\boldsymbol{s}} = [\hat{s}_1, \hat{s}_2, \dots \hat{s}_N]$ (Ideally x_i should belong to $\mathbb{I}_{\hat{s}_i}$ for $i = 1 \dots N$). If there exists a method to find the location of these most likely intervals $\mathbb{I}_k$ we will define the clean estimated signal $\hat{\boldsymbol{x}}$ as the sequence of the most likely interval centers, i.e.,

$$\hat{\boldsymbol{x}} = [I_c(\hat{s}_1), I_c(\hat{s}_2), \dots, I_c(\hat{s}_N)]^T \tag{8.38}$$

Let $\boldsymbol{s}$ be a path of N consecutive states, i.e.,

$$\boldsymbol{s} = [s_1, s_2, \dots, s_N]^T \tag{8.39}$$

We have now to define what kind of criterion should be used in order to find the best path among the N consecutive states. A quite natural criterion is to find $\hat{\boldsymbol{s}}$ which satisfies

$$p(\hat{\boldsymbol{s}}|\boldsymbol{y}) = \max_{\boldsymbol{s}} p(\boldsymbol{s}|\boldsymbol{y}) \tag{8.40}$$

Note that it implies again a maximum a posteriori (MAP) solution, as in Section 8.3.2.3. The same discussion with respect to the link between MAP and minimum variance applies.

Using Bayes' rules we have

$$p(\boldsymbol{s}|\boldsymbol{y}) = \frac{p(\boldsymbol{y}|\boldsymbol{s})p(\boldsymbol{s})}{p(\boldsymbol{y})} = \frac{p(\boldsymbol{s}, \boldsymbol{y})}{p(\boldsymbol{y})} \tag{8.41}$$

As the observation $\boldsymbol{y}$ is given, we have to find $\hat{\boldsymbol{s}}$ which satisfies

$$p(\hat{\boldsymbol{s}}, \boldsymbol{y}) = \max_{\boldsymbol{s}} p(\boldsymbol{s}, \boldsymbol{y}) = \max_{\boldsymbol{s}} p(\boldsymbol{y}|\boldsymbol{s})p(\boldsymbol{s}) \tag{8.42}$$

Let us define as $\boldsymbol{s}_m$ the first m states in $\boldsymbol{s}$, i.e.,

$$\boldsymbol{s}_m = [s_1, s_2, \dots, s_m]^T \tag{8.43}$$

What is essential to recognize first is that the "process" $\boldsymbol{s}_1^m$ is a first-order Markov process, i.e.,

$$p(\boldsymbol{s}_m) = p(s_m|s_{m-1})p(\boldsymbol{s}_{m-1}) \tag{8.44}$$

where $p(s_m|s_{m-1})$ is the discretized version of the FPO, the state transition probability from state s_{m-1} to state s_m.

Secondly, due to the i.i.d. nature of the noise we have that

$$p(y_m|\boldsymbol{s}_m) = \prod_{k=1}^{m} p(y_k|s_k) \tag{8.45}$$

$$= \prod_{k=1}^{m} p_w\left(y_k - I_c(s_k)\right) \tag{8.46}$$

where $p(y_k|s_k)$ is the conditional observation probability and p_w is the probability distribution of the noise. (The last equation in (8.45) is approximative; it is true if the number K of intervals is great enough.)

The probability $p(\boldsymbol{s}, \boldsymbol{y})$ can be therefore expressed as a standard product of state transition probabilities times conditional observation probabilities.

$$p(\boldsymbol{s}, \boldsymbol{y}) = \prod_{k=1}^{N} p(s_k|s_{k-1})p(y_k|s_k) \tag{8.47}$$

Note that the quality of the estimate depends on the resolution of the quantization, as does the computational complexity. This approach consequently allows us to control the tradeoff between precision and computational cost. The implementation of a noise filter based on quantization is presented in Section 8.3.4.3.

8.3.3.3 Parameterization

Another way to reduce the general problem is to use a parameterized approximation [19] as a base for the density space and find $\mathcal{O}$ and $\mathcal{P}$ correspondingly. Where applicable, this leads to a solution that is more computationally efficient than approximating by quantization. Consequently, the infinite dimensional problem of Equation (8.34) is reduced to a finite dimension equal to the number of parameters.

An application of this method to piecewise linear maps can be found in [19]. As a general rule, as explained there, it is difficult to find a common basis to represent both the (bounded) piecewise constant densities found for piecewise linear chaotic systems on one side and the (unbounded) Gaussian noise on the other side.

An advantage of the method is that no quantization errors (as in general for approximations based on a quantization) occur. However, other approximation errors depending on the ability of the chosen parameterized basis to represent the real densities can occur.

8.3.4 Implementation

In this section we give some implementation detail on the methods shown in Section 8.3. In particular, optimization schemes, that is, schemes that find a (local) minimum of cost functions with or without constraints are investigated, and an efficient algorithm for the discretized (quantized) approach are presented.

8.3.4.1 Gradient type methods

If optimization is used to find a solution to the noise reduction problem based on local properties, a suitable recursive method is needed to minimize the cost function $C(\boldsymbol{x}, \boldsymbol{y})$ (8.31) as introduced in Section 8.3.2.3. Our goal is to find a pseudo-trajectory

$$\hat{\boldsymbol{x}} = [\hat{x}_1, , \hat{x}_2, \ldots \hat{x}_N]^T \tag{8.48}$$

given the noisy observation $\boldsymbol{y} = [y_1, , y_2, \ldots y_N]^T$ which satisfies two conditions

A. $\hat{\boldsymbol{x}}$ is in the neighborhood of $\boldsymbol{y}$ according to some measure of closeness.

B. $\hat{\boldsymbol{x}}$ is as close as possible to a system true trajectory according to an another measure of closeness.

With the general cost function $C(\boldsymbol{x}, \boldsymbol{y})$ as defined by (8.31) (which takes into account these two conditions) it remains to derive a recursive method which finds a local minimum of $C(\hat{\boldsymbol{x}}, \boldsymbol{y})$. First-order methods (gradient methods) or second-order methods like Newton or Quasi-Newton methods can solve this problem. Let us define as $\hat{\boldsymbol{x}}(k)$ the recursive estimate of $\hat{\boldsymbol{x}}$. A very common method is to use a gradient descent method, i.e.,

$$\begin{aligned} \hat{\boldsymbol{x}}(0) &= \boldsymbol{y} \\ \hat{\boldsymbol{x}}(k) &= \hat{\boldsymbol{x}}(k-1) - \mu \left.\frac{\partial C(\hat{\boldsymbol{x}}, \boldsymbol{y})}{\partial \hat{\boldsymbol{x}}}\right|_{\hat{\boldsymbol{x}}=\hat{\boldsymbol{x}}(k-1)} \end{aligned} \tag{8.49}$$

where μ is a positive real number. Provided that μ is chosen small enough (in general μ is chosen as a decreasing function of k), $\hat{\boldsymbol{x}}(k)$ converges towards a local minimum of $\mathcal{C}(\hat{\boldsymbol{x}}, \boldsymbol{y})$. These kinds of methods have been proposed by Kostelich and Yorke [25] and Lee and Williams [28] among others. The approach is very simple and does not need any matrix inversion. Improvement in the speed of convergence can be obtained by using second order methods which require a matrix inversion of the Hessian matrix of the cost function. It is important to realize that the algorithm is a suboptimal algorithm which is intrinsically biased. The reason for this bias is obvious. Since the cost function $\mathcal{C}(\hat{\boldsymbol{x}}, \boldsymbol{y})$ is a tradeoff between two cost functions, the minimization of $\mathcal{C}(\hat{\boldsymbol{x}}, \boldsymbol{y})$ does not imply the simultaneous minimization of $\mathcal{C}_1(\hat{\boldsymbol{x}}, \boldsymbol{y})$ and $\mathcal{C}_2(\hat{\boldsymbol{x}})$. Therefore this kind of method does not force $\hat{\boldsymbol{x}}$ to be exactly on a true trajectory of the system. Even if numerical experiments remain to be done, we suspect the method to be inefficient for low signal-to-noise ratios (SNR).

8.3.4.2 Constrained optimization with Lagrangian methods

The main problem which limits the previous approach is the inability to deal properly with the trajectory constraints. It has been proposed by Farmer and Sidorowich [22] to consider the above stated problem as a problem of minimization of $C_1(\hat{\boldsymbol{x}}, \boldsymbol{y})$ with respect to $\hat{\boldsymbol{x}}$ subject to $N-1$ equality constraints

$$f(\hat{x}_n) = \hat{x}_{n+1} n = 1 \ldots N-1 \tag{8.50}$$

The problem becomes a standard optimization problem with equality constraints. Its solution $\hat{\boldsymbol{x}}_*$ is therefore a stationary point of the Lagrange function $L(\hat{\boldsymbol{x}}, \boldsymbol{\Lambda})$, i.e.,

$$L(\hat{\boldsymbol{x}}, \boldsymbol{\Lambda}) = \mathcal{C}_1(\hat{\boldsymbol{x}}, \boldsymbol{y}) + \sum_{n=1}^{N-1} \lambda_n^T \left[f(\hat{x}_n) - \hat{x}_{n+1} \right] \tag{8.51}$$

where λ_n is the D-dimensional Lagrange multiplier associated with the $n-th$ constraint and

$$\boldsymbol{\Lambda} = [\lambda_1, \lambda_2, \ldots, \lambda_{N-1}]^T \tag{8.52}$$

We have $2N-1$ unknowns which consist of the N pseudo-trajectory points and of the $N-1$ Lagrange multipliers. These unknowns can be determined by setting to zero the derivatives of $L(\hat{\boldsymbol{x}}, \boldsymbol{\Lambda})$ with respect to $\hat{\boldsymbol{x}}$ and $\boldsymbol{\Lambda}$. For instance, suppose that the cost function $\mathcal{C}_1(\hat{\boldsymbol{x}}, \boldsymbol{y})$ is given by

$$\mathcal{C}_1(\hat{\boldsymbol{x}}, \boldsymbol{y}) = \frac{1}{2} \sum_{n=1}^{N} ||\hat{x}_n - \boldsymbol{y}_n||^2 \tag{8.53}$$

Setting to zero the derivative of (8.51) with respect to $\hat{\boldsymbol{x}}$ we obtain

$$\begin{aligned} -(\boldsymbol{y}_1 - \hat{x}_1) + D_f^T(\hat{x}_1)\lambda_1 &= 0 \\ (\boldsymbol{y}_k - -\hat{x}_k) + D_f^T(\hat{x}_k)\lambda_k - \lambda_{k-1} &= 0, \qquad -k = 2 \ldots N-1 \\ (\boldsymbol{y}_N - \hat{x}_N) - \lambda_{N-1} &= 0 - \end{aligned} \tag{8.54}$$

where $D_f(\hat{x}_k)$ is the Jacobian matrix of the function f. Setting to zero the derivative of (8.51) with respect to $\boldsymbol{\Lambda}$ we obtain

$$f(\hat{x}_n) - \hat{x}_{n+1} = 0, \qquad n = 1 \ldots N-1 \tag{8.55}$$

Equalities (8.54) and (8.55) form a set of $2N-1$ nonlinear equations which can be solved by using an iterative Newton-Raphson technique. Starting from $\hat{\boldsymbol{x}}(0) = \boldsymbol{y}$ we can try to linearize the nonlinear equations around $\hat{\boldsymbol{x}}(0)$ and find the vector $\Delta\hat{\boldsymbol{x}}(0)$ we should add to $\hat{\boldsymbol{x}}(0)$ to fulfill the equalities (8.54) and (8.55).

We can proceed recursively until we find a fixed point for $\hat{\boldsymbol{x}}$ and $\boldsymbol{\Lambda}$. Linearizing (8.54) and (8.55) we find

$$
\begin{aligned}
-(\boldsymbol{y}_1 - \hat{x}_1 - \widehat{\Delta x}_1) + D_f^T(\hat{x}_1)\lambda_1 &= 0 \\
-(\boldsymbol{y}_k - \hat{x}_k - \widehat{\Delta x}_k) + D_f^T(\hat{x}_k)\lambda_k - \lambda_{k-1} &= -0, \quad k = 2\ldots N-1 \\
(\boldsymbol{y}_N - \hat{x}_N - \widehat{\Delta x}_N) - -\lambda_{N-1} &= 0
\end{aligned}
\tag{8.56}
$$

and

$$
f(\hat{x}_n) + D_f(\hat{x}_n)\widehat{\Delta x}_n - \hat{x}_{n+1} - \widehat{\Delta x}_{n+1} = 0, \quad n = 1\ldots N-1 \tag{8.57}
$$

Equations (8.56) and (8.57) can be cast into matrix form

$$
A \begin{bmatrix} \widehat{\Delta x}_1 \\ \lambda_1 \\ \widehat{\Delta x}_2 \\ \lambda_2 \\ \vdots \\ \widehat{\Delta x}_{N-1} \\ \lambda_{N-1} \\ \widehat{\Delta x}_N \end{bmatrix} = \underline{B} \tag{8.58}
$$

where B is the following $(2N-1)$ rows vector

$$
B = \begin{bmatrix} \boldsymbol{y}_1 - \hat{x}_1 \\ \hat{x}_2 - f(\hat{x}_1) \\ \boldsymbol{y}_2 - \hat{x}_2 \\ \hat{x}_3 - f(\hat{x}_2) \\ \vdots \\ \hat{x}_N - f(\hat{x}_{N-1}) \\ \boldsymbol{y}_N - \hat{x}_N \end{bmatrix} \tag{8.59}
$$

and A is $(2N-1)\times(2N-1)$ matrix deduced from (8.56) and (8.57), i.e.,

$$
A = \begin{bmatrix}
I & D_f^T(\hat{x}_1) & 0 & 0 & 0 & 0 \\
D_f(\hat{x}_1) & 0 & -I & 0 & 0 & 0 \\
0 & -I & I & D_f^T(\hat{x}_2) & 0 & 0 \\
0 & 0 & D_f(\hat{x}_2) & 0 & -I & 0 \\
\cdots & \cdots & \cdots & \cdots & \cdots & \cdots \\
\cdots & \cdots & \cdots & \cdots & \cdots & \cdots \\
0 & 0 & 0 & 0 & -I & I
\end{bmatrix} \tag{8.60}
$$

The optimal noise reduction technique obeys the following algorithm

1. Initialize $\hat{\boldsymbol{x}}(0) = \boldsymbol{y}$.

2. Form the matrix $A(0)$and vector $\underline{B}(0)$.

3. Compute $\widehat{\boldsymbol{\Delta x}}(0)$ and $\boldsymbol{\Lambda}(0)$ by computing $A(0)^{-1}\underline{B}(0)$.

4. Form $\hat{\boldsymbol{x}}(1) = \hat{\boldsymbol{x}}(0) + \widehat{\boldsymbol{\Delta x}}(0)$.

5. Proceed recursively until $\hat{\boldsymbol{x}}(k)$ converges towards a fixed point.

8.3.4.3 Viterbi-like methods

For an approach to noise reduction that involves discretization, the exact computation usually requires an entire tree of possible discrete combinations. However, for (hidden) Markov processes, it is not necessary to solve this problem of exponential complexity. Instead, the tree can be converted into a lattice (cf. Figure 8.4) on which the computational cost is essentially linear with N [3]. The same method also applies in the case where an approximation with a Markov process is valid.

We show the Viterbi algorithm, an efficient way to solve maximum likelihood problems for the quasi-Markov process arising from uniform quantization discussed in Section 8.3.3.2. Note that there are other algorithms to treat hidden Markov problems that we do not consider here. Also, here we do not consider the *natural* discretization from 8.3.2.2.

Using the Viterbi algorithm for maximizing (8.47) is a standard approach. For the enhancement of chaotic signals it was first introduced by Marteau and Abarbanel [30]. The Viterbi algorithm avoids making an exhaustive search among the lattice of K^N paths. Let δ_n^k be the following probability quantity

$$\delta_n^i = \max_{\boldsymbol{s}_{n-1}} p(\boldsymbol{s}_{n-1}, s_n = i, \boldsymbol{y}_n) \tag{8.61}$$

The Viterbi algorithm is based on the fact that

$$\delta_n^j = \max_i [\delta_{n-1}^i a_{ij}] b_j(y_n) \tag{8.62}$$

where $\mathbf{A} = \{a_{i,j}\}$ is the transition probability matrix (the discretized approximation of the FPO).

$$a_{ij} = p(s_n = j | s_{n-1} = i) \tag{8.63}$$

and

$$b_j(y_n) = p(y_n | s_n) = p_w\left(y_n - I_c(s_n)\right) \tag{8.64}$$

The Viterbi Algorithm works in two main passes, a forward one and a backward one (backtracking). During the forward pass Equation (8.62) is used to compute at time n the δ_n^j at the K lattice nodes $(n, j = 1, \ldots, K)$. Among the K paths which can link nodes $(n-1, k = 1 \ldots K)$ to node (n, j) only the best one is kept. After the termination $n = N$, we select the node (N, k^*) which gives the highest probability. To actually retrieve the state sequence, we need to keep track of the argument that maximized (8.62), for each n and j. This can be done via the array ϕ_n^j in which the optimal trajectory in node j at time n is stored. With the *a priori* probability π_i of y_1 to be in state i ($\pi_i = \frac{1}{K}$), the complete Viterbi procedure can be stated as follows

1. Initialization

$$\begin{aligned} \delta_1^i &= \pi_i b_i(y_1) 1 \leq i \leq K \\ \phi_1^i &= 0 \end{aligned}$$

2. Forward pass

$$\begin{aligned} \delta_n^j &= \max_{1 \leq i \leq K} [\delta_{n-1}^i a_{ij}] b_j(y_n) \\ &\quad 2 \leq n \leq N 1 \leq j \leq K \end{aligned}$$

$$\begin{aligned} \phi_n^j &= \arg \max_{1 \leq i \leq K} [\delta_{n-1}^i a_{ij}] b_j(y_n) \\ &\quad 2 \leq n \leq N 1 \leq j \leq K \end{aligned}$$

3. Termination

$$\hat{s}_N = \arg \max_{1 \leq i \leq K} [\delta_N^i]$$

4. Backward pass

$$\hat{s}_n = \phi_{n+1}^{\hat{s}_{n+1}} n = N-1, \ldots 1$$

Applying the Viterbi algorithm requires the knowledge of the probability distribution of the observation noise p_w. If the noise of observation is Gaussian, the only parameter to be estimated is the standard deviation. The standard deviation can be in general estimated off line. However we propose a recursive estimation of the noise variance. Starting from some *a priori* estimation of the noise variance one can launch the Viterbi algorithm and find the most probable sequence. Once this optimal sequence is found the re-estimation of the noise variance can be done. We launch again the Viterbi algorithm with this new value of the variance and so forth. The algorithm is halted when the noise variance converges. The recurrent application of the algorithm obeys the following procedure.

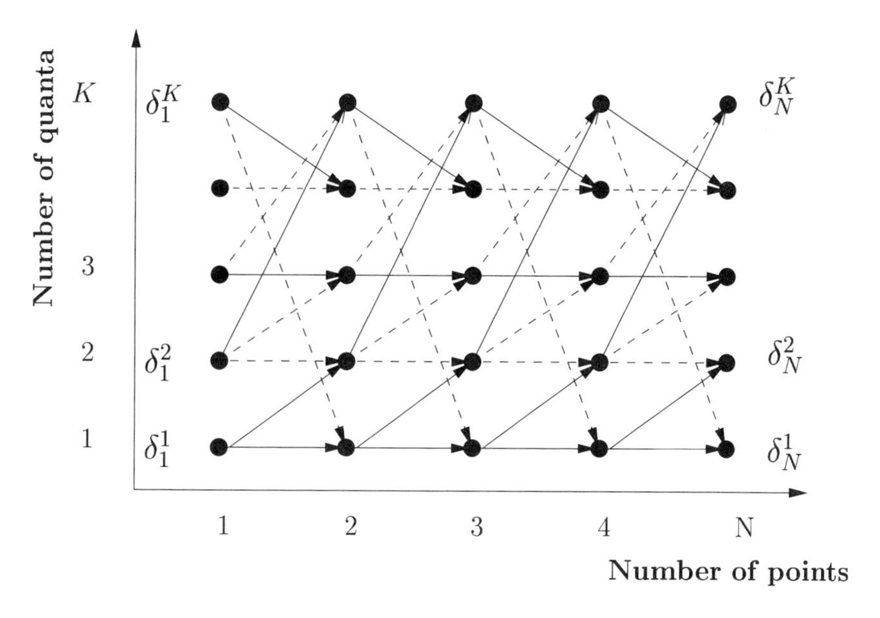

Figure 8.4: Lattice for the decoding of N points with K balls.

1. Find a crude approximation of σ_0 by some off line procedure. For instance choose

$$\sigma_0 = \sqrt{\sigma_y^2 - \sigma_x^2}$$

2. Initialize $i = 0$.

3. Repeat the Viterbi algorithm and find the optimal sequence $\hat{\boldsymbol{s}}$. The observation probabilities are computed by using

$$p_w(y) = \frac{1}{\sigma_i \sqrt{2\pi}} \exp^{-\frac{1}{2}\frac{y^2}{\sigma_i^2}}$$

4. Re-estimate the variance of the noise

$$\sigma_{i+1} = \sqrt{\frac{1}{N} \sum_{k=1}^{N} (I_c(s_k) - y_k)^2}$$

5. Go to 3. until $|\sigma_{i+1} - \sigma_i| \leq \epsilon$ where ϵ is some tolerance factor fixed in advance.

Experiments have shown that the convergence of the variance estimate is very fast even with bad estimations of σ_0; three to five iterations of this procedure are enough.

8.4 Explicit and Implicit Application

As discussed before (Section 8.2.3), we consider explicit noise reduction (noise pre-filtering) and implicit noise reduction (detection on internals of the noise pre-filter). The ultimate goal of applying a noise reduction scheme is the improvement of the BER to SNR performance. But it is also important to assess the noise reduction capabilities of the various sub optimal implementations with regard to the optimal noise filter.

8.4.1 Explicit Noise Pre-Filtering

The design criterion for the noise reduction filter has been the processing gain from Equation (8.21), or equivalently, the output noise variance for a given input noise variance for the noise filter. Consequently, we show the performance in these terms for some of the methods introduced in Section 8.3.

First, however, we need to define how to apply the general noise reduction scheme to particular chaos-based communication systems.

8.4.1.1 Noise reduction and signal sets

Considering chaos-based communication schemes as defined in Section 8.2.1, there are two ways to implement a noise reduction filter. Either a filter designed for all signal sets V_b simultaneously can be placed at the input of the receiver; or M filters specifically designed for each individual V_b can be placed in each of the M paths of the receiver (cf. Figure 8.5). We will consider the second case, as the noise reduction schemes developed in Section 8.3 then apply directly.

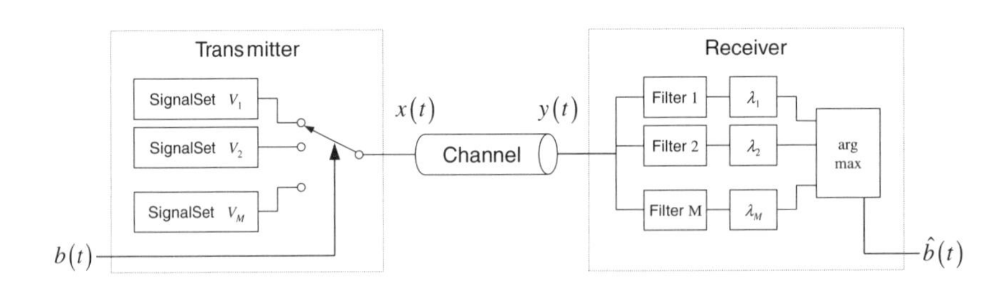

Figure 8.5: Communication system with M symbols and noise filters.

The individual noise filters take into account the specific initial densities $\rho_b(x_1)$ and the map determining the dynamical system $f_b(\cdot)$ as used for the definition of the signal sets in Section 8.2.1.

8.4.1.2 Performance of the discretized method for piecewise linear maps

In the case of piecewise linear maps, the estimator ϕ from Equation (8.26) can be calculated explicitly; for more details see [37]. It involves a discretization of the problem based on the symbolic dynamics of the map $f(\cdot)$. Consequently, as N grows, the computational complexity of the optimal estimator grows exponentially with the number of possible sequences in the symbolic dynamics. As an example we consider iterations of a skew tent map. Figure 8.6 presents noise variance diagrams obtained from numerical simulations, showing the variance per sample before and after the noise cleaning procedure for block lengths $N = 2, 4, 8$.

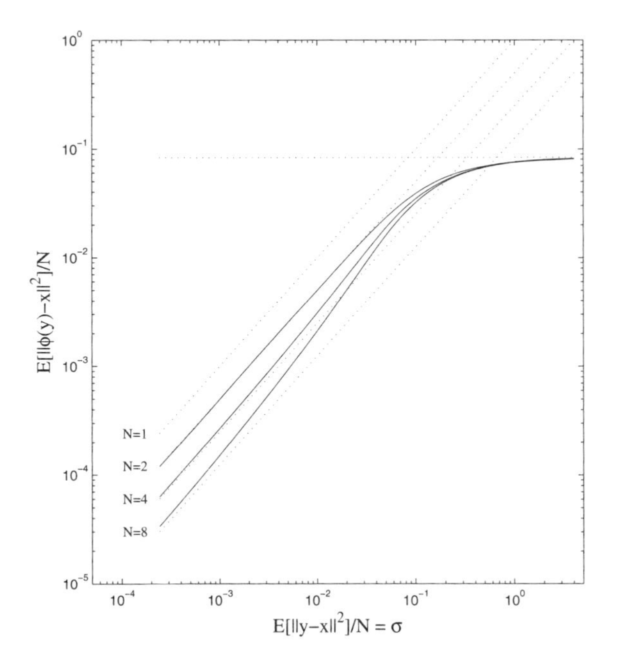

Figure 8.6: Unconstrained denoising, skew tent map, block lengths $N = 2, 4, 8$.

Note that, while the energy of $\hat{\boldsymbol{x}}$ is limited (due to the bounded $\boldsymbol{x}$), the noise energy grows without bound as σ grows. Consequently, the residual error saturates in the diagram. Further, for very small noise the performance tends to a $1/N$ noise reduction (as seen in classical systems). While for large σ the processing gain P_G will grow without bound, this does not mean that the estimate will get better and better then; instead, it will tend to be completely random, but bounded in variance. As a reference, the limit cases $\sigma \to 0$ and $\sigma \gg 1$ are represented as dotted lines.

Adding the constraint $\hat{x}_{n+1} = f(\hat{x}_n)$ to the estimator results in a suboptimal noise removal. As explained in detail in [15, 37], for piecewise linear maps the optimal estimator with constraint ϕ_c can be found explicitly.

Considering the same example as before, the iterated skew tent map, Figure 8.7 presents noise variance diagrams obtained from numerical simulations comparing the noise reduction in the unconstrained case with the constrained case for block length $N = 8$.

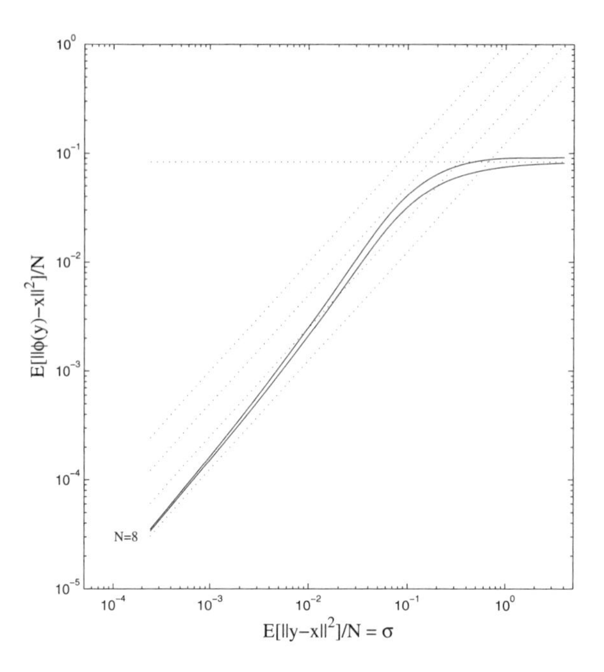

Figure 8.7: Constrained and unconstrained denoising - skew tent map, block length $N = 8$.

8.4.1.3 Noise reduction for CSK based on quantization

The methods from Section 8.3.3.2 (quantization) and 8.3.4.3 (Viterbi algorithm) have been applied to chaos shift keying (CSK), resulting in an efficient (computational cost is proportional to the block length N) and powerful noise reduction scheme. A performance increase both with respect to signal-to-noise ratio improvement and bit error rate reduction has been observed. For a comprehensive description of the method, please refer to [34].

8.4.1.4 DCSK with optimization (gradient descent) based noise reduction

Applying the methods from Section 8.3.2.3 (optimization) and 8.3.4.1 (gradient type methods) to differential chaos shift keying (DCSK) has been considered in [17]. The simulation results presented there suggest a considerable improvement of the signal-to-noise ratio but only a minor decrease of the bit error rate as a result thereof. The results demonstrate that a relatively ad-hoc method such as gradient type methods for noise reduction do not necessarily lead to a performance increase where the ultimate merit figure is the bit error rate and not the processing gain (signal-to-noise ratio improvement).

8.4.2 Implicit Noise Filtering

Given the structure of the presented noise reduction schemes, it is not always necessary (or, in fact, advisable) to perform explicit noise filtering and then apply the ad-hoc classifiers that have been proposed in the literature. Instead, an implicit approach can be taken.

Consequently, we present the two possible ways to exploit this approach. First, we demonstrate the detection based on internals of the noise pre-filter and second, the direct detection.

8.4.2.1 *Detection on internals of the noise pre-filter*

In the optimal noise reduction filters, typically there is already a (statistically motivated) likelihood information present. In the optimal parallel estimator (Section 8.3.2) this is simply the denominator of Equation (8.24). In the case of optimization methods (Section 8.3.2.3), it is the "final" value of the cost function $C(\hat{\boldsymbol{x}}, \boldsymbol{y})$ at the optimal point $\hat{\boldsymbol{x}}$. In case of the uniform quantization with the Viterbi solution (Section 8.3.4.3), with

$$\lambda_1 = p(\boldsymbol{s},\ \boldsymbol{Y}|M_1) = \prod_{k=1}^{N} p(s_k|\,(s_{k-1},\ M_1))\ p(y_k|s_k) \tag{8.65}$$

$$\lambda_2 = p(\boldsymbol{s},\ \boldsymbol{Y}|M_2) = \prod_{k=1}^{N} p(s_k|\,(s_{k-1},\ M_2))\ p(y_k|s_k) \tag{8.66}$$

the results of the forward pass of the algorithm directly yield likelihoods that can be used for the detection.

Thus, we can make the decision based on the implicit likelihood information already present in the noise filters (cf. Figure 8.8) and we can eliminate the ad-hoc likelihoods present in the receiver(s) as defined in 8.2.1. According to our experiments, this approach is to be preferred over an explicit noise pre-filtering, as the latter in general does not assure the compatibility of the noise reduction goal and the requirements of the receiver with respect to low error decisions.

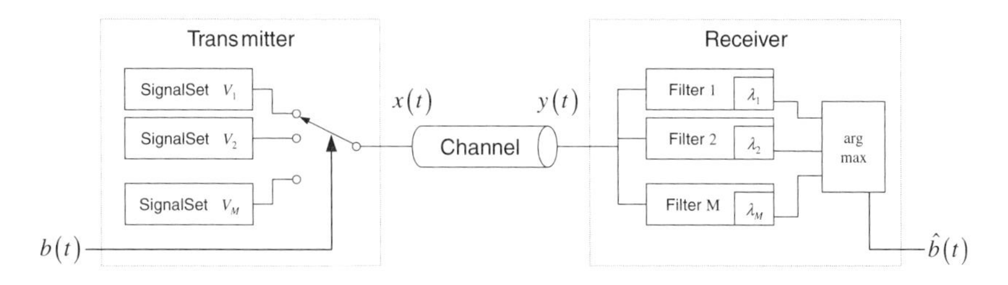

Figure 8.8: Communication system with M symbols and detection based on implicit likelihood information from noise filters.

8.4.2.2 Detection based on a direct classification

In the case of piecewise linear maps, the relation between the detection on internals of the noise pre-filter and the direct detection by an optimal classifier is particularly close. In fact, the two detection methods coincide, since the noise reduction filter for piecewise linear systems in its optimal version implicitly calculates the (probabilistically) correct likelihoods (while other ad-hoc likelihoods such as synchronization-based schemes would not lead to the same results). Here however, direct classification is not a central point of focus. We therefore refer the reader to [37, 46].

8.5 Conclusions

Noise filtering, be it performed explicitly or implicitly, has been shown to be an important tool for designing/implementing receivers for chaos-based communication systems that possess good performance characteristics. The ultimate performance criterion in use is the symbol error rate (bit error rate) given a certain noise level. An intermediate performance measure is the processing gain achieved by the noise reduction, typically in terms of noise energy before, versus after, applying the filter.

Here, for the example of reduction of additive white Gaussian noise, a framework has been presented which allows us to classify the noise reduction methods proposed in the literature. Chaos communication schemes have been introduced through the notion of signal sets which establishes a concise setting in which noise reduction can be introduced rigorously and that, on the other hand, allows a comparison with conventional communication schemes with respect to the specific advantages and drawbacks related to the proposed use of chaos.

The concept of optimal and suboptimal methods has been introduced, underlining the fundamental tradeoff between complexity (high for optimal methods) and noise reduction performance (reduced for suboptimal methods). It turns out that the value of the optimal methods (as far as they have an explicit solution, as in the example of piecewise linear systems) lies primarily in establishing the upper performance bound of a given system. Since we have considered the transmitter part of the system as given and fixed, the upper bound (determined by the optimal methods) characterizes primarily the appropriation of the (given) transmitter for the (given) communication channel. Note that this implies directly that the transmitter can be optimized with respect to these upper bound performance figures, but this has not been considered here. Unfortunately, optimal methods are in general not useful for direct implementation due to their computational cost (typically exponential in the block length N). Thus, efficient suboptimal methods such as the described methods based on quantization and the Viterbi algorithm, are of key importance for practical realizations. With the knowledge of the limit performance of optimal methods, the tradeoff can be assessed and quantified.

At this point, the main focus has been on discrete time models with additive Gaussian noise. In order to address the reality of many communication channels more effectively, more sophisticated channel models (in our framework this corresponds directly to noise models) can be incorporated into the framework, leaving its key elements in place. Concentrating primarily on noise reduction, direct methods for detecting digital information in transmissions using chaos have not been considered in detail here, even though they present an interesting alternative to the noise reduction based approach. For the specifics of direct methods, a number of references to the literature have been given.

The correct use of noise reduction increases the performance of classical receivers that were proposed for communication systems based on chaos, as their equivalent counterparts in traditional communication schemes.

8.6 Acknowledgments

The authors would like to thank Jörg Schweizer for many useful discussions on the subject. We acknowledge financial support of the Swiss National Science Foundation, grant no. 2000-56030.98.

References

[1] MacKay, D.J.C., *Information Theory, Inference, and Learning Algorithms*, Cambridge University Press, to appear 2001.

[2] Proakis, J.G., *Digital Signal Processing*, Macmillan Publishing, New York, 1992.

[3] Proakis, J.G., *Digital Communications*, MacGraw-Hill, New York, 1995.

[4] Abarbanel, H.D.I., *Analysis of Observed Chaotic Data*, Springer-Verlag, New York, 1996.

[5] Ott, E., Sauer T. and Yorke, J.A., *Coping with Chaos. Analysis of Chaotic Data and the Exploitation of Chaotic Systems*, Wiley Series in Nonlinear Science, Wiley, New York, 1994.

[6] Kantz, H. G. and Schreiber, T., *Nonlinear Time Series Analysis*, Cambridge University Press, Cambridge, 1997.

[7] Bai-Lin, H., *Elementary Symbolic Dynamics*. World Scientific, Singapore, 1989.

[8] Haykin, S., *Communication Systems*. John Wiley & Sons, New York, 1994.

[9] Cover, T.M. and Thomas, J.A., *Elements of Information Theory*. John Wiley & Sons, New York, 1991.

[10] Anderson, B.D.O. and Moore, J.B., *Optimal Filtering*, Prentice-Hall, Englewood Cliffs, 1979.

[11] Anderson, B.D.O. and Moore, J.B., *Optimal Control Linear Quadratic Methods*, Prentice-Hall, Englewood Cliffs, 1990.

[12] Poor, H.V., *An Introduction to Signal Detection and Estimation*, Springer-Verlag, New York, 1988.

[13] Lasota, A. and Mackey, M.C., *Chaos, Fractals and Noise*, Springer-Verlag, New York, 1994.

[14] Gershenfeld, N., An Experimentalist's Introduction to the Observation of Dynamical Systems, in *Directions in Chaos*, Hao Bai-Lin, ed. World Scientific, Singapore, pp. 311-382.

[15] Dedieu, H., Schimming, T. and Hasler, M., Separating a chaotic signal from noise and applications, in *Controlling Chaos and Bifurcations in Engineering Systems*, ed. G.R. Chen, CRC Press, Boca Raton, 1999.

[16] Sorenson, H.W., Recursive Estimation for Nonlinear Dynamical Systems, in *Bayesian Analysis of Time Series and Dynamic Models*, ed. J.C. Spall, Marcel Dekker, New York, 1988.

[17] Schweizer, J., *Application of Chaos to Communication*, Ph.D. thesis EPFL No. 1953, Lausanne, Switzerland, 1999.

[18] Richard, M.D., *Estimation and Detection with Chaotic Systems*, Ph.D. thesis, Massachusetts Institute of Technology, 1994.

[19] Götz, M., *Analyse des Frobenius-Perron-Operators und Korrelationstheorie stückweise linearer zeitdiskreter chaotischer Systeme (in German)*, Ph.D. thesis, TU-Dresden, Dresden, Germany, 1997.

[20] Dedieu, H., Kennedy, M.P. and Hasler, M., Chaos shift keying: modulation and demodulation of a chaotic carrier using self-synchronizing Chua's circuits, *IEEE Trans. Circ. Syst.*, Part II, vol. 40, No. 10, pp. 634-642, Oct. 1993.

[21] Broomhead, D.S., Huke, J.P. and Jones, R., Signals in chaos: A method for the cancellation of deterministic noise from discrete signals, *Physica D*, vol. D80, pp. 413-432, 1995.

[22] Farmer, J.D. and Sidorowich, J.J., Optimal shadowing and noise reduction, *Physica D*, vol. D47, pp. 373-392, 1991.

[23] Grassberger, P., Hegger, R., Kantz, H., Schaffrath, C. and Schreiber, T., On noise reduction methods for chaotic data, *Chaos*, vol. 3, No. 2, pp. 127-141, 1993.

[24] Haykin, S. and Li, X.-B., Detection of Signals in Chaos, *Proceedings of the IEEE*, Vol. 83, No. 1, pp. 95-122, 1995.

[25] Kostelich, E.J. and Yorke, J.A., Noise reduction in dynamical systems, *Physical Rev. A*, vol. 38, No. 3, pp. 1649-1652, 1988.

[26] Kostelich, E.J. and Yorke, J.A., Noise reduction: Finding simplest dynamical system consistent with the data, *Physica D*, vol. D41, pp. 183-196, 1990.

[27] Kostelich, E.J. and Schreiber, T., Noise reduction in chaotic time-series data: A survey of common methods, *Phys. Rev. E*, pp. 1752, 1993.

[28] Lee, C. and Williams, D.B., Generalized Iterative Methods for Enhancing Contaminated Chaotic Systems, *IEEE Trans. on Circuits and Systems-I*, Vol. 44, No. 6, June 1997.

[29] Hammel, S.M., Noise Reduction for Chaotic Systems, *Naval Surface Warfare Center*, Silver Spring, MD, 1989.

[30] Marteau, P.F. and Abarbanel, H.D.I., Noise Reduction in Chaotic Time Series Using Scaled Probabilistic Methods, *Journal of Nonlinear Science*, Vol. 1, pp. 313-343, 1991.

[31] Eckmann, J.P. and Ruelle, D., Ergodic Theory of Chaos and Strange Attractors, *Rev. Mod. Phys.*, Vol. 57, pp. 617-656, 1985.

[32] Heidari-Bateni, G. and McGillem, C.D., A Chaotic Direct-Sequence Spread-Spectrum Communication System, *IEEE Trans. on Communications*, 42 (2/3/4): 1524-1527, 1994.

[33] Kolumban, G., Kennedy, M.P. and Chua, L.O., The Role of Synchronization in Digital Communications using Chaos - Part I: Fundamentals of Digital Communications, *IEEE Trans. on Circuits and Systems*, Part I, vol. 44, pp. 927-936, Oct. 1997.

[34] Kisel, A., Dedieu, H. and Schimming, T., Maximum Likelihood Approaches for Noncoherent Communications with Chaotic Carriers, *IEEE Journal on Circuits and Systems*, Part I, to appear in 2000.

[35] Hasler, M., Engineering chaos for encryption and broadband communication, *Philosophical Transactions of the Royal Society of London*, A(353):115–126, 1995.

[36] Hasler, M. and Schimming, T., Chaos communication over a noisy channel, *Int. J. Bifurcation and Chaos, special issue*, to appear 2000.

[37] Schimming, T. and Hasler, M., Constrained and unconstrained noise reduction on chaotic trajectories, *Proc. NDES 1999*, Denmark, 1999.

[38] Götz, M. and Schwarz, W., Correlation Theory of a Class of N-Dimensional Piecewise Linear Systems, *Proc. ISCAS'97*, vol. II, pp. 1049–1052, Hong Kong, 1997.

[39] Schweizer, J., The Performance of Chaos Shift Keying: Synchronization versus Symbolic Backtracking, *IEEE International Symposium on Circuits and Systems*, Monterey, 1998.

[40] Hasler, M., Chaos Shift Keying in the Presence of Noise: A simple Discrete Time Example, *IEEE International Symposium on Circuits and Systems*, Monterey, 1998.

[41] Myers, C., Kay, S. and Richard, M.D., Signal Separation for Nonlinear Dynamical Systems, *ICASSP 1992*, pp. IV-129, IV-132.

[42] Abel, A., Götz, M. and Schwarz, W., Statistical Analysis of Chaotic Communication Schemes, *IEEE International Symposium on Circuits and Systems*, Monterey, 1998.

[43] Dedieu, H. and Schweizer, J., Noise Reduction in Chaotic Time Series - An Overview, *Proceedings of the 6th Int. Workshop on Nonlinear Dynamics of Electronic Systems, NDES'98*, Budapest, pp. 53-62.

[44] Dedieu, H. and Ogorzałek, M., Overview of Noise Reduction Algorithms for Systems with Known Dynamics, *International Symposium on Nonlinear Theory and Its Applications, NOLTA'98*, Crans-Montana, Switzerland, pp. 1297-1300.

[45] Schimming, T. and Schweizer, J., Chaos Communication from a Maximum Likelihood Perspective, *International Symposium on Nonlinear Theory and its Applications, NOLTA'98*, Crans-Montana, Switzerland, pp. 77-80.

[46] Schimming, T. and Hasler, M., Statistically Motivated Detection methods for chaos shift keying, *International Symposium on Nonlinear Theory and its Applications, NOLTA'99*, Hawaii, pp. 569-572.

[47] Khoda, T. and Fujisaki, H., Kalman's Recognition of Chaotic Dynamics in Designing Markov Information Sources, *International Symposium on Nonlinear Theory and Its Applications, NOLTA'98*, Crans-Montana, Switzerland, pp. 619-622.

[48] Kolumbán, G., Vizvari, K., Schwarz, W. and Abel, A., Differential chaos shift keying: A robust coding for chaos communication, *Proc. NDES 1996 Seville*, pp. 87–92, 1999.

Chapter 9

Statistical Analysis and Design of Chaotic Systems

Wolfgang Schwarz, Marco Götz, Kristina Kelber, Andreas Abel, Thomas Falk, and Frank Dachselt
Faculty of Electrical Engineering / IEE
Technical University Dresden
Mommsenstr. 13, D-01062 Dresden, Germany
schwarz@iee1.et.tu-dresden.de

9.1 Introduction

This chapter deals with the statistical description of chaotic signals and systems. Although chaotic systems are purely deterministic they can be modelled, analyzed, and designed by using probability measures and statistical characteristics, such as probability density functions and correlation functions, widely used in signal theory and in engineering applications.

The fundamental approach is to model the initial state of the chaotic system to be a random variable. Then the generated signal can be treated as a stochastic process. Ensembles of trajectories are regarded and modelled by the time development of their density functions.

This is done in Section 9.2 where the fundamentals of statistical analysis of chaotic systems are explained systematically. The theory is mainly developed for multidimensional systems and illustrated with numerous examples for the one-dimensional case. The section introduces and develops tools for the statistical analysis.

Section 9.3 grapples with the solution of the inverse problem, the synthesis and design of chaotic systems from prescribed statistical characteristics of the signals to be generated.

Statistical analysis and design are especially simple for piecewise linear system maps; the corresponding systems are a paradigm for chaos generator design and experiments. The practical realization of adjustable generator structures with piecewise linear maps is discussed in Section 9.4, which shows how system structures for piecewise linear maps can be constructed and implemented as electronic circuits.

Section 9.5 provides a statistical design case, the synthesis of a chaotic system for information encryption. A top-down design approach leads from the design objectives to system structures from which a simple example is provided up to its realization. The system then undergoes a crypto-analysis in order to evaluate its security performance.

Section 9.6 treats an extended problem of statistical analysis, the performance evaluation of chaotic signal processing schemes. This requires the introduction of an extended calculus because random processes (signals and noise) interact with chaotic signals. Several chaos communication schemes are analyzed and the results are presented in terms of performance criteria commonly used in communication engineering.

9.2 Statistical Analysis of Discrete Time Chaotic Systems

9.2.1 Function Space Description

Models. Each analysis requires a specific model of the reality tailored to the problem under consideration. For the analysis of nonlinear discrete-time systems different models or descriptions are in use – e.g., state space models, input-output models, and signal models in the time and frequency domains. The statistical analysis deals with statistical characteristics such as probability density functions and averages of the signals generated by chaotic systems. For this analysis the adequate description of the system behaviour is by means of density functions and their linear extensions, the so-called *L^1-function space description.* The basic idea is to describe the time evolution of measures of sets in the state space of the dynamical system. These measures are probability measures represented by the corresponding density functions. The evolution of an initial density function of the state of a chaotic system is uniquely determined by the state equation from which an operator can be derived which describes the density function evolution. This operator can be defined on a suitable chosen linear function space, the space L^1 of all integrable functions. It has the important property to be linear. Thus statistical analysis of chaotic systems turns out to be an application of functional analysis and linear operator theory. In this subsection the L^1-function space description of chaotic systems is developed starting from the state space description.

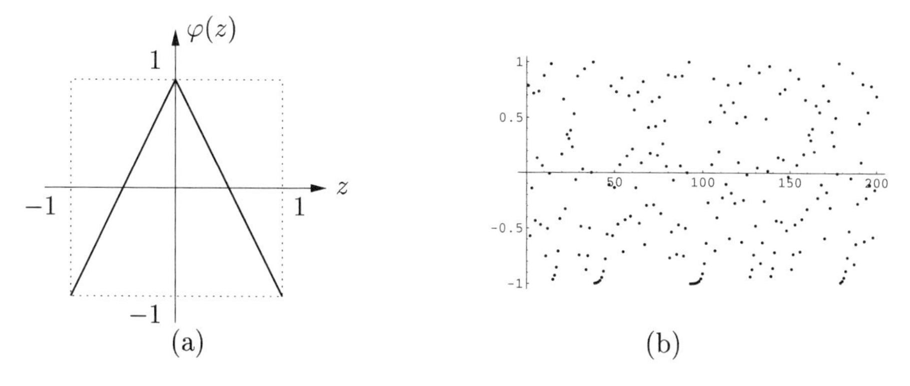

Figure 9.1: Graph of the tent map (a) and a generated state sequence $z(k)$ (b).

State space model. An autonomous nonlinear discrete-time system is defined by a state equation and an output map:

State Space Description

$$\begin{aligned} z(k+1) &= \varphi\left(z(k)\right) \\ y(k) &= h\left(z(k)\right) \end{aligned} \tag{9.1}$$

The integer variable k denotes the discrete time. We call Z the state space and $z(k)$ the state variable. Usually Z is a subset of the n-dimensional real vector space $\mathbb{R}^n$, but in some cases Z is better modelled as a manifold. For example, the state space Z of a circle map is a torus [3]. The map $\varphi(.)$ is called a system map. The output variable $y(k) \in \mathbb{K}$ usually is a complex or real value. The system generates a trajectory $c(z_0)$, which originates from the initial condition $z_0 := z(k_0)$. The trajectory is the sequence:

$$c(z_0) := \left(z_0, \varphi^1(z_0), \varphi^2(z_0), ..., \varphi^k(z_0), ...\right) \tag{9.2}$$

As a rule the initial time is set to zero, i.e., $k_0 = 0$. The trajectory depends on the system map as well as on the initial condition. The generated output signal depends on the initial condition too, i.e., we have

$$y(z_0) := \left(h(z_0), h \circ \varphi^1(z_0), h \circ \varphi^2(z_0), ..., h \circ \varphi^k(z_0), ...\right) \tag{9.3}$$

Example: Trajectory of the tent map. The tent map, given by

$$\varphi(z) = 1 - 2 \cdot |z|, \tag{9.4}$$

is depicted in Figure 9.1(a). A typical state sequence generated by this map is depicted in Figure 9.1(b). The reason for the complex system behaviour is that the complexity of the k-fold maps φ^k increases with k.

L^1-function space model. In statistical analysis we study ensembles of trajectories instead of single ones. For that we regard the initial state as a random variable and introduce a time-dependent probability measure P_k on the state space, which assigns a number between zero and one to measurable subsets A. Albeit $P_k(A)$ assigns merely a number between zero and one to A, one can interpret $P_k(A)$ as the probability that $z(k)$ is in the set A. With respect to the dynamical system (9.1) the measures $P_k, P_{k+1}, ...$ are related to each other by the measure conservation principle, which can be stated as:

Measure Conservation Principle

$$P_{k+1}(A) \quad = \quad P_k(\varphi^{-1}(A)) \tag{9.5}$$

This equation defines a dynamical discrete-time system on the set of measures P_k, which is uniquely related to the system map φ of the original system (9.1). If we choose an initial measure P_0, then a sequence of measures $P_0, P_1, .., P_t$ will be generated by (9.5). Let us assume that these measures can be represented by the Riemann integral:

$$P_k(A) = \int_A f(x,k)\,\mathrm{d}x, \tag{9.6}$$

where $f(z,k)$ is the density function of the state at time k. The conservation principle reads in terms of densities as

$$\int_{A_z} f(x,k+1)\,\mathrm{d}x = \int_{\varphi^{-1}(A_z)} f(x,k)\,\mathrm{d}x, \tag{9.7}$$

where $A_z := \{x \in Z : x_i \leq z_i\}$. Differentiating both sides by all state-vector components z_i we obtain:

$$f(z,k+1) = \frac{\partial^n}{\partial z} \int_{\varphi^{-1}(A_z)} f(x,k)dx \tag{9.8}$$

where the abbreviation

$$\frac{\partial^n}{\partial z} = \frac{\partial^n}{\partial z_1 \cdot \ldots \cdot \partial z_n} \tag{9.9}$$

has been used. By use of δ-functions Equation (9.8) can be rewritten as an integral operator

$$f(z,k+1) = \int_Z \delta(\varphi(x)-z) f(x,k)\,\mathrm{d}x, \tag{9.10}$$

where the kernel $\delta(\varphi(x)-z)$ represents the transition probability density function for two subsequent states. Equation (9.8) is linear and can be interpreted as a dynamical system on the densities $f(z,k)$. Since the density functions do not form a linear space we extend the equation to all integrable functions $g(.)$. These functions form the infinite-dimensional linear space L^1. In that way we get the function space model of the chaotic system given by

L^1 - Space Description

$$\begin{aligned} g(z,k+1) &= = \frac{\partial^n}{\partial z} \int_{\varphi^{-1}(A_z)} g(x)\,\mathrm{d}x \\ g(z,0) &= g_0(z) \end{aligned} \tag{9.11}$$

Example: L^1-space description of the tent map. First we determine the pre-image $\varphi^{-1}(A_z)$. This is especially simple for piecewise linear maps. For one-dimensional systems on $Z = [-1,1]$ we have $A_z = [-1,z]$. So for the tent map (9.4) we obtain:

$$\varphi^{-1}([-1,z]) = \left\{[-1, \frac{z-1}{2}] \cup [\frac{1-z}{2}, 1]\right\} \tag{9.12}$$

This pre-image is depicted in Figure (9.2). Inserting Equation (9.12) into (9.11)

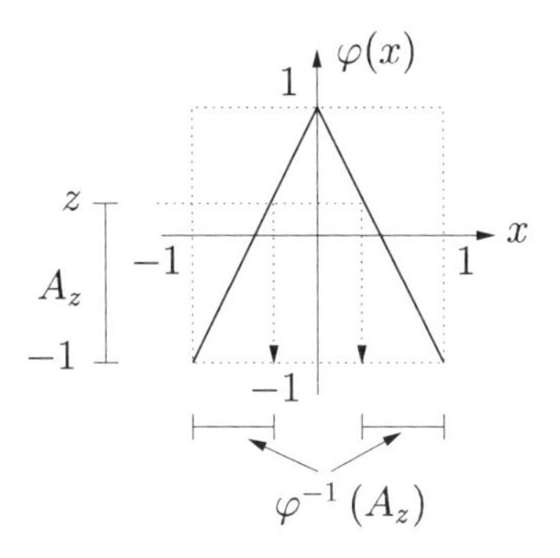

Figure 9.2: The pre-image $\varphi^{-1}(A_z)$ of the tent map .

we obtain after differentiation

$$g(z,k+1) = \frac{1}{2} \cdot \left(g\left(\frac{z-1}{2}, k\right) + g\left(\frac{1-z}{2}, k\right)\right). \tag{9.13}$$

A similar sum expression is obtained for any piecewise linear map. This is the most suitable representation for the L^1-function space description. It simplifies the calculation of sequences of functions from an initial $g_0(z)$ considerably. We show this with two examples for the tent map using (9.13). First, let the initial function be $g_0(z) = a \cdot z$. A simple calculation shows that $g(z,1) = 0$ and due to linearity we have $g(z,k) = 0$ for all $k \geq 1$. Second, let the initial function be a density function[1] $f_0(z) = \frac{z+1}{2}$. It has the properties $f_0(z) \geq 0$ and

[1] We use for the density functions in L^1 the special symbol f instead of g.

$\int_{[-1,1]} f_0(z) \cdot \mathrm{d}z = 1$. Inserting $f_0(z)$ in Equation (9.13) we get $f(z,1) = \frac{1}{2}$ which is again a density. In the next step we obtain $f(z,2) = \frac{1}{2}$ again and consequently $f(z,k) = \frac{1}{2}$ for all $k \geq 1$. Hence the density $f(z) = \frac{1}{2}$ is invariant under the transformation (9.13). As shown later invariant densities play a major role in statistical analysis.

9.2.2 Frobenius-Perron Operator

Definition and properties. The L^1-function space description (9.11) of the nonlinear system (9.1) defines a linear operator, which is the key tool in statistical analysis of nonlinear dynamic systems. This operator is called Frobenius-Perron operator (FPO). It is given by

Frobenius-Perron Operator (FPO)

$$\mathcal{P} : L^1 \to L^1$$
$$\mathcal{P}g(z) := \frac{\partial^n}{\partial z} \int_{\varphi^{-1}(A_z)} g(x)\,\mathrm{d}x \tag{9.14}$$

and has an integral representation in accordance with Equation (9.10):

$$\mathcal{P}g(z) = \int_Z \delta(\varphi(x) - z) g(x)\,\mathrm{d}x. \tag{9.15}$$

The FPO has the following important properties:

Linearity. This is the most important property of the FPO:

$$\mathcal{P}\left(a \cdot g_1(z) + g_2(z)\right) = a \cdot \mathcal{P}\left(g_1(z)\right) + \mathcal{P}\left(g_2(z)\right) \tag{9.16}$$

It results from the linearity of both the integration and the partial differentiation in Equation (9.14).

Invariance on densities. The FPO is a density operator. It is invariant on the subset of density functions in L^1. Thus if $f(z)$ is a density function with the property

$$\int_Z f(x)\,\mathrm{d}x = 1, \quad f(z) \geq 0 \tag{9.17}$$

the FPO maps it into another density function:

$$\int_Z \mathcal{P}f(x)\,\mathrm{d}x = 1, \quad \mathcal{P}f(z) \geq 0 \tag{9.18}$$

If the FPO maps a density function f to itself

$$\mathcal{P}f(z) = f(z) \tag{9.19}$$

we call f an *invariant density*.

Constraints on eigenvalues. A function g is called an *eigenfunction* of the FPO if it is mapped to a scaled version of itself:

$$\mathcal{P}g(z) = \lambda \cdot g(z) \tag{9.20}$$

The scaling value λ is the corresponding *eigenvalue.* Eigenfunctions and their corresponding eigenvalues can be complex valued. For instance the invariant density is a real valued eigenfunction with an eigenvalue equal to one. All eigenvalues are located inside or on the unit circle:

$$|\lambda| \leq 1 \tag{9.21}$$

Further properties of the FPO can be found in [2].

The dual operator of the FPO. An important concept of linear operator theory is *duality.* We will exploit dual spaces and the dual operator when defining and calculating averages. The dual space L^∞ is the linear space of all functionals on L^1 given by

$$\mathcal{F}_h(g) := \int_Z h(x) \cdot g(x)\,\mathrm{d}x \tag{9.22}$$

where h is a real- or complex-valued function of the state space Z. We say h represents the functional and write for short $h \in L^\infty$. Additionally we introduce for (9.22) the bracket notation:

$$\mathcal{F}_h(g) =: \langle h(z), g(z) \rangle \tag{9.23}$$

Let $\mathcal{U}$ be a linear operator on the dual space, i.e., $\mathcal{U} : L^\infty \to L^\infty$. We write $\mathcal{U}h$ for the image of the function h. The dual operator of the FPO is then uniquely defined by

$$\langle \mathcal{U}h(z), g(z) \rangle = \langle h(z), \mathcal{P}g(z) \rangle \tag{9.24}$$

for all $h \in L^\infty$. It is called the Koopman operator (KO) [2]

Koopman Operator (KO)

$$\mathcal{U} : L^\infty \to L^\infty$$
$$\mathcal{U}h(z) := h \circ \varphi(z) \tag{9.25}$$

The integral representation of the KO follows from that of the FPO (9.15) simply by interchanging the variables x and z in the kernel:

$$\mathcal{U}h(z) = \int_Z \delta(\varphi(z) - x) h(x)\,\mathrm{d}x \tag{9.26}$$

which directly yields $\mathcal{U}h(z) = h(\varphi(z))$. The KO has some important properties, useful for the application to statistical analysis:

Iterates. The τ-th action of the KO to a function h is given by

$$\begin{aligned} \mathcal{U}^{\tau}(h) &= h \circ \varphi^{\tau} \\ \mathcal{U}^{0}(h) &= h \end{aligned} \tag{9.27}$$

The τ-th iterate of the KO and the FPO are dual:

$$\langle \mathcal{U}^{\tau}(h), f \rangle = \langle h, \mathcal{P}^{\tau}(f) \rangle \tag{9.28}$$

Shift property. In the time domain the action of the KO can be interpreted as a time shift:

$$\mathcal{U}^{\tau}\left(h\left(z(k)\right)\right) = h\left(z(k+\tau)\right) \tag{9.29}$$

since $\mathcal{U}^{\tau}\left(h\left(z(k)\right)\right) = h \circ \varphi^{\tau}\left(z(k)\right) = h\left(z(k+\tau)\right)$.

2-Products. For products of two functionals the following shift rule holds true for all times satisfying $0 \leq \tau_1 \leq \tau_2$:

$$\begin{aligned} \mathcal{U}^{\tau_1}(h_1) \cdot \mathcal{U}^{\tau_2}(h_2) &= h_1 \circ \varphi^{\tau_1} \cdot h_2 \circ \varphi^{\tau_2} \\ &= h_1 \circ \varphi^{\tau_1} \cdot h_2 \circ \varphi^{\tau_1} \circ \varphi^{\tau_2 - \tau_1} \\ &= \mathcal{U}^{\tau_1}\left(h_1 \cdot \mathcal{U}^{\tau_2 - \tau_1}(h_2)\right) \end{aligned} \tag{9.30}$$

General products. For an ever increasing sequence of time moments τ_i by induction we obtain from (9.30) the general product rule:

$$\prod_{i=1}^{m} \mathcal{U}^{\tau_i}(h_i) = \mathcal{U}^{\theta_1}\left(h_1 \cdot \mathcal{U}^{\theta_2}\left(h_2 \cdot \ldots \cdot \mathcal{U}^{\theta_{n-1}}\left(h_{n-1} \cdot \mathcal{U}^{\theta_n}(h_n)\right) \ldots\right)\right) \tag{9.31}$$

where the positive time differences θ_i are given by

$$\theta_i = \tau_i - \tau_{i-1}, \quad k_0 = 0 \tag{9.32}$$

We will use this formula for calculating higher-order correlation functions.

Example: Action of the KO of the tent map. The KO of the map (9.4) maps the function $h(z) = z$ according to (9.25) to $\mathcal{U}h(z) = h \circ \varphi(z) = h(1 - 2|z|) = 1 - 2|z|$. This is depicted in Figure 9.3.

9.2.3 Conjugated System

Conjugacy is an important concept in dynamical system theory [3]. It is formulated in the state space. We briefly review it and extend conjugacy to the L^1-function space description.

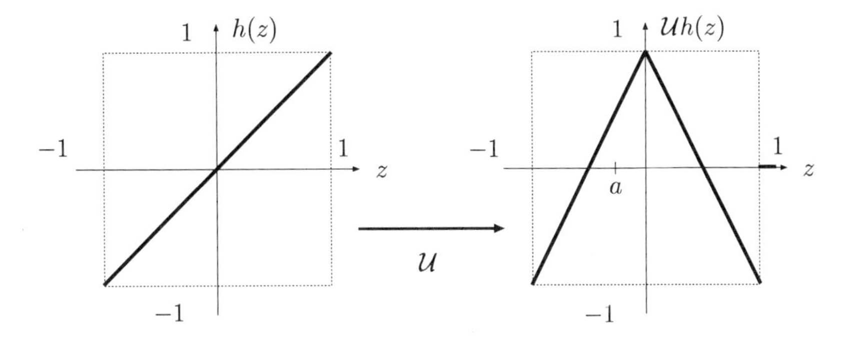

Figure 9.3: Action of the KO to $h(z) = z$.

State space conjugacy. Two discrete time systems

$$\begin{aligned} z_1(k+1) &= \varphi_1(z_1(k)) \\ z_2(k+1) &= \varphi_2(z_2(k)) \end{aligned} \tag{9.33}$$

are said to be equivalent if there exists a one-to-one map ϕ between their trajectories:

$$z_1(k) = \phi(z_2(k)) \tag{9.34}$$

If we know, say, the trajectory $z_2(k)$ of the second system we can determine the trajectory $z_1(k)$ of the equivalent system using ϕ exclusively. This is illustrated in the following diagram:

$$\begin{array}{ccc} Z_1 & \xrightarrow{\varphi_1} & Z_1 \\ \phi \uparrow\downarrow \phi^{-1} & & \phi \uparrow\downarrow \phi^{-1} \\ Z_2 & \xrightarrow{\varphi_2} & Z_2 \end{array} \tag{9.35}$$

The system maps of equivalent systems are called conjugated maps. From (9.33) and (9.34) we have

Conjugated Maps

$$\varphi_1(z) = \phi \circ \varphi_2 \circ \phi^{-1}(z) \tag{9.36}$$

Example: A conjugate of the tent map. Let the tent map (9.4) be conjugated to a system map φ_c by the conjugacy

$$\phi(z) = -\sin\left(\frac{\pi}{2} \cdot z\right). \tag{9.37}$$

Then the conjugated map is

$$\varphi_c(z) = \phi \circ \varphi \circ \phi^{-1}(z) = -\cos(2 \cdot \arcsin(z)) = 2 \cdot z^2 - 1. \tag{9.38}$$

This is the second-order Chebyshev polynomial.

Function space conjugacy. The two systems in (9.33) have a function space description based on the corresponding FPOs:

$$\begin{aligned} g_1(z_1, k+1) &= \mathcal{P}_1 g_1(z_1, k) \quad , \quad g_1(z_1, k) \in L_1^1 \\ g_2(z_2, k+1) &= \mathcal{P}_2 g_2(z_2, k) \quad , \quad g_2(z_2, k) \in L_2^1 \end{aligned} \tag{9.39}$$

In analogy to (9.34) there exists an invertible linear map $\mathcal{O} : L_2^1 \to L_1^1$ such that similar to (9.35) we have

$$\begin{array}{ccc} L_1^1 & \xrightarrow{\mathcal{P}_1} & L_1^1 \\ \mathcal{O} \uparrow\downarrow \mathcal{O}^{-1} & & \mathcal{O} \uparrow\downarrow \mathcal{O}^{-1} \\ L_2^1 & \xrightarrow{\mathcal{P}_2} & L_2^1 \end{array} \tag{9.40}$$

The map $\mathcal{O}$ connects the FPOs of conjugated maps and is given by

Conjugated FPOs

$$\mathcal{P}_1 g(z) = \mathcal{O} \circ \mathcal{P}_2 \circ \mathcal{O}^{-1} g(z)$$

$$\begin{aligned} \mathcal{O} g(z_1) &= |\det \mathrm{D}\phi^{-1}(z_1)| \cdot g \circ \phi^{-1}(z_1) \\ \mathcal{O}^{-1} g(z_2) &= |\det \mathrm{D}\phi(z_2)| \cdot g \circ \phi(z_2) \end{aligned} \tag{9.41}$$

Observe that the conjugacy ϕ uniquely determines the conjugacy $\mathcal{O}$ of the FPOs.

Invariance of eigenvalues. The eigenvalues of conjugated systems are equal and the eigenfunctions are transformed via the map $\mathcal{O}$. If $\xi_1(z_1)$ is an eigenfunction of $\mathcal{P}_1$ and λ_1 is the corresponding eigenvalue, then $\mathcal{P}_2$ also has an eigenvalue $\lambda_2 = \lambda_1$ and the corresponding eigenfunction is

$$\xi_2(z_2) = \mathcal{O}^{-1} \xi_1(z_2) = |\det \mathrm{D}\phi(z_2)| \cdot \xi_1 \circ \phi(z_2) \tag{9.42}$$

So, if we know the eigensystem of a particular FPO $\mathcal{P}$ we can determine the eigensystem of any other FPO conjugated to $\mathcal{P}$.

Example: Invariant density of the 2nd-order Chebyshev map. For the tent map the invariant density (eigenfunction of the FPO with eigenvalue 1) is constant and given by $f(z) = \frac{1}{2}$. The second-order Chebyshev polynomial is conjugated to the tent map via (9.37). Applying $\mathcal{O}^{-1}$ (9.41) we obtain the invariant density of the polynomial

$$f_c(z_c) = \left|\det \mathrm{D}\phi^{-1}(z_c)\right| \cdot \frac{1}{2} = \frac{1}{\pi \cdot \sqrt{1 - z_c^2}} \tag{9.43}$$

This example demonstrates the advantage of applying conjugacy. The direct calculation of the invariant density of the Chebyshev map would have been much more complicated.

9.2.4 Markov Maps – A Chaotic System Model Class

Conjugacy divides the set of dynamical systems into equivalence classes. In order to perform an eigensystem analysis, only the eigensystem of one particular member (representative) of the corresponding equivalence class has to be analyzed. A general method in statistical analysis is to analyze systems, which are easy to treat and to which other systems can be conjugated. Important examples are the *piecewise linear Markov maps.* Here we define this class and discuss their main properties.

Piecewise linear maps. A piecewise linear map (PWLM) is defined by a set of affine linear map segments on a partition of the state space:

Piecewise Linear Maps (PWLM)

$$\begin{aligned} &\text{map partition} &&: \; \Pi = \{I_1, ..., I_N\}, \quad Z = \cup_j I_j \\ &\text{map segments:} &&: \; \varphi_j(z) = A_j \cdot z + b_j \end{aligned}$$

where A_j are square matrices and b_j are vectors of appropriate dimensions.

(9.44)

We call a PWLM *regular* if all matrices A_j are invertible. Then the inverse of any map segment exists and is equal to

$$\varphi_j^{-1}(z) = A_j^{-1} \cdot (z - b_j) \tag{9.45}$$

Fully stretching maps. A PWLM is called fully stretching, if the image of each map segment is the whole state space, i.e., if we have

$$\varphi_j\,(I_j) = Z \tag{9.46}$$

These maps are the easiest to treat as we will see later.

Markov maps. A PWLM is said to be Markovian if a partition $\Pi_M = (I_{M1}, ..., I_{MN})$ exists such that the image of each interval I_{Mj} is a union of members of Π_M:

Piecewise Linear Markov Maps

$$X_{Mj} \;=\; \cup_{i \in O_j} I_{Mi} \tag{9.47}$$

where $X_{Mj} = \varphi\,(I_{Mj})$ and $O_j \subset \{1, .., N\}$ is an index set.

We refer to Π_M as the Markov partition of the map.

Example: Tent map. The tent map is piecewise linear and Markovian. Its Markov partition is $\Pi_M = \{I_{M1} = [-1,0), I_{M2} = [0,1]\}$ and we have $X_{M1} = X_{M2} = I_{M1} \cup I_{M2}$. The map segments are

$$\begin{aligned} \varphi_1(z) &= 2 \cdot z + 1, \quad z \in I_{M1} \\ \varphi_2(z) &= -2 \cdot z + 1. \quad z \in I_{M2} \end{aligned} \tag{9.48}$$

The tent map is also fully stretching. In general every fully stretching map is Markovian with the Markov partition equal to the map partition since $X_j = Z$ for all j.

FPO of piecewise linear Markov maps. The FPO is defined by its action on all partition elements. The action on the partition element I_{Mj} has the form

Action of $\mathcal{P}$ on $g(z) \cdot \mathbf{1}_{Mj}(z)$

$$\mathcal{P}\left(g(z) \cdot \mathbf{1}_{Mj}(z)\right) = \triangle_j \cdot g\left(A_j^{-1} \cdot (z - b_j)\right) \cdot \sum_{i=1}^{N} u_{ij} \cdot \mathbf{1}_{Mi}(z) \tag{9.49}$$

$$\triangle_j := \left|\det A_j\right|^{-1}$$

where $\mathbf{1}_{Mj}(z)$ is the indicator function of the interval I_{Mj}, i.e.,

$$\mathbf{1}_{Mj}(z) = \begin{cases} 1 & : \ z \in I_{Mj} \\ 0 & : \ else \end{cases} \tag{9.50}$$

and the u_{ij} are given by

$$u_{ij} = \begin{cases} 1 & : \ I_{Mj} \cap \varphi^{-1}(I_{Mi}) \neq \emptyset \\ 0 & : \ else \end{cases} \tag{9.51}$$

Observe that each function $g(z)$ can be uniquely represented by the base functions $g(z) \cdot \mathbf{1}_{Mj}(z)$. Using (9.49) and the FPO's linearity the image of each function $g(z)$ can be calculated.

Example: FPO of the tent map. For the tent map we have $\triangle_1 = \triangle_2 = \frac{1}{2}$ and $u_{ij} = 1, \quad i,j \in \{1,2\}$. From that we have by (9.49)

$$\begin{aligned} \mathcal{P}g(z) &= \mathcal{P}\left(g(z) \cdot \mathbf{1}_{M1}(z)\right) + \mathcal{P}\left(g(z) \cdot \mathbf{1}_{M2}(z)\right) \\ &= \tfrac{1}{2} \cdot g\left(\tfrac{z-1}{2}\right) + \tfrac{1}{2} \cdot g\left(\tfrac{1-z}{2}\right) \end{aligned} \tag{9.52}$$

in accordance with Equation (9.13).

9.2.5 Polynomial Eigenspace Analysis

The behaviour of a linear operator can be best understood by its eigenfunctions. The FPO of piecewise linear Markov maps has polynomial eigenfunctions. These functions form a finite dimensional subspace [1]. This simplifies the statistical analysis for this particular system class considerably. In the following we demonstrate this analysis for one-dimensional fully stretching maps.

One-dimensional fully stretching maps. A fully stretching map and its parameters are depicted in Figure 9.4.

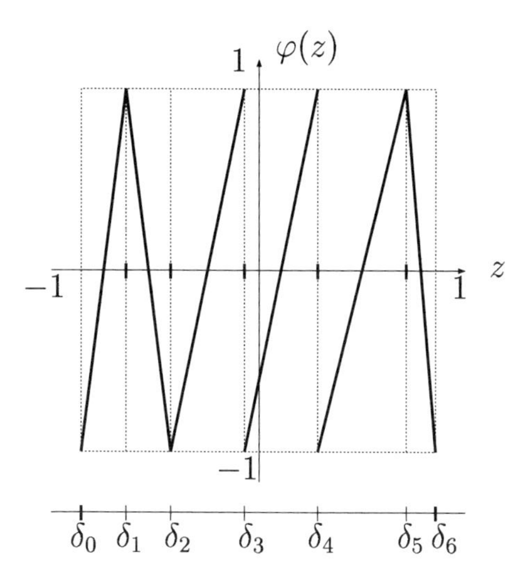

Figure 9.4: Fully stretching map with six segments.

FPO. The action of the FPO on a function $g(z)$ can be obtained from (9.49):

$$g(z) \to \mathcal{P}(g(z)) = \frac{1}{2}\sum_{i=1}^{N}(\delta_i - \delta_{i-1})g\left(s_i\frac{\delta_i - \delta_{i-1}}{2}z + \frac{\delta_i + \delta_{i-1}}{2}\right) \tag{9.53}$$

where the δ_i are the interval boundaries of the system partition, the $s_i \in \{-1, 1\}$ are the slope signs of the map segments, and $\triangle_i = \frac{\delta_i - \delta_{i-1}}{2}$.

Representation of the FPO on polynomial subspaces. Let us introduce the following subspaces of polynomial functions:

$$\begin{aligned} W_n &= \text{span}\{e_n(z)\} \\ e_n(z) &:= z^n \\ V_n &= W_0 \bigoplus \ldots \bigoplus W_n \end{aligned} \tag{9.54}$$

where $\bigoplus$ denotes the direct sum of the subspaces W_i. Thus V_n is the space of all polynomials up to order n. The subspaces W_i are one-dimensional, i.e., $\dim(W_n) = 1$ and thus $\dim(V_n) = n + 1$. It is easy to see from (9.53) that these subspaces are invariant under the action of the FPO. For instance, for the function $e_1(z) = z \cdot \mathbf{1}(z)$ the image is given by

$$\begin{aligned}
\mathcal{P}e_1(z) &= \tfrac{1}{2}\sum_{l=1}^{N} \delta_l - \delta_{l-1} \cdot \left(s_l \cdot \tfrac{\delta_l - \delta_{l-1}}{2} \cdot z + \tfrac{(\delta_l + \delta_{l-1})}{2}\right) \\
&= \tfrac{1}{4}\sum_{l=1}^{N} (\delta_l - \delta_{l-1})^2 \cdot s_l \cdot z \\
&= \tfrac{1}{4}\sum_{l=1}^{N} (\delta_l - \delta_{l-1})^2 \cdot s_l \cdot z \in W_1
\end{aligned} \tag{9.55}$$

because of $\sum_{l=1}^{N}(\delta_l^2 - \delta_{l-1}^2) = \delta_N^2 - \delta_{N-1}^2 + \delta_{N-1}^2 - \ldots - \delta_1^2 + \delta_1^2 - \delta_0^2 = \delta_N^2 - \delta_0^2 = 1 - 1 = 0$. Therefore we can represent the FPO on the space W_1 by means of a matrix with the structure:

$$G_1 = \begin{bmatrix} 1 & 0 \\ 0 & g_{11} \end{bmatrix} \tag{9.56}$$

According to (9.55) g_{11} is

$$g_{11} = \tfrac{1}{4} \cdot \sum_{i=1}^{N} (\delta_i - \delta_{i-1})^2 \cdot s_i. \tag{9.57}$$

The higher order polynomial spaces V_n are invariant too. There the FPO can be represented by the matrix

$$G_n = \begin{bmatrix} 1 & 0 & g_{02} & g_{03} & \cdots & g_{0n} \\ 0 & g_{11} & g_{12} & g_{13} & \cdots & g_{1n} \\ 0 & 0 & g_{22} & g_{23} & \cdots & g_{2n} \\ 0 & 0 & 0 & g_{33} & \cdots & g_{3n} \\ \cdots & \cdots & \cdots & \cdots & \cdots & \cdots \\ 0 & 0 & 0 & 0 & \ldots & g_{nn} \end{bmatrix} \tag{9.58}$$

and the matrix elements are given by:

$$\forall\, l \leq k,\ k \in \{1, \ldots, n\} : g_{lk} = \frac{k!}{l! \cdot (k-l)! \cdot 2^{k+1}} \cdot \sum_{i=1}^{N} (\delta_i + \delta_{i-1})^{k-l} \cdot (\delta_i - \delta_{i-1})^{l+1} \cdot s_i^l \tag{9.59}$$

Polynomial eigenvalues. For obvious reasons we call the eigenvalues of the matrix G_n polynomial eigenvalues. They can be ordered according to the subspace V_n and we obtain for the n-th eigenvalue

Polynomial Eigenvalues of $\mathcal{P}$

$$\lambda_n = \tfrac{1}{2^{n+1}} \cdot \sum_{i=1}^{N} (\delta_i - \delta_{i-1})^{n+1} \cdot s_i^n \tag{9.60}$$

All eigenvalues are real. They are located inside or on the unit circle. Furthermore they form a convergent zero sequence, i.e., $\lambda_n \to 0$ for $n \to \infty$. If the map parameters fulfill the following conditions

$$\begin{aligned} &(\mathbf{1}) \quad \forall\, k \in \{1, ..., N\} : \quad s_k = s \in \{-1, 1\} \\ &(\mathbf{2}) \quad \forall\, k \in \{1, ..., N\} : \quad (\delta_k - \delta_{k-1}) = \tfrac{2}{N} \end{aligned} \tag{9.61}$$

then the eigenvalues satisfy the equation

$$\lambda_n = \lambda_1^n. \tag{9.62}$$

Maps of this kind form the family of c-adic Renyi maps.

9.2.6 Expectation Values

Averages. The expectation value of a function $h : Z \to \mathbb{K}$ is defined by the formula

$$\begin{aligned} E[h] &= \int_Z h(x) \cdot f(x) \cdot \mathrm{d}x \\ &=: \langle h(z), f(z) \rangle \end{aligned} \tag{9.63}$$

The time average of $h\,(z(k))$ is defined as the limit

$$A\,[h, z_0] = \lim_{T \to \infty} \frac{1}{T} \cdot \sum_{k=0}^{T-1} h\,(z(k)) \tag{9.64}$$

This average depends on the initial value. But for ergodic systems the time average (9.64) is constant for almost all initial conditions and equal to the expectation value (9.63). In particular we have the following result from ergodic theory [4]:

Ergodicity

The time average of the function $h(z)$, given by

$A\,[h, z_0] = \lim_{T\to\infty} \frac{1}{T} \sum_{k=0}^{T-1} h\,(z(k))$

is equal to the expectation value

$E\,[h] = \langle h(z), f(z) \rangle$

for almost all state sequences having the unique distribution density $f(z)$. (9.65)

In the following we assume the system to be ergodic. The function h is called a test-function. It is an element of the dual space L^∞.

Static test-functions. We refer to a test-function as static if it does not imply a time shift on the trajectory. An example for such a test-function is the logarithm on the determinant of the differential of the map, i.e.,

$$h(z) = \log |\det \mathrm{D}\varphi(z)| \tag{9.66}$$

Expectation values with static test-functions can be calculated easily using the invariant density of the map only.

Example: Tent map. For the tent map we get for the test-function h defined by Equation (9.66) the value $E[h] = \log 2$. This is the Lyapunov exponent.

Dynamic test-function. We refer to a test-function as dynamic if it does imply a time shift on the trajectory. An example provides the correlation function. Here we have the test-function

$$h(z;\tau) = z \cdot \varphi^{\tau}(z) \tag{9.67}$$

where $\varphi^{\tau}(z)$ indicates the time shift. Note that τ is an parameter of the test-function. We can express this test-function in terms of a KO-product (9.27) and obtain

$$h(z;\tau) = z \cdot \mathcal{U}^{\tau} z \tag{9.68}$$

KO representation of dynamic test-functions. Dynamic test-functions which are sums of products of state values at different times can be expressed using the KO. We demonstrate by two examples an intuitive procedure to derive the KO representation of the test-function:

1. Suppose $x(k) = h(z(k))$ is a scalar valued output signal. We want to calculate the correlation function of $x(k)$. Here the following correspondence from the space of signals to the space of test-function is obvious:

$$x(k) \cdot x(k+\tau) \quad \rightarrow \quad h(z) \cdot h \circ \varphi^{\tau}(z) \quad \rightarrow \quad h(z) \cdot \mathcal{U}^{\tau} h(z) \tag{9.69}$$

2. For higher order correlations we have the correspondence:

$$\prod_{l=0}^{n} x(k+\tau_l) \quad \rightarrow \quad \prod_{l=0}^{n} h \circ \varphi^{\tau_l}(z) \quad \rightarrow \quad \prod_{l=0}^{n} \mathcal{U}^{\tau_l} h(z) \tag{9.70}$$

   where $\tau_0 = 0 \leq \tau_1 \leq \ldots \leq \tau_n$.

FPO representation of correlation functions. Using Equation (9.68) the correlation function can be expressed by means of the KO:

$$R(\tau) = \langle z \cdot \mathcal{U}^{\tau} z, f(z) \rangle \tag{9.71}$$

Because the KO is dual to the FPO we get a representation of $R(\tau)$ by means of the FPO

$$R(\tau) = \langle z, \mathcal{P}^{\tau} \left(z \cdot f(z) \right) \rangle \tag{9.72}$$

This is important, since the theoretical results from the spectral decomposition of the FPO can be applied to determine correlation functions. In order to calculate the correlation function we have to determine the action of the FPO on the function $z \cdot f(z)$. As demonstrated in Section 9.2.5 for fully stretching maps, this can be done by applying the results on polynomial eigenspaces. A FPO representation for higher order correlation functions can also be derived by applying the rule (9.31) to the test-function (9.70):

$$\begin{aligned} R\left(\tau_1, .., \tau_n\right) &= \left\langle h(z) \cdot \prod_{l=1}^{n} \mathcal{U}^{\tau_l} h(z), f(z) \right\rangle \\ &= \left\langle h(z), \left(\mathcal{P}^{\theta_n} h(z) \cdot ... \left(\mathcal{P}^{\theta_2} h(z) \cdot \left(\mathcal{P}^{\theta_1} \left(h(z) \cdot f(z)\right)\right)\right) ...\right)\right\rangle \end{aligned} \tag{9.73}$$

where $\theta_n = \tau_n - \tau_{n-1}$.

Example: Correlation function of the tent map. The invariant density of the tent map is uniform, i.e., $f(z) = \frac{1}{2}$ and we get

$$R(\tau) \;=\; \tfrac{1}{2} \cdot \langle z, \mathcal{P}^{\tau} z \rangle \tag{9.74}$$

Hence, in order to calculate the correlation value we have to determine the action of the FPO on the function $g(z) = z$. For the tent map we have already shown that $\mathcal{P}^{\tau} z = 0, \tau \geq 1$. Inserting this result in Equation (9.74) we obtain the result

$$R(\tau) = \frac{1}{3} \cdot \delta(\tau) \tag{9.75}$$

which indicates that the state values at different times are not correlated.

9.3 System Design

9.3.1 Prescribed Density Function

Synthesis problem. Given the density f of a real valued signal to be generated, find a system map φ and an output map h of a one dimensional system (9.1), such that the generated signal possesses the density f. This problem can be solved by the two approaches below. In approach 1 we select the output map $h(z) = z$ and design for φ. In approach 2 we choose a fixed φ and design for h. Both approaches apply uniform density generators, explained next.

Uniform generators. A chaotic system generating uniformly distributed states is called a uniform generator. For one-dimensional systems with the state space $Z = [-1, 1]$ that means the invariant density is equal to $f(z) = \frac{1}{2}$. We denote the system maps of this class of generators by φ_u.

Example: Fully stretching maps. Every chaotic system with a fully stretching system map is an uniform generator.

Approach 1: Designing φ. Given a uniform generator map φ_u and a conjugacy ϕ, from the conjugacy principle (Section 9.2.3) it follows that the conjugated map φ and its invariant density $f(z)$ is related to the uniform map and its uniform density by:

$$\begin{aligned} \varphi(z) &= \phi \circ \varphi_u \circ \phi^{-1}(z) \\ f(z) &= |\det \mathrm{D}\phi^{-1}(z)\ | \cdot \tfrac{1}{2}. \end{aligned} \tag{9.76}$$

Hence we can choose an arbitrary uniform map φ_u and the design problem reduces to that of finding a conjugacy ϕ which satisfies:

$$|\det \mathrm{D}\phi^{-1}(z)\ | = 2 \cdot f(z) \tag{9.77}$$

Note that this equation does not define a map ϕ uniquely. If we require $\phi : [-1, 1] \to [-1, 1]$ additionally to be increasing we get the explicit formula

$$\phi^{-1}(z) = 2 \cdot \int_{-1}^{z} f(z)\, \mathrm{d}z - 1 \tag{9.78}$$

for the calculation of ϕ from a given f.

Example: Triangular density. Let the prescribed density be $f(z) = \frac{1-z}{2}$. From formula (9.78) we get the conjugacy $\phi^{-1}(z) = \frac{1-z^2}{2} + z$ and its inverse $\phi(z) = 1 - \sqrt{2 - 2 \cdot z}$. One can choose an arbitrary φ_u and according to Equation (9.76) one obtains the system map.

Approach 2: Designing h. Given an arbitrary uniform generator, the uniform density of the state variable is transformed by an output map h into a density f according to

$$f(z) = |\det \mathrm{D}h^{-1}(z)\ | \cdot \tfrac{1}{2} \tag{9.79}$$

which is similar to (9.77). The output map h can be determined by formula (9.78) replacing ϕ by h only.

9.3.2 Prescribed Correlation Function

Synthesis problem. Given is the correlation function $R(\tau)$ of a signal to be generated. Find the system map φ of the generator. In order to solve this problem we have to give a suitable description of correlation sequences. A class of correlation functions satisfy the recursive equation [5]:

$$R(\tau) = \sum_{l=1}^{n} c_l \cdot R(\tau - l), \quad \tau \geq n \tag{9.80}$$

or, in implicit form

$$0 = \sum_{l=0}^{n} c_l \cdot R(\tau - l) \tag{9.81}$$

where $c_0 = -1$. This autoregressive model of correlation sequences is very popular in signal analysis especially in human voice modelling. In the following we concentrate on correlation sequences satisfying (9.81).

Fundamental design conditions. The correlation sequence of a one-dimensional chaotic map on state space $Z = [-1, 1]$ is given by (9.72)

$$R(\tau) = \langle z, \mathcal{P}^{\tau} (z \cdot f(z)) \rangle \tag{9.82}$$

where the bracket notation is defined in (9.63). Assume $R(\tau)$ fulfills (9.81). Inserting (9.82) into (9.81) we get

$$\begin{aligned} 0 &= \left\langle z, \sum_{l=0}^{n} c_l \cdot \mathcal{P}^{\tau - l} (z \cdot f(z)) \right\rangle \\ &= \left\langle \mathcal{U}^{\tau - n} z, \sum_{l=0}^{n} c_{n-l} \cdot \mathcal{P}^{l} (z \cdot f(z)) \right\rangle \end{aligned} \tag{9.83}$$

Note that $\mathcal{U}^{\tau - n} z$ is the $\tau - n$-iterated of the map φ (Section 9.2.2). Equation (9.83) holds for all $\tau - n \geq 0$. Therefore we have for all $k = 0, 1, ...$

$$0 = \left\langle \varphi^k(z), \sum_{l=0}^{n} c_{n-l} \cdot \mathcal{P}^{l} (z \cdot f(z)) \right\rangle \tag{9.84}$$

Note that the integral $\int_{[-1,1]} g \, dz$ vanishes if the integrand g is either identical zero or an odd function. For this reason there are the following conditions on the system map and its FPO in order to fulfill (9.84).

CI Even map condition: Function φ is even. Then φ^k is even for all k. In this case the function $\sum_{l=0}^{n} c_{n-l} \cdot \mathcal{P}^{l} (z \cdot f(z))$ has to be odd.

CII Odd map condition: Function φ is odd. Then φ^k is odd for all k. In this case the function $\sum_{l=0}^{n} c_{n-l} \cdot \mathcal{P}^l (z \cdot f(z))$ has to be even.

CIII Zero condition: Function φ is neither even nor odd. In this case we have the condition

$$0 = \sum_{l=0}^{n} c_{n-l} \cdot \mathcal{P}^l (z \cdot f(z)) \tag{9.85}$$

In the following we demonstrate some system design cases. For more details see [1].

White noise generators. A correlation function of white noise is obtained from (9.81) setting the parameters $n = 1, c_1 = 0$. Then Equation (9.84) reduces to

$$0 = \langle \varphi^k(z), z \cdot f(z) \rangle \tag{9.86}$$

Applying the conditions stated above we get

CI The function $z \cdot f(z)$ has to be odd. Because z is odd $f(z)$ has to be even.

CII The function $z \cdot f(z)$ has to be even. Because z is odd $f(z)$ has to be odd too.

CIII We can not fulfill condition (9.85) because

$$0 \neq z \cdot f(z) \tag{9.87}$$

Consequently the system map φ of a white noise generator has to be an odd or an even function.

Example: Even fully stretching maps. Choose any even fully stretching map, e.g., the symmetric tent map. Because f is constant and hence even it will generate white noise. As already shown (Section 9.2.3) the second-order Chebyshev polynomial, which is an even function, has an even density. Hence it will generate white noise too.

AR(1)-generators. The AR(1) correlation function is defined by $R(\tau) = R(0) \cdot c_1^{|\tau|}$. It is obtained by $n = 1, c_1 \neq 0$. We note that the corresponding spectrum is of low-pass type if $0 < c_1 < 1$ and of high-pass type if $-1 < c_1 < 0$. Equation (9.84) reduces to

$$0 = \langle \varphi^k(z), c_1 \cdot z \cdot f(z) - \mathcal{P}(z \cdot f(z)) \rangle \tag{9.88}$$

In order to keep off symmetry restrictions from φ we consider case three:

CIII We can fulfill condition (9.85) by satisfying

$$0 = c_1 \cdot z \cdot f(z) - \mathcal{P}\left(z \cdot f(z)\right) \tag{9.89}$$

Hence the function $z \cdot f(z)$ is required to be an eigenvector with eigenvalue $\lambda = c_1$. From this, according to Section 9.2.4, piecewise linear Markov maps can be constructed.

Example: 1D-fully stretching maps. Every non-even one-dimensional fully stretching map is an AR(1)-Generator. This is because the space spanned by the function z is invariant under the FPO and hence $z \cdot f(z)$ is an eigenfunction. The coefficient c_1 is equal to the eigenvalue λ_1 (see Equation (9.60)) and given by:

$$c_1 = \frac{1}{4} \cdot \sum_{i=1}^{N} (\delta_i - \delta_{i-1})^2 \cdot s_i^2 \tag{9.90}$$

For this reason infinitely many AR(1)-Generators can be find in this class for a given c_1.

9.4 Generator Implementation Based on Piecewise Linear Maps

In Section 9.2.4 the important role of piecewise linear maps was pointed out. In this section a structure for the implementation of discrete-time generators based on such maps is described. In comparison to other implementations of chaos generators, e.g., [6–8], the used map is widely adjustable, thus allowing us to realize a large set of maps simply by programming the circuit. The resulting circuit is suited for the generation of chaotic signals with prescribed statistical properties and can be used as a vehicle for chaos generator design and experiments.

The structure of the generator is developed in a systematic way. Finally a design example is described, which uses a digitally controlled analog signal processing structure.

9.4.1 System Structure

Figure 9.5 shows the conceptual block diagram of a discrete-time generator, which consists of a delay block T and a nonlinear transfer function φ parameterized by a parameter set p. Both components are connected in a feedback loop. The operation of this system is described by the state equation (9.1)

$$z(k+1) = \varphi\left(z(k)\right). \tag{9.91}$$

where z is the state vector and k the iteration index.

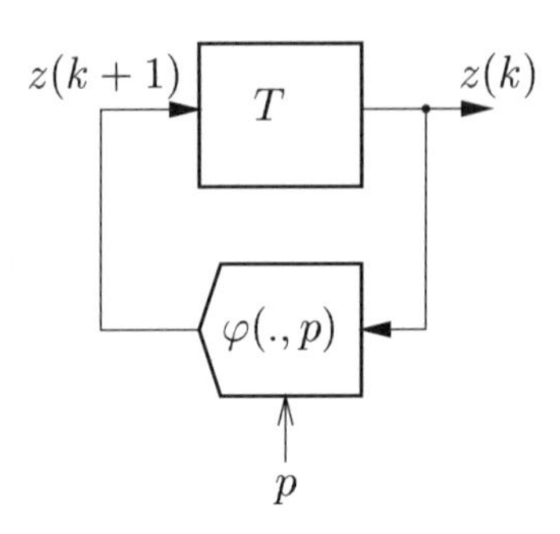

Figure 9.5: System structure.

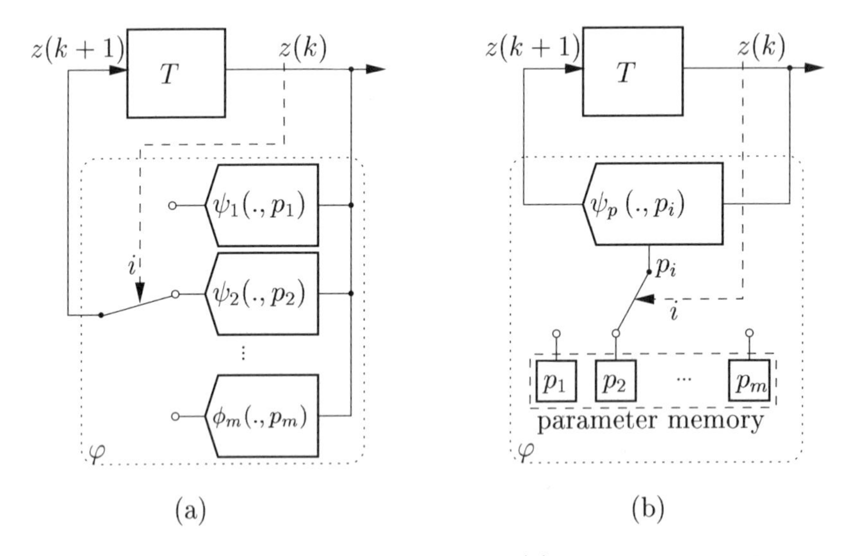

Figure 9.6: System with multiplexed maps ψ_i (a) and multiplexed parameters (b).

The main block in the system implementation is the piecewise defined map φ. This map can be interpreted as a multiplexed structure of separate maps ψ_i as depicted in Figure 9.6(a). If all ψ_i are of the same type, e.g. $\psi_i(z) = a_i\, z + b_i$, and have only different parameters then this structure can be simplified to the structure shown in Figure 9.6(b) where parameters p_i of a parameterized map ϕ_p are multiplexed. This results in a time-variant system where the recursion in each time step is calculated by an analog structure while the switching of the parameters is controlled digitally. The parameter memory is a separate building block in the structure and can be built as a digital memory. This allows us to easily adjust the parameters of the map and thus the characteristics of the generated signals. A pure digital realization of Equation (9.91) is also possible, but may lead to problems with limit cycles and fixed points caused by the finite precision of the arithmetic used.

In the following section the structures required for implementing the function φ will be described. The implementation of the delay blocks depends mainly on the technology used and will not be discussed here.

9.4.2 Synthesis of Piecewise Linear Maps

The function φ is defined on the state space $I \subset \mathbb{R}^n$ of the system. The state space is divided into m nonoverlapping *segments* $I_i \subset I$ $(1 \le i \le m)$. A piecewise defined map φ can be decomposed into

$$\varphi(z) = \psi(z)\,\delta(z) = \sum_{i=1}^{m} \psi_i(z)\,\delta_i(z) \tag{9.92}$$

where ψ is a vector of elementary affine functions $\psi_i(z)$ defined on the segments I_i and the vector δ consists of *indicator functions* $\delta_i(z)$ with

$$\delta_i(z) = \begin{cases} 1: & z \in I_i \\ 0: & \text{else} \end{cases} \tag{9.93}$$

Thus a piecewise defined function can be implemented as a vector product of the vector of the elementary functions and the vector of indicator functions. In the following the implementations of both vectors will be discussed.

Elementary functions ψ_i. In the case of a piecewise linear function φ elementary functions ψ_i of the form

$$\psi_i(z) = A_i\, z + b_i \tag{9.94}$$

have to be implemented. The multiplication of the analog values of z and the digital stored coefficients of the $n \times n$ matrix A_i and the n dimensional vector b_i can be implemented using multiplicating digital-to-analog converters (mDAC) where the signal z is the reference value of the converter, while the components of A_i and b_i control the digital inputs.

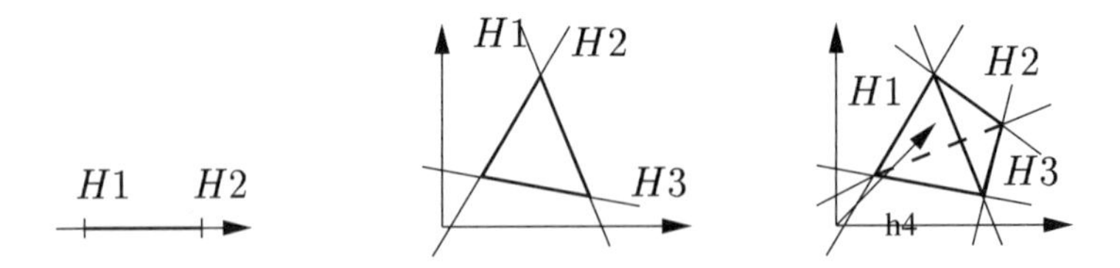

Figure 9.7: Segments in one-, two-, and three-dimensional spaces defined by hyper-planes.

Indicator functions δ_i. The functions δ_i map a vector z to the number i of the particular segment I_i containing the actual value of z. The segments are determined by $n+1$ hyperplanes which are each defined by

$$H_i: \; h_{i,0} + h_i \, z = 0 \qquad i = 1, \ldots, n+1. \tag{9.95}$$

where the h_i is the n-dimensional normal vector of the plane H_i and $h_{i,0}$ its the offset.

The resulting segments in different dimensions are simplexes and are shown in Figure 9.7. More complicated segments, e.g., squares in a two-dimensional space can be built from such simple segments. The segments can not be defined independent from each other because each hyper-plane defines the borders of different segments.

The assignment of a vector z to a segment is done by determining the orientation of the each of the planes corresponding to the point z as follows: Each of the hyper-planes H_i divides the state space into two half spaces. A point z can be assigned to one of these halves by

$$\lambda_i(z) = \mathrm{s}(h_{i,0} + h_i \, z) \qquad \text{where} \qquad \mathrm{s}(x) = \begin{cases} 1: & x \geq 0 \\ 0: & \text{else} \end{cases} \tag{9.96}$$

The ordered set $\{\lambda_i\}$ can be interpreted as a binary number and thus a value Λ

$$\Lambda(z) = \sum_i \lambda_i(z) \, 2^i \tag{9.97}$$

can be assigned to each segment resulting from a given set of hyper-planes. This value corresponds to an unique segment number and can be used to address the parameter memory. An example is shown in Figure 9.8. Figure 9.9 shows the structure realizing Equation (9.96). The resulting complete circuit structure is shown in Figure 9.10. On the left-hand side are the multiplying digital-to-analog converters for the maps ψ_i while on the right-hand side we have the circuitry for determining the segment numbers. There are two signal flows in this circuit: an analog one from the output of the delay blocks to the inputs of the mDAC's, and a digital signal flow from the outputs of the comparators to the data inputs of the mDAC's. Thus the calculation of the analog values of Equation (9.91) is done using analog components while the coefficients of the map f_i are switched by digital means.

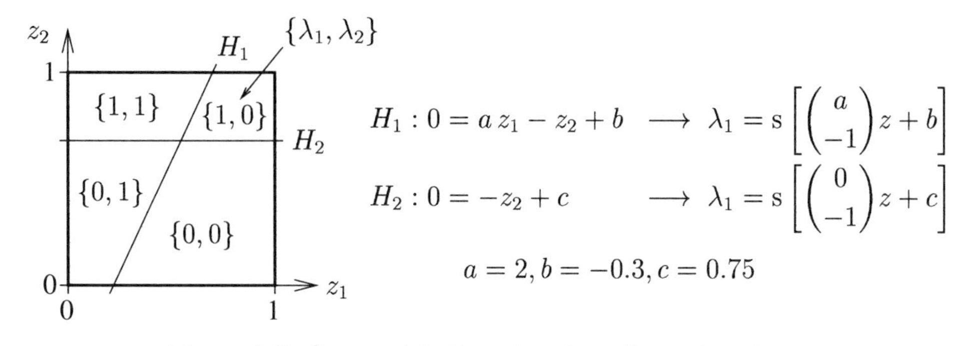

Figure 9.8: Segment indices in a two-dimensional space.

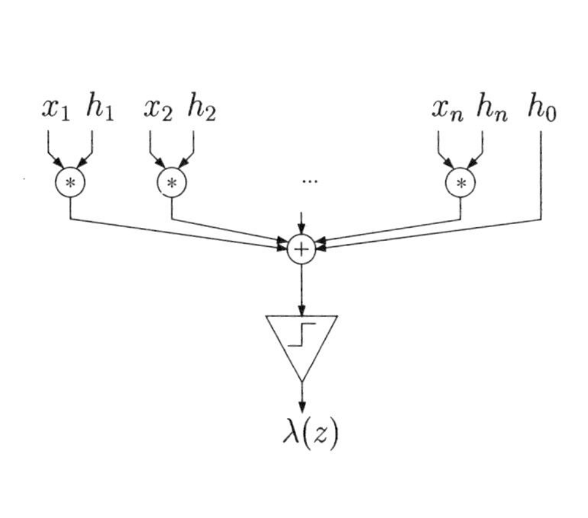

Figure 9.9: Basic structure for segment determination.

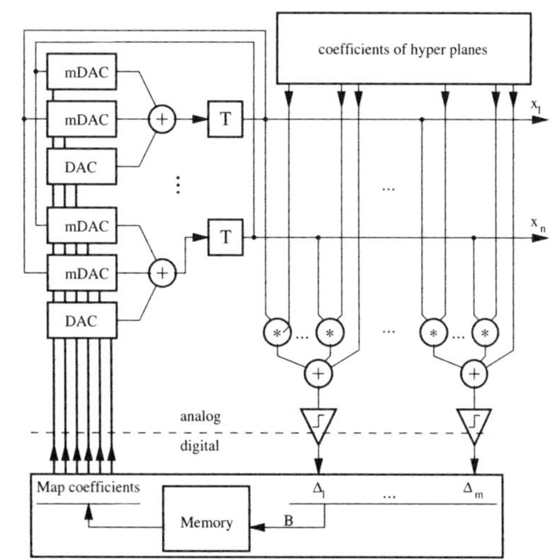

Figure 9.10: Block diagram of the general circuit structure.

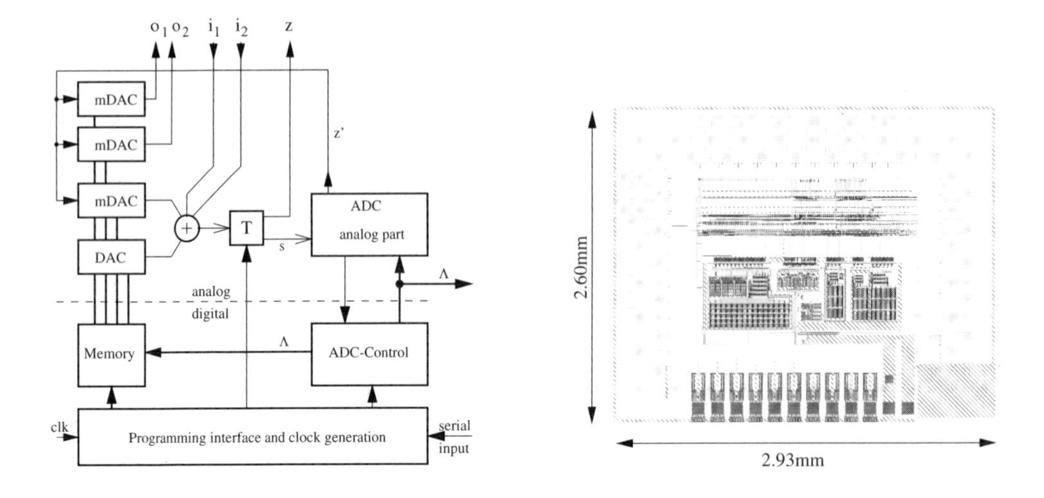

Figure 9.11: Block diagram and layout of the realized circuit.

9.4.3 Integrated Circuit Design Example

A circuit following the principle shown in Figure 9.10 has been implemented in a standard 0.8 μm CMOS technology using switched-current techniques. To reduce the costs some simplifications have been done starting with the structure in Figure 9.10. The reduced structure is shown in Figure 9.11. The determination of the number of the segment for a given z is done by simple analog-to-digital converters (DAC). Thus 8 fixed, rectangular segments per dimension are possible. The system shown in Figure 9.10 has been split up into a structure realizing a one-dimensional system in one circuit. Up to three circuits can be coupled together, thus realizing a three-dimensional system. This can be done with the signals o_j and i_j. The signal x is the output of this circuit. The memory for the coefficients of the map has been realized by an external circuit, giving more flexibility because its size depends on the dimension of the complete system. An interface for adjusting the coefficients using a PC has been included on the chip.

This circuit has been implemented in cooperation between the Technical University Dresden in Germany and the Institute of Microelectronics at the University of Seville in Spain.

9.5 Chaotic Encryption Systems

In this section a *systematic statistical approach* to the top-down design of discrete-time continuous-value information encryption systems is presented. The aim of the design is a basic structure for which an increase of privacy or secrecy can be achieved by techniques used in applied cryptography such as cascading. The design philosophy is explained and two main axioms for the coder design are derived. Based on these axioms the encoder cell is designed. Then a simple example and its crypto-analysis are presented.

9.5.1 Statistical Design Approach

In contrast to classical or applied cryptography which operates on finite spaces or modules with a finite number of symbols in chaotic cryptography the value range of the information signal as well as of the system parameters and states is considered to be continuous. Therefore an analytic approach to the encryption problem is necessary (Table 9.1).

	classical cryptography	*chaotic cryptography*
methods	algebraic, discrete-value, on finite spaces	analytic, continuous-value
digital realisation	integer arithmetic	non-integer arithmetic

Table 9.1: Comparison of classical and chaotic cryptography.

Design axioms. The design of the coder system is based on two main design axioms. The first one states desired transformations of the signal characteristics. A minimum requirement for public channel crypto-systems is to hide (at least some of) the characteristics of the information signal in the characteristics of the encoded signal. This can be achieved either by *diffusion* where the characteristics of the encoded signal $\boldsymbol{s}$ are independent of the information signal $\boldsymbol{i}$ or by *confusion* where the dependence is highly complex and cannot be analyzed in a simple way (see also [11]).

The objective of our design is to achieve a certain level of secrecy or privacy by *diffusion*. Therefore the encoder is considered to be a channel where an information signal $\boldsymbol{i}$ is transformed into an encoded signal $\boldsymbol{s}$. Since many crypto-analysis methods use statistical properties of the encoded signal we require the encoded signal $\boldsymbol{s}$ to have an n-th order probability distribution which is independent of the probability distribution of the information signal $\boldsymbol{i}$ (to provide a certain level of diffusion). Additionally, the n-dimensional distribution of the encoded signal $\boldsymbol{s}$ should be uniform to ensure the maximum entropy of $\boldsymbol{s}$. This is stated in Axiom 1.

- **Axiom 1. n-th order secrecy:** The conditional density $f_{\boldsymbol{s}|\boldsymbol{i}}$ of the signal vector $s_k = (s(k), s(k-1), ..., s(k-n+1))$ is independent of i_k for all $(k, k-1, ..., k-n+1)$ and uniformly distributed:

$$f_{\boldsymbol{s}|\boldsymbol{i}}(s_k|i_k) = \begin{cases} \frac{1}{vol(S^n)} & s_k \in S^n \\ 0 & else \end{cases} \tag{9.98}$$

A further advantage of the n-dimensional uniform distribution of $\boldsymbol{s}$ is its autocorrelation function $R(\tau)$ with $R(0) = 1$ and $R(\tau) = 0$ for $\tau = 1, \dots, n-1$.

The decoder has to retrieve the information signal, at least asymptotically. This can be achieved by the *inverse system approach* [12] (Figure 9.12). If the system Σ^{-1} is the inverse of Σ and if suitable initial conditions have been selected then Σ^{-1} is able to reproduce the input signal of Σ from its output signal at least asymptotically. This is stated in Axiom 2 and illustrated in Figure 9.12.

- **Axiom 2. Asymptotic invertibility:** For all initial conditions $(z(0), z^*(0)) \in Z$ (Z: state space of the encoder-decoder system) and all admissible information signals $\boldsymbol{i}$ the decoded signal $\boldsymbol{i}^*$ approaches $\boldsymbol{i}$ (at least) asymptotically, i.e., $\lim_{t\to\infty} |i^*(k) - i(k)| = 0$.

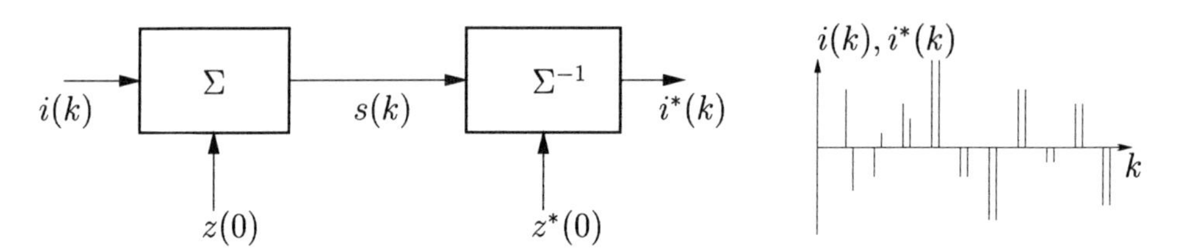

Figure 9.12: Inverse systems Σ and Σ^{-1}.

Basic structures. In the encoder the information signal $\boldsymbol{i}$ has to be encoded by the operation $\varphi(i, \kappa)$ using the key stream $\kappa(k)$ (Figure 9.13). In the simplest case this operation is static. The key stream $\kappa(k)$ is generated by a dynamical system $\Sigma_k(c)$, the so-called key stream generator, with the key parameter c.

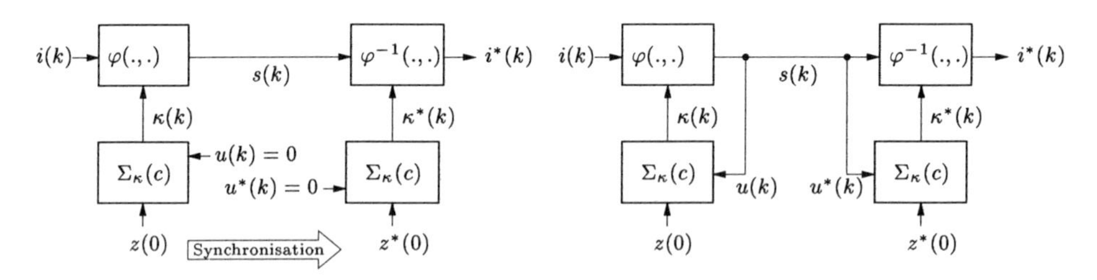

Figure 9.13: Synchronized and self-synchronizing scheme.

In the decoder the information signal has to be recovered by the inverse operation $\varphi^{-1}(s, \kappa)$. For that it is necessary to have identical key streams $\kappa(k)$ and $\kappa^*(k)$ in the encoder and the decoder, respectively. This identity can be achieved either by a directly setting the initial conditions of the two key stream generators using special synchronization frames (synchronized structure) or using a feed back feed forward control of the encoder and the decoder by the encoded signal $s(k)$ (self-synchronizing structure). The basic structures for both cases are depicted in Figure 9.13. Conditions for *self-synchronization* for various self-synchronizing systems are derived in [12]. Table 9.2 lists differences between both schemes.

	synchronized	*self-synchronizing*
$u(k)$	$u(k) = 0$	$u(k) = s(k)$
$\Sigma_\kappa(.)$	autonomous, complex behaviour	non-autonomous, unique asymptotic behaviour
initial condition	$z(0) = z^*(0)$	in general $z(0) \neq z^*(0)$
key	parameters c, $z(0)$	parameters c
realization	discrete-value only	discrete- or continuous-value

Table 9.2: Comparison of synchronized and self-synchronizing coders.

Subsequently, only self-synchronizing coder systems are considered because they do not need an extra synchronization signal. A disadvantage is that the signal $s(k)$ which controls the key stream generator is publicly accessible and thus can be used for a crypto-analysis [11].

In order to design an encryption system that satisfies Axioms 1 and 2 appropriate properties of the nonlinear operation $\varphi(i, \kappa)$ and the dynamical subsystems $\Sigma_\kappa(.)$ have to be derived. These properties serve as design rules. Without loss of generality we assume $i(k) \in I = [-1, 1), s(k) \in S = [-1, 1)$ and $\kappa(k) \in K \subset \mathbb{R}$.

Design of the nonlinear operation $\varphi(i, \kappa)$. From Axioms 1 and 2 the corresponding design requirements for the nonlinear operation $\varphi(i, \kappa)$ are deduced:

- **Requirement 1. Preservation of the uniform distribution:** Each element $\varphi(i, .) : K \to S$ preserves the uniform distribution:

$$f_\kappa(\kappa) = \begin{cases} \frac{1}{vol(K)} & \kappa \in K \\ 0 & else \end{cases} \quad \longrightarrow \quad \forall i \in I : f_{s|i}(s|i) = \begin{cases} \frac{1}{vol(S)} & s \in S \\ 0 & else \end{cases} .$$

- **Requirement 2. Invertibility of $\varphi(., \kappa)$:** Each element of the family of κ-parametric maps $\varphi(., \kappa)$ has to be invertible, i.e., for all $\kappa \in K$ $\varphi(., \kappa) : I \to S$ is bijective.

According to Requirements 1 and 2 the operation $\varphi(i, \kappa)$ has to preserve the uniform distribution along $K \to S$ and bijective along $I \to S$. Such an operation can be constructed as follows:

1. Assume $K \in [-1, 1)$ and design the map $\varphi(0, .)$ to fulfill Requirement 1, i.e., to preserve the uniform distribution. Rules for the design of uniform distribution preserving maps and numerous examples can be found in [14]. An example is shown in Figure 9.14. Based on such a map the nonlinear operation $\varphi(i, \kappa)$ is derived by shifting the map $\varphi(0, \kappa)$ according to the

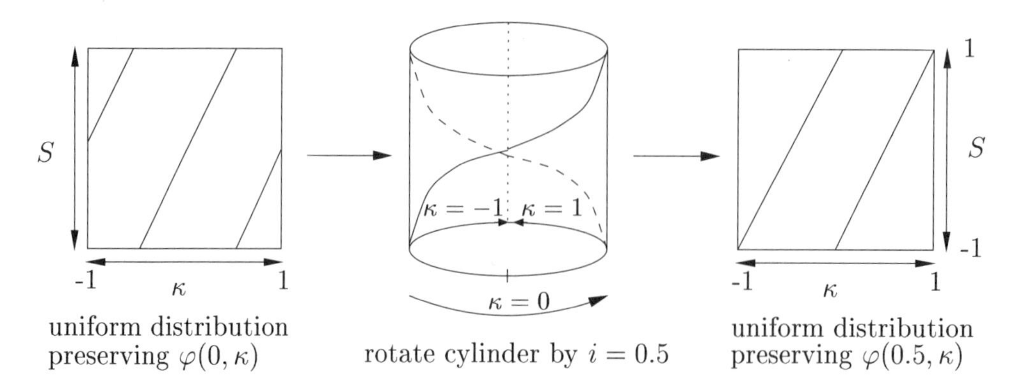

Figure 9.14: Generation of $\varphi(i,\kappa)$ by shifting.

information signal i. This is demonstrated in Figure 9.14 by rotation on a cylinder and can be mathematically described by $\mathrm{mod}\,(i(k)+\kappa(k))$, where

$$\mathrm{mod}\,(x) = x - 2 \cdot \left\lfloor \frac{x+1}{2} \right\rfloor \tag{9.99}$$

is the two's complement overflow nonlinearity. It can be proved that all these shifted maps $\varphi(i,\kappa)$ also preserve the uniform distribution. It follows

$$\varphi(i,\kappa) = h \circ \mathrm{mod}(i(k) + \kappa(k)), \tag{9.100}$$

with $h = \varphi(0,\kappa)$ (because of $\mathrm{mod}(\kappa) = \kappa$ for all $\kappa \in [-1,1)$).

In order to satisfy Requirement 2 (invertibility) the map $\varphi(.,\kappa)$ must be bijective. The mod-operation is bijective on each interval $[v1, v2)$, where $v2 - v1 = 2$. It follows that the map $h : [-1,1) \to [-1,1)$ must be one-to-one and therefore the absolute value of its derivation has to be everywhere equal to 1. The map h in Figure 9.14 is two-to-one and therefore does not fulfill Requirement 2.

2. Assume $K \neq [-1,1)$. Then it can be proved that K has to be equal to $[v_1, v_2)$, where $v_2 - v_1 = 2m$, $m \in \mathbb{Z}$, and $\varphi(0,\kappa)$ has to be periodic with period 2 as shown in Figure 9.15 for $\kappa \in [-2,2)$. The obtained structure of the nonlinear operation and its inverse are depicted in Figure 9.15.

Design of the dynamical system $\Sigma_\kappa(.)$. From Axioms 1 and 2 the corresponding design requirements for the dynamical system $\Sigma_\kappa(.)$ are deduced:

- **Requirement 3. Preservation of nth-order uniform distribution:** The dynamical input-output-system $\Sigma_\kappa(.)$ transforms the n-dimensional uniformly distributed signal $s(k)$ into the n-dimensional uniformly distributed signal $\kappa(k)$.

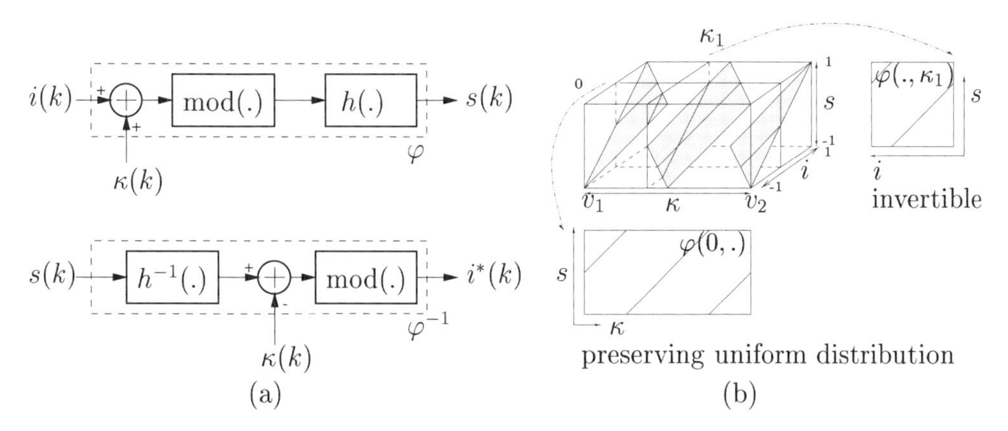

Figure 9.15: Structure of the nonlinear operation $\varphi(i, \kappa)$ and its inverse operation $\varphi^{-1}(s, \kappa)$ (a) and an example of $\varphi(i, \kappa)$ (b).

- **Requirement 4. Unique asymptotic behaviour of $\Sigma_\kappa(.)$:** For all pairs of initial conditions $(z(0), z^*(0)) \in Z$ and all signals $s(k)$ the key stream signal $\kappa^*(k)$ approaches (at least) asymptotically to $\kappa(k)$, i.e., $\lim_{t\to\infty} |\kappa^*(k) - \kappa(k)| = 0$ ($\kappa^*(k)$ and $\kappa(k)$: key signals belonging to the initial conditions $z(0)$ and $z^*(0)$, respectively).

Furthermore, we have to ensure that the autonomous encoder is *ergodic* in order to obtain the n-dimensional uniform distribution independent of the initial values $z(0)$.

The simplest structure of the dynamical system $\Sigma_\kappa(.)$ consists of one delay element only. It preserves the one-dimensional uniform distribution of $s(k)$. $\Sigma_\kappa(.)$ satisfies Requirements 3 and 4 but the autonomous encoder is not ergodic. For example, if $h(x) = x$ then it always reproduces the initial value, i.e., $s(k) = x(0)$ for all $k \geq 0$. Therefore we need the ability to stretch the range of the signal $\kappa(k)$ to $[v1, v2)$ (with $v_2 - v_1 = 2m$, $m \in \mathbb{Z}$). Hence, a uniform distribution preserving map $g(.) : [-1, 1) \to [v1, v2)$ is added after the delay element.

In the simplest case where $g(.)$ is just an amplification by c_g we obtain $s(k) = \text{mod}(c_g \cdot s(k-1))$. For $c_g = 2$ this system is similar to the well-known and well-analyzed Bernoulli system.

In order to construct a system $\Sigma_\kappa(.)$ which preserves the n-dimensional uniform distribution we extend the subsystem to a chain of n delay elements. However, the autonomous system is not ergodic. This can be demonstrated easily by choosing an arbitrary initial state $z(0)$ and considering its development in time. The coordinate $z_1(k+1)$ depends on $z_n(k)$ only and is independent of $z_i(k)$, $i = 1, \ldots, n-1$. So, the system is decomposable into n systems of first order.

In order to avoid this, the coordinates have to influence each other. This can be achieved by introducing operations $\phi(.,.)$ after the nonlinearity $g(.)$. These operations have to have the preservation property (Axiom 1), i.e., they have to preserve the uniform distribution independently from $y_m(k)$. In contrast to $\varphi(i, \kappa)$ they do not need to be invertible. So we obtain $\phi_m(y_m, u_m)$ by replacing

$h(.)$ in $\varphi(i, \kappa)$ by an arbitrary uniform distribution preserving map $g_m(.)$.

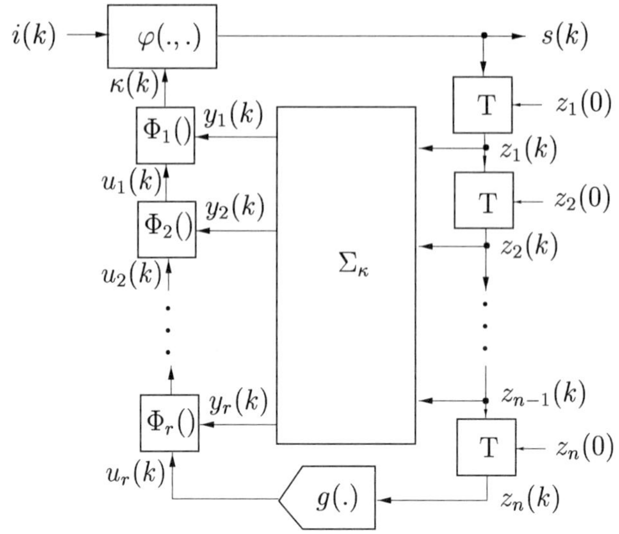

Figure 9.16: General structure of the encryption coder.

Finally, we connect the z-terminals with the y ones by a static or dynamical input-output-system $\Sigma_i(.)$. It is necessary and sufficient that $\Sigma_i(.)$ has unique asymptotic behaviour in order to ensure this property for the system $\Sigma_\kappa(.)$. The general structure obtained is shown in Figure 9.16.

It is obvious that this structure satisfies Axiom 2. With the encoded signal

$$s(k) = h \circ \mathrm{mod}(i(k) + \kappa(k)) \tag{9.101}$$

we directly obtain the decoded signal

$$i(k) = \mathrm{mod}(h^{-1}(s(k)) - \kappa(k)). \tag{9.102}$$

Fulfillment of Axiom 1 is more difficult to illustrate for the general case. Therefore we assume that $\Sigma_i(.)$ is a static system and $g(.)$ is a fully stretching Markov map. Then, for the autonomous system there exists a partition on the n-dimensional state space I^n. All elements of this partition are completely mapped into I^n by bijective maps (see [15] for details and the proof). It follows that the autonomous system preserves the n-dimensional uniform distribution. Furthermore it can be proved that the signal $\mathrm{mod}(i(k) + \kappa(k))$ is uniformly distributed on I independently of the distribution of $i(k)$ if $\kappa(k)$ is uniformly distributed on $[v_1, v2)$ [15]. So, the system also fulfills Axiom 1 for almost all signals $i(k)$.

9.5.2 A Simple Example

Several well-known and well-analyzed systems can be regarded as special cases of the general structure in Figure 9.16. One example is the class of systems based on AR-filter structures with two's complement overflow nonlinearity (9.99). Here the functions φ and ϕ_i are two's complement overflow additions and the systems Σ_i and g correspond to the weighting coefficients c_i, $i = 1, \ldots, n$, which are used as the encryption key.

Because of the special *modulo*-property[2] of the two's complement overflow nonlinearity (9.99) all two's complement overflow additions can be modelled by ordinary additions and one subsequent two's complement nonlinearity. This simplifies the continuous-value realization of such systems. Using this simplification the encoded and the decoded signal are described by

$$s(k) = \operatorname{mod}(i(k) + \sum_{j=1}^{n} c_j s(k-j)) \quad \text{and} \quad i^*(k) = \operatorname{mod}(r(k) - \sum_{j=1}^{n} c_j^* r(k-j)), \tag{9.103}$$

respectively. For this system the encoder and the decoder have been implemented on printed circuit boards using analog, discrete-time circuits [10]. The general scheme of the circuit is given in Figure 9.17. There are two additional function

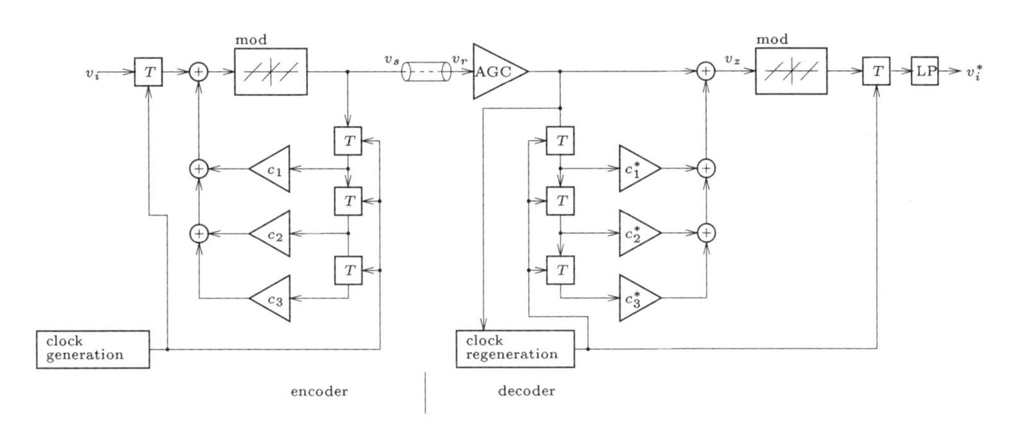

Figure 9.17: Configuration of the system.

blocks in the decoder: an automatic gain control (AGC) to compensate the attenuation in real channels (as the principle of the inverse system requires the amplitude of the encoder output also at the decoder input) and a clock regenerator (PLL structure) to reconstruct the clock of the encoder circuit from the transmitted signal.

Measurement results obtained for a sinusoidal input signal v_i (1 kHz, scaled to maximum input range) are depicted in Figure 9.18. Figure 9.18(a) confirms the recovery of the sinusoidal signal, Figure 9.18(b) shows the one-dimensional uniform distribution of the encoded signal. The signal-to-noise ratio of the decoded signal is about 45 dB.

[2] mod (mod $a \pm$ mod b) = mod ($a \pm b$)

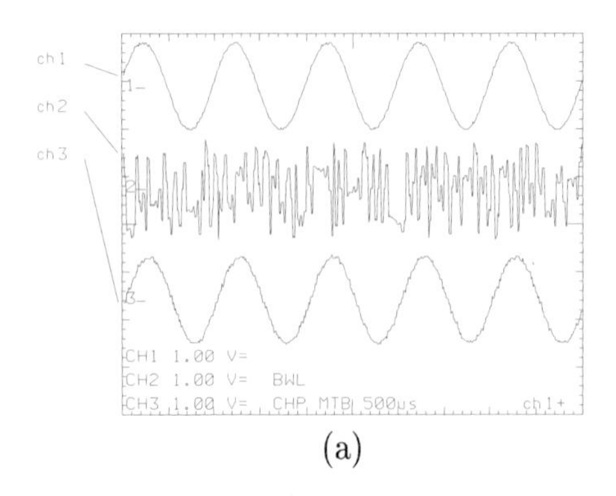

(a)

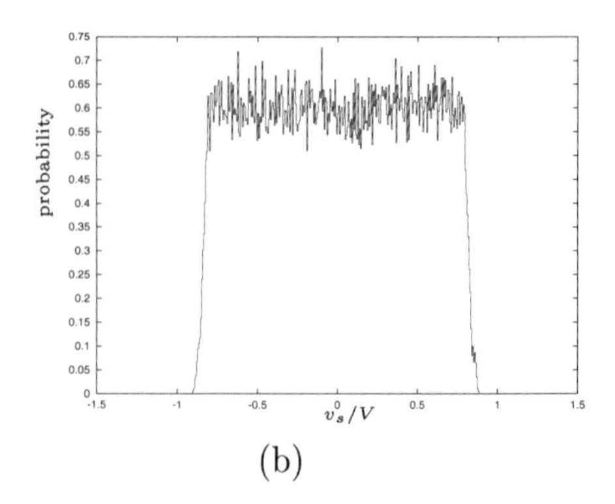

(b)

Figure 9.18: (a) Input signal v_i (ch1), transmitted signal v_s (ch2) and received signal v_i^* (ch3) versus *time*; (b) Measured density at the output of the encoder (Parameters: $c_1 = 2, c_2 = -2$).

9.5.3 Cryptographical Analysis

To evaluate the systems from a cryptographical point of view two ciphertext-only attacks [13] have been performed on the system discussed in Section 9.5.2. The first attack is based on the encoded signal, the second one on the decoded signal. Here we only show the general idea of the methods. The analysis itself is explained in more detail in [11].

It should be noted that the following analysis cannot give a positive evaluation of the cryptographical performance of the system. As with all successful analysis methods it only provides security bounds which are not exceeded by the system considered.

Encoded signal based analysis. The analysis is performed in two steps. First, a measurable dependence of a particular statistic $V_{\boldsymbol{s}}(c)$ of the encoded signal $\boldsymbol{s}$ on the key parameters $c_1, \ldots, c_n$ is used to generate a reference function. Second, for the actual analysis an estimation value v_E of this statistic is obtained from the time series of the observed encoded signal segment. By means of this estimation possible values for the key parameters are selected by determining the pre-image $V_{\boldsymbol{s}}^{-1}(v_E)$.

To derive a suitable dependence $V_{\boldsymbol{s}}(c)$ the structure in Figure 9.19(a) is considered. The independent random variables I and S_i represent the information signal $\boldsymbol{i}$ at time k with an arbitrary but known probability density function and the uniformly distributed encoded signal $\boldsymbol{s}$ at time $k - i$, respectively.

For this structure the covariance between S_i and $S_0 = \bmod(I + c_i S_i)$ is given as

$$V_{\boldsymbol{s}}(c_i) = \mathrm{Cov}(S_0, S_i) = \frac{1}{2}\int_{-1}^{1} s_i \,\mathrm{E}[\bmod(I + c_i s_i)]\,\mathrm{d}s_i, \qquad (9.104)$$

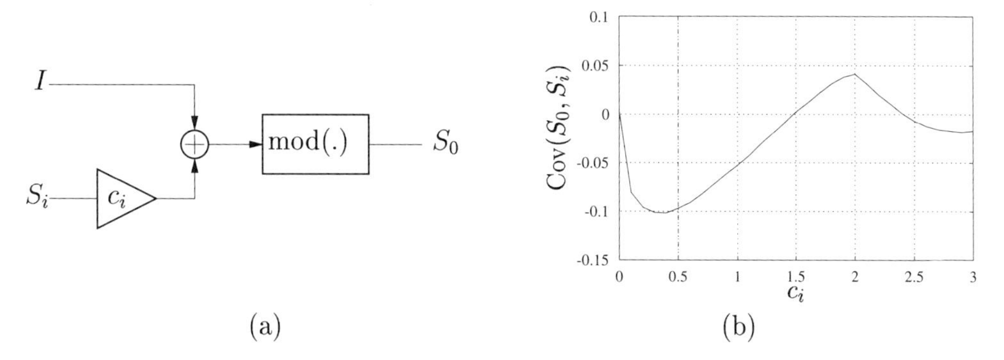

Figure 9.19: Structure for the derivation of a reference function (a), analytically determined reference function for a full-scale sinusoidal information signal (b).

where E[.] denotes the expectation. With the knowledge of the density of I, this relation allows to determine analytically or numerically the covariance as function of a single coefficient c_i. This function will be used as reference in the subsequent analysis step.

For the actual analysis we apply a set of transformations to the encoded signal $\boldsymbol{s}$ each of them consisting of a non-equidistant sampling of the vector $s(k) = (s(k) \cdots s(k-n))^T$. The sampling of $s(k)$ is carried out if the sum of the absolute values of $s(k-j)$, $j = 1, \ldots, n$, $j \neq i$, is less than ε, a suitable chosen small bound. This yields a dominant signal path involving the coefficient c_i, where the components $s(k)$ and $s(k-i)$ of the sampled vector, now denoted by $s_0(l_i)$ and $s_i(l_i)$, respectively, are realizations of S_0 and S_i. The sum of the remaining components $s_j(l_i)$, $j = 1, \ldots, n$, $j \neq i$, form the error signal $e_i(\varepsilon)$ with $|e_i(\varepsilon)| \leq (n-1) \cdot \varepsilon \cdot \max|c_j|$.

The n different transformations are chosen in such a way that each signal $s(k-i), i = 1, \ldots, n$ becomes once the dominant component of the sampled vector $s(l_i)$. Assuming a sufficient small value for ε, the influence of $e_i(\varepsilon)$ can be neglected. The n covariance estimates $\mathrm{Cov}_E(s_0(l_i), s_i(l_i))$ obtained from the respective dominant components of $s(l_i)$ allow us to find a set of possible coefficient tuples $(c_1, \ldots, c_n)$ by determining and combining the pre-images $V_{\boldsymbol{S}}^{-1}(\mathrm{Cov}_E(s_0(l_i), s_i(l_i)))$.

The analysis expense of this method is primarily given by the number of signal values $s(k)$ needed to be processed for a sufficient accuracy of the covariance estimates. An equivalent measure of the encryption security is therefore the estimation time t_E. Approximately, t_E increases exponentially with the system order n. The following table shows the typical estimation times for a system operating at a sampling frequency of 40 kHz and a required accuracy of 0.1 for the determination of the coefficients c_i:

system order n	2	3	4	5
estimation time t_E	25 s	8.3 min	4.2 h	3.5 d

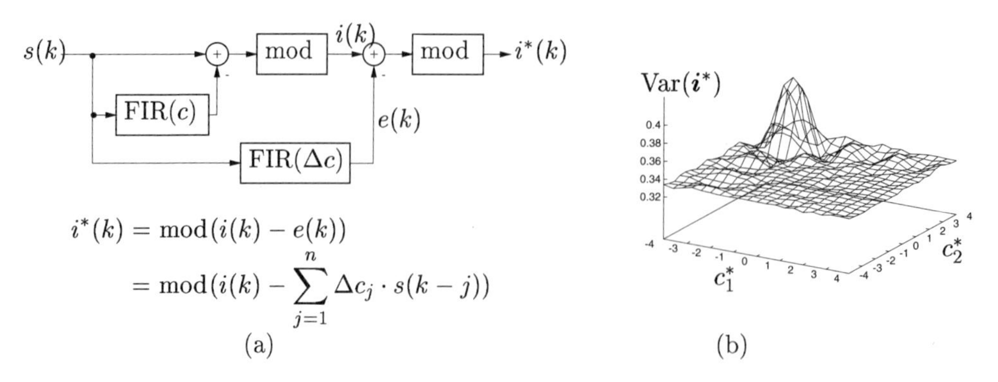

Figure 9.20: Model of the mismatched decoder (a), variance of the decoded signal $\boldsymbol{i}^*$ under parameter mismatch (b).

The described method does not allow a unique determination of the key parameter and has to be combined with other techniques such as a subsequent search among the n-tuples found before. The general behaviour of the chosen reference function (oscillating shape with decreasing envelope) also restricts the analysis to small coefficient values (e.g., $|c_i| \leq 3$ in our examples). The generalization to multi-parametric dependences allows a more time-efficient analysis [11].

Decoded signal based analysis. The second analysis takes advantage of the encoder property to produce the signal $\boldsymbol{s}$ with an n-dimensional uniform distribution independently of the information signal $\boldsymbol{i}$. We assume that $\boldsymbol{i}$ is not uniformly distributed as it holds for almost all information signals.

If the parameters of the encoder and the decoder are identical, then the signals $\boldsymbol{i}^*$ and $\boldsymbol{i}$ and also their distributions are exactly the same. By contrast, decoding the signal $\boldsymbol{s}$ using mismatched decoder parameters $c_j^* = c_j + \Delta c_j$, $j = 1, \ldots, n$, leads to a distribution of $\boldsymbol{i}^*$ which is different from the distribution of $\boldsymbol{i}$ and approaches a uniform distribution with increasing parameter mismatch Δc_j.

Because of the properties of the mod-operation (9.99), the mismatched decoder can be modelled according to Figure 9.20(a), where FIR(c) and FIR(Δc) denote FIR-filter structures with encoder parameter set c and set of parameter mismatches Δc, respectively. Corresponding to the conditions for the n-dimensional uniform distribution of $\boldsymbol{s}$, $\boldsymbol{i}^*$ is (one-dimensional) uniformly distributed on $[-1, 1)$ if there exists a $\Delta c_j \in \mathbb{Z}$ and $\Delta c_j \neq 0$. Outside this parameter region with increasing mismatch Δc_j the distribution of $\boldsymbol{i}^*$ approaches very rapidly the uniform shape. As an equivalent measure of this behaviour Figure 9.20(b) shows the variance Var($\boldsymbol{i}^*$) of the decoded signal $\boldsymbol{i}^*$ as a function in the two-dimensional parameter space.

Based on these prerequisites the system analysis consists of an adaptive-type algorithm. Using the known decoder structure, its aim is to find the parameter c set which corresponds to the absolute extremum of the distribution of i^*. Practically the first step is to determine roughly the position of the distribution peak by an extensive search in the parameter space at a suitable grid. Subsequently a classical hill-climbing algorithm can be applied to determine the parameter set with the desired precision.

The analysis expense of the second method mainly depends on the number of numerical operations and thus on the complexity of the applied algorithm. In contrast with the first method, the length of the necessary segment of $\boldsymbol{s}$ is rather uncritical. Once enough samples of the encoded signal have been collected they can be reused in each analysis step. Again the overall expense increases at least exponentially with the system order n which is typical for adaptive methods.

9.6 Statistical Analysis of Chaotic Signal Processing Schemes

This section describes the analysis of signal processing schemes (with emphasis on digital chaotic communications) which have chaotic and random input signals.

Signal processing in general may involve *random signals* of different kinds:

Random information signals: The exact content of the messages to be transmitted by a system is unknown. Statistical models can be used to describe these signals.

Random disturbances: Technical devices or a transmission over physical channels introduces disturbances (noise), originating from natural (e.g., thermal noise), but also technical sources (e.g., interfering signals on a transmission channel). Statistical signal models are suitable for the treatment of such signals.

Random auxiliary signals: These may be signals from artificial noise sources, chaotic signals, or signals with random parameters such as phase.

Random signal models can be employed and are useful if random information signals and random disturbances are treated in the same way.

Signal processing shall gain knowledge about the input signals and/or structural properties of the processing scheme from the random output signal. A part of the signals' randomness represents the information we are looking for, another stems from the random disturbances and leads to errors in the interpretation of the scheme's output. The knowledge of the impact of the random disturbances on these errors allows us to calculate performance figures for the evaluation and the comparison of different processing schemes.

Here we provide the necessary tools for a statistical analysis of signal processing schemes involving chaotic signals and demonstrate the usefulness and power of these methods by performing an example analysis of the Differential Chaos Shift Keying (DCSK) communication scheme and comparing it with classical communication systems.

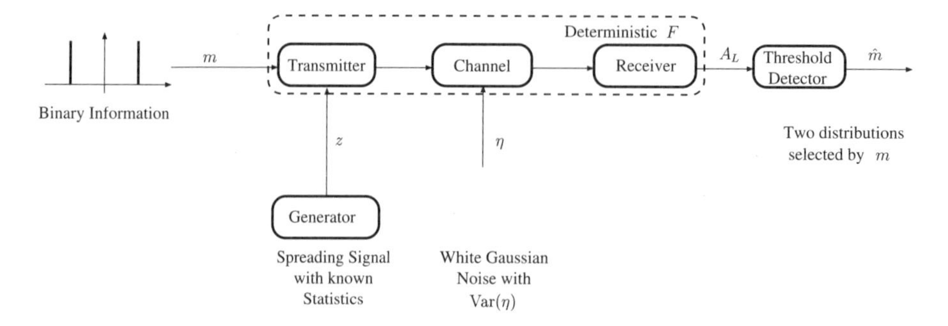

Figure 9.21: A communication system seen as a processing scheme for random signals.

9.6.1 Digital Chaotic Communications – The Processing of Random Signals

A general setup of a digital communication system is shown in Figure 9.21. A deterministic *nonlinear* functional F maps the information signal m, the channel noise η, and a carrier/spreading signal z into a certain random value A_L,

$$A_L = F(m, z, \eta). \tag{9.105}$$

A threshold detector gives an estimation $\hat{m}$ of the transmitted information. Usually A_L is represented by the value of a random signal at that particular time instant, when the detection happens. The functional F includes the complete processing chain of the transmission link. The channel introduces noise, damping, multi-path, fading, and other effects. The simplest non-ideal model is a channel with Additive White Gaussian Noise (AWGN) as the only disturbance.

If we know the statistics of A_L, we may calculate the statistical characteristics of the estimation $\hat{m}$ of m for a given threshold detector and thus characterize the quality of the transmission link. The statistics of A_L depend on the statistics of the processed random signals and on the structure of the functional F, i.e., on the way the statistics of the random input signals are transformed into the statistics of A_L. Normally, the calculation of the probability density function (PDF) of A_L, which would give a complete statistical description of A_L, turns out to be virtually impossible. However, there are equivalent descriptions via moments or cumulants which are obtained much more easily, as will be shown in the sequel.

We will consider two problems:

Analysis problem: Find the statistical characteristics of A_L for a given scheme in order to evaluate its behaviour.

Signal design problem: Find statistics of the spreading signal z which result in optimum performance for a given system. They may form statistical objectives for a chaos generator design. (The reader should refer to the sections on generator design in this chapter for appropriate methods.)

9.6.2 Statistics of Random Variables and Processes and Their Nonlinear Transformations

Characteristic functions, moments, and cumulants. A set of random variables $x_1, \ldots, x_n$ is characterized by the joint probability density function (PDF) $f(x_1, \ldots, x_n)$. To this probability density corresponds a characteristic function (moment generating function) [17, 18]

$$\begin{aligned} \Phi(\omega_1, \ldots, \omega_n) &= \int_{-\infty}^{\infty} f(x_1, \ldots, x_n)\, e^{j\omega_1 x_1 + \ldots + \omega_n x_n}\, dx_1 \cdots dx_n \\ &= \mathrm{E}\left[e^{j\omega_1 x_1 + \ldots + \omega_n x_n}\right], \end{aligned} \tag{9.106}$$

($E\,[.]$ denotes the expectation), which gives an equivalent and complete description of f. The moments α of the random variables $x_1, \ldots, x_n$ (i.e., expectations of products of the x_i)

$$\alpha_{\mathbf{x}}^{\mathbf{q}} = \alpha_{x_1,\ldots,x_n}^{q_1,\ldots,q_n} = \mathrm{E}\left[x_1^{q_1} \ldots x_n^{q_n}\right] = \int_{-\infty}^{\infty} x_1^{q_1} \ldots x_n^{q_n} f(x_1, \ldots, x_n) dx_1 \cdots dx_n \tag{9.107}$$

can be expressed using the characteristic function Φ [17, 18]

$$\alpha_{\mathbf{x}}^{\mathbf{q}} = (-j)^{q_1+\ldots+q_n} \frac{\partial^{q_1+\ldots+q_n}}{\partial\omega_1^{q_1} \ldots \partial\omega_n^{q_n}} \Phi(\omega_1, \ldots, \omega_n)\Bigg|_{\omega_1=\ldots=\omega_n=0} . \tag{9.108}$$

Obviously the moments represent the coefficients of a Taylor series expansion of the characteristic function Φ with respect to $\omega_1, \ldots, \omega_n$ at $\omega_1 = \ldots = \omega_n = 0$

$$\Phi(\omega_1, \ldots, \omega_n) = \sum_{q_1,\ldots,q_n=0}^{\infty} \frac{\alpha_{x_1,\ldots,x_n}^{q_1,\ldots,q_n}}{q_1! \ldots q_n!} (j\omega_1)^{q_1} \ldots (j\omega_n)^{q_n}. \tag{9.109}$$

There is a second characteristic function, the so-called cumulant generating function [17, 18]

$$\Psi(\omega_1, \ldots, \omega_n) = \ln \Phi(\omega_1, \ldots, \omega_n), \tag{9.110}$$

which in equivalence to (9.108) defines the cumulants $\kappa_{\mathbf{x}}^{\mathbf{q}}$ via its Taylor series expansion with respect to $\omega_1, \ldots, \omega_n$

$$\Psi(\omega_1, \ldots, \omega_n) = \sum_{q_1,\ldots,q_n=0}^{\infty} \frac{\kappa_{x_1,\ldots,x_n}^{q_1,\ldots,q_n}}{q_1! \ldots q_n!} (j\omega_1)^{q_1} \ldots (j\omega_n)^{q_n}, \tag{9.111}$$

with $\kappa_{x_1,\ldots,x_n}^{0,\ldots,0} = 0$.

Well-known cumulants are the mean value κ_x^1, the variance κ_x^2, the covariance $\kappa_{x,y}^{1,1}$, the skewness κ_x^3 (measuring the asymmetry of a PDF), and the curtosis κ_x^4 (describing the deviation of a PDF from a Gaussian PDF). Cumulants possess a set of special properties, which explain their wide exploitation:

- Gaussian PDF have cumulants of 1st and 2nd order only, i.e., the PDF is fully described by these cumulants.
- Joint cumulants of independent variables vanish (e.g., $\kappa_{x_1,x_2}^{q_1;q_2} = 0$ for independent x_1, x_2).
- the cumulant of a sum of independent variables is the sum of the respective cumulants of the variables (e.g $\kappa_{x+y}^3 = \kappa_x^3 + \kappa_y^3$).

Moments and cumulants can be calculated from each other, the n-th order moment of a random variable is a function of the cumulants of order up to n of this variable and vice versa

$$\begin{aligned} \alpha_{x_1}^1 &= \kappa_{x_1}^1 \\ \alpha_{x_1,x_2}^{1,1} &= \kappa_{x_1,x_2}^{1,1} + \kappa_{x_1}^1\kappa_{x_2}^1 \\ \alpha_{x_1,x_2,x_3}^{1,1,1} &= \kappa_{x_1,x_2,x_3}^{1,1,1} + \kappa_{x_1}^1\kappa_{x_2,x_3}^{1,1} + \kappa_{x_2}^1\kappa_{x_1,x_3}^{1,1} + \kappa_{x_3}^1\kappa_{x_1,x_2}^{1,1} + \kappa_{x_1}^1\kappa_{x_2}^1\kappa_{x_3}^1 \\ &\dots \end{aligned} \tag{9.112}$$

Further details on this can be found in the literature [17, 18].

Cumulant equations. For our analysis problem we are interested in the estimation of statistics of random variables depending in a nonlinear way from one or more other random variables

$$y = h(x_1, \dots, x_n). \tag{9.113}$$

where h is a test function in the sense of section 9.2.6 on page 267. The estimation of any moment $\alpha_y^q = \mathrm{E}[y^q]$, or of any cumulant κ_y^q (which may be expressed by moments of y too) requires us to calculate the expectation of some nonlinear function $\mathrm{E}[g(x_1, \dots, x_n)]$, which is usually not straightforward. The *cumulant equations* suggest a possibility to deal with these problems [17].

We are going to demonstrate the derivation of these equations for a nonlinear function of one variable. We know that [17, 18]

$$\mathrm{E}[g(x)] = \int_{-\infty}^{\infty} g(x) f(x) dx, \tag{9.114}$$

where the PDF f may be expressed as the inverse Fourier transform of the characteristic function $\Phi(\omega) = \mathrm{e}^{\Psi(\omega)}$

$$\mathrm{E}[g(x)] = \int_{-\infty}^{\infty} g(x) \frac{1}{2\pi} \int_{-\infty}^{\infty} \mathrm{e}^{-j\omega x} \mathrm{e}^{\Psi(\omega)} \, d\omega dx. \tag{9.115}$$

Using the Taylor series expansion (9.111) we obtain

$$\mathrm{E}[g(x)] = \int_{-\infty}^{\infty} g(x) \frac{1}{2\pi} \int_{-\infty}^{\infty} \exp\left(-j\omega x + \sum_{q_1=1}^{\infty} \frac{\kappa_x^{q_1}}{q_1!} (j\omega)^{q_1}\right) d\omega dx. \tag{9.116}$$

In order to see the influence of a particular cumulant on $\mathrm{E}[g(x)]$ we calculate the partial derivative

$$\begin{aligned} \frac{\partial^k \mathrm{E}[g(x)]}{\partial \kappa_x^{q\ k}} &= \frac{(-1)^{kq}}{(q!)^k} \int_{-\infty}^{\infty} g(x) \frac{1}{2\pi} \int_{-\infty}^{\infty} (-j\omega)^{kq} \cdot \\ &\cdot \exp\left(-j\omega x + \sum_{q_1=1}^{\infty} \frac{\kappa_x^{q_1}}{q_1!} (j\omega)^{q_1}\right) d\omega dx \\ &= \frac{(-1)^{kq}}{(q!)^k} \int_{-\infty}^{\infty} g(x) \frac{\partial^{kq}}{\partial x^{kq}} \left[\frac{1}{2\pi} \cdot \right. \\ &\left. \cdot \int_{-\infty}^{\infty} \exp\left(-j\omega x + \sum_{q_1=1}^{\infty} \frac{\kappa_x^{q_1}}{q_1!} (j\omega)^{q_1}\right) d\omega \right] dx \\ &= \frac{(-1)^{kq}}{(q!)^k} \int_{-\infty}^{\infty} g(x) \frac{d^{kq} f(x)}{dx^{kq}} dx. \end{aligned} \tag{9.117}$$

By partially integrating kq times and taking into account that $f(x)$ and all its derivatives vanish at infinity (since f is a PDF) the cumulant equations for the one-dimensional case are found [17]

$$\frac{\partial^k \mathrm{E}[g(x)]}{\partial \kappa_x^{q\ k}} = \frac{1}{(q!)^k} \mathrm{E}\left[\frac{d^{kq} g(x)}{dx^{kq}}\right]. \tag{9.118}$$

By the same principle the cumulant equations for the multi-dimensional case and for dependencies on several cumulants are derived [17]

$$\begin{aligned} &\frac{\partial^{k_1+\ldots+k_N} \mathrm{E}[g(\mathbf{x})]}{(\partial \kappa_{\mathbf{x}}^{\mathbf{q}_1})^{k_1} \ldots (\partial \kappa_{\mathbf{x}}^{\mathbf{q}_N})^{k_N}} = \\ &\frac{1}{(\mathbf{q}_1!)^{k_1} \ldots (\mathbf{q}_N!)^{k_N}} \mathrm{E}\left[\frac{\partial^{(q_{11}+\ldots+q_{1n})k_1+\ldots+(q_{N1}+\ldots+q_{Nn})k_N} g(\mathbf{x})}{\partial x_1^{q_{11}k_1+\ldots+q_{N1}k_N} \ldots \partial x_n^{q_{1n}k_1+\ldots+q_{Nn}k_N}}\right] \end{aligned} \tag{9.119}$$

with $\mathbf{q}_i! = q_{i1}! \cdots q_{in}!$.

Statistical analysis. Cumulant equations allow us to express $\mathrm{E}[g(\mathbf{x})]$ as a Taylor series expansion with respect to the cumulants of $\mathbf{x}$. Obviously (9.119) represents the coefficients of this expansion. A convenient reference point for the expansion is the point where *all* cumulants of $\mathbf{x}$ become zero, hence we have to calculate the partial derivatives in (9.119) at this particular point. Since we represent the coefficients via expectations of derivatives of g by the x_i, we need to know the particular PDF $f(x_1, \ldots, x_n)$ which has zero cumulants only. Clearly this is $\delta(\mathbf{x})$, i.e., the PDF of the constant $\mathbf{x} = 0$. The expectations of derivatives of g then are calculated by setting $\mathbf{x} = 0$ in the derivatives. All vanishing derivatives will result in a zero Taylor coefficient. Usually there are infinitely many cumulants of $\mathbf{x}$ to be regarded in the Taylor series. However, if f is a polynomial of finite order in $x_1, \ldots, x_n$ we observe the following simplifications:

- There are no non-zero derivatives of orders higher than the respective polynomial orders, i.e., the series expansion will be finite and we find an exact representation for $\mathrm{E}[g(x)]$.
- The summands in the polynomial expression have non-zero derivatives only for one particular derivation corresponding to a limited number of cumulants. For all other derivatives the corresponding Taylor coefficients vanish.

Further simplifications may be gained from the statistical properties of the input variables:

- If there exist independent pairs of variables, their joint cumulants become zero.
- If one or more variables are Gaussian, all corresponding cumulants of order higher than two vanish.

Example: Expectation of a polynomial. Assume a nonlinear function of the form

$$g(x_1, x_2, x_3) = x_1^3 + a x_2^2 x_3 \tag{9.120}$$

(a shall be some parameter), whose expectation shall be expressed in terms of the cumulants of x_1, x_2, and x_3. Using (9.119) we find the following nonzero coefficients of the Taylor series

$$\begin{aligned}
\frac{\partial^3 \mathrm{E}[g(x_1,x_2,x_3)]}{\partial {\kappa^1_{x_1}}^3} &= 6 & \frac{\partial^2 \mathrm{E}[g(x_1,x_2,x_3)]}{\partial \kappa^1_{x_1} \partial \kappa^2_{x_1}} &= 3 \\
\frac{\partial \mathrm{E}[g(x_1,x_2,x_3)]}{\partial \kappa^3_{x_1}} &= 1 & \frac{\partial^3 \mathrm{E}[g(x_1,x_2,x_3)]}{\partial {\kappa^1_{x_2}}^2 \partial \kappa^1_{x_3}} &= 2a \\
\frac{\partial^2 \mathrm{E}[g(x_1,x_2,x_3)]}{\partial \kappa^2_{x_2} \partial \kappa^1_{x_3}} &= a & \frac{\partial^2 \mathrm{E}[g(x_1,x_2,x_3)]}{\partial \kappa^{1,1}_{x_2,x_3} \partial \kappa^1_{x_2}} &= 2a \\
\frac{\partial \mathrm{E}[g(x_1,x_2,x_3)]}{\partial \kappa^{2,1}_{x_2,x_3}} &= a.
\end{aligned} \tag{9.121}$$

Combining them into the series we obtain

$$\begin{aligned}
\mathrm{E}[g(x_1,x_2,x_3)] = {} & {\kappa^1_{x_1}}^3 + 3\kappa^1_{x_1}\kappa^2_{x_1} + \kappa^3_{x_1} + \\
& + a\left({\kappa^1_{x_2}}^2 \kappa^1_{x_3} + \kappa^2_{x_2}\kappa^1_{x_3} + 2\kappa^{1,1}_{x_2,x_3}\kappa^1_{x_2} + \kappa^{2,1}_{x_2,x_3}\right)
\end{aligned} \tag{9.122}$$

(Zero-order parts of the cumulants are omitted, e.g., $\kappa^{3,0,0}_{x_1,x_2,x_3} = \kappa^3_{x_1}$.)

Cumulants of random processes. The concept of moments and cumulants is easily extended to random processes [17,18] by regarding a process as an entity of random variables with the time k as a parameter (or index). The first-order cumulant (or moment) then has the form

$$\kappa^1_{x(k)} = \kappa^1_x(k) \tag{9.123}$$

which is a function of the time index k. The second-order cumulant will be a function depending on two variables (or time indices) k_1 and k_2

$$\kappa^{1,1}_{x(k_1),x(k_2)} = \kappa^2_x(k_1, k_2) \tag{9.124}$$

and so on. If the process is *stationary*, i.e., if the statistics are invariant under time shifts, the m-th order cumulants and moments will depend on $m-1$ variables representing time differences [17, 18]

$$\begin{aligned} \kappa^1_x(k) &= \kappa^1_x \\ \kappa^2_x(k_1, k_2) &= \kappa^2_x(k_2 - k_1) \\ \kappa^3_x(k_1, k_2, k_3) &= \kappa^3_x(k_2 - k_1, k_3 - k_1) \\ &\dots \end{aligned} \tag{9.125}$$

If the stationary process is a signal, κ^1_x is its DC value, $\kappa^2_x(0)$ its AC power, and $\kappa^2_x(k)$ its auto-covariance function.

As for random variables, the set of cumulant or moment functions gives a complete statistical description of a random process. The joint cumulants of several random processes are defined equivalently.

9.6.3 Analysis of Communication Schemes

Discrete-time baseband representations. Communication is the transmission of information over a certain physical channel. This requires a suitable adaption of the transmitted information to the channel. The physical channel is usually a bandpass due to the particular physical effect (e.g., the propagation of electromagnetic waves) exploited for transmission, but often also due to administrative limitations (frequency and bandwidth assignments). So we have to transfer the information from a low-frequency range (the baseband) towards a high-frequency signal suitable for the given channel. This process is called modulation. For an analysis or simulation of the complete scheme the time scale is determined by the high-frequency part, albeit we are interested in the low frequency of the information and spreading signals only. This inefficiency is avoided by the so-called analytic signal approach [20], which transforms the bandpass channel and the corresponding signals into the baseband (low-pass equivalents). This is very well described in the literature for both classical and chaotic communications [20, 25, 27]. As the result a model in the baseband frequency range with complex-valued signals is obtained. By sampling it can be transferred to a discrete-time model having an I/O behaviour corresponding to the behaviour of the initial communication system model.

Measuring the performance. Let us consider a classical communication scheme: Direct-Sequence Spread Spectrum (DS-SS). In this scheme a sequence of L samples (from a pseudo-noise generator) is multiplied by the information bit ± 1 and then modulates the sinusoidal carrier [19]. The receiver correlates the received sequence with the original sequence (the receiver possesses a reference either in a matched filter or from a local synchronized generator). A threshold detection occurs once every L samples on the result of the correlation. If we have a channel with additive noise, the detector input at this moment appears as follows (complex baseband representation [27]):

$$A_L(k_D) = \frac{1}{L}\sum_{i=0}^{L-1}(mz(k_D - i) + \eta(k_D - i))\, z^*(k_D - i), \tag{9.126}$$

where $m \in \{-1, 1\}$ and * denotes complex conjugation. The first part in brackets is received from the channel, the complex conjugated part is the receiver's reference. Assume that z and η are statistically independent and stationary and that η is a zero-mean complex Gaussian-distributed white signal. By applying the cumulant equations as described above we obtain [27] (remember that for complex values the variance is $\kappa^{1,1}_{x,x^*} = \mathrm{E}[xx^*] - \mathrm{E}[x]\,\mathrm{E}[x^*]$)

$$\begin{aligned} \kappa^1_{A_L} &= m(\kappa^{1,1}_{z,z^*} + \kappa^1_z \kappa^1_{z^*}) = m\,\mathrm{E}[zz^*] = m\kappa^1_{zz^*} \\ \kappa^{1,1}_{A_L,A_L^*} &= \frac{\kappa^{1,1}_{\eta,\eta^*}\kappa^1_{zz^*}}{L}, \end{aligned} \tag{9.127}$$

Now we introduce the instant transmitter power $P(k) = z(k)z^*(k)$ having the expectation $\mathrm{E}[P] = \kappa^1_P = \kappa^1_{zz^*}$. This transforms (9.127) into

$$\begin{aligned} \kappa^1_{A_L} &= m\kappa^1_P \\ \kappa^{1,1}_{A_L,A_L^*} &= \frac{\kappa^{1,1}_{\eta,\eta^*}\kappa^1_P}{L}. \end{aligned} \tag{9.128}$$

Since η is zero mean, its variance $\kappa^{1,1}_{\eta,\eta^*}$ equals its power $\kappa^1_{\eta\eta^*}$.

$A_L(k_D)$ is a weighted sum of Gaussian random variables $(mz(k_D - i) + \eta(k_D - i))$ and as such again a Gaussian random variable, only having cumulants up to order 2. So with (9.128) we have the complete statistical description of A_L at the time instant k_D. To each m corresponds a PDF of A_L, whose mean is the power of the signal z multiplied by the information m and which has a variance as shown in (9.128). A_L can be split into a sum of its mean and a disturbance, whose power equals A_L's variance. This power is equally distributed between the real and imaginary parts of the disturbance, i.e., $\kappa^{1,1}_{A_L,A_L^*} = 2\kappa^2_{\Re(A_L)} = 2\kappa^2_{\Im(A_L)}$. Now we can introduce a signal-to-noise ratio (SNR) for the threshold detector input A_L. It is the ratio between the mean powers $\kappa^1_{A_L}\kappa^1_{A_L^*}$ (which is the power of our "signal," i.e., the part of A_L we are interested in) and the power of the disturbance, which is the mean-free part of A_L and has the variance of $\kappa^{1,1}_{A_L,A_L^*}$ as its power

$$SNR = \frac{\kappa^1_{A_L}\kappa^1_{A_L^*}}{\kappa^{1,1}_{A_L,A_L^*}} = \frac{L\kappa^1_P}{\kappa^{1,1}_{\eta,\eta^*}} = \frac{E_b}{N_0}. \tag{9.129}$$

E_b is the energy transmitted per bit (signal power times bit duration) and N_0 the AWGN power. In the continuous-time bandpass system E_b/N_0 is the ratio of bit energy and noise power spectral density [27].

If both possible values of m have equal probabilities, the optimum Bayesian detector [23] has its decision threshold at the imaginary axis, i.e., it determines the sign of the real part of A_L. We measure the performance of the system by the amount of decisions that lead to an estimation $\hat{m} \neq m$, i.e., the probability that an error occurs. Since we transmit discrete (here in particular binary) information which is measured in bits, this error probability is termed bit error rate (BER). For the Gaussian-distributed $\Re(A_L)$ with binary m it is calculated as [23]

$$BER = \frac{1}{2}\,\mathrm{erfc}\left(\sqrt{\frac{\left(\kappa^1_{A_L}\right)^2}{2\kappa^2_{\Re(A_L)}}}\right) = \frac{1}{2}\,\mathrm{erfc}\left(\sqrt{SNR}\right) = \frac{1}{2}\,\mathrm{erfc}\left(\sqrt{\frac{E_b}{N_0}}\right) \tag{9.130}$$

(erfc – complementary error function). This is well known from the communications literature [16, 22]. The formula shows a key aspect of a spread spectrum communication system: The BER is a function of the E_b/N_0 only. This means

- for a given N_0 the bit *energy* determines the system performance. It is irrelevant how this is achieved (either with high power and small bit length or with low power and large bit length).
- the signal can be "hidden" in the noise floor, i.e., such a system can cope with large noise levels, which is of interest if secrecy is to be achieved or if many users access a channel simultaneously (code-division multiple access).
- an increase in noise power can be compensated by a proportional increase of either the signal power or the bit duration.
- the system is scalable with respect to the power levels.

As a result, the diagram $BER = g(E_b/N_0)$ became the standard tool to characterize and compare the performance of communication schemes in AWGN. In the sequel we will call systems with such a BER characteristic *robust against the channel noise.*

Chaos Communications Example: DCSK. In this section we will analyze an example chaotic communication scheme – the Differential Chaos Shift Keying (DCSK), which was first presented in [24] and since then attracted much research

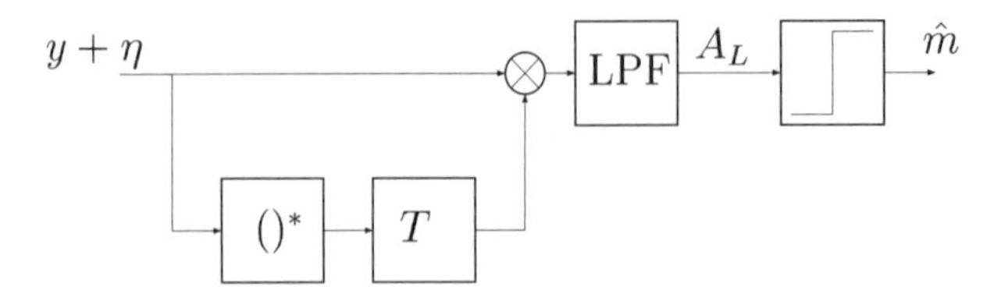

Figure 9.22: Baseband model of a DCSK receiver.

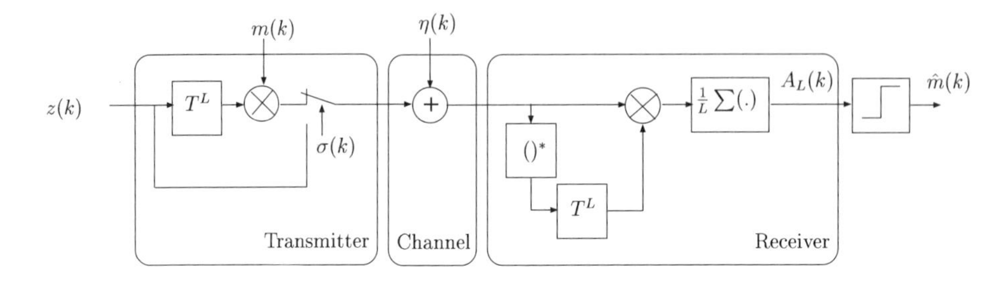

Figure 9.23: Model of a DCSK system suitable for statistical analysis.

interest (see, e.g., other parts of this book). The modulation scheme works as follows: The transmitter first sends a reference signal of a certain length T and then repeats this reference but modulated by the discrete information to be transmitted. The receiver delays the first part (the reference) by the time T and then correlates it with the information-carrying copy. At the output of the correlator a threshold detector makes a decision $\hat{m}$ of the transmitted information with period $2T$. Figure 9.22 shows the corresponding baseband model of the receiver (y – transmitted signal). DCSK is a variant of the so-called transmitted-reference (TR) principle. Such systems exploit correlation for the detection of the transmitted information. They transmit both the information-carrying signal and a reference for correlation via separate channels. Whereas the classical approach used two separate carrier frequencies, DCSK employs two channels separated in time.

For the analysis in the discrete-time domain we obtain the model of the complete DCSK processing scheme (representing (9.105)) as shown in Figure 9.23 [27]. In the figure σ is a switching function ensuring the selection of either the reference or its information-carrying copy. z is an arbitrary signal which may be chaotic. The transmission of one information bit requires $2L$ samples, L for the reference and L for the information carrying part.

In the time instant k_D of information detection the threshold detector input consists of exactly the L-sample long correlation sum of the two signal segments, both corrupted by noise. Since m only changes every $2L$ samples, we may consider it as a constant parameter for the statistical analysis of $A_L(k_D)$. We find

$$A_L(k_D) = \frac{1}{L}\sum_{i=0}^{L-1}(m\, z(k_D - L - i) + \eta(k_D - i))(z(k_D - L - i) + \eta(k_D - L - i))^*. \tag{9.131}$$

For the two signals η and z we make the same assumptions as in the analysis of the DS-SS example in Section 9.6.3. The cumulant analysis leads to [28]

$$\begin{aligned} \kappa^1_{A_L} &= m\kappa^1_P \\ \kappa^{1,1}_{A_L,A_L^*} &= \frac{1}{L}\left(\sum_{k=-L+1}^{L-1}(L - |k|)\,\kappa^2_P(k) + 2\kappa^{1,1}_{\eta,\eta^*}\kappa^1_P + \left(\kappa^{1,1}_{\eta,\eta^*}\right)^2\right). \end{aligned} \tag{9.132}$$

$\kappa^2_P(k)$ is the auto-covariance function of the signal power $P(k)$. The sum term in the formula describes the variance of A_L for the zero-noise case, i.e., the variance which is caused by the signal z itself. Describing a variance, the sum expression is always non-negative.

If we calculate the SNR at the threshold detector input and replace $2L\kappa^{1,1}_{z,z^*}/\kappa^{1,1}_{\eta,\eta^*}$ by E_b/N_0 (each bit requires $2L$ samples to be transmitted), we obtain

$$SNR = \left(\sum_{k=-L+1}^{L-1} 2L(L - |k|)\frac{\kappa^2_P(k)}{E_b^2} + 4\left(\frac{N_0}{E_b}\right) + 4L\left(\frac{N_0}{E_b}\right)^2\right)^{-1}. \tag{9.133}$$

Comparing this result with (9.129) we observe two principal differences:

- Beside E_b and N_0 the SNR is influenced by two other system parameters: L and the covariance function $\kappa^2_P(k)$ of the transmitted signal's power.
- The dependence of the SNR on N_0/E_b is not linear but quadratic, so we must expect an increase of the channel noise to have a higher impact on the SNR than in DS-SS.

A maximum for the SNR at any E_b, N_0, and L is found, if the sum term in (9.133) vanishes. We can achieve this by requiring $\kappa^2_P(k) = 0$ for all k. This holds if $P(k)$ is constant, i.e., if $z(k)$ is a signal with constant power or constant amplitude [28]. If z shall be chaotic then only its phase can be chaotic. In the real system this corresponds to a chaotically phase- or frequency-modulated carrier, which was found to be the most suitable one for DCSK [26]. There are no further restrictions to be placed on the chaotic phase. Any chaos generator can be used.

Another way to maximize the SNR is to minimize L. The smaller the L, the better the system will perform. In order to achieve a high SNR at the threshold detector, a high-power transmission leads to better results than a low-bitrate solution. Thus the property observed in DS-SS, that only the energy per bit E_b determines the performance at a given N_0, is lost for DCSK. This shall be

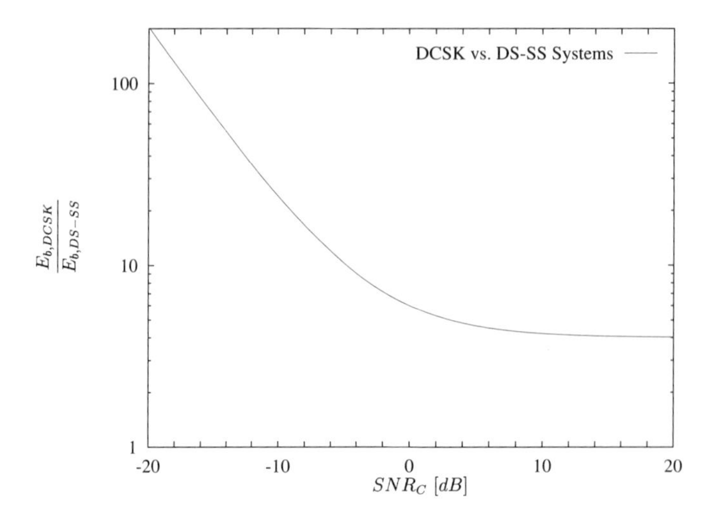

Figure 9.24: Bit energy of DCSK relative to the bit energy in DS-SS for the achievement of the same performance under equivalent channel conditions.

illustrated by comparing the efforts to be made (in terms of the bit energy E_b) in order to produce the same SNR at the threshold detector for both DCSK and DS-SS under equal channel conditions. By equating the two SNRs from (9.129) and (9.133) we can calculate the ratio of the bit energies as a function of the signal-to-noise ratio on the communication channel $SNR_C = \kappa_z^1 \kappa_{z^*}^1 / \kappa_{\eta\eta^*}^1 = \kappa_z^1 \kappa_{z^*}^1 / \kappa_{\eta,\eta^*}^{1,1}$ (here we already assume constant amplitude for the signal z)

$$\frac{E_{b,DCSK}}{E_{b,DS-SS}} = 4 + 2\,(SNR_C)^{-1} \tag{9.134}$$

shown in Figure 9.24. It is seen that with decreasing SNR_C an increasing effort in terms of the energy spent per transmitted bit is required, if we want to achieve an equivalent SNR at the threshold detector input. Since we use SNR_C as a parameter, a higher bit energy means a proportionally higher bit duration or a proportionally lower information transmission rate. This confirms the conclusion given above. This sensitivity of TR systems to channel noise is well known for the conventional TR with frequency division [16, 21], which can be shown to have exactly the same performance figures as DCSK by the statistical baseband analysis [27].

For the DS-SS system the BER follows directly from the SNR at the threshold detector input, since the distribution of $A_L(k_D)$ is Gaussian. Because A_L is obtained as an average, we may assume the distribution of A_L to converge to a Gaussian distribution for large L due to the central limit theorem [22]. (In this case (9.134) also compares the efforts for achieving the same BER in DS-SS and DCSK systems.) Under this assumption we find the DCSK error rate for the

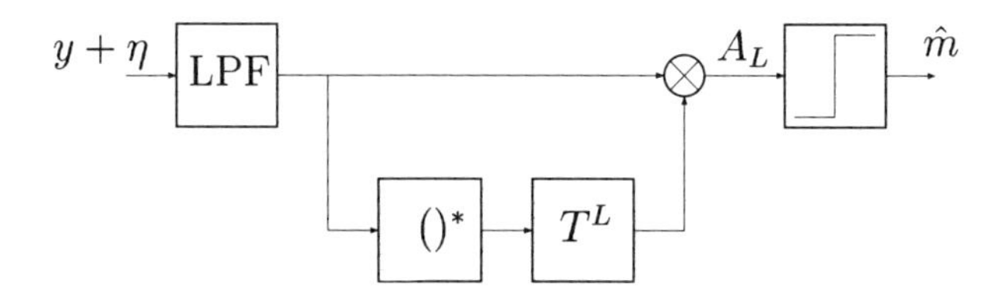

Figure 9.25: Baseband model of a DPSK receiver.

case of constant-amplitude signals as

$$BER = \frac{1}{2}\operatorname{erfc}\left(\left(4\left(\frac{N_0}{E_b}\right) + 4L\left(\frac{N_0}{E_b}\right)^2\right)^{-\frac{1}{2}}\right) \tag{9.135}$$

For the treatment of cases where L is small, in particular for the case $L = 1$, we will examine a classical communication scheme structurally similar to DCSK – the Differential Phase Shift Keying (DPSK). In DPSK the phase of a sinusoidal carrier is changed according to the bit to be transmitted: a 1 will result in no phase shift, a -1 in a $180°$ shift. The phase shift is analyzed by comparing the phase of two subsequent signal segments of one bit duration by a correlator. Since each transmitted signal segment serves as a reference for the following, the carrier signal has to be a periodic signal and the DPSK scheme is not suitable for chaotic communications. For noise cancellation DPSK exploits a band-pass filter at the receiver input, adjusted to the frequency of the carrier. In the baseband discrete-time model the sinusoidal carrier corresponds to a constant signal. The equivalent DPSK receiver structure is shown in Figure 9.25 [29]. Obviously it differs from DCSK only with respect to the position of the averaging device. The BER for DPSK is known from the literature [16, 19, 22] to be

$$BER = \frac{1}{2}\exp\left(-\frac{E_b}{N_0}\right) \tag{9.136}$$

This is clearly different from the DCSK result (9.135). DPSK does not share the drawback of DCSK and has a BER depending again on E_b/N_0 only. Here, as in DS-SS, we can trade power for bit duration. Contrary to DCSK the DPSK scheme is a stored reference (SR) system. The reference is contained in the band-pass filter, or the averaging device of the discrete-time baseband model. The noise cancellation, which is the aim of the averaging blocks in both DCSK and DPSK, occurs in DPSK at the very input of the receiver, whereas in DCSK it is performed after the correlators multiplier. The multiplier works (approximately) as a squaring device for the power of the disturbance [29], which gives an obvious explanation for the decrease in performance in DCSK at low signal-to-noise ratios on the channel.

For $L = 1$ the two receivers become exactly the same. However, since DCSK cannot use the encoding scheme of DPSK due to the aperiodicity of the carrier,

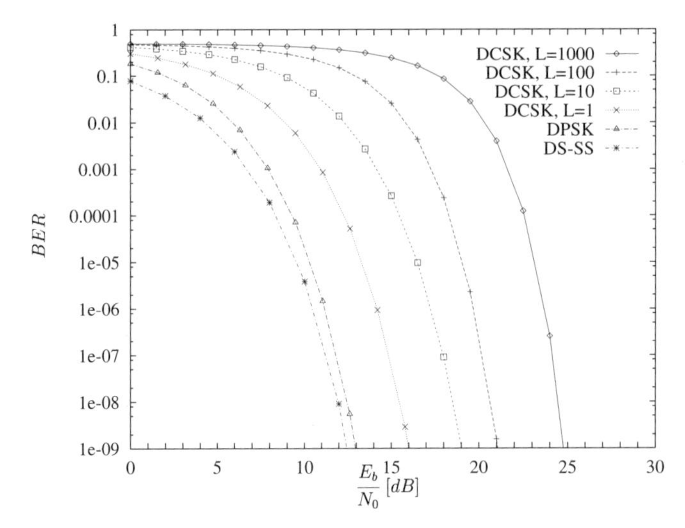

Figure 9.26: AWGN performance of several communication schemes.

DCSK has to spend twice the energy per bit. So from (9.136) we can easily derive the BER of DCSK in the limit case $L = 1$ (for which the PDF of $A_L(k_D)$ is not Gaussian)

$$BER = \frac{1}{2} \exp\left(-\frac{E_b}{2N_0}\right) \quad . \tag{9.137}$$

This is the performance limit in AWGN for the DCSK system.

Figure 9.26 displays the performance of all systems discussed here, based on the formulas (9.130), (9.135), (9.136), and (9.137). The DCSK results are confirmed by simulations [27], showing a good agreement of the curves, calculated under the assumption of a Gaussian PDF, with the numerical results already for $L = 10$.

A general criterion for noise robustness. For the analysis of the communication scheme in AWGN we looked at the dependence of the BER on the quotient E_b/N_0, which involves the communication system parameters κ^1_P, $\kappa^{1,1}_{\eta,\eta^*}$, and L. A dependence of the BER on only this quotient is the special case which we should aim at, since it results in noise robustness and the other beneficial properties listed in Section 9.6.3. For an arbitrary scheme we will face a dependence of the BER on $\kappa^1_P, \kappa^{1,1}_{\eta,\eta^*}, L$ (they do not have to appear always in form of the quotient E_B/N_0) and other cumulants of the signals z, η, m.

A sufficient criterion for noise robustness of a digital communication scheme then is *that all statistics of the threshold detector input signal A_L after a suitable normalization may be expressed by the quotient E_b/N_0 exclusively.* Then the BER of this scheme will depend solely on E_b/N_0.

For the communication schemes discussed here a suitable normalisation is the absolute value of the expectation of A_L, $|\kappa^1_{A_L}| = \kappa^1_P$. Then the first two cumulants of the normalized A'_L will be $\kappa^1_{A'_L} = m$ and $\kappa^{1,1}_{A_L,A^*_L} = SNR^{-1}$.

If additionally the distribution of A_L is Gaussian, the analysis of the cumulants up to second order will be sufficient to determine the robustness properties of a scheme. DS-SS is shown to be robust by (9.128) and (9.129). For DCSK (9.132) and (9.133) directly show that the system does not fulfill the robustness requirements, since we obtain further dependencies on system parameters such as L, or even κ_P^1 if the carrier signals are not properly chosen.

References

[1] M. Götz: *Analysis of the Frobenius-Perron operator and correlation theory of piecewise linear discrete time chaotic systems.* Ph.D. thesis, Dresden, 1997.

[2] A. Lasota and M.C. Mackey: *Chaos, Fractals, and Noise.* Applied Mathematical Sciences 97. Springer-Verlag, New York, 1994.

[3] A. Katok and B. Hasselblatt: *Introduction to the Modern Theory of Dynamical Systems.* Encyclopedia of Mathematics and Its Applications 54. Springer-Verlag, Cambridge, 1995.

[4] P. Walters, *Ergodic Theory-Introductory Lectures*, Springer-Verlag, Berlin, 1975.

[5] B. Porat, *Digital Signal Processing of Random Signals: Theory and Methods*, Prentice-Hall, Englewood Cliffs, 1994.

[6] A. Rodríguez Vázquez et al., "Switched Capacitor Broadband Noise Generator for CMOS VLSI," *Electronics Letters*, Vol. 27(21), pp. 1913-1915, Oct. 1991.

[7] M. Delgado-Restituto, F. Medeiro, and A. Rodríguez Vázquez, "Nonlinear Switched-Current CMOS IC for Random Signal Generation," *Electronic Letters*, Vol. 29(25), pp. 2190-2191, Dec. 1993.

[8] J. M. Cruz and L. O. Chua, "An IC Chip of Chua's Circuit," *IEEE Trans. on Circuits and Systems - Part II*, Vol. 40(10), pp. 614-625, Oct. 1993.

[9] M. Götz, K. Kelber, and W. Schwarz, "Discrete-Time Chaotic Encryption Systems, Part I: Statistical Design Approach," *IEEE Trans. Circ. & Systems-I*, vol. 44, no. 10, 963-970, 1997.

[10] K. Kelber, T. Falk, M. Götz, W. Schwarz, and T. Kilias "Discrete-Time Chaotic Coders for Information Encryption - Part 2: Continuous- and Discrete-Value Realisation," *Proc. Workshop on Nonlinear Dynamics of Electronic Systems*, 27-32, Seville, Spain, 1996.

[11] F. Dachselt, K. Kelber, and W. Schwarz, "Discrete-Time Chaotic Encryption Systems - Part III: Cryptographical Analysis," *IEEE Trans. Circ. & Systems - I*, vol. 45, no. 9, 883-888, 1998.

[12] U. Feldmann, M. Hasler, and W. Schwarz, "Communication by chaotic signals: The inverse system approach," *Int. J. Circ. Th. Appl.*, vol. 24, no. 5, pp. 551-579, 1996.

[13] D.R. Stinson, *Cryptography - Theory and Practice*, CRC Press, Boca Raton, 1995.

[14] T. Kilias, *One-dimensional discrete-time maps*, Ph.D. thesis, Dresden, 1992 (in German).

[15] K. Kelber, "N-dimensional Uniform Probability Distribution in Nonlinear Auto-Regressive Filter Structures," to appear in *IEEE Trans. Circ. & Systems-I*, 2000.

[16] M. K. Simon, J. K. Omura, R. A. Scholtz, and B. K. Levitt. *Spread Spectrum Communications Handbook*, McGraw-Hill, New York, 1994.

[17] A. N. Malakhov, *Cumulant Analysis of Random Non-Gaussian Processes and Their Transformations*, Sovetskoe Radio, Moscow, 1978 (in Russian).

[18] C. L. Nikias and A. P. Petropulu, *Higher-Order Spectra Analysis*, Prentice-Hall, Englewood Cliffs, NJ, 1993.

[19] S. Haykin, *Digital Communications*, Wiley, New York, 1988.

[20] S. Haykin, *Communication Systems*, 3rd ed., Wiley, New York, 1994.

[21] B. L. Basore, *Noise-Like Signals and Their Detection by Correlation.* Ph.D. thesis, Massachussetts Institute of Technology, Cambridge, MA, 1952.

[22] J. M. Wozencraft and I. M. Jacobs, *Principles of Communication Engineering*, Wiley, New York, 1965.

[23] H. V. Poor, *An Introduction to Signal Detection and Estimation*, Springer-Verlag, Berlin, 1994.

[24] G. Kolumbán, B. Vizvári, W. Schwarz, and A. Abel, "Differential Chaos Shift Keying: A Robust Coding for Chaos Communication," *Proc. Workshop on Nonlinear Dynamics of Electronic Systems*, pp. 87-92, Seville, Spain, 1996.

[25] G. Kolumbán, "Performance evaluation of chaotic communication systems: Determination of low-pass equivalent model," *Proc. Workshop on Nonlinear Dynamics of Electronic Systems*, pp. 41-51, Budapest, Hungary, 1998.

[26] G. Kolumbán, G. Kis, Z. Jákó, and M. P. Kennedy, "FM-DCSK: A robust coding for chaotic communication," *IEICE Trans.*, E81-A(9), pp. 1798-1802, Sept. 1998.

[27] A. Abel, M. Götz, and W. Schwarz, "Statistical Analysis of Chaotic Communication Schemes," *Proc. Int. Symposium on Circuits and Systems*, pp. IV-465 - IV-468, Monterey, CA, 1998.

[28] A. Abel, M. Götz, and W. Schwarz, "An Analysis Method for the Performance Comparison of Communication Systems," *Proc. Int. Conference on Electronics, Circuits and Systems*, pp. 1.107-1.110, Lisboa, Portugal, 1998.

[29] A. Abel, M. Götz, and W. Schwarz, "DCSK and DPSK – What makes them different?" *Proc. Int. Symposium on Nonlinear Theory and Its Applications*, pp. 69-72, Crans Montana, Switzerland, 1998.

Part III

CHAOS AT HARDWARE LEVEL

Chapter 10

Applications and Architectures for Chaotic ICs: An Introduction

Angel Rodríguez-Vázquez, Manuel Delgado-Restituto, Rocío del Río, and Belén Pérez-Verdú
Instituto de Microelectrónica de Sevilla
Centro Nacional de Microelectrónica CSIC
Edificio CICA-CNM, Avda. Reina Mercedes s/n, 41012 Sevilla, Spain
angel@imse.cnm.es, mandel@imse.cnm.es

10.1 Introduction [1]

It is well known that nonlinear electronic circuits may exhibit deterministic chaos; examples are found in [1–7]. Recently, it has been also demonstrated that the design of electronic circuits with customized, controllable chaotic behaviour is of interest for applications such as neural computation, instrumentation, analog signal processing and control, communication, etc. These circuits can be realized by interconnecting discrete IC (Integrated Circuit) component parts on a printed circuit board. However, whenever system miniaturization and power consumption are issues, chaotic circuits must be realized as monolithic integrated circuits, preferably in standard VLSI CMOS technologies where they can be embedded with other digital and analog circuitry. Consequently, different authors have focused on the development of architecture and circuits to design chaotic ICs.

The two basic ingredients of chaotic behaviour are dynamics and non-linearities. Hence, for the purpose of IC synthesis we must first search for achieving these ingredients, and then figure out systematic procedures to combine them to obtain controllable chaotic behaviours. Regarding dynamics, two different possibilities

[1]This research has been partially funded by Spanish CICYT under contract TIC99-08 26 and the EU through ESPRIT Project 31103, INSPECT.

arise: (a) using discrete-time dynamics, which leads to systems described by nonlinear finite-difference equations (FDEs); and (b) using continuous-time dynamics, which leads to systems described by nonlinear ordinary differential equations (ODEs). Actually, in recent years a few chaotic chips have been reported based on either of these approaches [8–15]. Most of them use discrete-time circuits to generate random signals with white or coloured spectra [8–10], or to emulate the behaviour of chaotic neurons [11,12]. The circuits presented in [13,15] and [16] use electrically-controllable continuous-time circuits to generate a bifurcation sequence to chaos, including several well-known attractors (Rössler, Chua's double-scroll, etc.). The circuits in [13,15] and the chaotic neurons have been also demonstrated for chaotic modulation with wired links [12,17].

This introductory chapter first reviews different scenarios where chaotic ICs might play a role. It then outlines general considerations pertaining to the implementation of both ODE- and FDE-based chaotic ICs.

10.2 Application Scenarios for Chaotic Integrated Circuits: An Overview

10.2.1 Emulation of Living Beings

It is nowadays well recognized that chaos is present in biological systems [18]. Biological chaos has been experimentally confirmed at various levels of system complexity: from the microscopic level, at individual biological neurons, to the macroscopic level, at highly complex systems such as the human brain [19–32].

Regarding microscopic level, it has been demonstrated with squid giant axons that chaos can be observed not only in oscillatory nerve membranes stimulated by sinusoidal strength [18,22], but also in resting membranes driven by pulse-train stimuli [23]. The latter observation is of particular relevance since most neurons in the brain seem to be at rest until stimulated by pulses of action potential coming from other neurons through the synaptic connections. A similar transition to chaos also has been measured in the onchidium pacemaker neuron [24]. In more complex systems, such as aggregate of chick heart cells, it has been shown that period-doubling bifurcations, intermittency, bistability, and chaotic dynamics arise when the neural ensemble is periodically stimulated [25].

At the macroscopic level, chaos also appears in many physiological processes, such as the electrical activity of the brain, the beating rate of the heart, or the dynamic neuromuscular contraction. In each case, evidence of chaotic behaviour is obtained by calculating indices such as the correlation dimension and Lyapunov exponents from experimental one-dimensional traces which are representative of the underlying dynamics [26]. Thus, for instance, the electrical activity of the brain is measured by EEG (Electroencephalogram) and MEG (Magnetoencephalogram) recordings. Seemingly, the dynamics of the heart are measured by ECG (Electrocardiogram) recordings, and the muscle's contraction dynamics are stored in EMG (Electromyograms) recordings.

Different interpretations suggest that chaotic dynamics seems to be the natural state of healthy or normal physiological processes, whereas disease, aging, and drug toxicity are characterized by a certain loss of complexity of the dynamics. This supports the idea that health is an information-rich state, characterized by a broadband spectrum, whereas the highly periodic behaviour observed in a variety of pathologies reflects a loss of physiological information and is represented by a relatively narrow-band spectrum [27]. Some experimental works confirming this view are given in [28–32].

One means to gain a thorough understanding of the seemingly relevant role of chaos in real biological systems and to diagnose and treat diseases correctly and efficiently is to study biologically inspired neural models embedded within artificial networks. To date, most of these studies are made through digital computer simulation, and thus require excessive processing time due to the complexity of the dynamics and the necessity to evaluate very long time series. This can be overcome by building dedicated computing hardware based on chaotic IC blocks [33].

10.2.2 Neural Computation

Aside from providing a better understanding of real biological systems, the design of neural networks with complex dynamics is also interesting for other practical processing tasks. In the following, two of the most promising applications of chaos in neural computation will be briefly reviewed: associative memory and optimization.

Many artificial neural networks are composed of very simple processing units whose function is to calculate the output of a sigmoid-like nonlinear activation function when driven by a weighted sum of stimuli. Fully recurrent structures made of these elements have been proven to serve as a nonlinear associative memory or content-addressable memory, the primary function of which is to retrieve a pattern stored as a stable equilibrium point of the network in response to the presentation of an incomplete or noisy version of that item [34, 35]. The name of content-addressable memory is justified because the writing and reading out of information are not based on the location of the memory, as in conventional digital computers, but on the content of the information itself. A major limitation of recurrent networks which store information as stable equilibrium points, such as the well-known Hopfield network, is their reduced storage capacity if we require that all the patterns be retrieved correctly. For instance, in a Hopfield network with N processing units and hard-limiting activation functions, the maximum number of fundamental memories which can be stored is around $N/(4 \log N)$, out of 2^N different equilibrium points [36].[2] If this bound is not respected, then the efficiency of the Hopfield Network as a content-addressable memory tends

[2] An increased storage capacity can be obtained by using nonmonotonic continuous activation functions instead of hard-limiting neurons [37]. Interestingly enough, this also allows the disappearance of spurious states and the emergence of chaotic behaviour when memory fails.

to decrease, because, on the one hand, we can no longer ensure the stability of all the fundamental memories, and on the other hand, it is possible that some spurious states arise leading to erroneous retrieval of stored information.

One way to overcome the limited storage capacity is to use other dynamical states, such as limit cycles, to encode the information patterns [38]. Regarding this, the introduction of chaotic dynamics into conventional neural network models seems to be a promising alternative based on the property of strange attractors to possess an infinite number of unstable periodic solutions [39]. Furthermore, since chaotic attractors are topologically transitive, i.e., all points in the attractor come arbitrarily close to every other point under the evolution of the flow or map, it is possible to use the onset of chaos to execute complex memory search tasks. This clearly favors that the network avoids becoming trapped in spurious states in the retrieval process.

Several neural network models are found in literature which exploit chaos to improve flexibility in information storage.[3] In [41], the neural network model is similar to a Hopfield network, but the synaptic matrix is asymmetric, and the connectivity (numbers of neurons connected to each unit) is a system parameter that determines the qualitative dynamical properties (stable or complex itinerancy) of the network. Patterns are stored in the form of limit cycles, and the synaptic matrix is formed by summing Kronecker products of pairs of consecutive patterns. When connectivity is large enough the network converges to one of the stored patterns (the closest to the initial state). However, if the connectivity is low enough, the network wanders chaotically in the state space even if one of the stored patterns is chosen as the initial state. Memory search is accomplished by adaptive control of the network connectivity.

Another neural network model which makes use of chaos for information storage has been proposed in [42]. The model is formally identical to a Hopfield network, in the sense that it consists of a fully recurrent structure which codifies input patterns into stable equilibrium points. Moreover, as in the Hopfield network, the stored patterns are binary vectors and the synaptic weights are learned by using the outer product (Hebbian) rule. The difference between the models is that the processing units are not threshold elements, but chaotic oscillators.[4] Such chaotic oscillators consist of two mutually coupled discrete maps, whose interconnection strength is a weighted sum of the neuron inputs. The output of the processing unit is binary and takes a "1" value if the coupled maps synchronize within some error bound, and a "0" value if there is no synchronization. It

[3]Neural networks are not the unique possibility for storing and retrieving information. Another interesting method has been presented in [40] and consists of using stable cycles of one-dimensional maps to codify information patterns. In this case, learning is not applied in the connection strengths, but in the shape of the return map.

[4]A similar approach of replacing conventional neural units by chaotic oscillators also has been applied in a Self-Organizing Feature-Mapping Network (Kohonen map) [43]. Again, the employed chaotic oscillator is a simple discrete map. The output of the neural units is obtained as the mean value of the orbit over a given interval of time T. Authors in [43], show how the introduction of chaotic units into the self-organizing map improves the ability of the network to cluster input patterns.

is shown that the model it is able to retrieve any of the stored patterns when sufficient partial information of the pattern itself is presented to the network.

The idea of using the property of synchronization/desynchronization between chaotic oscillators to codify binary patterns is also applied in the network model proposed in [44]. In this case, the network is composed of N fully connected autonomous continuous-time chaotic oscillators which interact among themselves by means of nonlinear coupling elements. The characteristics of such coupling elements are derived through Hebbian learning, according to the binary patterns to be memorized. A main difference with respect to the model in [42], is that the network acts dynamically and there are no global stationary states. When the network is stimulated by an arbitrary input pattern, it produces a sequence of mutual synchronization states, which corresponds to the fundamental memory most closely correlated with the input. On the other hand, if the pattern presented to the network is equally correlated with more than one fundamental memory, then the network produces an output sequence of synchronization states which intermittently points to all of the stored patterns resembling the input.

Another structure which uses the ability of synchronization among chaotic trajectories to perform associative memory is reported in [45], and it is based on previous works by Kaneko [46, 47]. The model consists of a network of chaotic one-dimensional discrete maps which are globally coupled through diffusive connections. The internal chaotic behaviour of each discrete map is controlled by a bifurcation parameter which defines the "strength" of chaos. For the application of information storage, all the bifurcation parameters, as well as the coupling strength, are adjusted so that the spatio-temporal characteristics of the system are in the ordered phase, which means that the network elements can exhibit only a small number of synchronization states (clusters) [47]. Contrary to the model in [44], the Hebbian rule is applied to the bifurcation parameters of the discrete maps, not to the coupling term which remain fixed during the learning process. During retrieval, the system realizes a chaotic memory search by adaptive control of the "strength" of chaos at each element. At the early stage of retrieval we can observe chaotic motion but, as time goes on, the motion becomes quiet and the system falls into an equilibrium state corresponding to the stored pattern most closely correlated with the input. Authors in [45] experimentally confirm that their model has superior storage capacity and better retrieval rate than the Hopfield network.

The last associative memory architecture that we will mention is the chaotic neural network proposed by Aihara, Takabe and Toyoda in [48]. This Hopfield-like neural network shares some common features with the above models. On the one hand, as occurs in [41], the spatio-temporal dynamics of the network can show global chaotic behaviour depending on the neuron parameters [49]. On the other hand, as occurs in [42], the stored patterns are binary vectors, the connection weights are derived using the Hebbian learning rule, and the processing units are chaotic oscillators. Interestingly enough, the chaotic neuron model used for such oscillators is biologically inspired, since it can qualitatively reproduce the alternating periodic-chaotic output sequences experimentally observed with normal squid axons [23]. Additionally, as occurs in [44], whenever the individual neurons oscillate, the overall system is always in a non-equilibrium state. The properties of this global oscillating pattern strongly depend on the external stimulation to the network. When the excitation to the neurons is uniformly distributed, the pattern dynamics is chaotic (or at least non-periodic) but we found a successive recall of stored patterns, with intervening unlearned patterns between retrievals. Because there is a transition continuous in time among stored patterns, we say that the neural network shows a dynamic association of memory.[5] A rather different situation happens when the external stimulation corresponds to one of the fundamental memories. In this case, chaotic wandering is still persistent, but the network recalls the stored pattern much more often and during longer times than others. Furthermore, the output pattern response of the network is almost periodic [51]. Conversely, if the external stimulation is very different from any of the stored patterns, the global oscillating pattern remains chaotic.

Another neural computation field where chaotic neural networks have proven to be useful is in solving complex combinatorial optimization tasks, whose archetype is the "travelling salesman problem" (TSP). The TSP can be formulated as follows: given the positions of a specified number of cities, assumed to lie in a plane, the question is to find the shortest tour that starts and finishes at the same city. Hopfield and Tank first demonstrated that the travelling salesman problem can be solved using a neural network paradigm [52]. They built a recurrent network of neurons with continuous nonlinear activation functions and determined the synaptic weights of the system from the distances between the cities. In this way, the energy (Lyapunov) function of the network can be regarded as a cost function for the lengths of the TSP tours. Given that the network is shown to be globally asymptotically stable, after a transient time, the system always converges to an equilibrium state which represents one possible solution of the TSP.

The main drawback of the Hopfield-Tank approach is that the neurodynamics of the system can get stuck in local minima of the energy function, and therefore, there is no guarantee that the network will find the optimal solution. In fact, it has been shown that the success of the optimization task strongly depends on the initial conditions of the network [53]. To cope with this shortcoming, optimization networks incorporating chaotic neurodynamics appear as a reliable alternative given that the topological transitivity of chaos can prevent the system state from being trapped in local minima, and may realize the global optimum solution. Some recent proposals of chaotic neural networks with the ability of solving the travelling salesman problem are presented in [42, 54, 55]. Computer simulations confirm that these networks have superior performance

[5] A similar result is obtained in [50] by using a stochastic neural network model. In this model, the retrieval process is governed by a one-dimensional discrete map, and because of it, the author talks about a dynamic link of memory, where the linking process is defined by a chaotic memory map.

both in success rate and computation speed compared with other commonplace extensions of the Hopfield-Tank scheme, such as the Boltzmann machines [56] and the mean-field theory neural networks [57].

10.2.3 Instrumentation Systems

Random signal generators are basic building blocks for instrumentation and communication systems. They are essential for the implementation of noise sources - both white and coloured - which are frequently used in speech processing [58], for testing dynamic behaviour of electronic systems [59], and in the identification of biological systems [60], among many other applications [61]. Additionally, random signal generators can also be used for the generation of secure non-repeatable pseudo-random number sequences [62]. A more novel application of random signal generators is in VLSI neural networks incorporating stochastic learning techniques [63]. In all of these applications, implementing on-chip random signal generators with reasonable area occupation is an ideal target.

Traditionally, there have been two main approaches for the electronic generation of random signals [64]: the exploitation of natural noise sources, such as Zener diodes, or the use of linear feedback shift registers.[6] An obvious difference between alternatives is in the very nature of the generated random signals. While in the first case the amplitude of the signals varies continuously over its range of activity (analog), in the second case, the amplitude of the signal only can take two possible discrete values (digital). This, however, does not necessarily imply non-overlapping application fields. For instance, analog signals can be obtained from LFSR's by low-pass filtering the output binary sequence, and digital sequences can be generated from analog noise sources by comparing the output signal with a given threshold value.

Another important distinction between the two approaches arises from their suitability for VLSI implementations. Random signal generators based on the intrinsic noise of electronic devices present three serious drawbacks for VLSI. First, very low power signals are obtained, thus requiring the use of large gain amplifiers. Second, it may produce spurious correlation in cases where multiple sources are needed. Third, appropriate sources may not be well suited for on-chip implementation in a given technology, thereby forcing the designer to consider

[6] A shift register of length m is a device consisting of m consecutive two-state memory stages (flip-flops) regulated by a single timing clock. At each clock pulse, the binary state of the flip-flops is shifted to the next state down the line. To prevent the shift register from emptying by the end of m clock pulses, a feedback block, which realizes a Boolean function of the states of the m memory stages, is applied to the first flip-flop of the shift register. When the feedback block is composed of modulo-2 adders to combine the outputs of the various memory stages, the resulting circuit is said to be a linear feedback shift register (LFSR). Some linear feedback structures have the ability to produce linear maximal binary sequences, i.e., periodic sequences with the largest possible period of value (the number of distinct m-binary words which can take the shift register, with the exception of the zero state). Such sequences present useful noise-like properties in terms of auto-correlation, spectrum, and probability density, and because of it, they are also referred to as pseudo-random or pseudo-noise sequences.

off-chip components. On the other hand, the use of linear feedback shift registers is ideally suited for the VLSI generation of random signals, because of their structural simplicity, easy programmability, and fast operation (current GaAs technologies allow digital processing well beyond the GHz range). Moreover, a single LFSR has the ability of generating multiple, arbitrarily shifted, linear maximal bit streams by tapping the outputs of selected memory stages and feeding the tapped responses through a set of exclusive-OR gates [63]. Thus, the LFSR approach offers the possibility of implementing multiple uncorrelated random sources with very small area overhead.

A main drawback of the LFSR approach with respect to the noise amplification technique is the inherent word-length limitation of the output signals, which can be a serious shortcoming in some applications. For instance, in random number generation the repeatability of the binary sequence implies reduced security [62]. Also, the spectrum of the LFSR signals is not continuous, but consists of a collection of spectral lines, which is an undesirable situation for noise sources. One possible solution is to increase the length of the shift register and consequently, the period of the maximal binary sequences. However, this may be inefficient for VLSI realizations.

Chaotic electronic circuits represent an efficient alternative for random signal generation which overcomes the principal limitations of the above techniques. On the one hand, because their underlying dynamic is deterministic, they are feasible for VLSI implementation. On the other hand, chaotic signals are aperiodic and exhibit broadband continuous spectrum, thus offering superior performance to LFSR bit sequences for random signal generation. In fact, it has been recently shown that linear feedback shift registers can be regarded as quantized versions of the standard baker's map, which is perhaps the simplest chaotic system [65].

10.2.4 Analog Signal Processing and Control

It is well recognized that most analog circuit designers feel very skeptical about the possibility of engineering circuits with chaotic behaviour, or in general, with any strongly nonlinear dynamic behaviour. After all, most designers have faced in their work problems like distortion, generation of harmonics, reduced dynamic range, etc., mainly caused by weak nonlinearities. In this scenario, it seems justified to argue that the stronger the nonlinear behaviour, the less likely the chance to obtain something meaningful.

No doubt, this negative vision of chaotic behaviour is valid in many situations. For example, in most electronic control systems where stable outputs are required, the onset of chaos is an undesirable phenomenon which may lead to excess noise, false frequency locking, phase slipping, if not to an irremediable malfunction of the system. To avoid these situations, several techniques [66, 67] have emerged in the last years whose purpose is to control chaos, i.e., to influence the system operating in chaotic regime in order to achieve a desired dynamic behaviour (usually a fixed point or a periodic orbit, depending on the particular application). Interestingly enough, application of chaos control techniques is

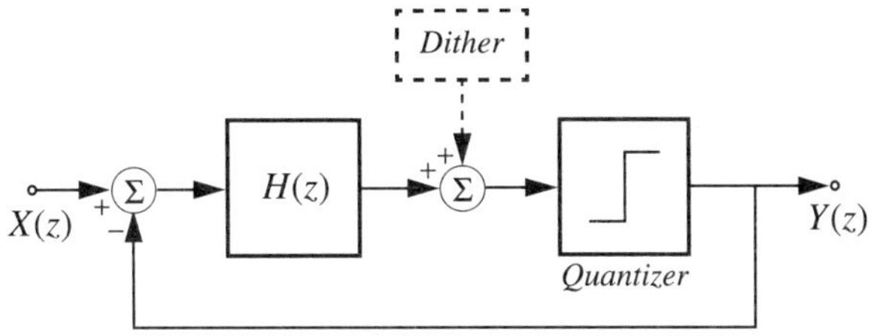

Figure 10.1: Discrete-time model of a Sigma-Delta modulator.

not reduced to the stabilization of electronic systems; they are also exploitable in other disciplines. These include neural signal processing [21, 28, 68]; biology and medicine (chaos control could mean elimination of heart arhythms and fibrillation [69], or epileptic attacks [70]); control of lasers [71]; thermodynamics (for laminarizing the flow in a thermal convection loop [72]); digital communications [73]; etc.

Aside from the situations where chaos can be seen as a disturbing mechanism, there are other signal processing tasks for which chaos can be beneficial. These applications include the linearization of Sigma-Delta ($\Sigma\Delta$) modulators, which are widely used for analog-to-digital and digital-to-analog signal conversion. Figure 10.1 shows a discrete-time model of a $\Sigma\Delta$ modulator which consists of a negative feedback loop embedding a one-bit quantizer and a discrete-time filter with transfer function $H(z)$. Structural stability of the feedback loop is ensured whenever the poles of $H(z)$ remain inside the unit circle in the z-domain [74].

Performance of stable $\Sigma\Delta$ modulators is degraded by the fact that for low amplitude input signals, the output waveform is corrupted by spurious tones, which are introduced during quantization and which would be audible in voice-band applications [75]. To avoid this situation, it is necessary to randomize these spurious tones, presumably at the expense of certain dynamic range reduction. Techniques that have been proposed to suppress or reduce the influence of spurious tones in $\Sigma\Delta$ modulators include the application of dither signals before the one-bit quantizer (see Figure 10.1), and the introduction of chaotic behaviour by moving the poles of $H(z)$ outside the unit circle [76, 77]. Preliminary results indicate that linearization of $\Sigma\Delta$ modulators by dithering leads to a lower dynamic range penalty than by making the circuit operate in a chaotic regime [78]. Regardless, an in-depth study is still lacking.

10.2.5 Communication Systems

The surprising fact that two chaotic oscillators could achieve synchronization without using an external driving signal has raised the possibility of exploiting chaos for communication purposes [79–83]. Chaotic signals present some particular features which make them interesting to convey information:

- Chaotic signals are not predictable at long term.

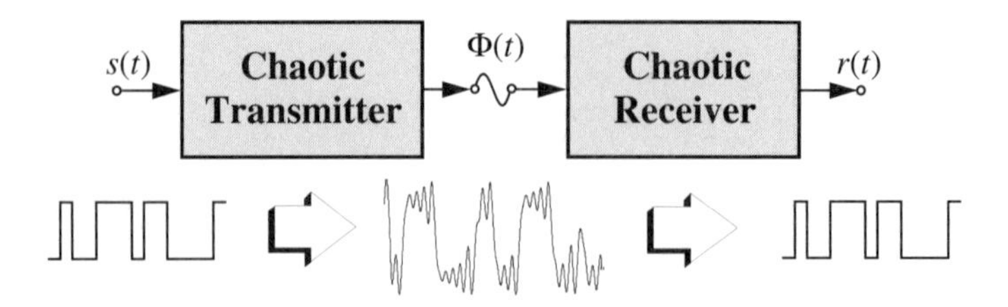

Figure 10.2: Basic Concept of the Communication Systems Utilizing Chaos.

- Estimation of future values from a segment of the signal is ill-conditioned.
- The two above properties make the communication system secure against jamming attacks.
- Chaotic signals appear noise-like, which render their usage well suited to hide information (chaotic coding). Chaotic coding can be viewed as a sort of analog cryptography where the parameters of the system act as the secret-key of the communication link.
- As chaotic signals are the homogenous response of certain nonlinear dynamical systems, they have more structure than stochastic signals. Clearly this is the basis which allows the transmitter and receiver to synchronize their trajectories.
- Chaotic signals exhibit broadband spectrum, which results in the frequency spreading of the transmitted information. This is advantageous because communication is thus endowed with robustness against interference, low disturbance from other users (because of the low power spectral density of the transmitted signal), and low probability of interception.

Figure 10.2 illustrates the basic concept behind communication systems based on chaos. An information bearing signal $s(t)$ consisting of an arbitrary pulse train is codified by a chaotic system which acts as the transmitter. The output signal $\Phi(t)$ generated by this block (assumed scalar in Figure 10.2) is then injected into another chaotic system which acts as the receiver. If the transmitter and receiver systems verify some conditions [84], and the parameters of both systems are nearly identical, chaotic synchronization can be achieved, and then the original information can be recovered. More precisely, signal $r(t)$ will reproduce asymptotically in time the original information $s(t)$. Since synchronization is only possible if the parameters of transmitter and receiver are close enough, they can be seen as the cryptographic key of the transmission.

The use of chaos in communication systems defines one of the most appealing applications of chaotic signals developed thus far. Actually, much of our recent research efforts have been devoted to probing the technical feasibility of the idea from an IC perspective. Note that the use of chaotic signals for hiding information provides simultaneous coding and spreading during data transmission, which are basic requisites for spread spectrum communications [85].[7] Most of the current spread spectrum techniques typically use pseudo-random sequences for widening the transmission band. Since such sequences can be understood as quantized versions of chaotic behaviour (they can have only a finite number of states), it is theoretically plausible that chaos could further improve the privacy performance of the communication.

10.3 Mathematical Models for Chaotic Circuits in ICs

The first step towards synthesizing chaotic ICs is identifying suitable mathematical models. Basic model features required for this purpose include: (a) possibility of obtaining a wide variety of dynamic behaviours by changing a reduced set of parameter values; (b) low parameter spreading, together with ease of controllability; (c) feasible nonlinearities; and (d) robust behaviour. In many occasions the original models proposed in literature must be modified to render them suitable for IC implementation [14, 15].

Two major approaches can be considered depending on whether the selected model is discrete-time, described by a vectorial finite-difference equation [86, 87],

$$\boldsymbol{x}(n+1) = \boldsymbol{f}[\boldsymbol{x}(n), \boldsymbol{p}], \qquad n = 0, 1, 2, \ldots \tag{10.1}$$

or continuous-time, described by a nonlinear differential equation [88],

$$\frac{d}{dt}\boldsymbol{x}(t) = \boldsymbol{f}[\boldsymbol{x}(t), \boldsymbol{p}], \qquad t > 0 \tag{10.2}$$

where $\boldsymbol{x}^T = [x_1, x_2, \ldots, x_N]$ represents a state variable vector; $\boldsymbol{f}(\cdot)$ is a nonlinear vector function; $\boldsymbol{p}$ is a parameter vector; t is the continuous-time variable; n is the discrete-time variable; and $\boldsymbol{x}(n)$ represents the value of the state vector at the n-th discrete time instant.

10.3.1 Systems Based on Finite-Difference Equations: Discrete Maps

Systems described by Equation (10.1) are also called discrete maps. In many discrete maps, changing the parameter vector $\boldsymbol{p}$ in Equation (10.1) obtains a rich and wide variety of dynamic behaviours with the same nonlinear function: stable fixed points, periodic orbits of different periodicity, and chaotic regimes [86, 87].

[7] Spread spectrum techniques are ideally suited for applications in satellite communications (low power spectral density to interfere with ground-based systems), mobile phones (high tolerance against multipath effects), military communications (low probability of interception), and emergency communications (below the noise floor of currently allocated frequency bands).

Consider for illustration purposes the widely known logistic map, given by [86,87]

$$x(n+1) = px(n)[1 - x(n)] \qquad n = 0, 1, \ldots \tag{10.3}$$

with parameter p comprised between 0 and 4. If $x(0) \in [0, 1]$, the above equation defines an endomorphism which transforms the unit interval into itself; thus, starting from such a value, a sequence $\{x(k)\}$ is obtained where the successive iterates are all comprised in this interval. Depending on the actual value of p, these sequences may have quite different properties. For $0 \leq p \leq 1$, they evolve towards a stable fixed point, located at $x = 0$. For $1 < p \leq 3$, such equilibrium state turns unstable, but another stable fixed point arises at $x = (p-1)/p$. As p increases beyond 3, this fixed point also becomes unstable and the system shows a period-2 limit cycle. This oscillatory behaviour remains stable if and only if $3 < p \leq 1 + \sqrt{6}$. If we increase p further, we find that $\{x(n)\}$ converges to orbits of periods equal to a successive power of 2. This phenomenon is called period doubling. Although the period doubling produces an infinite sequence of cycles, the width of their windows (intervals of p for which a particular stable periodic attractor exists) progressively diminishes. In the limit, a critical value p_∞ is reached, such that for $p_\infty < p < 4$ there is an infinite number of cycles, as well as an uncountable number of initial points that give bounded but totally aperiodic trajectories. This situation is generally referred to as chaos.

This example shows that chaotic behaviours can be obtained with rather simple mathematical models. Actually, the logistic map only involves a very simple quadratic law - feasible for IC implementation [89]. Also, it must be noted that the above mentioned bizarre behaviour is not a pathological characteristic of the logistic map. Rather, chaos has been observed in numerous dynamical systems modelled by discrete maps with simple mathematical structure. A set of discrete maps suitable for electronic implementation was reported in [89]; many others can be found in the reprint volumes [90,91].

10.3.2 Systems Based on Ordinary Differential Equations

Systems described by Equation (10.2) also exhibit a rich and wide variety of dynamic behaviours under changes of the parameter vector $\boldsymbol{p}$. However, in contrast to discrete maps, such chaotic behaviours are only exhibited by systems with at least third-order dynamics. Consider for illustration purposes the system [14],

$$\begin{aligned}
\tau\frac{d}{dt}x_1 &= m_1x_1 + \frac{m_0 - m_1}{2}(|x_1 + B_p| - |x_1 - B_p|) + \alpha x_2 \\
\tau\frac{d}{dt}x_2 &= \alpha(x_1 - x_3) - \gamma x_2 \\
\tau\frac{d}{dt}x_3 &= \beta x_2
\end{aligned} \tag{10.4}$$

whose parameterized behaviour is detailed in [14] and [15]. This model is more complex than that in Equation (10.3), but still feasible for IC implementation

[15]. Its chaotic behaviour is determined by six parameters. However, not all of them need to be changed; α, β, and γ can be set to fixed values, namely $\alpha = 3$, $\beta = 4$, and $\gamma = 1$. Then, by changing the others, a period-doubling bifurcation sequence is obtained including periodic windows and Rössler and double-scroll attractors. Examples of other chaotic ODEs can be found in the list of references at the end of this chapter.

10.4 Architectures and Concepts for Chaotic ICs

For the purpose of circuit synthesis, Equations (10.1) and (10.2) can be recast, respectively, as:

$$\boldsymbol{x}(n+1) = \boldsymbol{p}_0 + \boldsymbol{P}_L\boldsymbol{x}(n) + \boldsymbol{g}[\boldsymbol{x}(n), \boldsymbol{p}_{NL}] \tag{10.5}$$

$$\boldsymbol{\tau}^T\frac{d}{dt}\boldsymbol{x}(t) = \boldsymbol{p}_0 + \boldsymbol{P}_L\boldsymbol{x}(t) + \boldsymbol{g}[\boldsymbol{x}(t), \boldsymbol{p}_{NL}] \tag{10.6}$$

where $\boldsymbol{p}_0$ and $\boldsymbol{p}_{NL}$ are parameter vectors, $\boldsymbol{P}_L$ is a parameter matrix, and $\boldsymbol{\tau}^T = [\tau_1, \tau_2, \ldots, \tau_N]$ is a time constant vector. From here the updating of the generic k-th state variable is expressed, respectively, as

$$\begin{aligned} x_k(n+1) &= p_{0k} + \sum_{j=1,N} p_{Lkj}x_j(n) + g_k[\boldsymbol{x}(n), \boldsymbol{p}_{NL}] \\ \tau_k\frac{d}{dt}x_k(t) &= p_{0k} + \sum_{j=1,N} p_{Lkj}x_j(t) + g_k[\boldsymbol{x}(t), \boldsymbol{p}_{NL}] \end{aligned} \tag{10.7}$$

which are implemented by using the block diagrams of Figure 10.3.

Figure 10.3 allows us to identify the basic operations needed for the implementation of chaotic ICs, namely integration, delay, summation, signal weighting, and non-linear function generation. The IC building blocks needed to realize these operations cannot be covered with sufficient detail in this introduction. Hence, we will restrict ourselves to presenting some general ideas and concepts. More details can be found in [89, 92, 103, 104, 110] as well as in the works cited in these references. A key point for efficient realization of analog chaotic ICs is exploiting all functional features of the basic CMOS primitive: the MOS transistor. Compact descriptions of MOST operation including definitions and typical values of significant parameters can be found in [109, 110]. This latter reference discusses also possible usage and limitations of the MOST as well as other CMOS primitives.

10.4.1 Signal Weighting for ODE-Based Chaotic ICs

Because monolithic inductors are only feasible at very high frequencies, capacitors are the basic dynamic primitives of ODE-based chaotic ICs. State variables

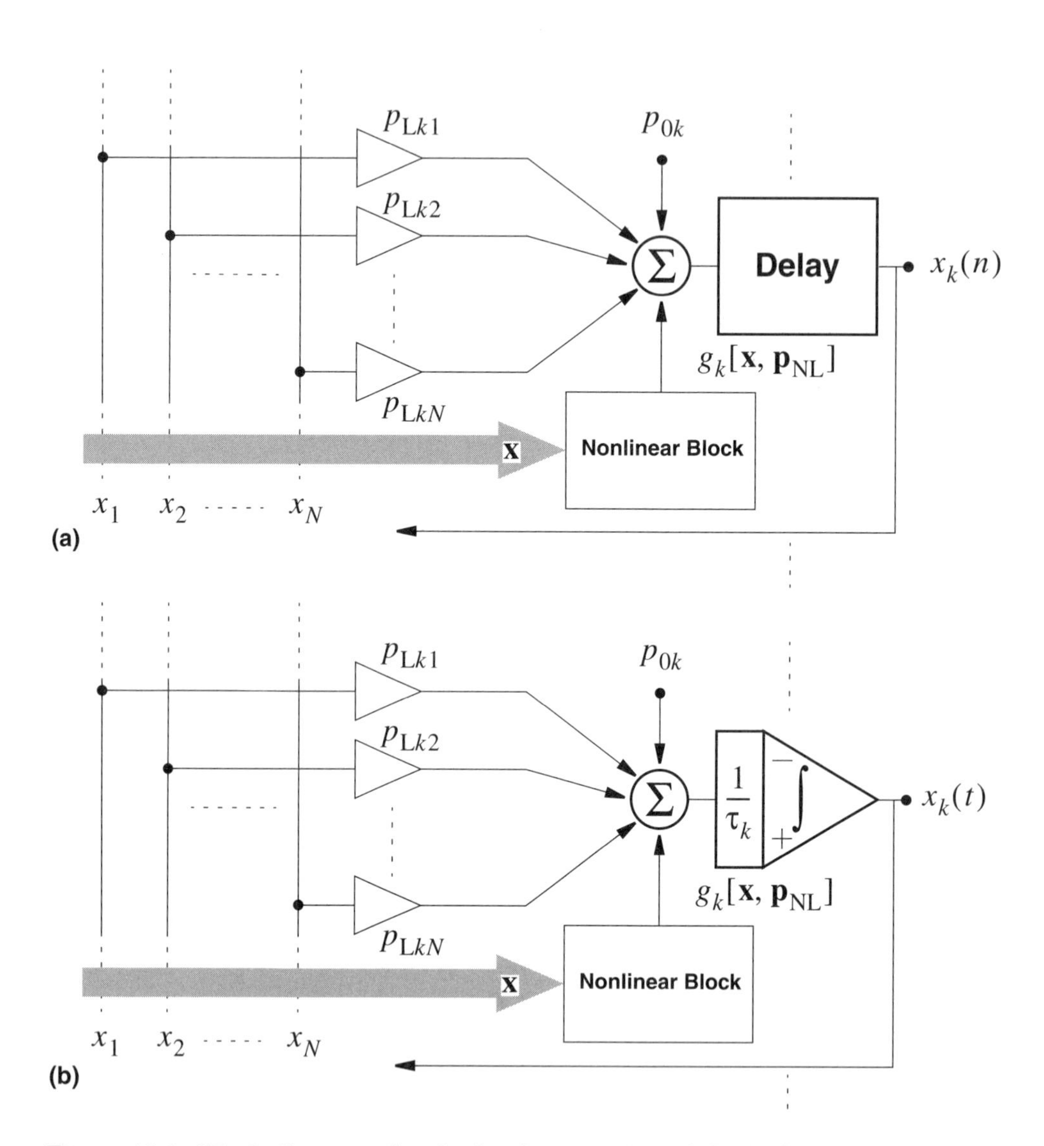

Figure 10.3: Block diagrams for the implementation of the k-th state variable of a chaotic circuit.

are hence voltages and the basic scaling operation is voltage-to-current transformation. Figure 10.4 illustrates several possible uses of a single n-channel MOST for voltage-to-current transformation. Note that this can be achieved by either using the MOST to realize a voltage-controlled current source or a resistor. Obviously the actual behaviour is nonlinear in all cases covered in the figure, which corresponds to different MOST operating regions [109, 110]; conditions for these regions are given in the figure inset. Expressions enclosed in the figure capture only the low-order relationships among incremental variables. These incremental variables are measured around a quiescent point; data pertaining to this quiescent point are in upper case and indicated by the subscript Q so that, for example, $y = \Delta y + Y_Q$. Actual voltage-to-current transformation circuits used in chaotic ICs contain more than just one transistor. However, their operation relies on the principles above. Important issues related to the design of these transformation circuits are:

- programmability
- linearity
- scaling factor accuracy
- parasitic resistances
- parasitic dynamics

10.4.1.1 Programmability

It basically refers to the possibility of scaling incremental transconductances through electrical control variables. From Figure 10.4, the incremental transconductance $G = (\beta/n_P)(X_Q - V_{T0})$ of MOST-based transformation circuits depend on the large-signal transconductance factor β, and on the values of biasing variables. Two possible controlling scenarios hence arise:

- Taking advantage of the dependence of β on transistor geometry,

$$\beta \propto \frac{W}{L} \tag{10.8}$$

  where W and L and are the transistor width and length, respectively, and of the MOST operation as an analog switch, to realize digitally controlled β-values.

- Taking advantage of the dependence of G with biasing to realize analog-controlled parameter values.

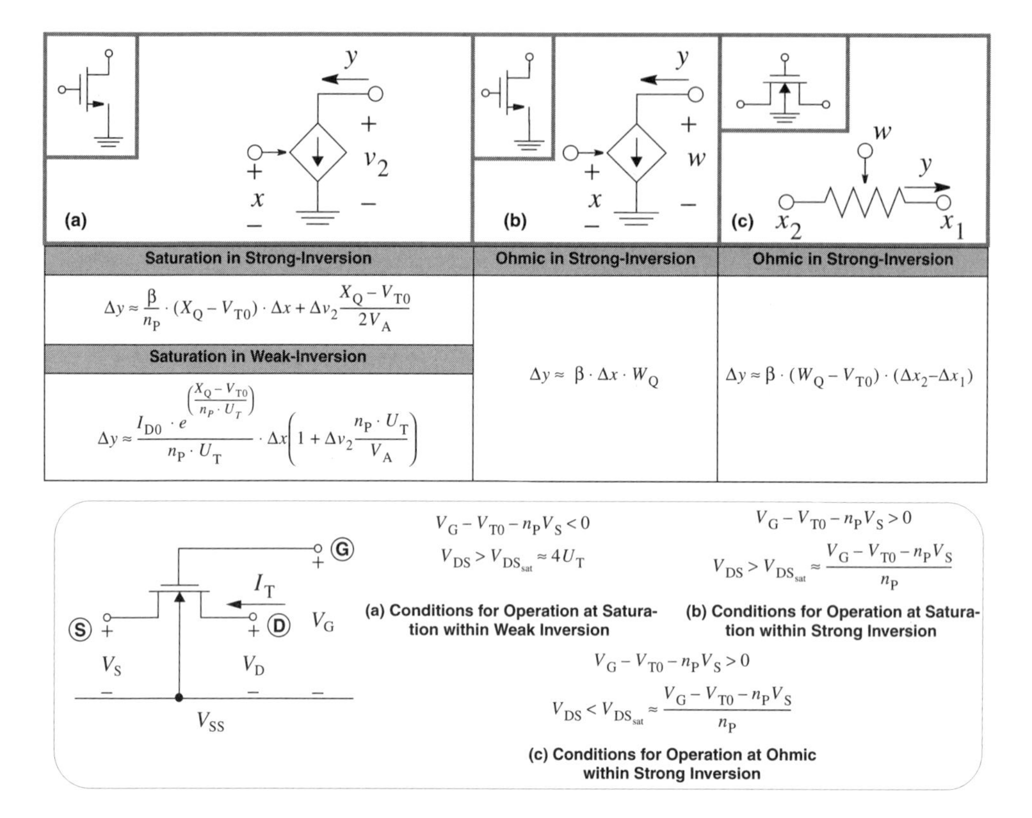

Figure 10.4: Using a single NMOST for voltage-to-current transformation. Only first-order terms are included in the behavioral equations displayed. β is the large-signal transconductance factor, a parameter proportional to W/L where W is the transitor width and L is the transistor length. V_{T0} is the zero-bias MOST threshold voltage. n_P is the slope factor in weak inversion. V_A is the equivalent Early voltage and U_T is the thermal voltage. I_{D0} is an asymptotic current proportional to β.

10.4.1.2 Linearity

Another important issue regarding scaling coefficients is the guarantee of linearity of the overall input-output characteristics. For instance, in the case of the MOST operation as a VCCS based on the saturation regime within strong inversion, one obtains by neglecting Early effects,

$$\Delta y \simeq \frac{\beta}{n_P}(X_Q - V_{T0})\Delta x \left[1 + \frac{1}{2(X_Q - V_{T0})}\Delta x\right] \tag{10.9}$$

which is a nonlinear function. Nonlinearities also appear in ohmic regime within strong inversion and in weak inversion. On the other hand, for transistors subjected to large fields, due to large bias voltages and/or small sizes, linearity may be worsened due to second-order phenomena such as mobility degradation, velocity saturation, and the like [109, 110]. Parasitic nonlinear terms can be attenuated by using differential configurations, by applying feedback, through inverse function nonlinear cancellation, etc. Some of these strategies are found in [92, 94, 97].

10.4.1.3 Scaling Factor Accuracy

Scaling factor accuracy is subjected to a threefold influence: (a) errors in technological parameter values that affect the transconductance factor β; (b) errors in the transistor sizes; and (c) biasing errors. Scaling factor accuracy has two faces: absolute accuracy and ratio accuracy. The former refers to exactness in absolute values of incremental transconductances and resistances which by itself cannot be guaranteed. Typical tolerances can be around 30%. However, absolute accuracy is important only in cases where timing is relevant; for example, for applications requiring synchronization among different chaotic oscillator units [15]. In those cases, tuning [94] is required to reduce absolute value tolerances. On the other hand, ratio accuracy refers to exactness of the ratios between transconductance or resistance values of devices of the same type (matching). Fortunately, devices of the same type can be made to match with quite good accuracy, up to 0.1%, depending on the device areas, shapes, and distances [110].

10.4.1.4 Parasitic Resistances

Because the MOST gate is isolated from the channel in dc, the configurations of Figure 10.4(a) and (b) have infinite input resistance. However, Figure 10.4(c) does not have large input resistance. Actually, its incremental input resistance, that measured at nodes x_1 and x_2, coincides with the inverse of the transconductance,

$$R_{gi} = G^{-1} = \frac{1}{\beta(X_Q - V_{T0})} \tag{10.10}$$

so that it may not be large, and cannot be controlled independently of the transconductance value. Regarding output resistance, for Figure 10.4(a) it is

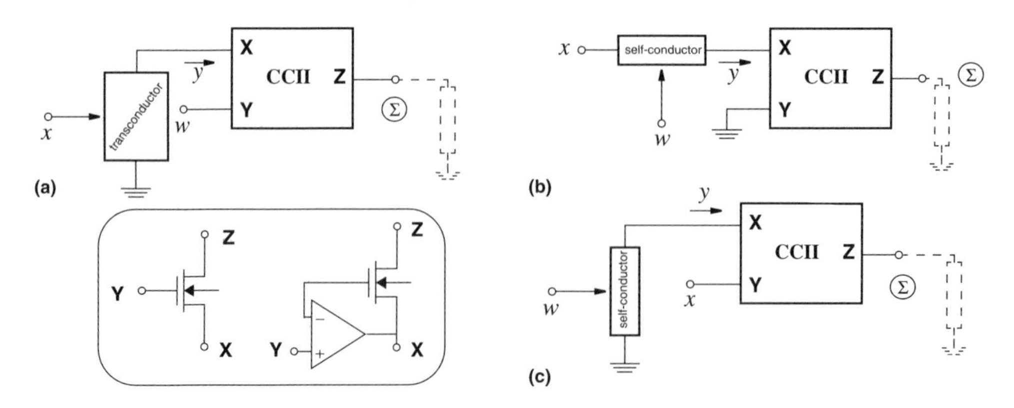

Figure 10.5: Basic structures for voltage-to-current conversion.

determined by the equivalent MOST Early voltage V_A. By properly setting this parameter, the output resistance can be fixed to a large value, independent of the transconductance value. However, the output resistance of Figure 10.4(b) is $R_{go} = (\beta\Delta x)^{-1}$, i.e., signal-dependent and not large by construction. Similarly, the output resistance of Figure 10.4(c), given also by Equation (10.10), is neither large by construction.

For voltage-to-current transformation circuits having low input resistance, the only way to attenuate loading errors [92] is by driving the input node with low output resistance. On the other hand, for those transformation circuits having low output resistance, the loading problems can be attenuated by resorting to one of the structures of Figure 10.5. They employ a conceptual block called current conveyor whose ideal behaviour is described by

$$\begin{bmatrix} i_y \\ v_x \\ i_z \end{bmatrix} = \begin{bmatrix} 0 & 0 & 0 \\ 1 & 0 & 0 \\ 0 & \pm 1 & 0 \end{bmatrix} \begin{bmatrix} v_x \\ i_x \\ v_z \end{bmatrix} \tag{10.11}$$

On the one hand, it creates a virtual ground between the terminals X and Y. On the other, it realizes a current follower operation between the terminals X and Z. Current conveyors can be realized through quite simple circuit configurations such as those depicted in the inset of Figure 10.5. In the simplest case a single cascode transistor may be sufficient; it exploits the operation of this transistor as a voltage follower. For "stronger" virtual ground, feedback can be applied around this transistor. More involved conveyor configurations can be found at [96].

Figure 10.5(a) is well suited for Figure 10.4(b). There, the output node is clamped at a fixed value w, thus annulling spurious current contributions to y due to node voltage fluctuations. On the other hand, Figure 10.5(b) and (c) are appropriate for Figure 10.4(c) and, in general, whenever the voltage-to-current transformation is realized by exploiting the self-conductance of either an active resistor, i.e., composed of MOSTs, or a passive resistor. Similar to Figure 10.4(b), in the case of Figure 10.5(b) the output node is at virtual ground and the current circulating through the output resistance is annulled. On the other hand, in the

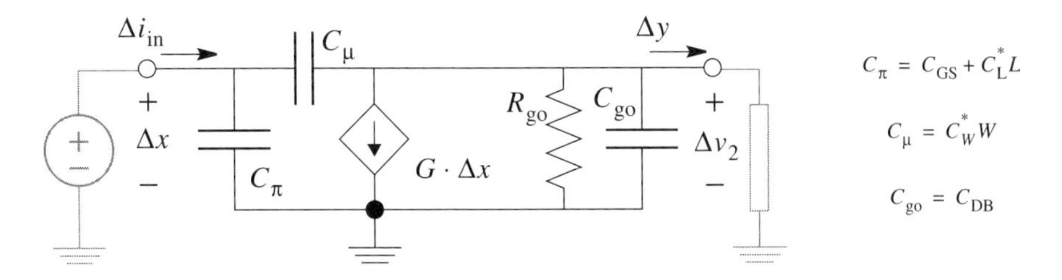

Figure 10.6: Small-signal hybrid-π MOSFET model.

case of Figure 10.5(c) the output node is forced to track the input signal x, thereby making the output current independent of the load connected to node Σ.

10.4.1.5 Parasitic Dynamics

Parasitic dynamics is due to MOST parasitic capacitances and must be assessed to guarantee that their associated time constants are much smaller than the nominal ones in (10.7). Figure 10.6 shows a hybrid-π model that captures the most significant linear parasitics. A useful figure of merit to quantify the frequency limitations is the transition frequency f_T [92]. It is defined as the frequency at which the magnitude of the current gain of Figure 10.6 with short-circuit load drops to unity. It can be calculated as:

$$2\pi f_T \simeq \frac{G}{C_\pi + C_\mu} \tag{10.12}$$

For instance, in the strong-inversion saturation region this can be evaluated as:

$$2\pi f_T \simeq \frac{3\mu(X_Q - V_{T0})}{L^2\left[3n_P - 1 + \frac{3n_P}{C^*_{ox}}\left(\frac{C^*_W}{L} + \frac{C^*_L}{W}\right)\right]} \tag{10.13}$$

where μ is electron mobility, C^*_{ox} is the gate transistor per unit area, and C^*_W, C^*_L are parasitic capacitances per unit length [110]. This expression shows that f_T increases as $X_Q - V_{T0}$ increases and L decreases.

For instance, in a 0.5μm technology, assuming W/L=1, $n_P = 1.3$, one obtains $f_T \simeq 12.8$GHz for $X_Q - V_{T0} = 1$V. In practice, however, the measured f_T is smaller. Among other reasons, such reduction is due to the fact that the transconductance of short-channel devices is not linear with $X_Q - V_{T0}$ but saturates at large fields due to phenomena such as mobility degradation and because (10.13) does not include extra parasitic capacitances that further reduce f_T. In any case, the range to allocate the chaotic IC natural frequencies can be quite large.

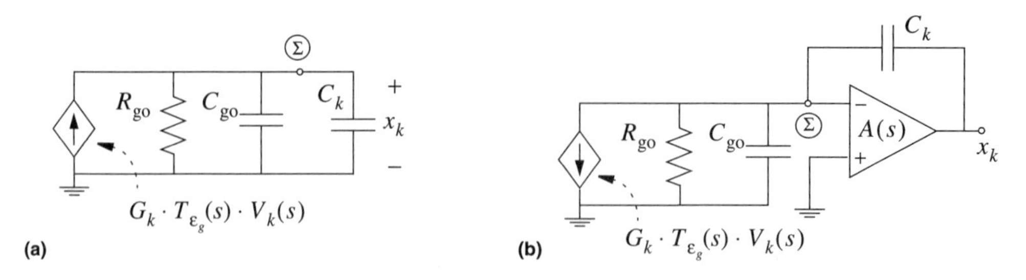

Figure 10.7: (a) Open Loop Integrator Concept; (b) Generalized Miller Integrator Concept.

10.4.2 Integrators

In actual circuit implementations of Figure 10.3(a) summation is performed by routing all voltage-to-current transformation circuit outputs to a common node, labelled Ⓢ in Figure 10.7, and letting Kirchoff Current Law work. Then integration is realized by simply using the aggregated current to drive a capacitor. Figure 10.7 shows two alternative implementations of the integration operation. There, the VCCS models the total aggregated current with three main parasitics:

- output resistance $G_{go} \equiv R_{go}^{-1}$ - it can be very large because it is obtained as the results of aggregating output conductances of all transformation circuits that drive node Ⓢ,
- output capacitance C_{go} - similar to the output conductance,
- frequency-dependent transconductance gain $G_k(s) = G_k T_{\epsilon_g}(s)$.

the two former parasitics can be attenuated by using a "glue" current conveyor in front of node Ⓢ(see Figure 10.5).

For discussion of pros and cons of both integrator structures, readers are referred to [92]. In summary, the open loop integrator of Figure 10.7(a) is preferable for high-frequency applications, despite requiring pre-distortion to compensate for time constant errors. Conversely, the Miller integrator of Figure 10.7(b) is more appropriate for low and medium frequency ranges, requiring no pre-distortion. However, note that degradation of frequency response in the Miller structure is mainly a consequence of using internally-compensated, general-purpose opamps. High-frequency advantages of the open-loop structure are not so evident if simpler, custom opamps without internal compensation are used [93]. In addition, frequency response of Miller structure may perhaps be enhanced by ingenious exploitation of active compensation techniques [94] to properly shape the integrator high-frequency response and, thus, combine the features of accuracy, small losses, and large frequency bandwidth in a single structure.

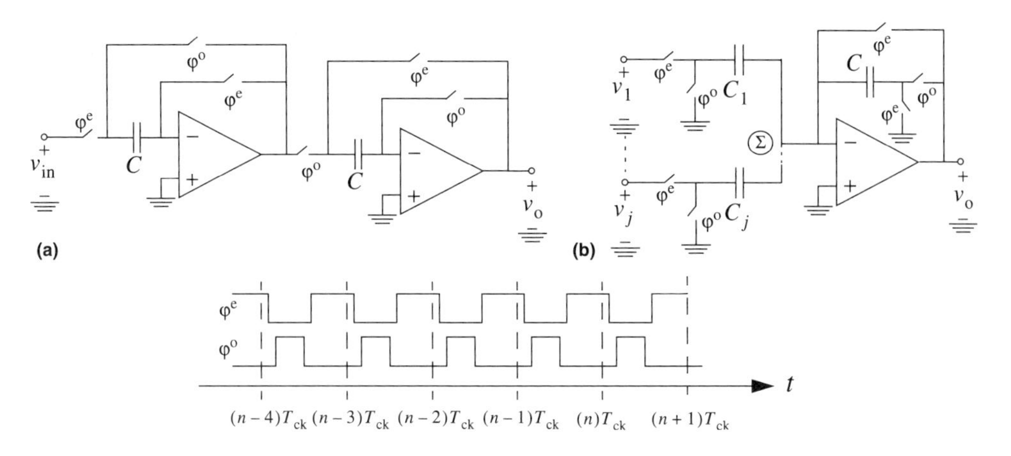

Figure 10.8: Switched-Capacitor Delay, Summation and Scaling.

10.4.3 Scaling and Delay for FDE-Based Chaotic ODEs

Two basic implementation styles are identified:

- switched-capacitor [8, 12, 89]
- switched-current [9–11]

Significant variables of the former case are voltages, and for the latter, currents. Because both styles are covered in greater detail in Chapter 11 of this volume, we only outline them below.

In both cases, scaling and delay operations are closely interrelated. Delays are realized by using sample-and-hold structures, which in the case of switched-capacitors employ opamps, capacitors, and analog switches. Figure 10.8(a) shows a possible implementation of a full-cycle delay stage which is insensitive to opamp offset and parasitic capacitances, and independent of capacitance ratios [95]. There each stage realizes a half delay. By replacing the first stage by that shown in Figure 10.8(b) weighted summation is achieved, the scaling factors being given by capacitor ratios,

$$j - \text{th Scale Factor} = \frac{C_j}{C} \tag{10.14}$$

Because the scaling factors are given as ratios of components of the same type, tolerances can be quite low, down to 0.1%, depending again on the device areas, shapes, and distances [110]. On the other hand, timing accuracy is guaranteed by the clock used to control the switches. Programmability can be quite easily incorporated by using digitally-controlled capacitor arrays [8].

Consider now switched-current techniques. They employ only transconductors and switches to process currents [96, 97]. Figure 10.9(a) shows a switched-current summer where each input current is weighted by a scale factor G_{mj}/G_m, and the aggregated current has a half-period delay with respect to the inputs.

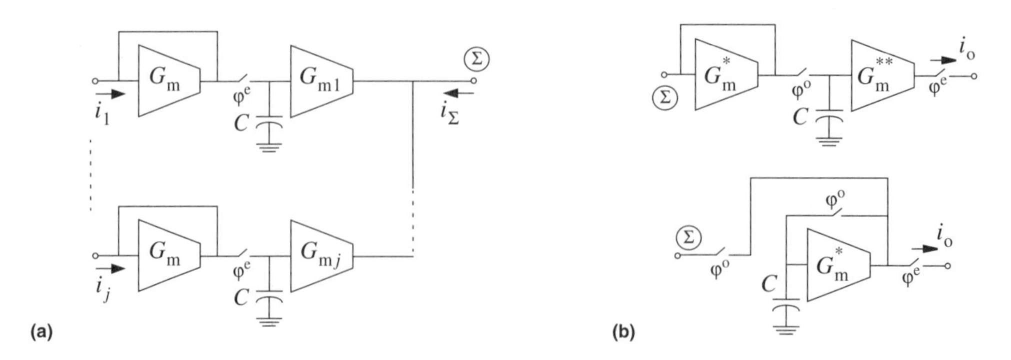

Figure 10.9: Scaling, summation and delay with switched-current circuits.

The transconductors can be realized by using a single transistor or a composite structure as discussed in [111]. On the other hand, the capacitor has only second-order influence on circuit performance; it can even be nonlinear provided that the clock period is long enough to guarantee that steady-state is reached at each clock phase. Hence, it can be implemented by taking advantage of parasitic capacitances at the transconductor input node.

The other half delay can be obtained by connecting node Ⓢ of Figure 10.9(a) to the corresponding one in either of the circuits of Figure 10.9(b). For that at the top (a first generation current memory cell [96]), the aggregated current is further scaled by G_m^{**}/G_m^*; for that at the bottom the scale factor is unity.

As for switched-capacitors, scaling factors are given as ratios among components of the same type and can hence be set with high accuracy. Timing is also precise because it is determined by the clock. On the other hand, similar to ODE-based chaotic ICs, programmability can be achieved through either analog control or digital control of transconductance values [11].

One of the most important dynamic limitations of discrete-time delayers (and in general, of any sampled-and-hold structure) is the so-called charge injection error [97–101]. Bear in mind that correct operation of these circuits relies on the capacitor's ability to hold the input voltage when the analog switch turns off. Assume that analog switches are implemented by single MOS transistors which are driven between its ohmic (closed state) and cut-off (open state) regions. In the turn-off process, the mobile charge in the inversion layer of the MOS switch (generated during the closed state) flow out of its drain, source, and substrate. As a consequence, a fraction of the charge released by the switch is added to the charge already stored on the hold capacitor, resulting in (charge injection) errors on the memorized voltage. Another error source that occurs during the switching-off process is the clock feedthrough effect. The rapidly changing gate voltage of the switch causes charges to flow from the gate diffusion overlap capacitances out of the MOS drain and source terminals. These charges also accumulate in the hold capacitor, thus leading to stored voltage deviations. The magnitude of these voltage errors is inversely proportional to the holding capacitance. For this reason, charge injection and clock feedthrough errors are particularly problematic in switched-current circuits, where hold capacitors consist of parasitics at the input of the transconductors; their capacitances are small if transconductors with reduced sizes are used. Charge injection errors can be attenuated with several techniques. They include dummy switch compensation, fully differential architectures, algorithmic cancellation, and adaptive timing of the clock signals.

10.4.4 Nonlinearities

Systematic procedures and circuits to realize nonlinear functions using IC components are described in [102–108], among others. Particularly, [104,107,108] are focused on the implementation of nonlinear behaviours using CMOS components and circuits. The design route towards the implementation of these behaviours basically comprises four steps:

- Identification of the intrinsic nonlinearities available at the design primitives; in the CMOS case, the nonlinearities available at the MOS transistor.
- Construction of basic circuit building blocks to realize elementary nonlinear operations such as logarithm, exponent, squaring, sign, etc.
- Interconnection of these blocks and the primitives themselves to realize nonlinear functions such as multiplication, division, absolute value, etc.
- Realization of nonlinear tasks through the proper interconnection of all the circuit blocks above.

Focusing on the first point above, nonlinear behaviours available at MOSFETs can be classified for design purposes in two groups: continuous (exponential, powers, etc.) and discontinuous or piecewise (rectification).

10.4.5 Piecewise-Linear Behaviours

They are due to the abrupt nonlinear dependence of MOSFET channel conductance with current polarity, as it explained for a NMOS with the help of Figure 10.10(a) and based on [109] and [110]. There, for $I_s = 0$, voltages V_+ and V_- are both equal to V_{DD} and consequently larger than the transistor pinch-off voltage $V_P \simeq (V_G - V_{T0})/n_P = (V_{DD} - 2V_{T0})/(2n_P)$. Consequently, the whole channel is in weak inversion and hence has very low conductivity. Now, if a positive input current $I_s > 0$ is applied, the voltage V_+ increases and the transistor current is approximately calculated as:

$$I_T \simeq I_{D0} e^{\dfrac{-0.8V_{DD} - V_{T0}}{1.3U_T}} \tag{10.15}$$

where U_T is the thermal voltage. Because the exponent is a large negative number, the transistor current is negligible (the channel conductivity remains very low) and practically all the input current is used to charge the parasitic capacitance at node V_+; this charging process lasts until either the compliance limit of the current source device is reached or the pn junction between the transistor diffused region and the substrate breaks down. On the other hand, if a negative input current $I_s < 0$ is applied, the voltage V_+ decreases, thereby increasing the channel conductivity. When V_+ in steady-state becomes smaller than the pinch-off voltage, all the input current circulates through the transistor in accordance to the following law

$$|I_s| \simeq \beta_{sat}(V_G - V_{T0} - n_P V_+)^2 \tag{10.16}$$

Similar, but complementary, behaviour is observed for the PMOST. Consequently, by connecting a PMOS and a NMOS transistor as shown in Figure 10.10(b) and driving the common node through a current source, positive currents will circulate through the PMOST while negative ones will circulate through the NMOST. Based on this a whole catalogue of piecewise-linear functions can be realized in CMOS [103, 108].

10.4.6 Continuous Nonlinear Behaviours

They can basically be in accordance to two different mathematical laws:

- Exponential
- Power

The former is observed for transistors operating in weak inversion. Particularly whenever the conditions in the inset of Figure 10.4 are fulfilled, the current is given by:

$$I_T \simeq I_{D0} e^{\dfrac{V_G - V_{T0} - n_P V_S}{n_P U_T}} \tag{10.17}$$

which is valid for low current levels such that $I_T \ll \beta U_T^2$. On the other hand, if the conditions in Figure 10.4 are fulfilled, one obtains

$$I_T \simeq \beta_{sat}(V_G - V_{T0} - n_P V_S)^2. \tag{10.18}$$

Circuits that exploit these nonlinear behaviours to realize nonlinear operations and functions are presented in [103, 104, 107]. For example, in the case powers, multiplication can be achieved by exploiting algebraic properties of the square function, namely,

$$z = \frac{1}{4}[(x + y)^2 - (x - y)^2] = xy \tag{10.19}$$

For transistors subjected to large electrical fields, however, deviations from square function law can be significant. On the other hand, in the case of exponentials, multiplication can be achieved based on,

$$\begin{aligned} z' &= \log(x) + \log(y) = \log(xy) \\ z &= e^{z'} = e^{\log(xy)} = xy \end{aligned} \tag{10.20}$$

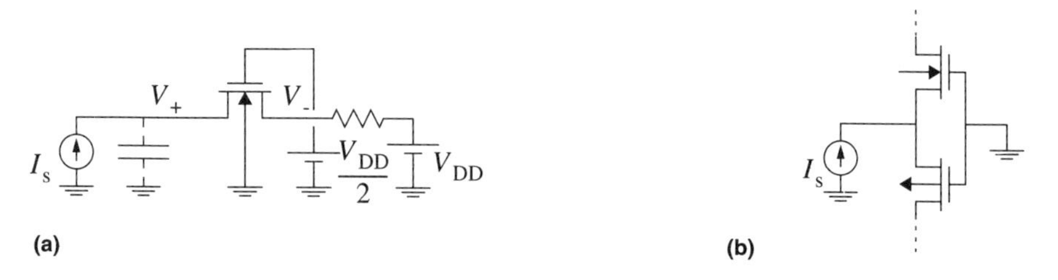

Figure 10.10: Current Rectification with MOSTs.

10.5 Summary

Using proper design it is possible to design compact and accurate chaotic ICs using analog CMOS circuits, as demonstrated by the design of several chips with different technologies. Robustness of the operation of these chips has been validated by using them in applications like noise generation and chaotic synchronization.

Because noise generation constitutes the most likely natural scenario for future chaotic ICs with many applications in fields like testing, system modelling, non-coherent communication, etc., it will be the topic of Chapter 11 of this volume.

References

[1] L.O. Chua (guest editor), "Special Issue on Chaotic Systems," *Proceedings of the IEEE*, vol. 75, August 1987.

[2] R.N. Madan (guest editor), *Chua's Circuit: A Paradigm for Chaos*, World Scientific, Singapore, 1993.

[3] L.O. Chua and M. Hasler (guest editors), "Special Issue on Chaos in Nonlinear Electronic Circuits. Part A: Tutorials and Reviews," *IEEE Transactions on Circuits and Systems-I*, vol. 40, October 1993.

[4] L.O. Chua and M. Hasler (guest editors), "Special Issue on Chaos in Nonlinear Electronic Circuits. Part B: Bifurcation and Chaos," *IEEE Transactions on Circuits and Systems-I*, vol. 40, November 1993.

[5] L.O. Chua and M. Hasler (guest editors), "Special Issue on Chaos in Nonlinear Electronic Circuits. Part C: Applications," *IEEE Transactions on Circuits and Systems-II*, Vol. 40, October 1993.

[6] L.O. Chua (guest editor), "Special Issue on Nonlinear Waves, Patterns and Spatio-Temporal Chaos in Dynamic Arrays," *IEEE Transactions on Circuits and Systems-I*, vol. 42, October 1995.

[7] M.P. Kennedy and M.J. Ogorzalek (guest editors), "Special Issue on Chaos Synchronization and Control: Theory and Applications," *IEEE Transactions on Circuits and Systems-I*, vol. 44, October 1997.

[8] A. Rodriguez-Vazquez, M. Delgado-Restituto, S. Espejo, and J. L. Huertas, "Switched Capacitor Broadband Noise Generator for CMOS VLSI," *Electronic Letters*, vol. 27(21), pp. 1913-1915, October 1991.

[9] M. Delgado-Restituto, F. Medeiro, and A. Rodriguez-Vazquez, "Nonlinear Switched-Current CMOS IC for Random Signal Generation," *Electronic Letters*, vol. 29(25), pp. 2190-2191, December 1993.

[10] A. Rodriguez-Vazquez and M. Delgado-Restituto, "Generation of Chaotic Signals using Current-Mode Techniques," *J. of Intelligent and Fuzzy Systems*, vol. 2(1), pp. 15-37, January 1994.

[11] M. Delgado-Restituto and A. Rodriguez-Vazquez, "Switched-Current Chaotic Neurons," *Electronic Letters*, vol. 30(5), pp. 429-430, March 1994.

[12] H. Horio and K. Suyama, "Experimental Verification of Signal Transmission Using Synchronized SC Chaotic Neural Networks," *IEEE Transactions on Circuits and Systems - Part I*, vol. 42(7), 393-395, July 1995.

[13] A. Rodriguez-Vazquez and M. Delgado-Restituto, "CMOS Design of Chaotic Oscillators Using State Variables: A Monolithic Chua's Circuit," *IEEE Transactions on Circuits and Systems - Part II*, vol. 40(10), pp. 596-613, October 1993.

[14] M. Delgado-Restituto and A. Rodriguez-Vazquez, "Design Considerations for Integrated Continuous-Time Chaotic Oscillators," *IEEE Transactions on Circuits and Systems - I* , vol. 45, pp. 481-495, April 1998.

[15] M. Delgado-Restituto, M. Linan, J. Ceballos, and A. Rodriguez-Vazquez, "Bifurcations and Synchronization using an Integrated Programmable Chaotic Circuit," *International Journal of Bifurcations and Chaos*, vol. 7, pp. 1737-1773, November 1997.

[16] J. M. Cruz and L. O. Chua, "An IC Chip of Chua's Circuit," *IEEE Transactions on Circuits and Systems - Part II*, vol. 40(10), pp. 614-625, October 1993.

[17] M. Delgado-Restituto, A. Rodriguez-Vazquez, R. L. Ahumada, and M. Linan, "Experimental Verification of Secure Communication Using Monolithic Chaotic Circuits," *Proceedings of the 1995 European Conference on Circuit Theory and Design*, vol. I, pp. 471-474, August 1995.

[18] H. Degn, A.V. Holden, and L.F. Olsen (editors), *Chaos in Biological Systems*, Plenum Press, New York, 1987.

[19] C.A. Skarda and W.J. Freeman, "How Brains Make Chaos in Order to Make Sense of the World," *Behavioral and Brain Science*, vol. 10, pp. 161-195, 1987.

[20] E. Basar (editor), *Chaos in Brain Function*, Springer-Verlag, New York 1990.

[21] W.J. Freeman, "Tutorial on Neurobiology: From Single Neurons to Brain Chaos," *International Journal of Bifurcation and Chaos*, vol. 2, n. 3, pp. 451-482, 1992.

[22] H. Hayashi, M. Nakao, and K. Hirakawa, "Chaos in the Self-Sustained Oscillation of an Excitable Biological Membrane under Sinusoidal Stimulation," *Physics Letters*, vol. 88A, pp. 265-266, 1982.

[23] G. Matsumoto, K. Aihara, Y. Hanyu, N. Takahashi, S. Yoshizawa, and J. Nagumo, "Chaos and Phase Locking in Normal Squid Axons," *Physics Letters A*, vol. 123, pp. 162-166, August 1987.

[24] H. Hayashi, S. Ishizuka, and K. Hirakawa, "Transition to Chaos via Intermittency in the Onchidium Pacemaker Neuron," *Physics Letters*, vol. 98A, n. 8, pp. 474-476, November 1983.

[25] M.R. Guevara, L. Glass, and A. Shrier, "Phase Locking, Period-Doubling Bifurcations, and Irregular Dynamics in Periodically Stimulated Cardiac Cells," *Science*, vol. 214, pp. 1350-1353, February 1981.

[26] G. Mayer-Kress, F.E. Yates, L. Benton, M. Keidel, W. Tirsch, S.J. Poppl, and K. Geist, "Dimensional Analysis of Nonlinear Oscillations in Brain, Heart and Muscle," *Mathematical Bioscience*, vol. 90, pp. 155-182, 1988.

[27] R. Pool, "Is It Healthy to Be Chaotic?" *Science*, vol. 243, pp. 604-607, February 1989.

[28] J.S. Nicolis, *Chaos and Information Processing*, World Scientific, Singapore, 1991.

[29] W.J. Freeman, "Strange Attractors that Govern Mammalian Brain Dynamics Shown by the Trajectories of Electroencephalographic (EEG) Potential," *IEEE Transactions on Circuits and Systems*, vol. 35, pp. 781-783, July 1988.

[30] A. Babloyantz and A. Destexhe, "Low-Dimensional Chaos in an Instance of Epilepsy," *Proceedings of National Academy of Sciences USA*, vol. 83, pp. 3513-3517, 1986.

[31] A.L. Goldberger, D.R. Rigney, and W.J. West, "Chaos and Fractals in Human Physiology," *Scientific American*, vol. 262 (February), pp. 42-49, 1990.

[32] M. Bodruzzaman, "Chaotic Classification of Electromyographic Signals via Correlation Dimension Measurement," *IEEE Proceedings Southeastcon'92*, vol. 2, pp. 95-101, 1992.

[33] P. Richert, B.J. Hosticka, M. Kesper, M. Scolles, and M. Schwarz, "An Emulator for Biologically-Inspired Neural Networks," *Proceedings of the 1993 International Joint Conference on Neural Networks*, vol. I, pp. 841-844, 1993.

[34] J.J. Hopfield, "Neural Networks and Physical Systems with Emergent Collective Computational Abilities," *Proceedings of National Academy of Sciences USA*, vol. 79, pp. 2554-2558, 1982.

[35] J.J. Hopfield, "Neurons with Graded Response have Collective Computational Properties like those of Two-State Neurons," *Proceedings of National Academy of Sciences USA*, vol. 81, pp. 3088-3092, 1984.

[36] D.J. Amit, *Modeling Brain Function: The World of Attractor Neural Networks*, Cambridge University Press, New York, 1989.

[37] S. Yoshizawa, M. Morita, and S.I. Amari, "Capacity of Associative Memory using a Nonmonotonic Neuron Model," *Neural Networks*, vol. 6, pp. 167-176, 1993.

[38] P. Thiran and M. Hasler, "Information Processing Using Stable and Unstable Oscillations: A Tutorial," *Proceedings of the Third IEEE International Workshop on Cellular Neural Networks and Their Applications*, pp. 127-136, 1994.

[39] B. Baird, M.W. Hirsch, and F. Eeckman, "A Neural Network Associative Memory for Handwritten Character Recognition Using Multiple Chua Characters," *IEEE Transactions on Circuits and Systems-II*, vol. 40, pp. 667-674, 1993.

[40] Y.V. Andreyev, A.S. Dmitriev, L.O. Chua, and C.W. Wu, "Associative and Random Access Memory Using One-Dimensional Maps," *International Journal of Bifurcation and Chaos*, vol. 2, n. 3, pp. 483-504, 1992.

[41] S. Nara, P. Davis, and H. Totsuji, "Memory Search Using Complex Dynamics in a Recurrent Neural Network Model," *Neural Networks*, vol. 6, pp. 963-973, 1993.

[42] M. Inoue and A. Nagayoshi, "A Chaos Neuro-Computer," *Physics Letters A*, vol. 158, pp. 373-376, 1991.

[43] A. Dingle, J.H. Andreae, and R.D. Jones, "A Chaotic Neural Unit," *Proceedings of the 1993 IEEE International Conference on Neural Networks*, vol. I, pp. 335-340, 1993.

[44] S. Jankowski, A. Londei, C. Mazur, and A. Lozowski, "Synchronization and Association in a Large Network of Coupled Chua's Circuits," *Proceedings of the 1994 International Workshop on Nonlinear Dynamics of Electronic Systems*, pp. 15-20, 1994.

[45] S. Ishii, K. Fukumizu, and S. Watanabe, "Associative Memory using Spatiotemporal Chaos," *Proceedings of the 1993 International Joint Conference on Neural Networks*, vol. I, pp. 2638-2641, 1993.

[46] K. Kaneko, "Pattern Dynamics in Spatiotemporal Chaos: Pattern Selection, Diffusion of Defect and Pattern Competition Intermittency," *Physica D*, vol. 34, pp. 1-41, 1989.

[47] K. Kaneko, "Clustering, Coding, Switching, Hierarchical Ordering and Control in a Network of Chaotic Elements," *Physica D*, vol. 41, pp. 137-172, 1990.

[48] K. Aihara, T. Takabe, and M. Toyoda, "Chaotic Neural Networks," *Physics Letters A*, vol. 144, pp. 333-340, March 1990.

[49] T. Ikeguchi, M. Adachi, and K. Aihara, "Chaotic Neural Networks and Associative Memory," in *Artificial Neural Networks*, A. Prieto, ed., pp. 17-24. Springer-Verlag, Berlin, 1991.

[50] I. Tsuda, "Dynamic Link of Memory - Chaotic Memory Map in Nonequilibrium Neural Networks," *Neural Networks*, vol. 5, pp. 313-326, 1992.

[51] M. Adachi, K. Aihara, and M. Kotani, "An Analysis of Associative Dynamics in a Chaotic Neural Network with External Stimulation," *Proceedings of the 1993 International Joint Conference on Neural Networks*, vol. I, pp. 409-412, 1993.

[52] J.J. Hopfield and T.W. Tank, "Neural Computation of Decisions in Optimization Problems," *Biological Cybernetics*, vol. 52, pp. 141-152, 1985.

[53] G.V. Wilson and G.S. Pawley, "On the Stability of the Traveling Salesman Problem Algorithm of Hopfield and Tank," *Biological Cybernetics*, vol. 58, pp. 63-70, 1988.

[54] H. Nozawa, "A Neural Network Model as a Globally Coupled Map and Applications Based on Chaos," *Chaos*, vol. 2, pp. 377-386, March 1992.

[55] T. Yamada, K. Aihara, and M. Kotani, "Chaotic Neural Networks and the Travelling Salesman Problem," *Proceedings of the 1993 International Joint Conference on Neural Networks*, vol. II, pp. 1549-1552, 1993.

[56] E. Aarts and J. Korst, *Simulated Annealing and Boltzmann Machines: A Stochastic Approach to Combinatorial Optimization and Neural Computing*, John Wiley, New York, 1989.

[57] C. Peterson, "Mean Field Theory Neural Networks for Feature Recognition, Content Addressable Memory and Optimization," *Connection Science*, vol. 3, pp. 3-33, 1991.

[58] R. Gregorian and G.C. Temes, *Analog MOS Integrated Circuits for Signal Processing*, John Wiley, New York, 1986.

[59] J. Shoukens, "Survey of Excitation Signals for FFT Based Signal Analyzers," *IEEE Transactions on Instrumentation and Measurements*, vol. 37, pp. 342-352, September 1988.

[60] V.Z. Marmarelis, "A Family of Quasi-White Random Signals and Its Optimal Use in Biological System Identification," *Biological Cybernetics*, vol. 27, pp. 49-56, 1977.

[61] M. Gupta, "Applications of Electrical Noise," *Proceedings of the IEEE*, vol. 63, pp. 996-1010, July 1975.

[62] G.M. Bernstein and M.A. Lieberman, "Secure Random Number Generation Using Chaotic Circuits," *IEEE Transactions on Circuits and Systems*, vol. 37, pp. 1157-1164, September 1990.

[63] J. Alspector, J.W. Garnett, S. Haber, M.B. Porter, and R. Chu, "A VLSI Efficient Technique for Generating Multiple Uncorrelated Noise Sources and its Application to Stochastic Neural Networks," *IEEE Transactions on Circuits and Systems*, vol. 38, pp. 109-123, January 1991.

[64] P. Horowitz and W. Hill, *The Art of Electronics* (second edition), Cambridge University Press, New York, 1989.

[65] A.C. Davies, "Relating Pseudorandom Signal Generators to Chaos in Non-Linear Dynamical Systems," *Proceedings of the 1993 European Conference on Circuit Theory and Design*, vol. II, pp. 1287-1292, 1993.

[66] M.J. Ogorzalek, "Taming Chaos - Part II: Control," *IEEE Transactions on Circuits and Systems-I*, vol. 40, pp. 700-706, October 1993.

[67] G. Chen and X. Dong, "From Chaos to Order - Perspectives and Methodologies in Controlling Chaotic Dynamical Systems," *International Journal of Bifurcation and Chaos*, vol. 3, No. 4, pp. 1363-1409, 1994.

[68] G.J. Mpitsos and R.M. Burton, Jr., "Convergence and Divergence in Neural Networks: Processing of Chaos and Biological Analogy," *Neural Networks*, vol. 5, pp. 605-625, 1992.

[69] A. Garfinkel, M.L. Spano, W.L. Ditto, and J.N. Weiss, "Controlling Cardiac Chaos," *Science*, vol. 257, pp. 1230-1235, 1992.

[70] S.J. Schiff, K. Jeger, D.H. Duong, T. Chang, M.L. Spano, and W.L. Ditto, "Controlling Chaos in the Brain," *Nature*, vol. 370, pp. 615-620, 1994.

[71] R. Roy, T.W. Murphy, Jr., T.D. Maier, Z. Gills, and E.R. Hunt, "Dynamical Control of a Chaotic Laser: Experimental Stabilization of a Globally Coupled System," *Physical Review Letters*, vol. 68, pp. 1259-1262, 1990.

[72] J. Singer, Y.Z. Wang, and H.H. Bau, "Controlling a Chaotic System," *Physical Review Letters*, vol. 66, pp. 1123-1125, 1991.

[73] J. Schweizer, Methods of Control and Synchronization of Chaotic Systems. Ph.D. thesis, University College Dublin, 1994.

[74] O.C. Feely, An Analytical Study of the Nonlinear Dynamics of Sigma-Delta Analog-to-Digital Conversion. Ph.D. thesis, University of California at Berkeley, May 1992.

[75] P.R. Gray, "Oversampled Sigma-Delta Modulation," *IEEE Transactions on Communications*, vol. 35, pp. 481-488, May 1987.

[76] R. Schreier, Noise-Shaped Coding. Ph.D. thesis, University of Toronto, 1991.

[77] S. Hein, "Exploiting Chaos to Suppress Spurious Tones in General Double-Loop $\Sigma\Delta$ Modulators," *IEEE Transactions on Circuits and Systems-II*, vol. 40, pp. 651-659, October 1993.

[78] C. Dunn and M. Sandler, "Linearizing Sigma-Delta Modulators Using Dither and Chaos," *Proceedings of the 1995 IEEE International Symposium on Circuits and Systems*, pp. 625-628, 1995.

[79] T. Carroll and L.M. Pecora, "Synchronizing Chaotic Signals," *IEEE Transactions on Circuits and Systems*, vol. 38, pp. 453-456, April 1991.

[80] M.J. Ogorzalek, "Taming Chaos - Part I: Synchronization," *IEEE Transactions on Circuits and Systems-I*, vol. 40, pp. 693-699, October 1993.

[81] M. Hasler, "Synchronization Principles and Applications," *Proceedings of the 1994 IEEE International Symposium on Circuits and Systems - Tutorials*, Chapter 6.2, pp. 314-327, 1994.

[82] A.V. Oppenheim, G.W. Wornell, S.H. Isabelle, and K.M. Cuomo, "Signal Processing in the Context of Chaotic Signals," *Proceedings of the 1992 IEEE International Conference on Acoustics, Speech and Signal Processing*, vol. IV, pp. 117-120, 1992.

[83] L.J. Kocarev, K.S. Halle, K. Eckert, U. Parlitz, and L.O. Chua, "Experimental Demonstration of Secure Communications via Chaotic Synchronization," *International Journal of Bifurcation and Chaos*, vol. 2, pp. 709-713, 1992.

[84] C.W. Wu and L.O. Chua, "A Unified Framework for Synchronization and Control of Dynamical Systems," *International Journal of Bifurcation and Chaos*, vol. 4, n. 4, pp. 979-998, 1994.

[85] D.T. Magill, F.D. Natali, and G.P. Edwards, "Spread-Spectrum Technology for Commercial Applications," *Proceedings of the IEEE*, vol. 82, pp. 572-584, April 1994.

[86] H.G. Schuster, *Deterministic Chaos: An Introduction* (2nd edition), VCH Verlagsgesellschaft, Weinheim, 1988.

[87] R.L. Devaney, *An Introduction to Chaotic Dynamical Systems*, Benjamin/Cummings, Menlo Park, 1986.

[88] J. Guckenheimer and P. Holmes, *Nonlinear Oscillations, Dynamical Systems and Bifurcations of Vector Field*, Springer-Verlag, Heidelberg, 1983.

[89] A. Rodriguez-Vazquez, et al., "Chaos from Switched-Capacitor Circuits: Discrete Maps," *Proceedings of the IEEE*, vol. 75, pp. 1090-1106, August 1987.

[90] H. Bai-Lin, *Chaos*, World Scientific, Singapore, 1984.

[91] P. Cvitanovic, *Universality in Chaos* (2nd edition), Adam Hilger, Bristol, U.K., 1989.

[92] A. Rodriguez-Vazquez, M. Delgado-Restituto, et al., "On the Implementation of Linear and Nonlinear Interaction Operators Using CNNs," Chapter 4 in Towards the Visual Microprocessor, T. Roska and A. Rodriguez-Vazquez, eds., John Wiley & Sons, Chichester, 2000.

[93] K.D. Peterson, A. Nedungadi, and R.L. Geiger, "Amplifier Design Considerations for High Frequency Monolithic Filters," *Proceedings of the 1987 European Conference on Circuit Theory and Design*, pp. 321-326, September 1987.

[94] R. Schaumann, M.S. Ghausi, and K.R. Laker, *Design of Analog Filters*, Prentice-Hall, Englewood Cliffs, 1990.

[95] G.C. Temes and K. Haug, "Improved Offset Compensation Schemes for Switched-Capacitor Circuits," *Electronics Letters*, vol. 20, pp. 508-509, June 1984.

[96] C. Toumazou, F.J. Lidgey, and D.G. Haigh (editors), *Analogue IC Design: The Current-Mode Approach*, Peter Peregrinus Ltd., London, 1990.

[97] C. Toumazou, J.B. Hughes, and N.C. Battersby (editors), *Switched-Currents: An Analogue Technique for Digital Technology*, Peter Peregrinus Ltd., London, 1993.

[98] J.H. Shieh, M. Patil, and B.J. Sheu, "Measurement and Analysis of Charge Injection in MOS Analog Switches," *IEEE Journal of Solid-State Circuits*, vol. 22, pp. 277-281, April 1987.

[99] G. Wegmann, E.A. Vittoz, and F. Rahali, "Charge Injection in Analog MOS Switches," *IEEE Journal of Solid-State Circuits*, vol. 22, pp. 1091-1097, December 1987.

[100] C. Eichenberger, *Charge Injection in MOS-Integrated Sample-and-Hold and Switched-Capacitor Circuits*, Hartung-Gorre Verlag, Konstanz, 1989.

[101] S. Espejo, R. Dominguez-Castro, F. Medeiro, and A. Rodriguez-Vazquez, "Tunable Feedthrough Cancellation in Switched-Current Circuits," *Electronics Letters*, vol. 30, pp. 1912-1914, November 1994.

[102] D. H. Sheingold, *Nonlinear Circuits Handbook*, Analog Devices Inc., Norwood, 1976.

[103] A. Rodriguez-Vazquez, M. Delgado-Restituto, J.L. Huertas, and F. Vidal, "Synthesis and Design of Nonlinear Circuits," Chapter 32 in *The Circuits and Filters Handbook*, W.K. Chen, ed., CRC Press, Boca Raton, 1995.

[104] M. Delgado-Restituto, A. Rodriguez-Vazquez, and F. Vidal, "Nonlinear Synthesis Using ICs," in *Encyclopedia of Electrical and Electronics Engineering*, J.G. Webster, ed., John Wiley & Sons, New York, 1999.

[105] E. Seevinck, *Analysis and Synthesis of Translinear Integrated Circuits*, Elsevier, Amsterdam, 1988.

[106] B. Gilbert, "Current-Mode Circuits from a Translinear Viewpoint: A Tutorial," Chapter 2 in *Analogue IC Design: The Current-Mode Approach*, C. Toumazou, F.J. Lidgey, and D.G. Haigh, eds., Peter Peregrinus Ltd., London, 1990.

[107] R. Wiegerink, *Analysis and Synthesis of MOS Translinear Circuits*, Kluwer Academic, Boston, 1993.

[108] A. Rodriguez-Vazquez, R. Dominguez-Castro, F. Medeiro, and M. Delgado-Restituto, "High-Resolution CMOS Current Comparators: Design and Applications to Current-Mode Function Generation," *Analog Integrated Circuits and Signal Processing*, vol. 7, pp. 149-166, March 1995.

[109] C.C. Enz, F. Krummenacher, and E.A. Vittoz, "An Analytical MOS Transistor Model Valid in ALL Regions of Operation and Dedicated to Low-Voltage and Low-Current Applications," *Analog Integrated Circuits and Signal Processing*, vol. 8, pp. 83-114, July 1995.

[110] A. Rodriguez-Vazquez, et al., "CMOS Analog Design Primitives," Chapter 3 in *Towards the Visual Microprocessor*, T. Roska and A. Rodriguez-Vazquez, eds., John Wiley & Sons, Chichester, 2000.

[111] A. Rodriguez-Vazquez, S. Espejo, R. Dominguez-Castro, and J.L. Huertas, "Current Mode Techniques for the Implementation of Continuous and Discrete-Time Cellular Neural Networks," *IEEE Transactions on Circuits and Systems*, vol. 40, pp. 132-146, March 1993.

Chapter 11

Chaos-Based Noise Generation in Silicon

Manuel Delgado-Restituto and Angel Rodríguez-Vázquez
Instituto de Microelectrónica de Sevilla
Centro Nacional de Microelectrónica CSIC
Edificio CICA-CNM, Avda. Reina Mercedes s/n, 41012 Sevilla, Spain
mandel@imse.cnm.es, angel@imse.cnm.es

11.1 Introduction [1]

On-chip random signal generators (RSG's) are essential components in many of today's mixed-signal very large scale integration (VLSI) electronic circuits for signal or information processing. They are aimed at providing random electrical excitations, either in analog or digital format, to the embedding integrated system, usually for one of the following purposes: to support testing and characterization of the hardware, to improve performance of other circuits in the system, or as part of the implemented algorithms.

An example of the first application scenario is the mixed-signal VLSI built-in self-test (BIST) methodology in which test functions are included in the integrated circuit itself, thereby facilitating at-speed test and substantially reducing the test application time [1–3]. In the second application scenario, RSG's are used, for example, for dithering $\Sigma\Delta$ modulators to suppress unwanted idle-tones and to whiten the noise floor [4], or for improving the linearity performance of Nyquist-rate analog-to-digital [5] and digital-to-analog data converters [6]. Most commonly, RSG's are found in the third scenario, i.e., as part of the implemented algorithms. For example, RSG's are efficiently used in some neural networks for simulated annealing optimization or stochastic model-free learn-

[1]This research has been partially funded by Spanish CICYT under contract TIC99-08 26 and the EU through ESPRIT Project 31103, INSPECT.

ing [7, 8]. Also, RSG's are basic elements for random number generation in computation [9] and cryptographic systems [10]. Yet another application area is that of spread-spectrum techniques for wideband digital communications and ranging systems [11].

To fit the purposes previously highlighted, on-chip RSG's have been conventionally implemented following either of two major approaches which can be referred to as physical and algorithmic. Physical generators are those which directly or indirectly exploit a random phenomenon of the physical world. On the other hand, algorithmic generators are those based on well-defined results from number theory, finite elements mathematics, computation science, and the like.

Two examples of physical random signal generation techniques are direct noise amplification and phase noise sampling. In the first case, a high gain amplifier is used to amplify the small ac voltage produced by a thermal or shot noise source [12]. Thereafter, an analog-to-digital converter can be used to obtain a random output digital word. The challenge of this approach is to control the sen- sitivity to physical parameters such as temperature, and to minimize correlation effects due to unavoidable power supply coupling and non-random substrate noise, particularly in mixed-signal chips or in system architectures requiring multiple random sources.

The second approach for physical random signal generation exploits the frequency instability, or jitter, of free-running oscillators - a by-product of transistor thermal noise - as the fundamental source of randomness [13]. In a typical embodiment, the outputs from two high frequency oscillators are sampled and exor-ed at a much lower clock rate, ideally producing a random bit per sample. An analog random source can be obtained from the output bit-stream through digital-to-analog conversion and low-pass filtering. Jitter-based RSG's are commonly combined with other random generation techniques for improved randomness of the synthesized signals, thus leading to large area occupation.

Algorithmic RSG's are the most popular solutions for random signal generation. They use either linear feedback shift registers (LFSR's) [9] or linear cellular automata registers (LCAR's) [14] yielding compact and scalable parallel VLSI architectures. As in the jitter-based approach, analog random signals can be obtained from properly selected taps of the register through digital-to-analog conversion and low-pass filtering. Because of the deterministic, infinitely precise and limited resolution features of the digital approach, generated sequences are guaranteed to repeat themselves. Sequences synthesized by using LFSR's (or LCAR's) are referred to as pseudo-random. These repeatable pseudo-random sequences are essential in applications which require exact code regeneration by a tracking system as, for example, to de-spread the desired information in code-division multiple-access (CDMA) communication networks, or to extract the enciphered message in some cryptographic systems. However, for many application fields, repeatability is not necessary; it may even be considered an inconvenience, for example, for private key generation or neural computation. Another drawback of the approach, particularly for analog random generation as required in mixed-signal BIST techniques, is the need to sub-sample the output bit-streams from the register to eliminate sequential correlations over time of the analog samples.

An alternative approach for random signal generation which overcomes many of the above difficulties is based on the exploitation of the noise-like properties of chaotic dynamics. Similar to the LFSR approach, chaotic systems are deterministic and, hence, described by equations, so that they represent an algorithmic method for random signal generation. However, because of the limited accuracy and infinite resolution features of their analog representation, and the intrinsic randomness of additive noise in their physical implementation, chaotic systems exhibit drastically different random properties than do the purely digital RSG's counterparts. On the one hand, chaotic signals are non-periodic and wideband and, hence, stochastic-like as the random signals obtained by direct noise amplification, but not requiring extra amplification. On the other hand, the divergence of trajectories associated with chaos combined with the injected physical noise renders the waveforms generated by chaos-based RSG's unpredictable and non-reproducible even if begun from identical initial conditions.

Based on the above properties, chaotic RSG solutions have been already proposed in many application areas as discussed in Chapter 5. They include, for instance, random number generation and cryptography [15], neural computation for associative dynamics and optimization [16, 17], ranging systems [18], and, fundamentally, digital communications. Some of these applications not only use chaos-based RSG's as an efficient means for random signal generation, but also take advantage of the inherent structure behind chaotic signals and some of their properties such as, for example, the capability of synchronization.

The applications mentioned above have undeniable market projections; the trend in all these areas is towards complete system integration, prompted by the convenience of reducing the size and power consumption of electronic realizations. Hence, to fully exploit the inherent potential of chaos-based RSG's, they must be realized using VLSI compatible techniques as done in [19–29]. Accordingly, the main objective of this chapter is to present a survey of integrated circuit (IC) techniques, from model formulation to silicon design, for the monolithic implementation of chaos-based RSG's. More concretely, the chapter deals with the realization of autonomous chaotic circuits based on recursive finite-difference equations (FDE's), also called discrete maps. Readers interested in the IC details involved in the implementation of chaotic circuits based on ordinary-differential equations (ODE's) are referred to [28, 29] and to the reference section of Chapter 5 of this volume.

The chapter is organized as follows. Section 11.2. illustrates, through an exemplary FDE system, some of the most relevant features of the dynamic models covered in this chapter. It also introduces the family of piecewise-linear (PWL) discrete maps as a simple, yet powerful, model for synthesizing chaotic behavior. This section identifies the operators required for the implementation of discrete maps, whose implementation is covered in Sections 11.3 and 11.4. The former is devoted to linear operators while the latter covers implementation of the nonlinear ones. Both switched-capacitor (SC) and switched-current (SI) techniques are covered in the two sections. Finally, Section 11.5 shows the experimental results obtained from two monolithic chaotic circuits, which confirm the possibility of generating chaotic behaviors on-chip in a robust and controlled manner.

11.2 Discrete-Time Chaos Generators

Any autonomous discrete map can be generically described by the following -th order -dimensional recursive equation,

$$\boldsymbol{x}(k+n) = \boldsymbol{F}[\boldsymbol{x}(k+n-1), \ldots, \boldsymbol{x}(k); \boldsymbol{P}] \tag{11.1}$$

where $\boldsymbol{x}^T = [x_1, x_2, \ldots, x_m]$ represents the state vector of the system; $\boldsymbol{F}(\cdot; \boldsymbol{P})$ is a nonlinear vector field which depends on a parameter set $\boldsymbol{P}$; $k = 0, 1, 2, \ldots$ symbolizes the discrete-time variable; and $\boldsymbol{x}(i)$ is the state vector $\boldsymbol{x}$ at the i-th discrete time instant. In some cases, the system is also endowed with an output equation defined in terms of the r most recent states of the system as

$$\boldsymbol{y}(k+n) = \boldsymbol{G}[\boldsymbol{x}(k+r), \ldots, \boldsymbol{x}(k+1); \boldsymbol{Q}] \tag{11.2}$$

where $\boldsymbol{y}(i)$ is the p-dimensional output vector of the discrete map at the i-th instant; and $\boldsymbol{G}(\cdot; \boldsymbol{Q})$ is a function, in general, nonlinear and parameterized by a vector $\boldsymbol{Q}$. The dynamics of the discrete map formed by Equations (11.1) and (11.2) is essentially defined by the former, as the output equation exclusively performs a transformation of the signals generated by the recursive system. From this, and taking into account the similarity of the mathematical representations of Equations (11.1) and (11.2) (identical design techniques can be employed for their electronic implementations), we will exclusively focus on system (11.1).

The dynamic response of the system (11.1), referred to as orbit or trajectory, strongly depends on the nonlinearity of the map, as well as on the particular configuration of the parameters vector. Additionally, the qualitative behavior of the trajectories may also depend on the initial state value $\boldsymbol{x}(0) = \boldsymbol{x}_0$ presented to the system. As for signal generation, we will assume that system (11.1) is characterized by an invariant set J under $\boldsymbol{F}(\cdot, \boldsymbol{P})$, such that any trajectory starting in J remains confined to it, thus avoiding the onset of divergence and, thereby, the saturation of the outputs in the electronic realization.

In spite of the structural simplicity of (11.1), the dynamic behavior of many nonlinear discrete maps is extremely rich and complicated. They can exhibit stable fixed points, periodic orbits of different periodicity, or chaotic regimes, depending upon the non-empty set in the parameter space in which vector $\boldsymbol{P}$ falls.

Consider, for illustration purposes, the Bernoulli family of discrete maps defined by the following first-order 1-D FDE system [30–32],

$$x(k+1) = \begin{cases} Bx(k) + A & x(k) < 0 \\ Bx(k) - A & x(k) > 0 \end{cases} \quad (11.3)$$

where A and B are real parameters. For $0 < B < 2$, the dynamic law (11.3) maps the interval $J = [-A, A]$ onto itself. In this sense, system (11.3) can be considered as a signal generator which provides a sequence of sampled values $\{x(k)\}$, $k = 0, 1, \ldots$, after initialization by a leading state $x(0) = x_0$. Assuming that x_0 lies inside J, it can easily be shown that parameter B solely determines the dynamic properties of the generated signal, whereas parameter A acts as mere scale factor [30, 31]. For $B < 0$ or $B > 2$, orbits diverge regardless of the initial state, leading to unstable operation. For $0 \leq B < 1$, trajectories evolve towards an oscillatory pattern between points $x_1^* = A/(1+B)$ and $x_2^* = -A/(1+B)$, conforming a stable period-2 orbit. For $1 \leq B < 2$, the map enters in chaotic regime. In this regime, there exists a countable infinite number of unstable periodic orbits of all possible periods, along with an uncountable number of aperiodic solutions that remain confined inside J. Further, if x_0 is not a periodic point of (11.3), then x_0 is not an asymptotically periodic point. Hence, any two trajectories either coincide with one another or are separated forever. In such a case the map is said to exhibit sensitive dependence on initial conditions because small perturbations of the initial conditions are largely amplified in a few iterations. A major consequence of this property is that signals generated in the chaotic regime decorrelate rapidly with themselves, i.e., the autocorrelation function has a large peak at zero and decays exponentially with time. In the frequency domain, this implies that chaotic signals exhibit a continuous noise-like broad power spectrum. Within the Bernoulli family, all the above properties become even more noticeable as the parameter B approaches 2. For instance, it can be shown that for $\sqrt{2} < B < 2$, and with the exception of a countable set of periodic points, any arbitrarily small sub-interval inside J is reached regardless of the initial state, i.e., the map is ergodic and topologically transitive. Indeed, for almost all solutions, iterations of (11.3) are uniformly distributed in J when B tends to 2.

Figure 11.1 illustrates the above characteristics of the Bernoulli family assuming $A = 1.0$, an arbitrary $x_0 \in J$ and a sampling frequency $f_s = 20$MHz. In all the figures, parameter $B = 1.95$, with the exception of Figure 11.1(b) where B is swept over the range $[0, 2)$. Figure 11.1(a) displays the first return map of the system, discovering the deterministic structure of the iterations in accordance to the nonlinearity of (11.3). Figure 11.1(b) represents the bifurcation diagram of the map in which the location of the points of a long enough sequence $\{x(k)\}$ are plotted versus the sweeping (bifurcation) parameter B. Observe that both the periodic and ergodic behaviors of the discrete map are readily appreciated. Figure 11.1(d) shows a normalized histogram of the map trajectory, as

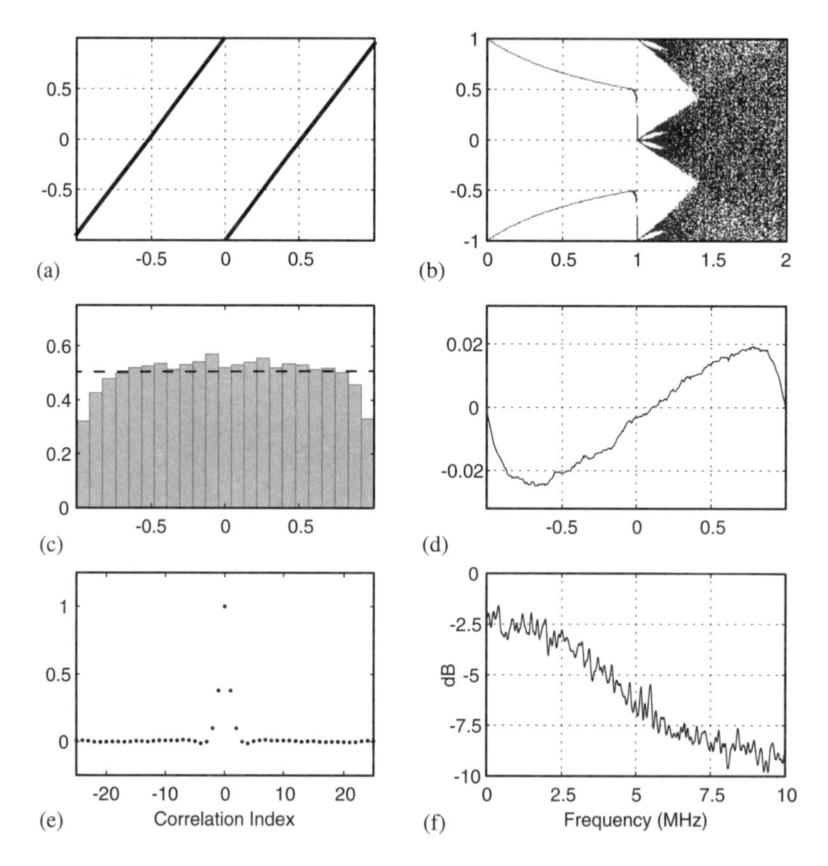

Figure 11.1: Properties of the Bernoulli map: (a) return map for $B = 1.95$ and $A = 1.0$; (b) bifurcation diagram for $B \in [0, 2)$; (c) estimation of the power spectral density for the parameters in (a); (d) deviation from an ideal uniform distribution function; (e) autocorrelation function; (f) power spectrum.

an estimation of its invariant probability density function.[2] As can be seen, the histogram approaches a constant value (uniform distribution) in the definition interval J. This is confirmed in Figure 11.1(d) which indicates a maximum deviation of 2.5% between the probability distribution function of the iterations and an ideal uniform distribution defined in the same interval. Finally, Figures 11.1(e, f) display, respectively, the autocorrelation function and power spectrum of signal $\{x(k)\}$. The rapidly decaying feature of the former and the continuous noise-like characteristics of the latter can be easily observed.

An important conclusion which can be drawn from the above is that a very simple mathematical model is able to generate chaotic behavior. In fact, the Bernoulli family is not a pathological case; rather, chaos has been observed in numerous dynamical systems modeled by discrete maps. Some of these maps (only 1-D and 2-D examples for illustration purposes) are collected in Table 3.2.1, together with their definition intervals and the ranges of parameters which lead to chaotic regime. Other FDE systems able to exhibit chaos can be found in reprint volumes [34, 35] and the specialized literature.

Besides the ubiquity of chaotic dynamics in several nonlinear models, discrete maps also exhibit some interesting features for VLSI realization:

- As can be noted from Table 11.1, chaos can be generated by first-order 1-D discrete maps of the form,

$$x(k+1) = f[x(k); \boldsymbol{P}] \tag{11.4}$$

  such as those listed in th first eight rows of Table 11.1. Conversely, autonomous chaotic systems based on ODE's require at least third-order equations, i.e., three state variables. Thus, simpler monolithic realizations can be expected from the use of discrete maps.

- Also, for low and medium-frequency operation, sampled-data techniques feature integrated circuits with low sensitivity to parasitics and well-controlled dynamics. This is clearly advantageous as compared to continuous-time techniques whose natural frequencies require tuning for accurate setting. It is also important in many digital communication systems based on chaos, which demand a precise timing control on the generated signals.

- Last but not least, there is a wide catalog of chaotic discrete maps where the nonlinear mapping is piecewise-linear. As will be stressed in the following sections, this kind of discrete map leads to simple and accurate monolithic realizations, requiring no bulky and drift-prone analog multipliers.

Figure 11.2(a) shows a simple yet completely general conceptual block diagram of a first-order 1-D discrete map, comprising a linear block and a parameterized nonlinear transfer function, connected in a feedback loop [36]. The linear

[2] The ergodicity of the map guarantees that the invariant density function is unique regardless of the initial value of the trajectory (except a set of measure zero) [33].

Map	Definition	Chaotic regime
Logistic	$x(k+1) = f_{log}(x(k); A, B) = B(A^2 - x(k)^2) - A$	$x(k) \in [-A, A]$ $3/2 < AB < 2$
Quadratic	$x(k+1) = f_{sqr}(x(k); A, B) = B - Ax(k)^2$	$x(k) \in [-2/A, 2/A]$ $3/4 < AB < 2$
Exponent	$x(k+1) = f_{exp}(x(k); A, B) = x(k)\exp(B(A - x(k)))$	$x(k) \in [0, \frac{\exp(AB-1)}{B}]$ $AB > 2$
Sine Circle	$\theta(k+1) = f_{cir}(\theta(k); A, B) = \theta(k) + A - B\sin(\theta(k))$	$\theta(k) \in [0, 2\pi]$ $0 < A < 2\pi;\ B > 0$
Bernoulli	$x(k+1) = f_{ber}(x(k); A, B) = \begin{cases} Bx(k) + A & x(k) < 0 \\ Bx(k) - A & x(k) > 0 \end{cases}$	$x(k) \in [-A, A]$ $0 < B < 2$
Tent	$x(k+1) = f_{tnt}(x(k); A, B) = A - B\lvert x(k)\rvert$	$x(k) \in [A(1-B), A]$ $0 < B < 2$
Congruent	$x(k+1) = f_{mod}(x(k); A, B) = \begin{cases} Bx(k) - C & x(k) > A \\ Bx(k) & \lvert x(k)\rvert \leq A \\ Bx(k) + C & x(k) < -A \end{cases}$	$x(k) \in [-C, C]$ $1 < B < 2$ $C = 2A$
Hopping	$x(k+1) = f_{hop}(x(k); A, B, D) = \begin{cases} D(x(k) - A) + C & x(k) > A \\ Bx(k) & \lvert x(k)\rvert \leq A \\ D(x(k) + A) - C & x(k) < -A \end{cases}$	$x(k) \in [-C, C]$ $B, -D > 1\ \ C = BA$ $f_{hop}(-C) \in (0, C)$ $f_{hop}(C) \in (-C, 0)$
Henon	$x(k+1) = C + y(k) - Ax(k)^2$ $y(k+1) = Bx(k)$	$x(k), \frac{y(k)}{B} \in [-\frac{2}{A}, \frac{2}{A}]$ $A > \frac{(1-B)^2}{4C}, \lvert B\rvert < 1$ $3/4 < AC < 2$
Standard	$\rho(k+1) = \theta(k) + \rho(k)$ $\theta(k+1) = \theta(k) - B\sin(\rho(k))$	$\rho(k), \theta(k) \in [0, 2\pi]$ $B > 0$
Lozi	$x(k+1) = C + y(k) - A\lvert x(k)\rvert$ $y(k+1) = Bx(k)$	$0 < B < 1$ $B + 1 < A < 2 - B/2$ $C > 1/2$
Arnold	$x(k+1) = f_{mod}(x(k) + y(k); A, 1)$ $y(k+1) = f_{mod}(x(k) + By(k); A, 1)$	$x(k), y(k) \in [-C, C]$ $1 < B < 2$ $C = 2A$
Baker	$x(k+1) = f_{mod}(x(k); A, B)$ $y(k+1) = \begin{cases} By(k) + A & x(k) < 0 \\ By(k) - A & x(k) > 0 \end{cases}$	$x(k) \in [-C, C]$ $y(k) \in [-A, A]$ $1 < B < 2$ $C = 2A$

Table 11.1: Short catalog of chaotic discrete maps.

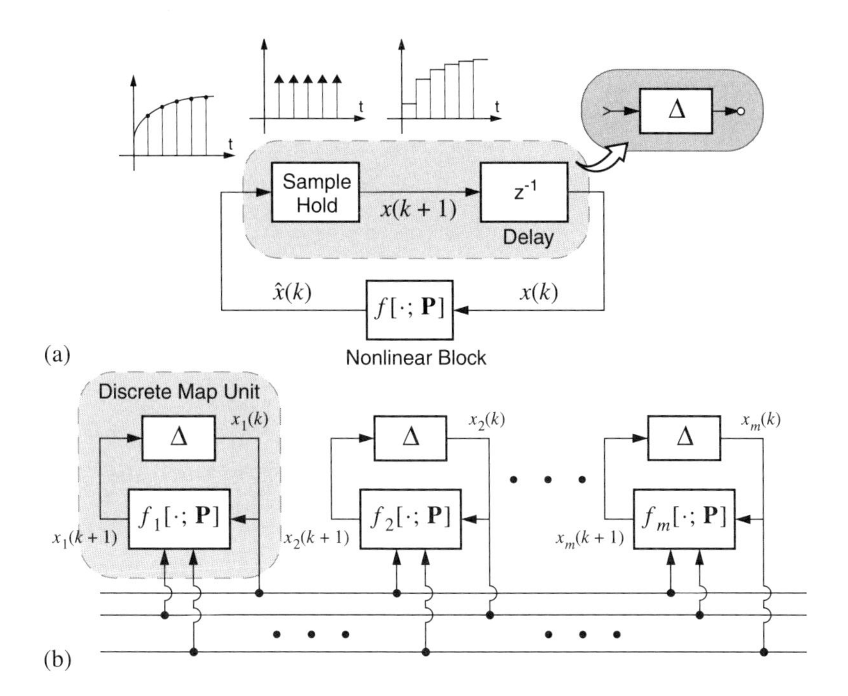

Figure 11.2: Conceptual diagrams for (a) 1-D and (b) m-D first-order discrete maps.

block performs sample-and-hold (S/H) and delay operations, and its implementation basically relies on accuracy and speed considerations, regardless of the discrete map model to be synthesized. Usually, such implementation consists of a single electronic device which, hereafter, will be represented by the symbol in the inset of Figure 11.2(a) and denoted as a delay element. Only the nonlinear feedback block, implementing $f(\cdot;\boldsymbol{P})$ in (11.4), needs to be tailored to the particular first-order 1-D discrete map at hand. In Figure 11.2(a), the value of the signal x at the k-th time interval is transformed into the signal $\hat{x}$. The output of the nonlinear block is first sampled-and-held and delayed, and the resulting signal is fed back to the input of the nonlinearity. The clock signal fixing the sampling period determines the iteration pace for the feedback loop. The dynamics of this conceptual block diagram is therefore exactly described by (11.4).

Extension to the first-order m-dimensional case can be achieved by simply combining 1-D discrete map units as shown in Figure 11.2(b). For each state variable x_j, $j = 1, 2, \ldots, m$, there is associated one discrete map unit which consists of a delay element and a nonlinear block, implementing the multi-valued function $f_j[\cdot;\boldsymbol{P}]$. This obtains the recursive equation

$$\boldsymbol{x}(k+1) = \boldsymbol{F}[\boldsymbol{x}(k);\boldsymbol{P}] \tag{11.5}$$

where $\boldsymbol{F}(\cdot)^T = [f_1(\cdot), f_2(\cdot), \ldots, f_m(\cdot)]$ is a m-dimensional nonlinear vector field.

Figure 11.3 shows the generic block diagram of a n-th order 1-D discrete map comprising a n-variate function block and a delay line, formed by the cascade

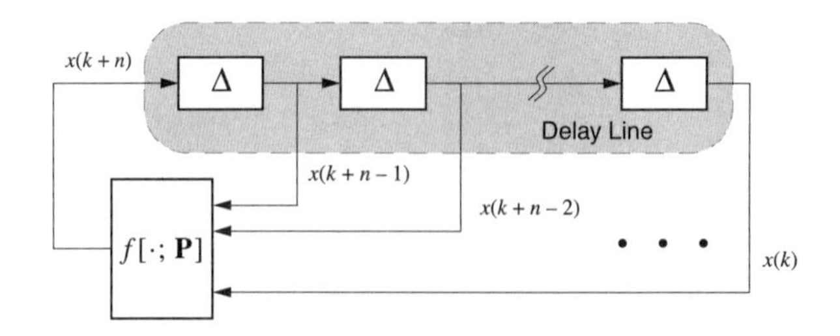

Figure 11.3: Conceptual diagram of generic a n-th order 1-D discrete map.

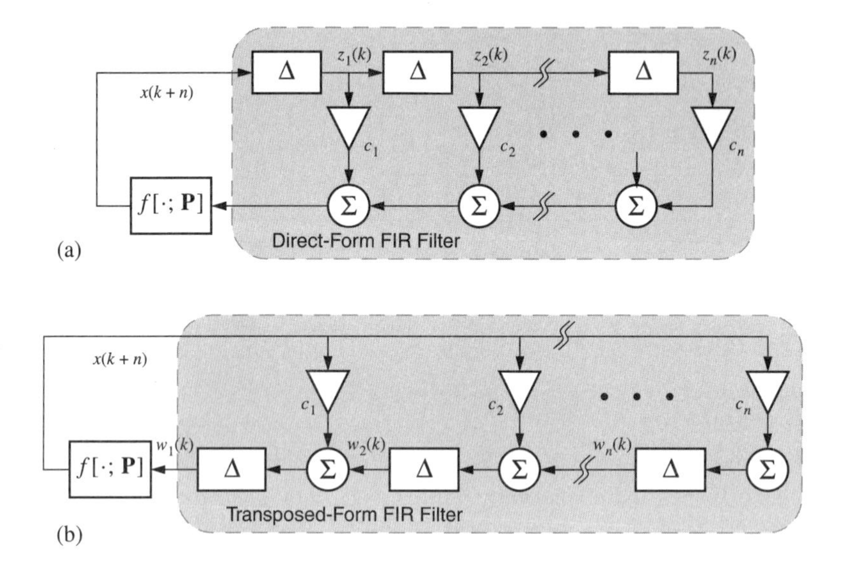

Figure 11.4: Conceptual diagrams for synthesizing n-th order 1-D discrete maps using (a) direct form and (b) transposed FIR structures.

connection of n delay elements. Extension to the m-dimensional case is obtained by combining m of the above stages in a similar way as done in Figure 11.2(b), therefore synthesizing the general FDE system of (11.1).

From this generic framework, some special cases deserve special attention because of their relevant application field and/or particular implementation style. Among them, Figure 11.4 shows different configurations of the block diagram in Figure 11.3 in which dynamics is determined by the linearly averaged history of the state variable. Such structures have been found well suited for the synthesis of RSG's with programmable statistical properties [37].

The conceptual diagrams of Figures 11.4(a, b) comprise a single nonlinear block, which implements the function $f[\cdot;\boldsymbol{P}]$, and a linear finite impulse response (FIR) filter, formed by a tap weighted delay line and adders. They differ on the chosen architecture for the FIR filter. Figure 11.4(a) uses a direct-form FIR

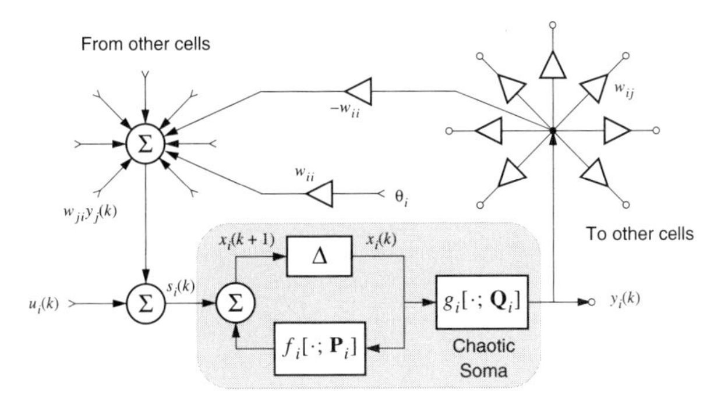

Figure 11.5: Analog computer concept of a chaotic neuron unit.

structure and obtains,

$$x(k+n) = f\left[\left(\sum_{i=1}^{n} c_i z_i(k)\right); \boldsymbol{P}\right] \tag{11.6}$$

where $z_i(k) = x(k+n-i)$, for $i = 1, 2, \ldots, n$, and c_i are real numbers. On the other hand, Figure 11.4(b) uses a transposed-form FIR structure which gives

$$x(k+n) = f[w_1(k); \boldsymbol{P}] \tag{11.7}$$

where $w_1(k)$ is defined recursively by the equation $w_i(k) = c_i x(k+n-1) + w_{i+1}(k)$, with $w_{n+1}(k) = 0$. Both (11.6) and (11.7) result in the same n-th order discrete map model with linear average; the choice between one structure or the other being mainly dictated by electronic implementation issues. Generalizations of these basic architectures with additional programming capabilities can be found in [37].

Another special case within the framework of (11.1) is related to the paradigm of chaotic neural networks [16, 17]. According to Figure 11.5, chaotic neural networks can be described in terms of scalar variables as

$$\begin{aligned} x_i(k+1) &= f_i[x_i(k); \boldsymbol{P}_i] + \left(\sum_{j=1, j\neq i}^{m} w_{ji} y_j(k)\right) + u_i(k) - w_{ii}(y_i(k) - \theta_i) \\ y_i(k) &= g_i[x_i(k); \boldsymbol{Q}_i] \end{aligned} \tag{11.8}$$

where $x_i(k)$, $u_i(k)$, and $y_i(k)$ are, respectively, the internal state, the external input, and the activation function output of the i-th chaotic neuron at the k-th sampling instant; w_{ji} represents the connection weight from neuron j to neuron i; θ_i is the self-recurrent bias of neuron i; and $f_i[\cdot; \boldsymbol{P}_i]$, $g_i[\cdot; \boldsymbol{Q}_i]$ are functions used, respectively, to define the next internal state and the output of neuron i. Usually, $g_i[\cdot; \boldsymbol{Q}_i]$ presents a sigmoid characteristic, whereas $f_i[\cdot; \boldsymbol{P}_i]$ is often replaced by a linear amplifier.

The relevance of the above architecture arises from its suitability for dynamical associative pattern classification tasks, as demonstrated in [16]. Further, if the self-feedback connection weights w_{ii} of the neurons are recurrently damped according to $w_{ii}(k+1) = (1-\beta)w_{ii}(k)$ (with $0 < \beta < 1$), it can be shown that the network, after a chaotic transient, eventually converges to a stable equilibrium point which is (almost always) the global minimum of the computational energy function defined by the network [17]. This global searching ability makes chaotic neural networks ideally suited for solving complex combinatorial optimization problems.

The block diagrams of Figures 11.4 and 11.5 show that the linear part of some discrete maps require the extra use of linear amplifiers and adders, together with the necessary delay elements for inducing the circuit dynamics. In fact, these three linear operators are the only fundamental for the implementation of the linear block of integrated discrete maps; any other operator, such as, for instance, the FIR filters of Figure 11.4, can be regarded as derived from the above. The following section presents a building block-based approach for the synthesis of both fundamental and derived linear operators using sampled-data techniques.

11.3 Linear Operators for Discrete Maps

Before giving detailed circuit realizations of the different linear operators involved in discrete maps - delay elements, adders and linear amplifiers - it should be beneficial to provide some general considerations which can help readers to understand their synthesis; this should be particularly useful to those readers unfamiliar with IC techniques.

The implementation of delay elements for discrete maps will invariably rely on analog sampled-data techniques based on capacitors, for storing and retrieving information in the form of voltages: equivalently charge packets; switches, for charging and/or discharging capacitors in response to a control signal; and active devices (basically opamps[3]) for defining the conditions of charge transfer - equivalently, current flow - among capacitors at each state of the control signals of the switches. Still, two major sampled-data design approaches can be distinguished, their merits and drawbacks mainly dictated by the features of the capacitors available in the chosen fabrication technology.

A first sampled-data approach is the Switched-Capacitor (SC) technique which requires opamps, as active devices, and linear capacitors, as holding elements [38–40]. Such capacitors are available in analog and mixed-signal oriented technologies, whose fabrication process incorporates specific steps for the imple-

[3]This term is used here to refer to a differential amplifier with very large voltage gain and very large input resistance. Such amplifiers are basically intended to be used in feedback configurations, and thereby to create virtual grounds. However, depending on their output impedance, these amplifiers can be classified in two different categories: on the one hand, operational voltage amplifiers (the typical opamps) have very low output resistance and thereby are capable of providing large voltage gain even when loaded with low resistive loads; on the other hand, operational transconductance amplifiers (OTAs) have large output resistance and are only capable of obtaining large voltage gain when driving either very large resistive loads or purely capacitive loads. In what follows the term opamp basically relates to OTAs.

mentation of parallel-plates capacitor structures - usually realized with polysilicon layers - which achieves high linearity, as well as reduced voltage and temperature dependence, and good matching properties [41–44]. The exploitation of these technological features and of SC techniques has been, in recent years, the basic tool for the implementation of state-of-the-art functional blocks in filtering [45] and data conversion [46] applications, among others [47]. Unfortunately, in pure digital CMOS technologies, the most standard and cheapest for VLSI design, these high-quality structures no longer exist and capacitors are commonly implemented by exploiting the capacitive effect between the gate and channel of MOS transistors operated in strong inversion or accumulation [48]. MOS-based capacitors usually exhibit larger capacitance per unit area and better matching than parallel-plate structures. However, they suffer from significant non-linearities and parasitic capacitances, and must be conveniently biased to guarantee a low resistivity conducting layer under the gate. As a consequence, SC circuits built on digital technologies have inevitably poorer performance than those implemented on analog-oriented processes, in aspects such as harmonic distortion or spurious-free dynamic range.

An alternative sampled-data approach which avoids the need for highly-linear capacitors is the Switched-Current (SI) technique [49]. In this technique, capacitors are simply formed by the input parasitics of transconductors,[4] thus making the approach especially appealing for standard digital technologies. SI circuits are a priori conceptually simpler than SC counterparts. Unfortunately, these notable simplifications are at the expense of performance degradations. Thus, in practice, and whenever high-performance is an issue, SI circuits can become much more complex than SC counterparts. However, for applications requiring moderate accuracy, the complexity - and thus the area consumption - of SI circuits is generally lower than that of SC circuits performing the same function, which makes SI technique a fall-back alternative when low-cost fabrication is mandatory. Nevertheless, since SC and SI techniques have at this time well-defined design scenarios, both alternatives will be considered next for the implementation of discrete-maps.

Regarding the aggregation and scaling operations, they constitute static operations which ideally act instantaneously on the driving signals. Accordingly, they can be implemented without resorting to any external timing control, as opposed to the delay operation. Nevertheless, we can take advantage of the holding operation exercised on the discrete-map variables and realize the required adders and linear amplifiers by also using SC or SI techniques that correspond. As will be shown, this leads to simple realizations, with a low number of circuit components, which also avoids the use of impedance matching techniques.

Let us first consider the implementation of the aggregation operator. Sum-

[4]The simplest transconductor is just one transistor acting as a voltage-controlled current source. However, a single transistor has terminal impedances and linearity which are far from ideal. Thereby, practical transconductors are formed by interconnecting several transistors to improve terminal impedances and/or linearity.

ming circuits are always based on the Kirchoff's current law (KCL) and generally implemented by first transforming the additive signals into currents or incremental charges, routing such intermediate currents to a common node with an output branch, and transforming the KCL-summed current flowing through such output branch back to the signal domain of the original data. This clearly gives an edge to SI summing circuits (signals are directly represented by currents) over SC ones (signals are in the form of voltages) because there is no need for data transformation in the former, while the second requires additional voltage-to-charge and charge-to-voltage converters. Regardless of the implementation technique, if the number of components is large, the output impedance of the driving nodes not large enough, and/or the input impedance of the load not small enough, signal aggregation will encompass significant loading errors due to variations of the voltage at the summing node. This is overcome by clamping the voltage of this node using a virtual ground, which in practical circuits is realized by using either the input port of an opamp, or the input terminals of a current conveyor.

A final design consideration, this time referred to as the scaling operation, is that its implementation must preferably rely on ratios of like quantities associated with similar integrated components. In other words, on-chip design of scaling operators must be based, not on absolute values of quantities, but only on relative values of certain electrical properties such as, for example, capacitances, resistances, or transconductances. This is because absolute values of on-chip quantities suffer from severe uncontrollable deviations (which may amount up to $\sim 30\%$ of the nominal value) as a result of statistical variances in technological parameters, temperature variations, and aging [41]. Conversely, relative ratio accuracies can be high if proper layout techniques are used [41–44, 50–54]. For example, for a modern technology the capacitance mismatch can be as low as 0.013% for capacitors of 2pF, which allows us to obtain resolutions of about 12 bits equivalent without the need of calibration. SC techniques take advantage of these excellent matching properties among capacitors to implement highly accurate voltage scaling operators. Similarly, SI techniques exploit the relative ratio accuracy among transconductors for current scaling operations, based on the current mirror concept [55].

The following provides a catalog of circuit realizations for the linear operators required by discrete maps. They have been grouped according to the sampled-data technique used, SC or SI. At the same time, their respective performances and limitations are highlighted, so that they can be chosen by a designer in accordance with the global specifications demanded for the discrete map implementation.

11.3.1 Switched-Capacitor Linear Operators

Consider first the delay operator. It is based on the sample-and-hold (S/H) function. Figure 11.6(a) shows a simple SC S/H circuit consisting of a capacitor C_S, an opamp, and three analog switches controlled by a clock with two non-overlapping phases [56]. Switches labelled ϕ_1 (respectively, ϕ_2) turn ON in

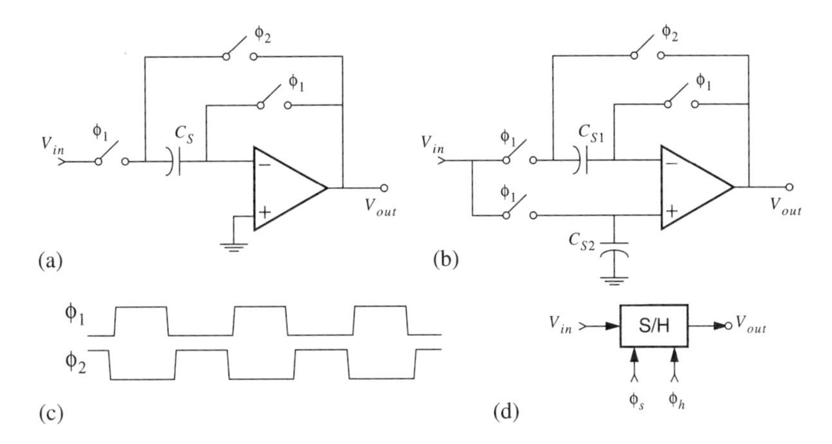

Figure 11.6: Offset and parasitic-free sample-and-hold SC circuits with (a) return-to-zero and (b) non-return-to-zero output during the sampling phase; (c) clock timing diagram; (d) symbol for sample-and-hold circuits; ϕ_s and ϕ_h represent, respectively, the sampling and holding phases.

synchronization with the first (respectively, second) clock phase.[5] The circuit operates as follows. In the acquisition phase, switches labelled ϕ_1 are ON (respectively, the switch driven by ϕ_2 is OFF) and the opamp is configured as a unity-gain amplifier. Assuming that the opamp has infinite gain and bandwidth and null offset, this creates a virtual ground at the inverting terminal of the opamp and then capacitor C_S tracks the input voltage V_{in}. In the holding mode, phase ϕ_2 is ON, and the left plate of the sampling capacitor C_S is connected to the opamp output. Since the right plate of C_S remains connected to the inverting input of the opamp, the output voltage during the holding mode is exactly equal to the previously sampled input. Hence, assuming an ideal opamp, the circuit operation can be described by the following recursive equations,

$$\begin{aligned} V_{out}(k+1) &= V_{in}(k+1/2) \\ V_{out}(k+1/2) &= 0 \end{aligned} \tag{11.9}$$

thus providing half-cycle delay of the input voltage during the holding phase and null output during the acquisition - reset - phase. In high clock rates applications, periodic resetting of the opamp output imposes demanding slew-rate and settling-time specs. To relax them, a slight modification of Figure 3.2.6(a), illustrated in Figure 11.6(b), can be used [57]. In this circuit, during the acquisition phase, the opamp is also configured as a unity-gain buffer, but its non-inverting terminal is connected to the input signal instead of ground. This makes $V_{out}(k+1/2) = V_{in}(k+1/2)$ and, hence, the output voltage ideally does not

[5]By convention, any arbitrary signal $s(\cdot)$, when observed during the first (respectively, second) clock phase will be denoted as $s(k+1/2)$, (respectively, $s(k)$), indicating that the observation time interval is given by $((k+1/2)T, (k+1)T)$ (respectively, $(kT, (k+1/2)T)$), where T is the clock signal period.

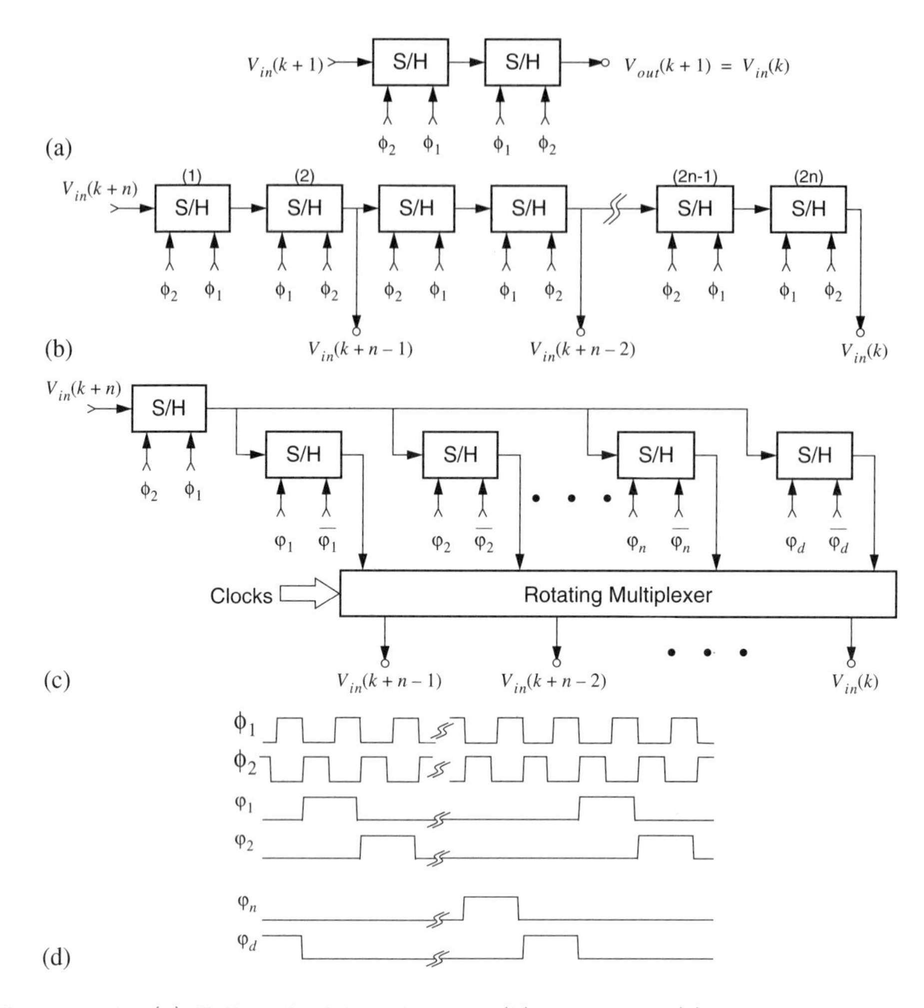

Figure 11.7: (a) Full-cycle delay element; (b) serial and (c) parallel approaches for analog delay-line with taps; (d) timing diagram for (c).

change during autozeroing. Other circuits that avoid the output resetting problem, and therefore operate much faster than Figure 11.6(a), are found in [58–60].

Full-cycle delay, as required by the circuits of Figures 11.2-11.5, is realized by cascading two half delay blocks with interchanged sampling/holding clock phases - shown at the conceptual level in Figure 11.7(a) where the S/H block is defined in Figure 11.6(d). Connecting $2n$ S/H circuits in a cascade with alternating clocking phases we realize an analog delay line with n full-delay tapped outputs, which can be used for the implementation of n-th order discrete maps (see Figure 11.7(b)). It suffers from two significant shortcomings. First, it requires two S/H circuits per tap, thus leading to large area and power consumption. Second, errors that arise in real S/H circuits (finite amplifier gain, noise, offset, among others) are propagated and accumulated from stage to stage, thus limiting the practical length of the delay chain. A more convenient solution relies on the

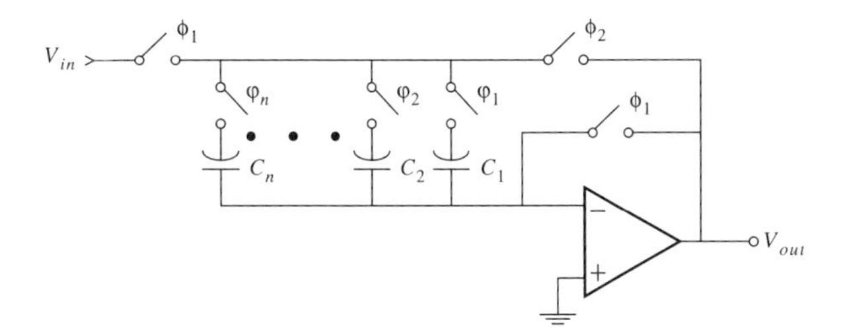

Figure 11.8: SC scanning delay-line.

use of parallel structures, where errors are only added once at the sampling instant [61–63]. Figure 11.7(c) shows the basic architecture. An array of parallel S/H circuits is used to sample the input voltage. The number of S/H's must be at least $n+1$, so that, at any time, there are available (in hold mode) n delayed outputs. The extra S/H unit is represented in Figure 11.7(c) by the circuit driven by the clock signal φ_d and its complement. The S/H control clocks cycle around the array in a circular manner as shown in Figure 11.7(d). The rotating multiplexer, which is also controlled by the S/H clocks, serves to appropriately arrange the S/H outputs so that they are ordered according to the number of clock cycles by which the input sample has been delayed.

The circuits of Figures 11.7(b, c) provide in parallel the complete and most recent history of the input voltage. If only the n-th previous sample must be considered for processing, a simpler circuit can be used. An example is shown in Figure 11.8 [64, 65]. It can be considered as an extension of the S/H circuit of Figure 11.6(a) with n sampling paths instead of only one. Selection of a specific path for sequentially read (hold mode) and write (acquisition mode) input voltage samples can be achieved with a timing control similar to that of Figure 11.7(d). Other solutions have been published elsewhere [66, 67].

In addition to delay, the implementation of the blocks diagrams in Figures 11.2-11.5 requires the operations of summation and signal weighting. As for the latter, consider the SC circuit of Figure 11.9 without switch S_R and assume the opamp is ideal. During phase ϕ_1, capacitor C_S charges to $V_{in}(k+1/2)$ and then, as a result of the virtual short circuit at the input terminals of the opamp, the stored charge is fully delivered to capacitor C_I during the next phase ϕ_2. By the charge conservation principle, it follows that the output voltage at the end of phase ϕ_2 satisfies

$$C_i[V_{out}(k+1) - V_{out}(k)] = C_S V_{in}(k+1/2) \tag{11.10}$$

from which

$$V_{out}(k+1) = V_{out}(k) + \frac{C_S}{C_I} V_{in}(k+1/2) \tag{11.11}$$

and during phase ϕ_1, we have $V_{out}(k+1/2) = V_{out}(k)$.

Note that one of the contributions to the output in (11.11) is a scaled version of the input voltage sampled at the previous phase. Its associated scaling factor

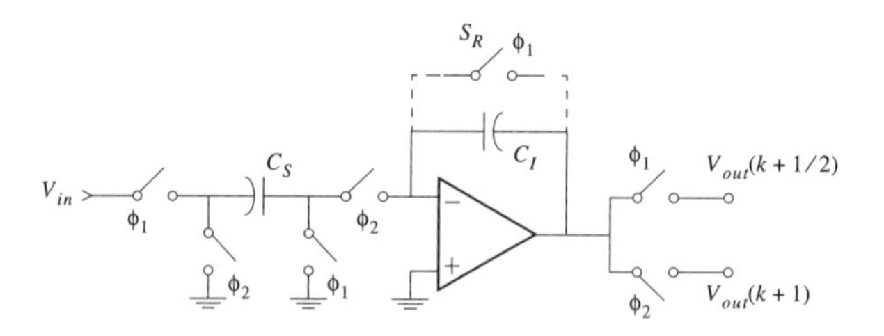

Figure 11.9: Stray-insensitive non-inverting SC integrator/amplifier circuit. The dotted lines indicate the realization of the amplifier.

is determined by the ratio of two capacitances and, hence, highly accurate. This contribution accumulates to a previous value of the output voltage stored in C_I - the circuit implements a discrete-time integration [38,39]. If only the scaled input voltage contribution is of interest, a simple solution to cancel the remaining term consists of discharging C_I during the acquisition phase. This can be done by short-circuiting C_I by means of an additional switch (S_R in Figure 11.9), that ideally gives at the end of ϕ_2,

$$V_{out}(k+1) = \frac{C_S}{C_I} V_{in}(k+1/2) \tag{11.12}$$

so that the circuit performs a non-inverting discrete-time amplification of the input signal with a gain factor C_S/C_I. During ϕ_1, the unity-gain feedback configuration of the opamp ideally gives $V_{out}(k+1/2) = 0$.

Scaling operation can be exploited to add flexibility to the implementation of first-order discrete maps, for example, by relaxing the specifications of the nonlinear block. To illustrate this point, Figure 11.10 shows different topologies for the realization of 1-D first-order discrete maps, each associated to a particular configuration of the linear operator. Each of these structures is composed of a parameterized nonlinear block $h[\cdot;\boldsymbol{P}']$; an SC integrator (Figures 11.10(a-d)) or a gain stage (Figures 11.10(e, f)); and an elementary delay unit, labeled $z^{-j/2}$, $j = 0, 1, 2$, which is inserted after the nonlinear block to guarantee a full-cycle delay in the feedback loop (depending on whether j is 1, 2, or 3, the block $z^{-j/2}$ introduces no-, half-, or full-cycle delay). We will assume that the nonlinear block $h[\cdot;\boldsymbol{P}']$ in Figure 11.10 provides valid output immediately after its driving switch (connected to the opamp output) turns ON. When this switch is in the OFF state, the output of the nonlinear block is connected to a convenient low impedance node.

In the case of the integrator-based circuits of Figures 11.11(a) and (c) analysis assuming ideal opamps obtains

$$x(k+1) = x(k) + \frac{C_S}{C_I} h[x(k);\boldsymbol{P}'] \tag{11.13}$$

which is equivalent to (11.4) upon choosing

$$f[x(k);\boldsymbol{P}] \equiv \frac{C_S}{C_I} h[x(k);\boldsymbol{P}'] - x(k) \tag{11.14}$$

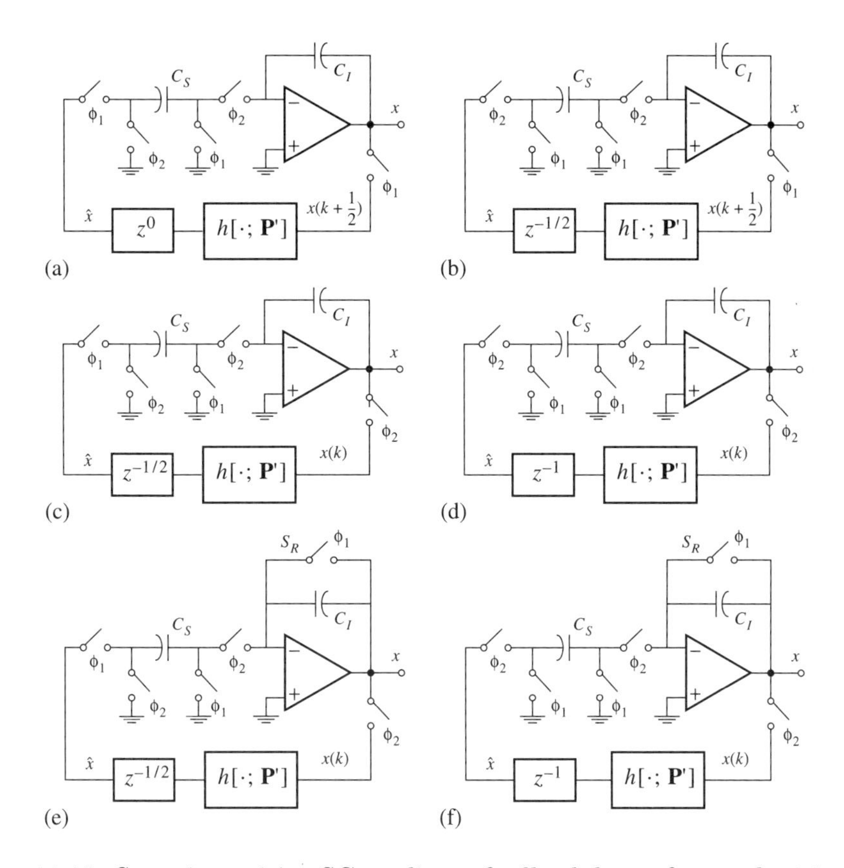

Figure 11.10: Stray insensitive SC nonlinear feedback loops for synthesizing 1-D first-order discrete maps, using: (a),(c) non-inverting integrators; (b),(d) inverting integrators; (e) a non-inverting amplifier; (f) an inverting amplifier.

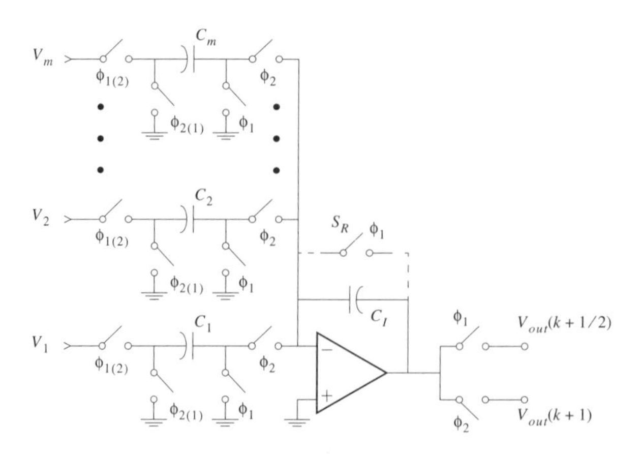

Figure 11.11: Stray-insensitive SC summing integrator/amplifier. The dotted lines indicate the realization of the amplifier configuration. Alternative phasing of the switches is shown in parentheses.

On the other hand, for Figsures 11.11(b) and (d),

$$x(k+1) = x(k) - \frac{C_S}{C_I} h[x(k); \boldsymbol{P}'] \tag{11.15}$$

which is equivalent to (11.4) by setting

$$f[x(k); \boldsymbol{P}] \equiv -\frac{C_S}{C_I} h[x(k); \boldsymbol{P}'] - x(k) \tag{11.16}$$

From (11.14) and (11.16) it is seen that the use of SC integrators provides an additional degree of freedom for the implementation of function $f[x(k); \boldsymbol{P}]$ as a consequence of the designer controllable factor $\pm C_S/C_I$, which affects the nonlinear block $h[x(k); \boldsymbol{P}']$. The same result can be also accomplished by using inverting or non-inverting SC gain stages, instead of integration, as illustrated in Figures 11.10(e, f). For these circuits it can easily shown that

$$x(k+1) = \pm\frac{C_S}{C_I} h[x(k); \boldsymbol{P}'] \tag{11.17}$$

where the positive (respectively, negative) sign applies for the circuit of Figure 11.10(e) (respectively, Figure 11.10(f)). Note that (11.17) is equivalent to (11.4) upon choosing

$$f[x(k); \boldsymbol{P}] \equiv \pm\frac{C_S}{C_I} h[x(k); \boldsymbol{P}'] \tag{11.18}$$

i.e., $f[x(k); \boldsymbol{P}]$ is obtained as a scaled version of the nonlinearity of the loop.

The circuit in Figure 11.9 can be easily transformed into a SC summing integrator/amplifier circuit. This can be accomplished by adding multiple SC sampling branches to the block, as shown in Figure 11.11. Assuming ideal components and clock phases outside parenthesis, the output voltage of the SC summing integrator (ignoring switch S_R) is given by:

$$V_{out}(k+1) = V_{out}(k) + \sum_{j=1}^{m} \frac{C_j}{C_I} V_j(k+1/2) \tag{11.19}$$

and for the phases inside parentheses,

$$V_{out}(k+1) = V_{out}(k) - \sum_{j=1}^{m} \frac{C_j}{C_I} V_j(k+1) \tag{11.20}$$

in both cases $V_{out}(k+1/2) = V_{out}(k)$.On the other hand, the output voltage of the SC summing amplifier (including switch S_R in Figure 11.11) is also given by (11.19) (phases outside parentheses) or (11.20) (phases inside parentheses) but without the term $V_{out}(k)$ at the right-hand side of the equations. Because of the integrating capacitor resetting during phase ϕ_1, we have $V_{out}(k+1/2) = 0$.

All SC circuits above are insensitive to stray capacitances to ground, unavoidable in monolithic realizations [38, 68]. This is because every node in the circuits verifies one of the following properties:

- It is permanently connected to low impedance nodes or to virtual ground.
- It is switched between two low impedance nodes.
- It is switched between ground and virtual ground, so that the voltage across the stray capacitance is always constant.

In this way, assuming ideal opamps, parasitic capacitances do not affect charge redistribution and the functionality of the block is not degraded, i.e., the discrete-time difference equation of circuit operation remains unaltered.

Stray capacitances are examples of parasitic effects in SC circuits which, as many other nonidealities, must be minimized or cancelled by appropriate design techniques for avoiding severe degradations of the circuit performance, or even a complete malfunction of the chip. Such nonideal effects are consequences of second-order phenomena exhibited by real SC circuit components (opamps, capacitors, and switches). Table 11.2 resumes the different non-idealities associated with such SC circuit components, as well as their effects on the linear part performance of discrete maps. Addressing all these second-order effects is beyond the scope of this chapter. Readers are referred to the bibliography attached to each row in Table 11.2. Such references provide in-depth analysis of corresponding effects and, whenever available, propose design strategies for their elimination, compensation, or reduction. For a general review, readers are referred to [38–40, 81].

For illustration purposes we find it convenient to comment on the effect of the opamp finite amplifier gain and offset and, more specifically, on improved SC

SC component	Second-order phenomena	Effects
OTA	Offset	Characteristic shifting [69, 74–78]
	Finite and nonlinear DC gain	Magnitude and phase errors [70–73] Harmonic distortion [79–81]
	Finite bandwidth	Incomplete settling/charging errors [70–73] Noise aliasing [82–84]
	Slew rate	Settling errors and harmonic distortion
	Limited output swing	Parasitic fixedpoints Harmonic distortion
	thermal noise	Folded-back noise [82–84]
Switches	Finite ON resistange	Incomplete settling/charging errors Thermal noise
	Charge Injection Clock feedthrough	Offset Harmonic distortion [85–94]
Capacitors	Mismatch	Magnitude and phase errors
	Nonlinearity	Harmonic distortion

Table 11.2: Second-order phenomena and their effects on SC circuits.

circuit structures which can be used to compensate for their influence. Let us use the stray insensitive non-inverting integrator of Figure 11.9 as an illustration vehicle. It was previously shown that the output of this block is ideally described in time-domain by (11.11). However, if we take into account the finite dc gain $A_0 = 1/\mu$, and offset voltage V_{OS}, of real opamps, such an equation should be modified to [69–73]

$$V_{out}(k+1) = \alpha p V_{in}(k+1/2) + \beta V_{out}(k) + \gamma V_{OS} \tag{11.21}$$

where $p = C_S/C_I$ is the ideal integrator gain (see (11.11)); and α, β, and γ are parameters which account for deviations. They are given by:

$$\alpha = \frac{1}{1+(1+p)\mu} \tag{11.22}$$

$$\beta = \frac{1+\mu}{1+(1+p)\mu} \tag{11.23}$$

$$\gamma = \frac{p}{1+(1+p)\mu} \tag{11.24}$$

As can be seen, the finite dc gain of the opamp introduces losses in the integrator gain ($\alpha < 1$) and modifies its pole location ($\beta < 1$). Both effects can be attenuated by increasing the opamp dc gain (μ tends to zero); however, this makes the offset effect more noticeable (γ tends to p). All the SC circuits presented so far, except Figure 11.6(b), also suffer from this drawback. To cope

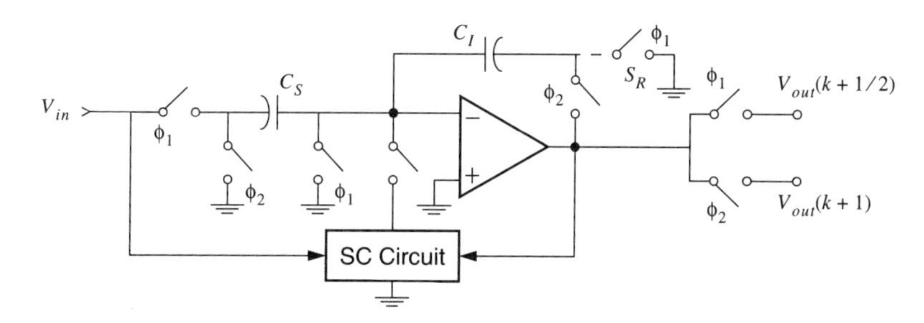

Figure 11.12: Generic stray-insensitive non-inverting offset-compensated SC integrator/amplifier. The dotted lines indicate the realization of the amplifier configuration.

with it, different offset-compensated structures have been proposed, the general scheme of which is shown in Figure 11.12. The basic idea is that capacitors C_S and C_I are not switched away from V_{OS} so that the offset term is cancelled (at least during the hold phase). This is the so-called correlated double-sampling (CDS) technique [40]. The block labelled SC circuit in Figure 11.12 is used to provide a feedback path between the input and output of the integrator/amplifier. Otherwise, the opamp will be in open loop during the acquisition interval, to define a low impedance path to ground for the inverting input of the opamp (otherwise, leakage currents will charge the parasitic capacitance at such node and force the opamp into saturation) and to keep the output voltage change as small as possible during the transition from the sampling to the holding mode of operation. Table 11.3 shows different offset- and dc-compensated SC integrators/amplifiers. For each block, the name of the first author is assigned to the circuit. In all cases, performance characteristics are evaluated for the integrator configurations (excluding the reset switch S_R) during phase ϕ_2. The first two rows in Table 11.3 correspond to non-inverting structures, while the remaining three rows are inverting ones. As can be seen, parameter γ is always proportional to μ, indicating that the offset effect can be reduced by increasing the dc gain of the OTA. The SC circuit proposed in [74] is the simplest scheme in Table 11.3; however, it suffers from the inconvenience that the output voltage must switch (slew) back and forth between the desired output and V_{OS} every half cycle. This is overcome in the other structures where transitions in the output voltage can be made approximately equal to V_{OS}.

11.3.2 Switched-Current Linear Operators

An alternative to the implementation of discrete maps is using switched-current (SI) techniques [49, 95] which require only transconductors and switches. Prior to addressing the implementation of linear SI operators it is useful to review the basic characteristics of real transconductors.

Ideally, transconductors behave as voltage-controlled current sources (VCCS's) however, in practice, their static transfer characteristics are only approximately

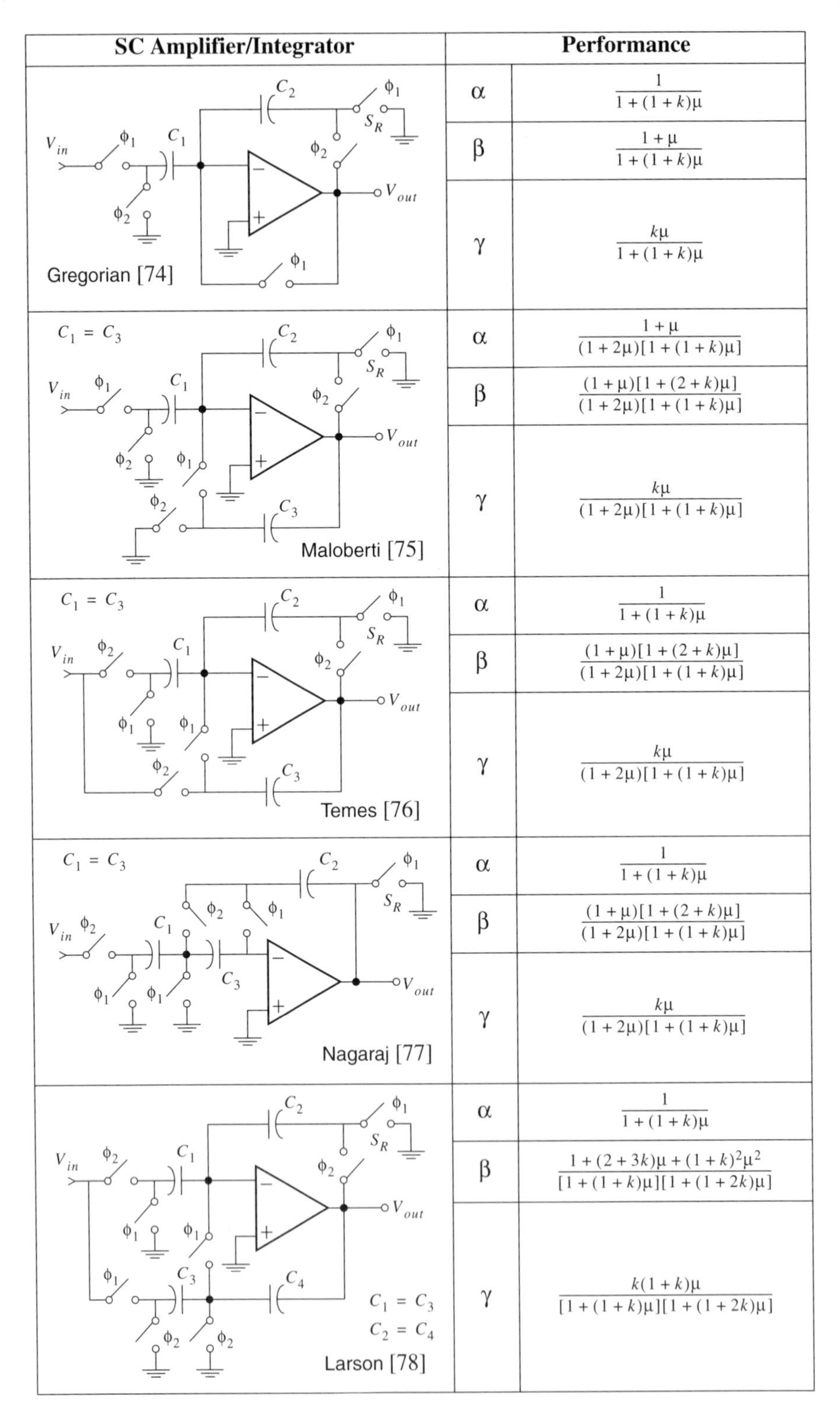

SC Amplifier/Integrator	Performance	
Gregorian [74]	α	$\frac{1}{1+(1+k)\mu}$
	β	$\frac{1+\mu}{1+(1+k)\mu}$
	γ	$\frac{k\mu}{1+(1+k)\mu}$
Maloberti [75] ($C_1 = C_3$)	α	$\frac{1+\mu}{(1+2\mu)[1+(1+k)\mu]}$
	β	$\frac{(1+\mu)[1+(2+k)\mu]}{(1+2\mu)[1+(1+k)\mu]}$
	γ	$\frac{k\mu}{(1+2\mu)[1+(1+k)\mu]}$
Temes [76] ($C_1 = C_3$)	α	$\frac{1}{1+(1+k)\mu}$
	β	$\frac{(1+\mu)[1+(2+k)\mu]}{(1+2\mu)[1+(1+k)\mu]}$
	γ	$\frac{k\mu}{(1+2\mu)[1+(1+k)\mu]}$
Nagaraj [77] ($C_1 = C_3$)	α	$\frac{1}{1+(1+k)\mu}$
	β	$\frac{(1+\mu)[1+(2+k)\mu]}{(1+2\mu)[1+(1+k)\mu]}$
	γ	$\frac{k\mu}{(1+2\mu)[1+(1+k)\mu]}$
Larson [78] ($C_1 = C_3$, $C_2 = C_4$)	α	$\frac{1}{1+(1+k)\mu}$
	β	$\frac{1+(2+3k)\mu+(1+k)^2\mu^2}{[1+(1+k)\mu][1+(1+2k)\mu]}$
	γ	$\frac{k(1+k)\mu}{[1+(1+k)\mu][1+(1+2k)\mu]}$

Table 11.3: Offset- and dc gain-compensated SC amplifiers/integrators.

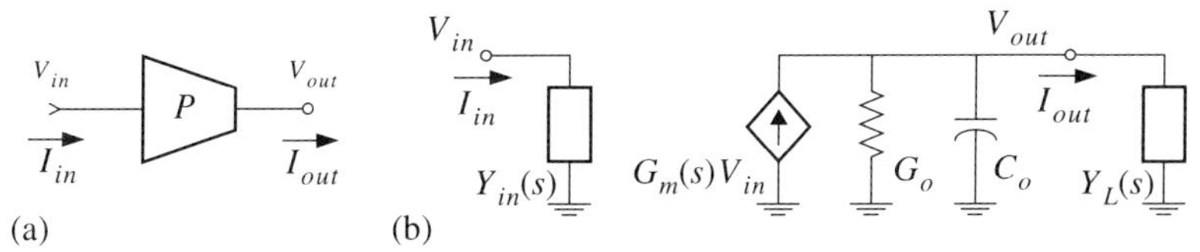

Figure 11.13: (a) Symbol and (b) general macromodel of a real transconductor.

linear. To account for this deviation, we define a real transconductor as an electronic device whose static transfer function can be defined as

$$I_{out} = Pn(V_{in}, V_{out}) \tag{11.25}$$

where P is a designer controllable scaling factor; and the (generally nonlinear) function $n(V_{in}, V_{out})$ is assumed invertible at least in V_{in}. Figure 11.13(a) shows the symbol used for real transconductors which clearly indicates the associated scaling factor P. In the figure, the variables at the input and output ports can be single-ended or differential. Invertibility of $n(V_{in}, V_{out})$ at least in V_{in} implies that there exists a set of functions $w(I, V)$ parameterized by V, such that

$$n(w(I,V), V) = I \tag{11.26}$$

Clearly, from (11.25), we have

$$V_{in} = w(I_{out}/P, V_{out}) \tag{11.27}$$

Also, the following two conditions are generally expected from transconductor blocks,

$$\begin{gathered} \frac{\partial I_{out}}{\partial V_{out}} \ll \frac{\partial I_{out}}{\partial V_{in}} \\ I_{in} \ll I_{out} \end{gathered} \tag{11.28}$$

The first condition indicates a weak dependency of the output current with the output voltage, and more concretely, that the output conductance must be much lower than the transconductance of the device. On the other hand, the second condition, together with the invertibility of function $n(V_{in}, V_{out})$, provides the necessary properties for the construction of current-mirrors, a well-known and widely used structure in analog design that yields linear current scaling and replication from nonlinear transconductors [55].

Figure 11.13(b) shows a general macromodel of a real transconductor including the most dominant AC parasitics, and where we have implicitly assumed that the block is loaded by a frequency-dependent admittance $Y_L(s)$. Of course, this model is a simplification of what can be obtained by substituting each MOS transistor in the transconductor with its small-signal equivalent model. In general, a larger number of internal nodes can result, causing a more complex behavior, and capacitive coupling between the input and output nodes is usually present.

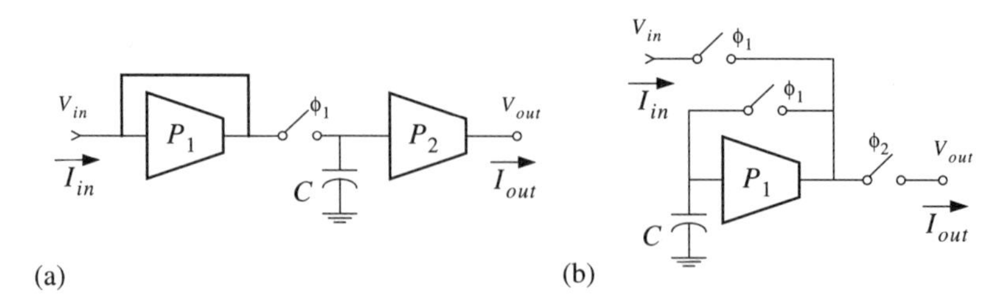

Figure 11.14: Switched current memory cells: (a) first generation; (b) second generation.

These limitations should be taken into account whenever greater accuracy is required, especially when modeling the transient and frequency response of the transconductor.

Figure 11.14 shows two switched-current sample-and-hold circuits, which are usually known as first- and second-generation current memory cells [49]. Assuming the same clock diagram as in Figure 3.2.6(c) and ideal transconductors, the functionality of the first-generation memory cell is given by

$$\begin{aligned} I_{out}(k+1) &= \frac{P_2}{P_1} I_{in}(k+1/2) \\ I_{out}(k+1/2) &= \frac{P_2}{P_1} I_{in}(k+1/2) \end{aligned} \tag{11.29}$$

and the functionality of the second-generation memory cell is defined by

$$\begin{aligned} I_{out}(k+1) &= I_{in}(k+1/2) \\ I_{out}(k+1/2) &= 0 \end{aligned} \tag{11.30}$$

thus providing, in both cases, half-cycle delay of the input current during the holding phase. It is worth noting that the second-generation memory cell has unity gain by construction.

Note that the grounded capacitor C in both implementations does not affect their respective functionalities, whenever the clock rate is low enough to allow the capacitor to reach a steady-state voltage during each phase. Consequently, there is no need for linear capacitors, and, in the majority of applications, C can be made only of the parasitic capacitance at the input of the transconductors. This is an important advantage over switched-capacitor realizations, since switched-current circuitry is fully compatible with digital technologies.

Full-cycle switched-current delayers can be readily obtained by cascading two of the current memories in Figure 11.14 with complementary phase clocks, as shown in Figure 11.15. The circuit in Figure 11.15(a) comprises two first-generation current memories, and operates as follows:

$$\begin{aligned} I_{out}(k+1) &= -\frac{P_2 P_4}{P_1 P_3} I_{in}(k+1/2) \\ I_{out}(k+1/2) &= -\frac{P_2 P_4}{P_1 P_3} I_{in}(k-1/2) \end{aligned} \tag{11.31}$$

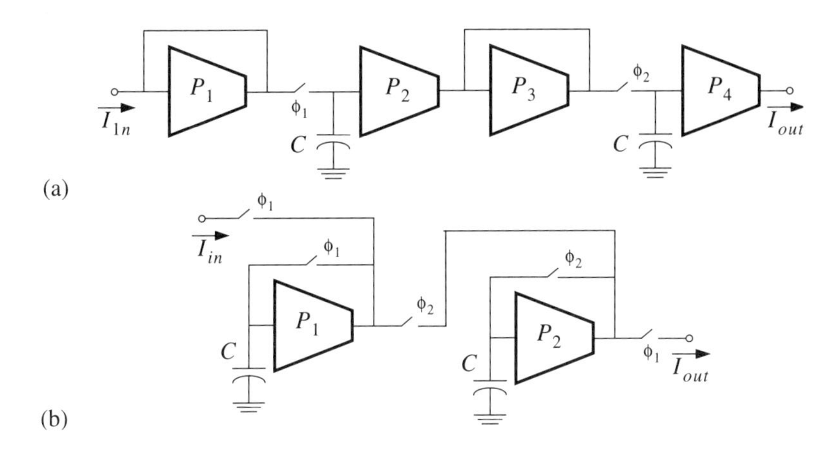

Figure 11.15: First- (a) and second-generation (b) full-cycle switched-current delayers.

where we notice that delay is concurrent with signal scaling, determined by transconductance ratios. An alternative for the delay block is using a cascade of second-generation current memories (see Figure 11.16(b)) which yields

$$\begin{aligned} I_{out}(k+1) &= 0 \\ I_{out}(k+1/2) &= -I_{in}(k-1/2) \end{aligned} \tag{11.32}$$

with no scaling. A final alternative is using a first generation current memory followed by a second generation one; thus scaling is achieved at the front stage. Obviously in practice transconductor blocks must be replaced by actual transistor circuits. For illustration purposes, Table 11.4 shows different alternative circuits together with expressions for their relevant terminal impedances.

One of the most important dynamic limitations of discrete-time delayers (and in general, of any sampled-and-hold structure) is the so-called charge injection error [85–89]. Bear in mind that correct operation of these circuits relies on the capacitor's ability to hold the input voltage when the analog switch turns off. Assume that an analog switch is implemented by a single MOS transistor which is driven between linear (closed state) and cut-off (open state) regions. In the turn-off process, the mobile charge in the inversion layer of the MOS switch (generated during the closed state) flow out of its drain, source, and substrate. As a consequence, a fraction of the charge released by the switch is added to the charge already stored on the hold capacitor, resulting in charge injection errors on the memorized voltage. Another error source also occurring during the switching-off process is the clock feedthrough effect. This is caused by the changing gate voltage of the switch, which makes charges flow from the gate diffusion overlap capacitances out of the MOS drain and source terminals. These charges also accumulate in the hold capacitor, thus leading to stored voltage deviations. The magnitude of these voltage errors is inversely proportional to the holding capacitance. For this reason, charge injection and clock feedthrough errors are particularly problematic in switched-current circuits, where hold capacitors con-

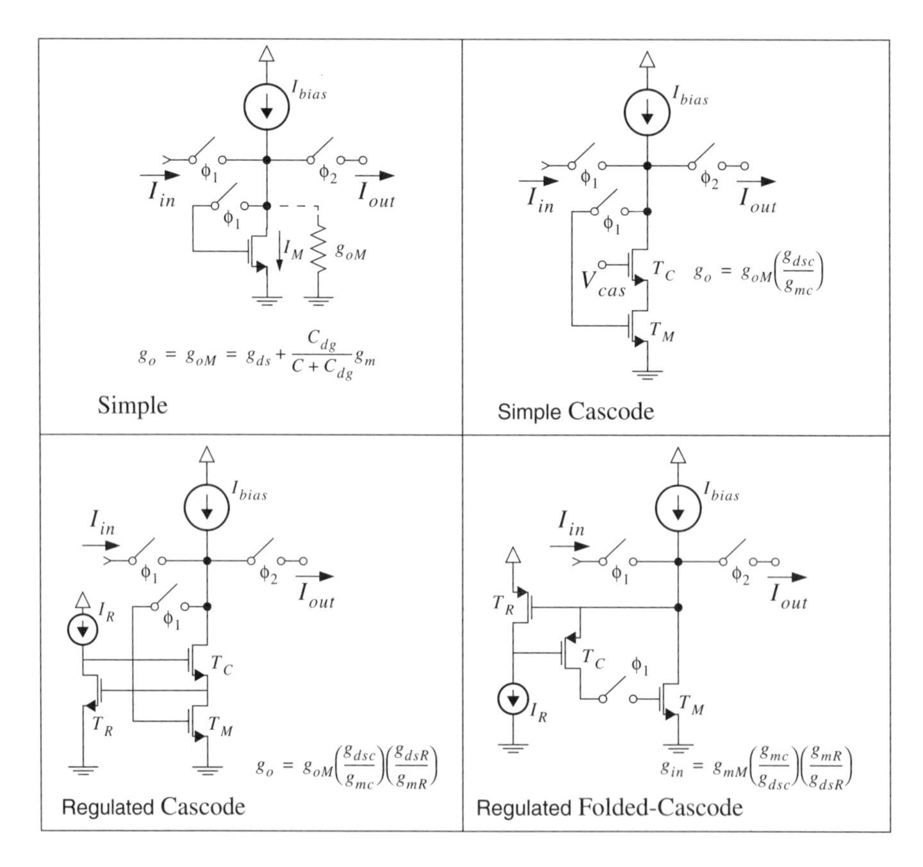

Table 11.4: Memory cell circuits and associated conductance errors.

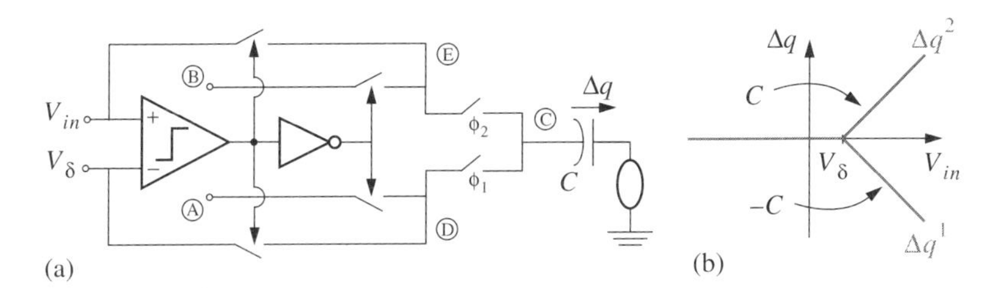

Figure 11.16: Circuit for rectification in voltage-charge domain.

sist of parasitics at the input of the transconductors –their capacitances are small if transconductors with reduced geometries are used. Charge injection errors can be attenuated with several techniques. They include dummy switch compensation, fully differential architectures, algorithmic cancellation, and adaptive timing of the clock signals.

11.4 Piecewise-Linear Operators for Discrete Maps

Piecewise-linear nonlinearities are of special interest for noise generation because they can be easily implemented in a robust manner and because they provide a wide catalog of useful behaviors. The first step towards PWL implementation is decomposing the global function into a sum of elementary functions. Then these elementary functions must be realized through circuits. Issues related to function decomposition are well documented in the literature and will not be repeated here [96–99].

11.4.1 PWL Shaping of Voltage to Charge Transfer Characteristics

The realization of PWL relationships among sampled-data signals is based on nonlinear voltage-to-charge transfer and uses analog switches and voltage comparators. Figure 11.16(a) shows a circuit structure, where one of the capacitor terminals is connected to virtual ground and the other to a switching block. Assume that nodes A and B are both grounded. Note that for $V_{in} - V_\delta > 0$ the switch arrangement set node D to V_δ, while node E is set to V_{in}. For $V_{in} - V_\delta < 0$, nodes D and E are both grounded. Consequently, voltage at node C in this latter situation does not change from one clock phase to the next and consequently, the incremental charge becomes null for $V_{in} - V_\delta < 0$. On the other hand, for $V_{in} - V_\delta > 0$, the voltage at node C changes from one clock phase to the next, and generates an incremental charge,

$$\begin{aligned} \Delta q(k+1) &= C[V_{in}(k+1) - V_\delta] \\ \Delta q(k+1/2) &= -C[V_{in}(k) - V_\delta] \end{aligned} \tag{11.33}$$

which enables obtaining negative and positive slopes using the same circuit. Thus, if the input voltage remains constant during each full clock cycle, the circuit in Figure 11.16(a) realizes a concave extension [96] operator in the voltage-charge domain, which can either be positive or negative depending on the clock phase. These characteristics are represented in Figure 11.16(a), and defined by

$$\begin{aligned} \Delta q^2 &= C\Delta^+(V_{in}, V_\delta) \\ \Delta q^1 &= -C\Delta^+(V_{in}, V_\delta) \end{aligned} \tag{11.34}$$

To synthesize the convex extension operator [96] and, therefore, to make the characteristics null for $V_{in} - V_\delta > 0$, it suffices to interchange the comparator inputs. Also, the technique is easily extended to the absolute value operation by connecting terminal A to V_{in}, and terminal B to V_δ. The realization of the Hermite linear basis function is straightforward, and can be found in [98].

Other approaches to the realization of PWL switched-capacitor circuitry use series rectification of the circulating charge through a comparator-controlled switch [98, 100].[6] Figure 11.17(a) shows an implementation of the concave extension operator using this technique. As for the circuit in Figure 3.2.16(a), convex rectification is easily obtained by swapping the comparator inputs. An interesting property of this approach is the possibility of truncating the voltage-to-charge transfer characteristics by just replacing the voltage comparator in Figure 11.17(a) by a window detector, as shown in Figure 3.2.17(b). Window detectors are electronic devices which provide a high- or low-level output binary signal depending on whether the input voltage falls within a specified voltage band or not. They can be easily implemented with two voltage comparators and logic gates, as indicated in Figure 11.18.

In the realization of Figure 11.17(b), truncation is applied to a concave extension operator. Observe that the truncated segment of the characteristic may exhibit a vertical offset charge by making $V_\delta < V_{\delta 1}$. Truncation on a convex extension operator is obtained by swapping the clock phase controls of the input switches and, if a vertical offset charge is needed, reference voltages must verify that $V_\delta > V_{\delta 2}$. In the case of Figure 11.17(b), assuming that V_{in} does not change from phase 2 to phase 1, the voltage-to-charge transfer characteristic can be expressed as:

$$\begin{aligned} \Delta q^2 &= C(V_{in} - V_\delta)\mathrm{wdw}(V_{in}, V_{\delta 1}, V_{\delta 2}) \\ \Delta q^1 &= -C(V_{in} - V_\delta)\mathrm{wdw}(V_{in}, V_{\delta 1}, V_{\delta 2}) \end{aligned} \tag{11.35}$$

where, wdw$(\cdot)$ is the window function defined as

$$\mathrm{wdw}(V_{in}, V_{\delta 1}, V_{\delta 2}) = \frac{1}{2}[\mathrm{sgn}(V_{in} - V_{\delta 1}) + \mathrm{sgn}(V_{\delta 2} - V_{in})] \tag{11.36}$$

[6] The former also discusses exploitation of these switched-capacitor circuits to realize continuous-time driving-point characteristics, the associated transformation circuits, and the dynamic problematics.

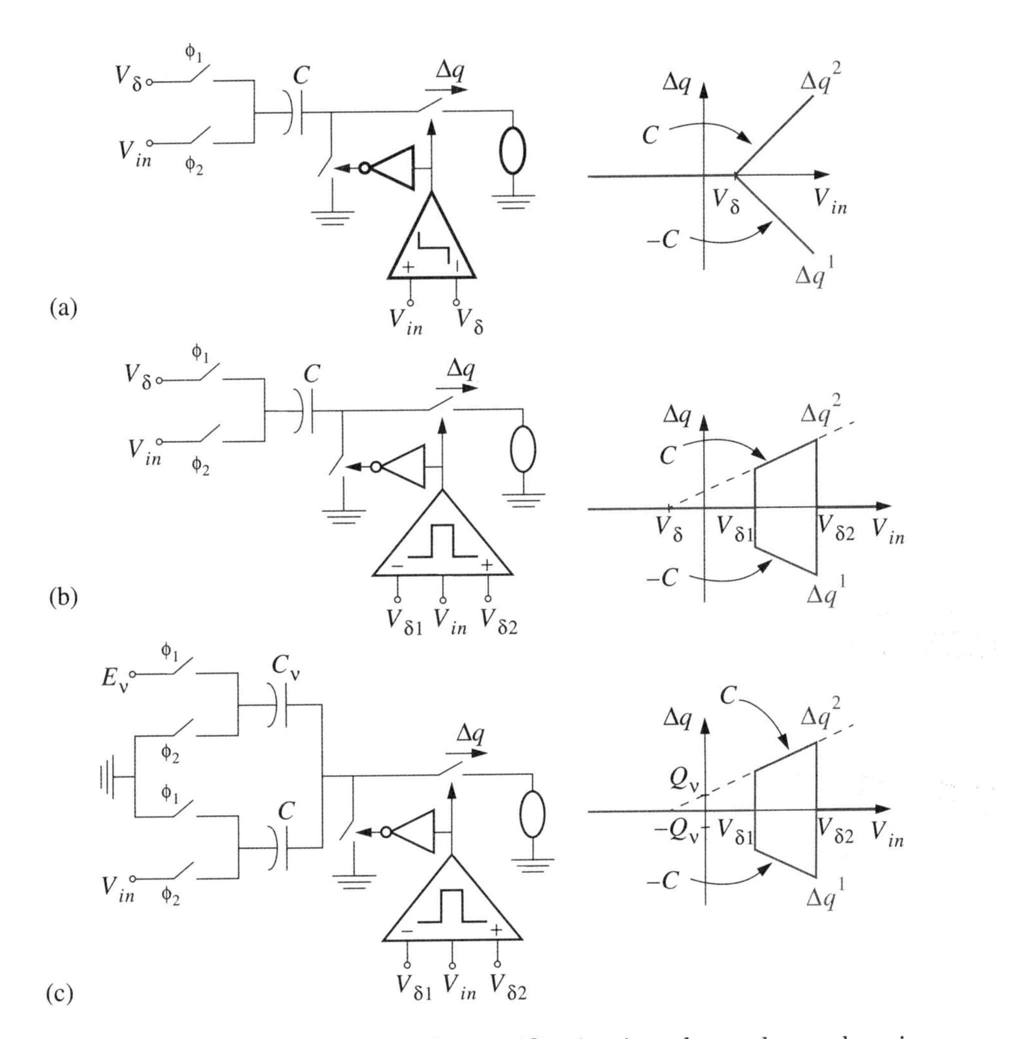

Figure 11.17: Alternative circuits for rectification in voltage-charge domain.

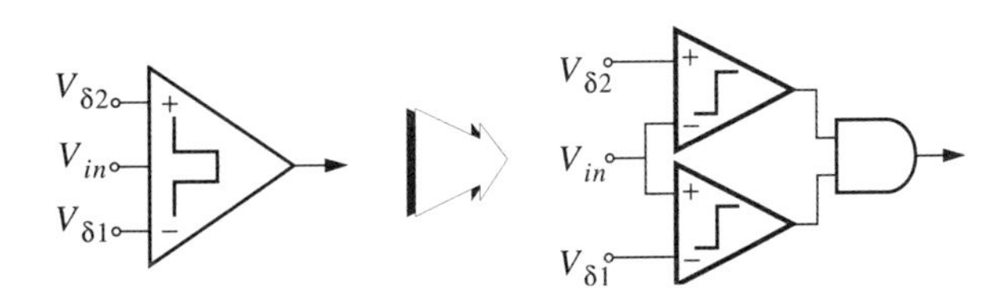

Figure 11.18: Window comparator.

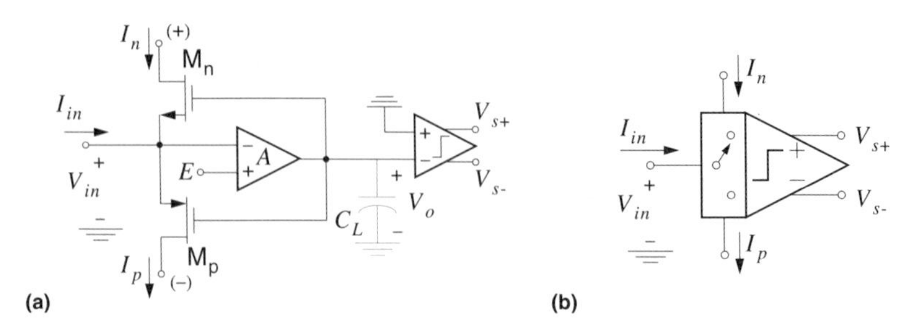

Figure 11.19: Current switch rectifier: (a) circuit concept; (b) symbol.

An inconvenience of the circuit in Figure 11.17(b) is the large value which may take V_δ if the capacitance C is small. This can be overcome with the circuit of Figure 11.17(c), where it has been assumed that $E_\nu < 0$. Simple analysis shows that

$$\begin{aligned} \Delta q^2 &= (CV_{in} + Q_\nu)\text{wdw}(V_{in}, V_{\delta 1}, V_{\delta 2}) \\ \Delta q^1 &= -(CV_{in} + Q_\nu)\text{wdw}(V_{in}, V_{\delta 1}, V_{\delta 2}) \end{aligned} \tag{11.37}$$

where the value of Q_ν is implicitly defined by the capacitance C_ν and the reference voltage E_ν, by the equation $Q_\nu = |C_\nu E_\nu|$. Clearly, the circuit in Figure 11.17(c) presents a problematic completely dual to that of Figure 11.17(b); i.e., it may be unsuitable for large values of C.

11.4.2 PWL Shaping in Current-Mode Domain

11.4.2.1 Rectification Operator

Rectification always relies on the capability to allow or preclude a current signal to flow through a circuit branch according to the polarity of a control variable. This suggests, for design simplicity, the use of current-mode techniques for the implementation of extension operators and derived elementary PWL functions, thus avoiding the need for any additional transformation circuitry.

The circuit of Figure 11.19(a), called current switch rectifier, is a versatile building block for rectification operations [101,102]. It consists of a voltage-mode amplifier and a nonlinear resistor (formed by transistors M_n and M_p) arranged in a negative feedback loop. We will assume that the drain terminals of M_n and M_p are connected to low impedance nodes whose voltage levels are such that transistors M_n and M_p operate in saturation under strong inversion when they are in the ON state.

The circuit exhibits three modes of operation depending on the input current level. For positive current flows, the incoming current I_{in} is integrated in the input parasitic capacitor and the input voltage V_{in} is pulled high. The voltage difference $E - V_{in}$ is then amplified by A (the gain of the voltage amplifier), causing V_o to go low and forcing M_n to the OFF state, so that $I_n = 0$. In addition, transistor M_p turns ON and a negative feedback loop is formed around the amplifier. This feedback configuration reduces the input resistance of the current switch and obtains $I_p = I_{in}$. Further, the input node voltage becomes

$$V_{in+}(I_{in}) = E\left(\frac{A}{1+A}\right) + \frac{V_{sgp}(I_{in})}{1+A} \tag{11.38}$$

which, for A large enough, is close to the reference voltage E of the amplifier, and the output node voltage is

$$V_{o-}(I_{in}) = \left(\frac{A}{1+A}\right)[E - V_{sgp}(I_{in})] \tag{11.39}$$

where $V_{sgp}(I_{in})$ is the source-to-gate voltage required by transistor M_p to supply the input current I_{in}.

A dual situation occurs for negative input currents. In this case, the input voltage V_{in} is pulled down and V_o goes up, turning M_n ON and M_p OFF, so that $I_p = 0$. The voltage amplifier is fed back by transistor M_n, thus reducing the input resistance of the rectifier. As a consequence, M_n completely draws the input current ($I_n = I_{in}$). The input node voltage becomes

$$V_{in-}(I_{in}) = E\left(\frac{A}{1+A}\right) - \frac{V_{gsn}(I_{in})}{1+A} \tag{11.40}$$

and the output node voltage is

$$V_{o+}(I_{in}) = \left(\frac{A}{1+A}\right)(E + V_{gsn}(I_{in})) \tag{11.41}$$

where $V_{gsn}(I_{in})$ is the gate-to-source voltage required by transistor M_n to supply the input current I_{in}.

In summary, the circuit of Figure 11.19(a) routes the input current to either the upper or the lower output terminal, depending on its sign, i.e.,

$$I_p = \begin{cases} I_{in} & I_{in} > 0 \\ 0 & I_{in} < 0 \end{cases} \qquad I_n = \begin{cases} 0 & I_{in} > 0 \\ I_{in} & I_{in} < 0 \end{cases} \tag{11.42}$$

and, thus, it generates simultaneously the concave (I_p output) and convex (I_n output) rectified versions of the input current. In addition, the output voltage abruptly swings from V_{o-} in (11.39) to V_{o+} in (11.41) when the input current goes from positive to negative values, and the opposite for the dual transition. This output voltage can be further extended to the rails using an additional amplifier stage (see Figure 11.19(a)), thus generating logical signals V_{s+} and V_{s-} according to the input current sign. For this purpose, a cascade of CMOS inverters can be used whenever the values of V_{o+} and V_{o-} are included, respectively, in the positive and negative noise margins of the first inverter. In this case, the output V_{s+} is "1" for $I_{in} > 0$, and "0" for $I_{in} < 0$, and the opposite for V_{s-}, thus implementing the comparison operation.

It is worth noting that due to the capacitive impedance of the circuit during the transitions around $I_{in} = 0$, it exhibits very high resolution (only limited by leakage currents), insensitive to transistor mismatch. Also, the feedback current-switch is advantageous as compared to using the inherent rectification property at the input node of a current mirror [103]. On the one hand, it achieves concave and convex curves with no input current replication needed. On the other hand, since it provides paths for both negative and positive input currents, it precludes that parasitics charge be accumulated at the input node, and hence has potential for greater operation speed [104].

Additionally, the feedback loop created by the amplifier allows the voltage excursions at the input node to remain small for a large input current range, thus alleviating the loading errors of the circuit with the environment. Indeed, according to (11.38) and (11.40), the benefits from the feedback for reducing the interstage error are increased by increasing the amplifier gain.

The current switch rectifier of Figure 11.19(a), however, exhibits a noticeable transient limitation which precludes its use for high-frequency applications (above 10MHz in a 0.8μm CMOS technology). This limitation arises from the Miller effect created around the overlapping capacitor C_f which connects input and output terminals of the amplifier in Figure 11.19(a), significant even for minimum-sized feedback transistors, in particular for low current levels.

Assume that the input current steps at $t = 0$ from a negative overdrive level $-J_-$ up to a positive overdrive level J_+. It can be shown [102] that the transient behavior of the current switch rectifier is dominated by the time (when $I_{in} \simeq 0$) needed to change V_o from $V_{o+}(0) = \alpha(E + V_{Tn})$ to $V_{o-}(0) = \alpha(E - |V_{Tp}|)$, where $\alpha = A/(1 + A)$ and V_{Tn}, V_{Tp} are, respectively, the threshold voltages of M_n and M_p. The delay time T_D involved in this transition, during which transistors M_n and M_p are OFF, can be approximated as [102]

$$T_D \simeq \frac{C_f}{J_+}(V_{Tn} + |V_{Tp}|) \tag{11.43}$$

which shows that the time delay is independent of A due to the Miller effect created around C_f. Therefore, though the high resolution property of the current switch rectifier can be improved by increasing A, the transient behaviour remains limited by (11.43) independent of A.

The next subsection presents an improved comparator structure which overcomes dynamic limitations of Figure 11.19, using a different current feedback mechanism. Although this new circuit is better suited for pure comparison purposes, it does not preserve the current rectification properties, and hence, does not qualify directly for current-mode function generation.

11.4.2.2 Comparison Operator

Improved transient response of the feedback switch comparator requires using circuit strategies to decouple input and output nodes of the amplifier and, hence, reduce Miller capacitance. On the one hand, cascode transistors should be used

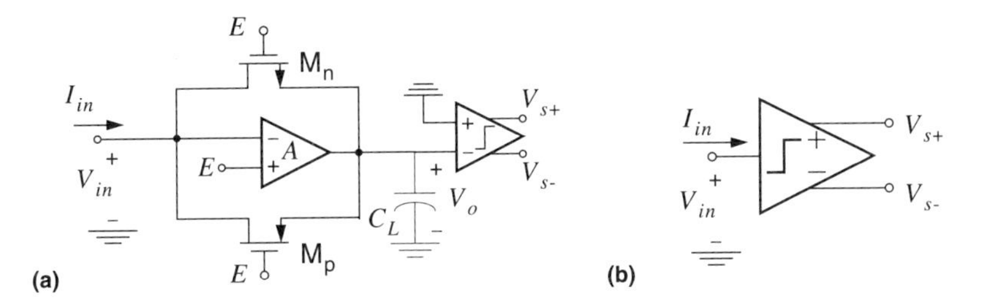

Figure 11.20: (a) Enhanced current comparator schematic; (b) symbol.

at the amplifier input; on the other hand, the feedback mechanism used to obtain the nonlinear driving-point characteristics should be modified. The circuit in Figure 11.20(a), whose symbol is drawn in Figure 11.20(b), incorporates a strategy to this end. Its static operation follows principles similar to that of Figure 11.19(a). At the center point ($V_{in} = E$, $I_{in} = 0$) both transistors are OFF and the circuit yields capacitive-input behaviour. Positive currents ($I_{in} > 0$) integrate in the input capacitor increasing V_{in}, and consequently decreasing V_o until the transistor M_n becomes conductive, absorbing the input current and stabilizing the voltage. The same occurs for negative currents, where M_p is the conductive transistor. Regarding the transient behaviour, analysis confirms that it is dominated by a quadratic, rather than a linear, term. Consequently, this circuit obtains a delay time,

$$T_D \simeq \sqrt{\frac{2\alpha(V_{Tn} + |V_{Tp}|)C_a C_b}{J_+ g_m}} \tag{11.44}$$

where C_a, C_b, and g_m are, respectively, the input capacitance, the output capacitance, and the transconductance gain of the opamp. For typical parameter values (11.44) is much smaller than (11.43). As a matter of fact, calculations with $\alpha = 1$ and $V_{Tn} = |V_{Tp}| = 1\text{V}$ obtain two orders of magnitude improvement.

11.5 Integrated Discrete Maps

In this section, we present different sampled-data IC designs which implement very simple chaotic discrete maps. They are the Bernoulli map and the tent map. These prototypes must be considered only as demonstration vehicles, because design aspects such as high operation speed, low area, and power consumption, etc., have been regarded as secondary objectives. Actually, the only severe restriction which has been imposed on the different prototypes is a minimum accuracy of around 7-8 bits on their analog operations. In this framework, design of the delay blocks, which are often the most critical elements as they suffer from charge injection effects, is largely simplified. In all the prototypes, holding capacitors, either implemented with parallel-plate structures (switched-capacitor) or based on the input parasitics of transconductors (switched-current), have been made

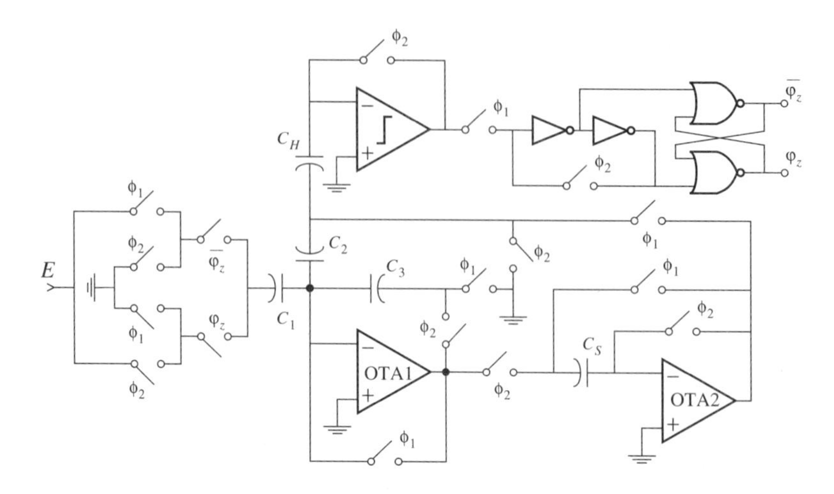

Figure 11.21: SC schematics for the Bernoulli map.

about three orders of magnitude larger than the equivalent capacitor of MOS switches. Clearly, this avoids resorting to complicated compensation mechanism for the delay blocks to fulfill the accuracy requirements, and simple structures can be used instead. A major drawback of this approach is the maximum frequency achievable, because of the need to guarantee that circuits evolve into steady state for each clock interval.

As will be shown next, the suitability of the prototypes for noise generation have been confirmed by experimental measures in the laboratory. Interestingly enough, they can serve as basic building blocks for generating colored noise with a prescribed power spectral density and probability density function, following the approaches proposed in [105, 106].

11.5.1 Bernoulli Map

Let us consider the implementation of the Bernoulli map in (11.3). Recall that, for $0 < B < 2$, trajectories remain confined in the interval $J = [-A, A]$ whenever the initial state lies inside J. If this not the case, two situations can be clearly distinguished within the chaotic regime. For $x_0 > A/(B-1)$ or $x_0 < A/(1-B)$ all orbits diverge (to $+\infty$ or $-\infty$, respectively). Conversely, for $A < x_0 < A/(B-1)$ (equivalently, $A/(1-B) < x_0 < A$), orbits evolve into the interval J, and remain confined to it. Both subintervals are interesting for practical implementations, since they provide safety zones where saturation characteristics of active devices must be located to avoid the corresponding circuit becoming locked at parasitic stable points. This is achieved by selecting a proper value of parameter A.

Let us first consider implementation of (11.3) by SC techniques [19]. Figure 11.21 shows an offset-free parasitics-insensitive SC schematic for the Bernoulli map. OTA1 and related capacitors perform the weighted summation in (11.3) and introduce a half-cycle delay in its output signal. OTA2 is used to implement the remaining half-delay stage and complete the concept of Figure 11.2(a). Pa-

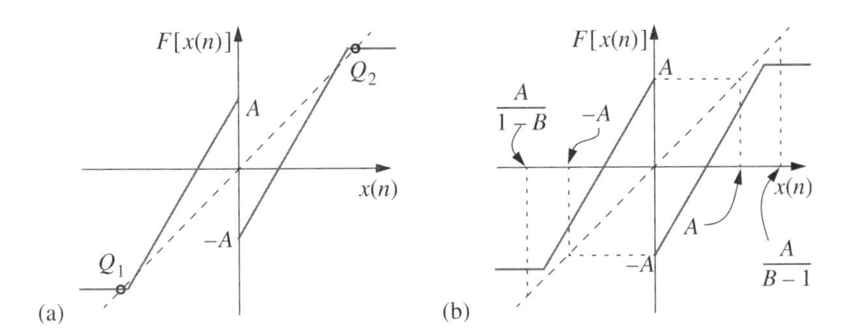

Figure 11.22: (a) Onset of parasitic stable points in the Bernoulli map due to improper setting of A; (b) strategy to avoid locked states.

rameters in the map are controlled by the capacitors C_1, C_2 and C_3, and the dc voltage E, as follows:

$$A = \frac{C_1}{C_3} E \qquad B = \frac{C_2}{C_3} \tag{11.45}$$

The nonlinearity is realized via a phase-reverser switch arrangement controlled by a dynamic comparator. Depending on the value of φ_z, this arrangement either adds or subtracts E at the input of the opamp OTA1, thereby yielding the change of sign in (11.3). The comparator consists of an input offset cancelled amplifier followed by a regenerative sense amplifier, made of two inverters in feedback loop, and a NOR-based latch. The latch is used to avoid any possible overlapping beween the phases of signals Z and $\bar{Z}$.

Operation of the circuit in Figure 11.21 is described by (11.3) whenever OTA's work in their linear region. If any of the amplifiers enters in saturation, the circuit no longer implements (11.3), and becomes locked at parasitic stable points close to the power rails. This undesirable situation can be avoided by properly setting parameter A. The design criteria is illustrated in Figure 11.22, where we have considered two different values of A for the same value of B ($1 \leq B \leq 2$) and drawn the corresponding open-loop transfer characteristics of Figure 11.21 including the influence of the opamp voltage saturation. In Figure 11.22(a) parasitic stable points, Q_1 and Q_2, appear at the crossing points between the transfer function characteristics and the 45° line. In this case, the circuit will evolve until locked either at Q_1 or Q_2, hence not yielding chaotic signals. On the other hand, for Figure 3.2.22(b) no spurious equilibria appear and chaotic signals will be obtained. The following must be fulfilled for the transfer function to be like the one in Figure 11.22(b),

$$A < V_{s+} < A/(B-1) \qquad A/(1-B) < V_{s-} < -A \tag{11.46}$$

where V_{s+} (V_{s-}) denotes the opamp positive (negative) saturation level.

Figure 11.23 shows a microphotograph of a programmable prototype of Figure 11.21, fabricated in a 3μm CMOS n-well double-poly double-metal technology.

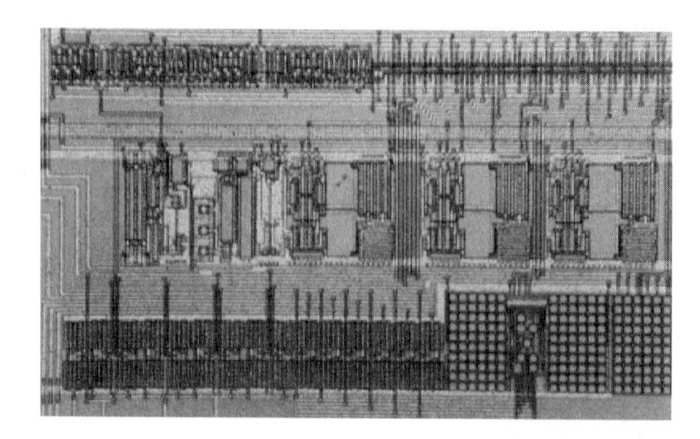

Figure 11.23: Microphotograph of the switched-capacitor Bernoulli map prototype.

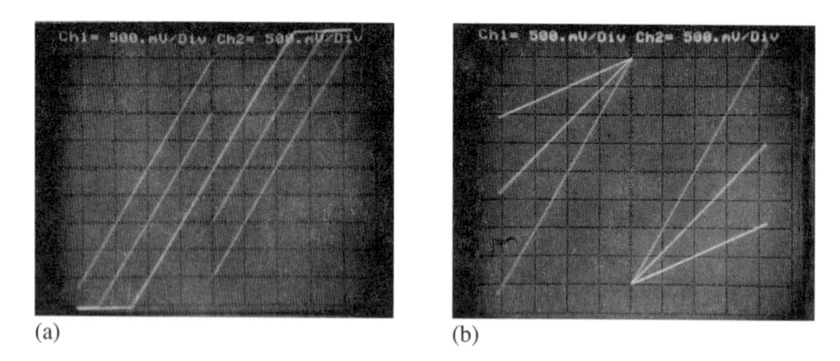

Figure 11.24: Measured open-loop transfer characteristic of the SC Bernoulli map: (a) for different values of A; (b) for different values of B_1 and B_2.

A slightly modified version of Figure 11.21 was considered where the value of the slope of (11.3) for $x(n) < 0$ (which we call B_1) can be controlled, independent of the corresponding value $x(n) > 0$ (denoted as B_2). Two binary weighted capacitors (with six control bits each) were used to separately adjust B_1 and B_2 over the chaotic regime, $1 \leq B_1, B_2 \leq 2$. Also a control bit was added to selectively open or close the feedback loop. All measurements performed on the prototype were done at a clock frequency $f_c = 200\text{kHz}$.

Figure 11.24 shows the open-loop transfer characteristic of the chip. Figure 11.24(a) shows a family of curves for different values of voltage E, and slopes B_1 and B_2 fixed at 2. Conversely, Figure 11.25(b) shows a set of transfer characteristics obtained for different values of B_1 and B_2, with E chosen to yield $A = 2\text{V}$. Measurements in closed loop were made for all possible combinations of B_1 and B_2 values inside the chaotic interval. Pictures in Figure 11.25 show the spectra obtained for some of these combinations. No periodicity was observed in any case, as demonstrated by the broadband characteristic of the pictures in Figure 11.25. Flat spectra were obtained for the cases where $B_1 = B_2 = B$. This is because the double symmetry[7] of the Bernoulli map for these settings makes the

[7]A map is said to be doubly symmetric if both the dynamic law $F(\cdot)$ and the invariant

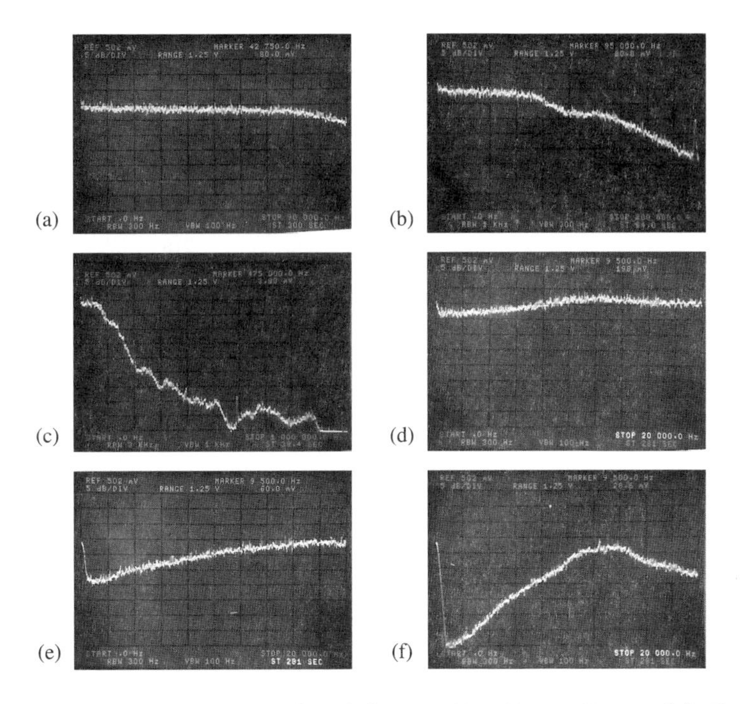

Figure 11.25: Measured spectra for different B_1, B_2 settings: (a) $B_1 = B_2 = 61/32$; (b) $B_1 = B_2 = 47/32$; (c) $B_1 = 47/32$ $B_2 = 39/32$; (d) $B_1 = 61/32$ $B_2 = 55/32$; (e) $B_1 = 61/32$ $B_2 = 47/32$; (f) $B_1 = 61/32$ $B_2 = 39/32$.

circuit generate nearly δ-correlated sequences, which in turn, makes the power spectral density approximately independent of the operating frequency. This is illustrated in Figures 11.25(a) and (b), obtained for $B = 61/32$ and $B = 47/32$, respectively. Observe that the flat spectrum band increases as the value of B tends to 2, i.e., as the Bernoulli map approximates the shift map. In particular, the spectrum in Figure 11.25(a) is flat up to 75kHz (35% of the clock frequency) with a maximum deviation of 1dB, which renders the circuit well-suited for white noise generation. On the other hand, for $B_1 \neq B_2$, because symmetry is broken, generated noise becomes colored. This is illustrated in Figures 11.25(c-f) which were obtained for different B_1 and B_2 settings.

Now let us consider implementation of (11.3) in current-mode domain [20]. The scaled delay operation can be realized as a cascade of two track-and-hold switched-current stages with complementary phase clocks, following the general architecture in Figure 11.15(a). On the other hand, following the design principle described in Section 11.4.2.1, Figure 11.26(a), which consists of two current sources (realized in practice by current mirror output branches), four transistors and two digital inverters, shows a conceptual schematic for the realization of the PWL characteristics of Figure 11.26(b). Transistors M_{CSn} and M_{CSp} in Figure 11.26(a) operate as a current-controlled-current-switch, while transistors M_{VSn}

---

measure (probability density function) are symmetric with respect to the origin [106].

Figure 11.26: Current-mode realization of the Bernoulli map nonlinearity.

and M_{VSp} operate as voltage-controlled current switches. Any positive input current increases the input voltage, turning M_{CSp} device ON, and since both devices in the current switch have the same gate voltage, M_{CSn} OFF. Simultaneously, the voltage at the second inverter output evolves to the high logic state, turning M_{VSn} ON and M_{VSp} OFF. Thus, a current $Bx(n) - A$ (obtained by KCL at node N_1) is directed to the output node through the transistor M_{VSn}; the right-hand piece of Figure 11.26(b) is implemented in this manner. Similarly, negative input currents turn M_{CSn} and M_{VSp} ON, so that a current $Bx(n) + A$ circulates through M_{VSp} to the output node.

As already mentioned, since current discrimination in the proposed circuit relies on integration function performed at the input node, resolution in the discrimination function is very high, not influenced by transistor mismatches (measurements from the CMOS prototype display resolution of 12pA's using minimum size transistors). Operation speed is also very high, limited mainly by nonlinear transients in the transistors that implement the current sources used to drive nodes N_1 and N_2. Also, the feedback created by inverter INV_1 yields significant reduction of the dead-zone exhibited by the driving point characteristics measured at the input node, which is proportional to $(V_{Tn} + |V_{Tp}|)/(1 + K)$, where V_{Tn} and V_{Tp} are the threshold voltages for the transistors and K is the inverter dc gain at $x(n) = 0$. This is an appealing feature that enables reduction of interstage loading errors caused by finite equivalent MOS transistors Early voltages.

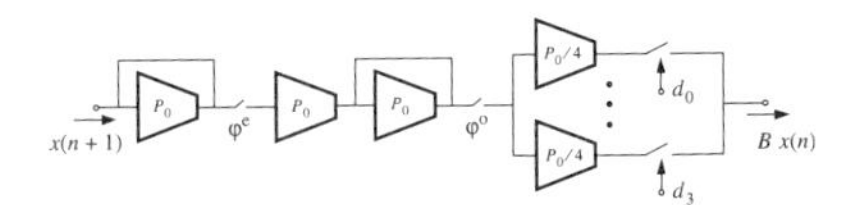

Figure 11.27: Microphotograph of the switched-current Bernoulli map prototype.

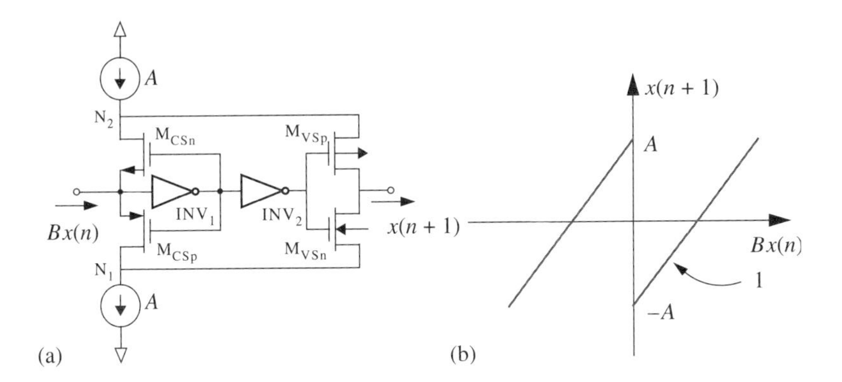

Figure 11.28: Programmable current-mode scaled delay block.

Figure 11.27 shows a microphotograph of a prototype fabricated in a 1.6μm CMOS double-metal single-poly n-well technology. In this prototype, regulated cascode transconductors have been used to implement the delay block in Figure 11.28. The output transconductor of the second track-and-hold stage has been binary-weighted (with four bits) for allowing digital control of parameter B[8]. Also, some extra miscellaneous circuitry has been added to enable testing the output current and the possibility to either open or close the feedback loop. Dummy switches were also added to reduce the influence of clock feedthrough. Bias current I_B for the delay stages was set to 50μA with $A = 20\mu$A.

The characteristics of Figure 11.29 have been measured in open loop, using

[8] In fact, the scaling factor of the MSB transconductor is slightly less than $P_0/4$ which makes parameter B lower than one. This avoids divergent orbits that may arise as a consequence of nonideal effects, such as clock feedthrough or variations on the technological parameters.

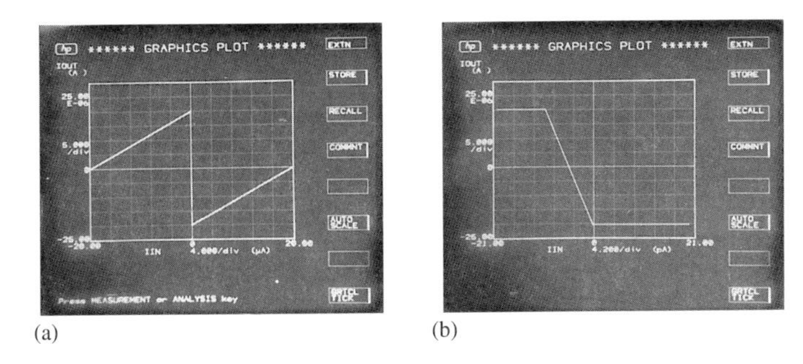

Figure 11.29: (a) Measured characteristic of the nonlinear block, and (b) detail of the discrimination function.

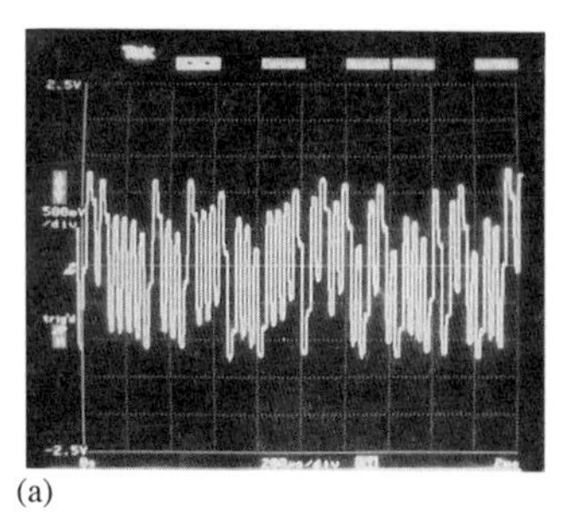

(a)

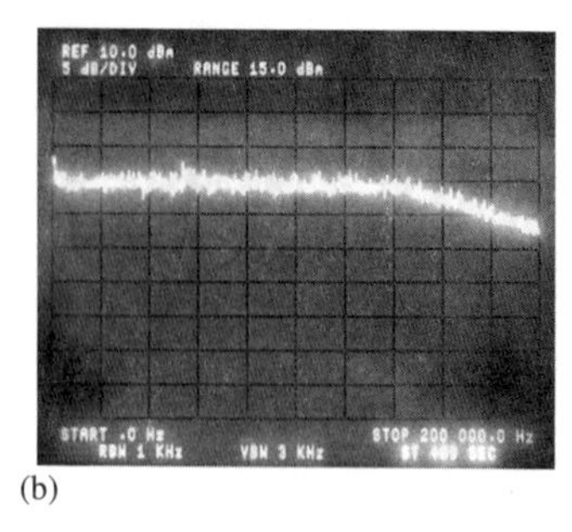

(b)

Figure 11.30: (a) Measured current waveform, and (b) power density spectrum.

the HP4145 semiconductor analyzer. Figure 11.29(a) ranges from -20μA to 20μA and shows the actual global PWL current transfer characteristics displayed by the prototype. Measured deviation from linearity in this range is less than 0.2%. Figure 11.29(b) shows a detail of the global characteristics. In this case the input current changes from -21pA to 21pA. It is intended to illustrate resolution achieved in the current discrimination which, as already anticipated above, amounts to few pA's.

Figure 11.30 illustrates closed loop operation of the prototype for a clock frequency of 500kHz. Figure 11.30(a) shows the measured current waveform at the output of the delay block for $B = 2$ (actually a slightly lower value to avoid divergent orbits), while Figure 11.30(b) shows its associated power density spectrum. The waveform of Figure 11.30(a) shows that apparently coincident values of $x(n)$ result in quite different values after few iterations, thereby confirming the expected unpredictably feature. Regarding Figure 11.30(b), detailed measurements show a very flat spectrum from dc up to about 30% of the clock frequency (deviation in this range was of less than 1dB).

It is useful to compare performance of this circuit to that of the SC circuit Figure 11.21. Area occupation of the switched-current prototype is about one order of magnitude smaller than that of the SC prototype. Also, for half the power consumption, the speed of the switched-current prototype is about three times greater than that obtained from the SC prototype.

11.5.2 Tent Map

Another simple FDE system which exhibits chaotic behavior is the tent map (sixth row of Table 3.2.1) [30,31]. This map has properties similar to those of the Bernoulli family; the main difference is that the definition interval generated by the dynamic law depends on both parameters A and B. Namely, for $0 < B < 2$, such interval is defined as $[A(1 - B), A]$. As for the Bernoulli map, parameter A does not influence the qualitative dynamic behavior, but just acts as a scale factor. However, different dynamic properties arise depending on the value of parameter B [30,31].

- For $B < 0$ or $B > 2$, all orbits diverge leading to unstable operation.

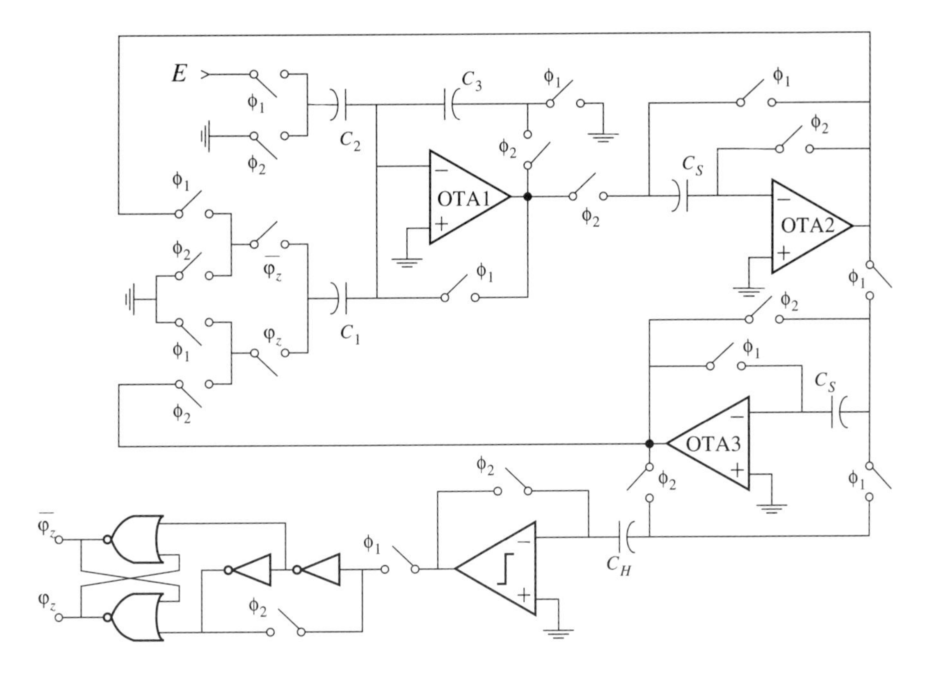

Figure 11.31: SC schematics for the tent map.

- For $0 \leq B < 1$, trajectories evolve towards a stable equilibrium point located at $x^* = A/(1+B)$ no matter which the initial point is.
- For $1 \leq B \leq 2$, the system is in the chaotic regime, and it is characterized by the same properties previously drawn for the Bernoulli map.

Figure 11.31 shows a stray insensitive SC schematic for the tent map [107]. The circuit uses design principles similar to those applied in the SC Bernoulli circuit of Figure 11.21. Observe, for instance, that the nonlinearity is implemented by using a dynamic comparator and a phase reverser switch arrangement. Parameters in the map are also defined by (11.45).

For proper operation of the circuit in Figure 11.31, parameter A must be adequately defined to preclude the appearance of parasitic stable points caused by the saturation of the active devices. The design criteria is illustrated in Figure 11.32, where a saturated tent map has been represented for two different values of A. For both figures, the same value of B ($1 \leq B \leq 2$) has been used. If parameter A is selected as shown in Figure 11.32(a), the circuit will evolve until locked at the parasitic stable point Q, located at the crossing point between the transfer function characteristics and the 45° line. Hence, after a transient, the circuit will fail to produce chaotic signals. On the other hand, if parameter A is set as indicated in Figure 11.32(b) no spurious equilibria appear and chaotic signals will actually be obtained. From Figure 11.32(b), the following conditions for the positive and negative saturation levels of the opamps can be drawn:

$$A < V_{s+} < A/(B-1) \qquad A/(1-B) < V_{s-} < A(1-B) \tag{11.47}$$

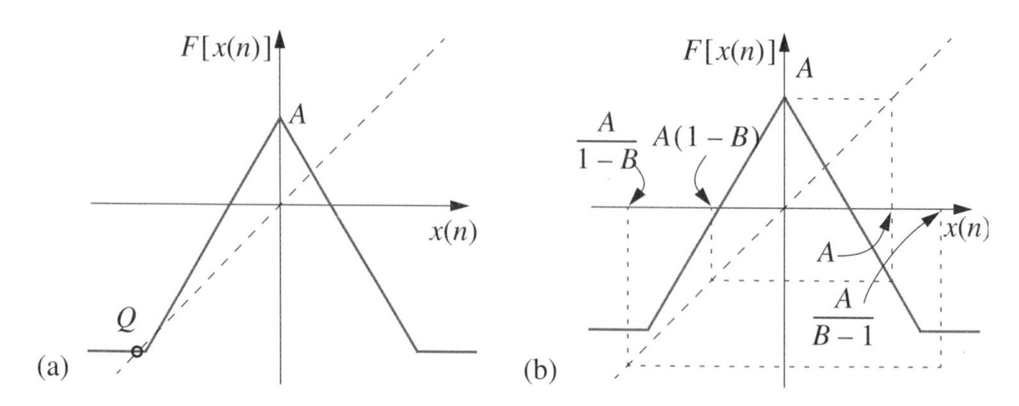

Figure 11.32: (a) Onset of parasitic stable points in the tent map due to improper setting of A; (b) Strategy to avoid locked states.

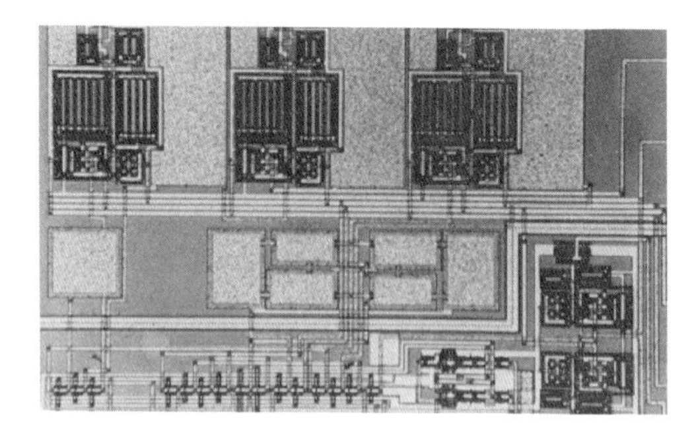

Figure 11.33: Microphotograph of the reconfigurable SC Bernoulli/tent map prototype.

which guarantee that the circuit in Figure 11.31 self-start to generate chaotic sequences in response to the power ON transient.

A prototype of Figure 11.31 has been fabricated in a 2.4μm CMOS n-well double-poly double-metal technology [108]. It is in fact a reconfigurable chip which can either implement the Bernoulli or the tent map via a digital control input. In addition, a binary weighted (six control bits) capacitor array is included to externally control the value of parameter B within the chaotic regime of both maps, namely $1 \leq B \leq 2$. Figure 11.33 shows a microphotograph of the prototype. All measurements performed on the prototype were done at a clock frequency $f_c = 200$kHz.

Figure 11.34 shows the open-loop transfer characteristic of the chip when configurated for the tent map. Figure 11.34(a) shows a family of curves for different values of voltage E, and slope B fixed at 2. Conversely, Figure 11.34(b) shows a set of transfer characteristics obtained for different values of B, with E chosen as to yield $A = 2$V.

Measurements in closed loop were made for all possible values of B. Figure 11.35 shows the power density spectrum obtained for the case $B = 2$. As can be seen, it is nearly flat from dc to about 75kHz, and hence, the circuit with this

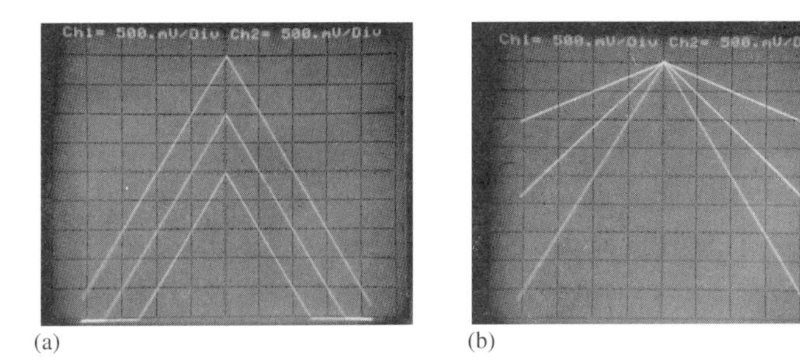

Figure 11.34: Measured open-loop transfer characteristic of the SC circuit in Figure 11.33 configurated as a tent map: (a) for different values of A; (b) for different values of B.

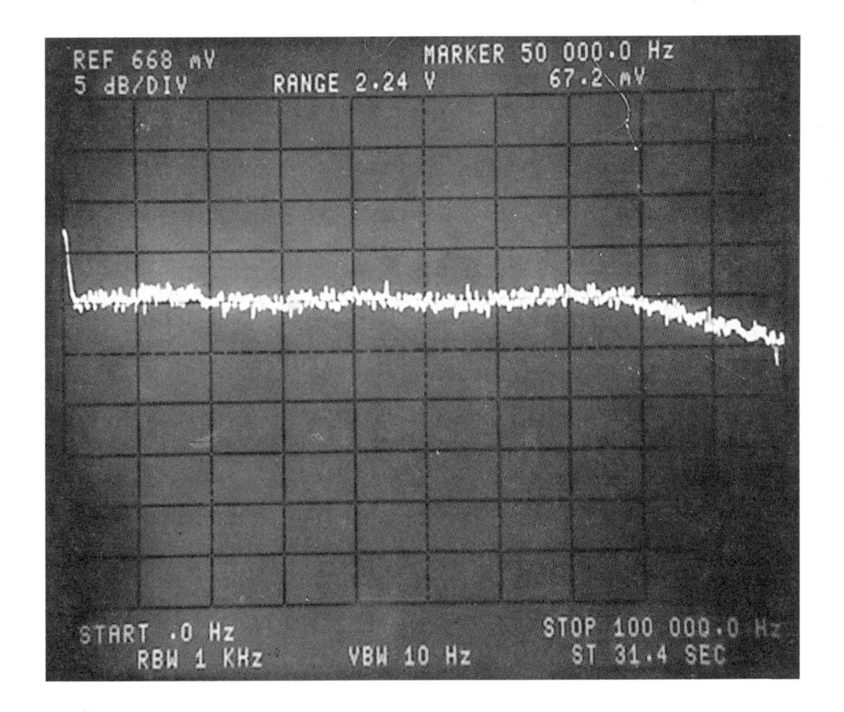

Figure 11.35: Power density spectrum for $B = 2$.

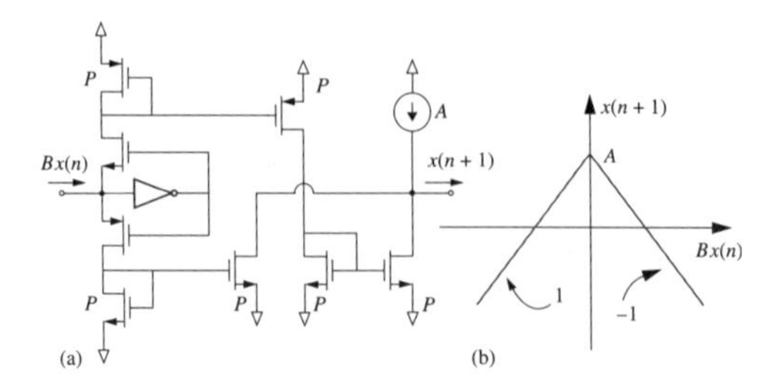

Figure 11.36: Current-mode realization of the tent map nonlinearity.

configuration can be regarded as a white noise generator. For any value of B other than 2, measured spectra show colored noise. This is because, though the map is symmetric with respect to the origin, its invariant measure is not uniform for $B < 2$, and therefore, the double symmetry property required for δ-correlated sequence generation is lost.

Let us now consider the realization in current-mode domain of the tent map. A possible implementation would consist of the feedback connection of a scaled delay block, like that in Figure 11.28, and the nonlinear circuit shown in Figure 11.36. Note that the last circuit is simply a full-wave rectifier based on a feedback current comparator, whose output current has been shifted by the biasing source A. Alternatively, the nonlinear function can be implemented by exploiting the rectifying characteristics of current mirrors.

References

[1] Milor, L. C., "A Tutorial Introduction to Research on Analog and Mixed-Signal Circuit Testing," *IEEE Trans. Circ. Sys. II*, 45(10), 1389, 1998.

[2] Ohletz, M. J., "Hybrid Built-In Self-Test (HBIST) for Mixed Analog/Digital Integrated Circuits," *Proc. 1991 Euro. Test Conf.*, 307, 1991.

[3] Pan, C.-Y. and Cheng, K.-T., "Pseudo-Random Testing and Signature Analysis for Mixed-Signal Circuits," *Proc. ICCAD*, 102, 1995.

[4] Norsworthy, S. R., "Quantization Errors and Dithering in ΣΔ Modulators," Chapter 3 in *Delta-Sigma Data Converters*, Norsworthy, S. R., Temes, G. C. and Schreier, R., eds., IEEE Press, Piscataway (NJ), 1997.

[5] Jewett, R., Poulton, K., Hsieh K.-C. and Doernberg, J., "Linearizing a 128 Msamples/s ADC," in *Analog Circuit Design*, van de Plassche, R. J., Huijsing, J. H. and Sansen, W. M. C., eds., Kluwer Academic, Dordrecht (the Netherlands), 1997.

[6] Carley, L. R., "A Noise Shaping Coder Topology for 15+ Bit Converters," *IEEE J. Solid-State Circ.*, 24(2), 267, 1989.

[7] Haykin, S., *Neural Networks: A Comprehensive Foundation*, Macmillan, New York, 1994.

[8] Alspector, J. et al., "A VLSI-Efficient Technique for Generating Multiple Uncorrelated Noise Sources and its Application to Stochastic Neural Networks," *IEEE Trans. Circ. Sys.*, 38(1), 109, 1991.

[9] Golomb, S. W., *Shift Registers Sequences*, Aegan Park Press, Laguna Hills (CA), 1982.

[10] Zheng, K., Yang, C.-H., Wei, D.-Y., and Rao, T. R. N., "Pseudorandom Bit Generators in Stream-Cipher Cryptography," *IEEE Comp. Mag.*, 8, Feb. 1991.

[11] Dixon, R. C., *Spread Spectrum Systems with Commercial Applications*, 3rd edition, John Wiley & Sons, New York, 1994.

[12] Holman, W. T., Conelly, J. A., and Dowlatabadi, A. B., "An Integrated Analog/Digital Random Noise Source," *IEEE Trans. Circ. Sys. I*, 44(6), 521, 1997.

[13] Letham, L., Hoff, D. and Folmsbee, A., "A 128K EPROM Using Encryption of Pseudorandom Numbers to Enable Read Access," *IEEE J. Solid-State Circ.*, 21(5), 881, 1986.

[14] Hortensius, P. D., McLeod, R. D., Pries, W., Miller, D. M. and Card, H. C., "Cellular Automata-based Pseudorandom Number Generators for Built-In Self-Test," *IEEE Trans. Computer-Aided Design*, 8(8), 842, 1989.

[15] Bernstein, G. M. and Liberman, M. A., "Secure Random Number Generation Using Chaotic Circuits," *IEEE Trans. Circ. Sys.*, 37(9), 1157, 1990.

[16] Adachi, M. and Aihara, K., "Associative Dynamics in a Chaotic Neural Network," *Neural Networks*, 10(1), 83, 1997.

[17] Chen, L. and Aihara, K., "Global Searching Ability of Chaotic Neural Networks," *IEEE Trans. Circ. Sys. I*, 46(8), 974, 1999.

[18] Walker, W. T., "Radar System Utilizing Chaotic Coding," U.S. Patent 5,321,409, Jun. 1994.

[19] Rodríguez-Vázquez, A., Delgado-Restituto, M., Espejo, S., and Huertas, J. L., "Switched Capacitor Broadband Noise Generator for CMOS VLSI," *Elec. Lett.*, 27(21), 1913, 1991.

[20] Delgado-Restituto, M., Medeiro, F., and Rodríguez-Vázquez, A., "Nonlinear Switched-Current CMOS IC for Random Signal Generation," *Elec. Lett.*, 29(25), 2190, 1993.

[21] Cruz, J. M. and Chua, L. O., "An IC Chip of Chua's Circuit," *IEEE Trans. Circ. Sys. II*, 40(10), 614, 1993.

[22] Delgado-Restituto, M. and Rodríguez-Vázquez, A., "Switched-Current Chaotic Neurons," *Elec. Lett.*, 30(5), 429, 1994.

[23] Horio, H. and Suyama, K., "Experimental Verification of Signal Transmission Using Synchronized SC Chaotic Neural Networks," *IEEE Trans. Circ. Sys. I*, 42(7), 393, 1995.

[24] Varrientos, J. E. and Sánchez-Sinencio, E., "A Low-Power Fully-Differential Current-Mode Synchronous Chaotic Oscillator," *IEEE Proc. 38th Midwest Symp. Circ. Sys.*, 1042, Aug. 1995.

[25] Pham, C.-K., Korehisa, M., and Tanaka, M., "Chaotic Behavior and Synchronization Phenomena in a Novel Chaotic Transistors Circuit," *IEEE Trans. Circ. Sys. I*, 43(12), 1006, 1996.

[26] Petrie, C. S. and Connelly, J. A., "A Noise-Based Random Bit Generator IC for Applications in Cryptography," *IEEE Proc. Int. Symp. Circ. Sys.*, TAA14-3, May-Jun. 1998.

[27] Cauwenberghs, G., "Delta-Sigma Cellular Automata for Analog VLSI Random Vector Generation," *IEEE Trans. Circ. Sys. II*, 46(3), 240, 1999.

[28] Rodríguez-Vázquez, A. and Delgado-Restituto, M., "CMOS Design of Chaotic Oscillators Using State Variables: A Monolithic Chua's Circuit," *IEEE Trans. Circ. Sys. II*, 40(10), 596, 1993.

[29] Delgado-Restituto, M. and Rodríguez-Vázquez, A., "Design Considerations for Integrated Continuous-Time Chaotic Oscillators," *IEEE Trans. Circ. Sys. I*, 45(4), 481, 1998.

[30] Schuster, H. G., *Deterministic Chaos: An Introduction*, 2nd edition, VCH Verlagsgesellschaft, Weinheim (Germany), 1988.

[31] Devaney, R. L., *An Introduction to Chaotic Dynamical Systems*, Benjamin/Cummings, Menlo Park (CA), 1986.

[32] Lasota, A., and Mackey, M.C., *Chaos, Fractals and Noise. Stochastic Aspects of Dynamics*, Springer-Verlag, New York, 1994.

[33] Boyarsky, A., and Scarowsky, M."On a Class of Transformations which have Unique Absolutely Continuous Invariant Measures," *Trans. Amer. Math. Soc.*, 255, 243, Nov. 1979.

[34] Bai-Lin, H., *Chaos*, World Scientific, Singapore, 1984.

[35] Cvitanovic, P., *Universality in Chaos*, 2nd edition, Adam Hilger, Bristol (UK), 1989.

[36] Rodríguez-Vázquez, A., Huertas, J. L., Rueda, A., Pérez-Verdú, B., and Chua, L. O., "Chaos from Switched-Capacitor Circuits: Discrete Maps," *Proc. IEEE*, 75(8), 1090, 1987.

[37] Gotz, M., Kelber, K. and Schwarz, W., "Discrete-Time Chaotic Encryption Systems. Part I: Statistical Design Approach," *IEEE Trans. Circ. Sys. I*, 44(10), 963, 1997.

[38] Gregorian, R. and Temes, G. C., *Analog MOS Integrated Circuits for Signal Processing*, John Wiley & Sons, New York, 1986.

[39] Unbehauen, R. and Cichocki, A., *MOS Switched-Capacitor and Continuous-Time Integrated Circuits and Systems*, Springer-Verlag, Heidelberg (Germany), 1989.

[40] Enz, C. C. and Temes, G. C., "Circuit Techniques for Reducing the Effects of Op-Amp Imperfections: Autozeroing, Correlated Double Sampling, and Chopper Stabilization," *Proc. IEEE*, 84(11), 1584, 1996.

[41] Allstot, D. J. and Black, W. C., "Technological Considerations for Monolithic MOS Switched-Capacitor Filtering Systems," *Proc. IEEE*, 71(8), 967, 1983.

[42] McCreary, J. L., "Matching Properties, and Voltage and Temperature Dependence of MOS Capacitors," *IEEE J. Solid-State Circ.*, 16(6), 608, 1981.

[43] Shyu, J., Temes, G. C. and Krummenacher, F., "Random Errors Effects in Matched MOS Capacitors and Current Sources," *IEEE J. Solid-State Circ.*, 19(12), 948, 1984.

[44] McNutt, M. J., LeMarquis, S., and Dunkley, J. L., "Systematic Capacitance Matching Errors and Corrective Layout Procedures," *IEEE J. Solid-State Circ.*, 29(5), 611, 1994.

[45] Moschytz, G. S. (ed.), *MOS Switched-Capacitor Filters. Analysis and Design*, IEEE Press, Piscataway (NJ), 1984.

[46] Razavi, B., *Principles of Data Conversion System Design*, IEEE Press, Piscataway (NJ), 1995.

[47] Hosticka, B. J., Brockherde, W., Kleine, U., and Schweer, R., "Design of Nonlinear Analog Switched-Capacitor Circuits Using Building Blocks," *IEEE Trans. Circ. Sys.*, 31(4), 354, 1984.

[48] Behr, A. T., Schneider, M. C., Filho, S. N., and Montoro, C. G., "Harmonic Distortion Caused by Capacitors Implemented with MOSFET Gates," *IEEE J. Solid-State Circ.*, 27(10), 1470, 1992.

[49] Toumazou, C., Hughes, J. B., and Battersby, N. C. (eds.),*Switched-Currents: An Analogue Technique for Digital Technology*, Peter Peregrinus Ltd., London, 1993.

[50] Vittoz, E. A., "The Design of High Performance Analog Circuits on Digital CMOS Chips," *IEEE J. Solid-State Circ.*, 20(6), 657, 1985.

[51] Lakshmikumar, K., Hadaway, R. A., and Copeland, M., "Characterization and Modeling of Mismatch in MOS Transistors for Precision Analog Design," *IEEE J. Solid-State Circ.*, 21(12), 1057, 1986.

[52] Pelgrom, M. J. M., Duinmaijer, A. C. J., and Welbers, A. P. G., "Matching Properties of MOS Transistors," *IEEE J. Solid-State Circ.*, 24(10), 1433, 1989.

[53] Maloberti, F., "Layout of Analog and Mixed Analog-Digital Circuits," Chap. 11 in Design of *Analog Digital VLSI Circuits for Telecommunications*, Franca, J. E. and Tsividis, Y. P., eds., Prentice-Hall, Englewood Cliffs (NJ), 1994.

[54] Tsividis, Y. P., *Mixed Analog-Digital VLSI Devices and Technology*, McGraw-Hill, New York, 1996.

[55] Gray, P. R. and Meyer, R. G., *Analysis and Design of Analog Integrated Circuits*, 3rd edition, John Wiley & Sons, New York, 1994.

[56] Haque, Y. A., Gregorian, R., Blasco, R. W., Mao, R. A., and Nicholson, W. E., "A Two Chip PCM CODEC with Filter," *IEEE J. Solid-State Circ.*, 14(6), 961, 1979.

[57] Di Cataldo, G., Palmisano, G. and Palumbo, G., "Gain-Compensated Sample-and-Hold Circuit for High Frequency Application," *Elec. Lett.*, 29(15), 1347, 1993.

[58] Watanabe, K. and Ogawa, S., "Clock-Feedthrough Compensated Sample/Hold Circuits," *Elec. Lett.*, 24(19), 1226, 1988.

[59] Wang, F.-J. and Temes, G. C., "A Fast Offset-Free Sample-and-Hold Circuit," *IEEE J. Solid-State Circ.*, 23(5), 1271, 1988.

[60] Temes, G. C., Huang, Y., and Ferguson Jr., P. F., "A High-Frequency Track-and-Hold Stage with Offset and Gain Compensation," *IEEE Trans. Circ. Sys.*, 42(8), 559, 1995.

[61] Kiriaki, S., Viswanathan, L., Feygin, G., Staszewski, B., Pierson, R., Krenik, B., de Wit, M., and Nagaraj, K., "A 160-MHz Analog Equalizer for Magnetic Disk Read Channels," *IEEE J. Solid-State Circ.*, 32(11), 1839, 1997.

[62] Wang, X. and Spencer, R. R., "A Low-Power 170-MHz Discrete-Time Analog FIR Filter," *IEEE J. Solid-State Circ.*, 33(3), 417, 1998.

[63] Lee, Y.-S. and Martin, K. W., "A Switched-Capacitor Realization of Multiple FIR Filters on a Single Chip," *IEEE J. Solid-State Circ.*, 23(2), 536, 1988.

[64] Matsui, K., Matsuura, T., Fukasawa, S., Izawa, Y., Toba, Y., Miyake, N., and Nagasawa, K., "CMOS Video Filters Using Switched Capacitor 14-MHz Circuits," *IEEE J. Solid-State Circ.*, 20(6), 1096, 1985.

[65] Carmona-Galán, R. Rodríguez-Vázquez, A., Espejo-Meana, S., Domínguez-Castro, R., Roska, T., Kozek, T., and Chua, L. O., "An 0.5-mm CMOS Analog Random Access Memory Chip for TeraOPS Speed Multimedia Video Processing," *IEEE Trans. Multimedia*, 1(2), 121, 1999.

[66] Nishimura, K. A. and Gray, P. R., "A Monolithic Analog Video Comb Filter in 1.2-μm CMOS," *IEEE J. Solid-State Circ.*, 28(12), 1331, 1993.

[67] Haller, G. M. and Wooley, B. A., "A 700-MHz Switched-Capacitor Analog Waveform Sampling Circuit," *IEEE J. Solid-State Circ.*, 29(4), 500, 1994.

[68] Martin, K., "Improved Circuits for the Realization of Switched-Capacitor Filters," *IEEE Trans. Circ. Sys.*, 27(4), 237, 1980.

[69] Ki, W.-H. and Temes, G. C., "Offset-Compensated Switched-Capacitor Integrators," *Proc. ISCAS*, 2829, 1990.

[70] Temes, G. C., "Finite Amplifier Gain and Bandwidth Effects in Switched-Capacitor Filters," *IEEE J. Solid-State Circ.*, 15(3), 358, 1980.

[71] Martin, K. and Sedra, A. S., "Effect of the Op Amp Finite Gain and Bandwidth on the Performance of Switched-Capacitor Filters," *IEEE Trans. Circ. Sys.*, 28(8), 822, 1981.

[72] Fisher, G. and Moschytsz, G., S., "On the Frequency Limitations of SC Filters," *IEEE J. Solid-State Circ.*, 19(4), 510, 1984.

[73] Haug, K., Maloberti, F., and Temes, G. C., "Switched-Capacitor Integrators with Low Finite-Gain Sensitivity," *Elec. Lett.*, 21(24), 1156, 1985.

[74] Gregorian, R. and Amir, G., "An Integrated Single-Chip Switched-Capacitor Speech Synthesizer," *Proc. ISCAS*, 733, 1981.

[75] Maloberti, F., "Switched-Capacitor Building Blocks for Analogue Signal Processing," *Elec. Lett.*, 19(7), 263, 1983.

[76] Temes, G. C. and Haug, K., "Improved Offset Compensation Schemes for Switched-Capacitor Circuits," *Elec. Lett.*, 20(12), 508, 1984.

[77] Nagaraj, K., Singhal, K., Viswanathan, T. R., and Vlach, J., "Reduction of Finite-Gain Effect in Switched-Capacitor Filters," *Elec. Lett.*, 21(15), 644, 1985.

[78] Larson, L. E., Martin, K. W., and Temes, G. C., "GaAs Switched-Capacitor Circuits for High-Speed Signal Processing," *IEEE J. Solid-State Circ.*, 22(6), 971, 1987.

[79] Lee, K. and Meyer, R. G., "Low Distortion Switched-Capacitor Filter Design Techniques," *IEEE J. Solid-State Circ.*, 20(12), 1103, 1985.

[80] Sansen, W. M. C., Qiuting, H., and Halonen, K. A. I., "Transient Analysis of Charge Transfer in SC Filters – Gain Error and Distortion," *IEEE J. Solid-State Circ.*, 22(2), 268, 1987.

[81] Medeiro, F., Pérez-Verdú, B. and Rodríguez-Vázquez, A., *Top-Down Design of High-Performance Modulators*, Kluwer Academic, Boston (MA), 1998.

[82] Hsieh, K.-C., Gray, P. R., Senderowicz, D. and Messerschmitt, D. G., "A Low-Noise Chopper-Stabilized Differential Switched-Capacitor Filtering Technique," *IEEE J. Solid-State Circ.*, 16(6), 708, 1981.

[83] Fischer, J. H., "Noise Sources and Calculation Techniques for Switched Capacitor Filters," *IEEE J. Solid State Circ.*, 17(4), 742, 1982.

[84] Gobet, C.-A. and Knob, A., "Noise Analysis of Switched Capacitor Networks," *IEEE Trans. Circ. Sys.*, 30(1), 37, 1983.

[85] Sheu. B. J. and Hu, C., "Switch-Induced Error Voltage on a Switched Capacitor," *IEEE J. Solid-State Circ.*, 19(8), 519, 1984.

[86] Wilson, W. B., Massoud, H. Z., Swanson, E. J., George R., Jr., and Fair, R. B., "Measurement and Modelling of Charge Feedthrough in n-Channel MOS Analog Switches," *IEEE J. Solid-State Circ.*, 20(12), 1206, 1985.

[87] Shieh, J.-H., Patil M. and Sheu, B. J., "Measurement and Analysis of Charge Injection in MOS Analog Switches," *IEEE J. Solid-State Circ.*, 22(4), 277, 1987.

[88] Wegmann, G., Vittoz, E. A., and Rahali, F., "Charge Injection in Analog MOS Switches," *IEEE J. Solid-State Circ.*, 22(12), 1091, 1987.

[89] Eichenberger, C. and Guggenbuhl, W., "On Charge Injection in Analog MOS Switches and Dummy Switch Compensation Techniques," *IEEE Trans. Circ. Sys.*, 37(2), 256-264, 1990.

[90] Van Peteghem, P., "On the Relationship Between PSRR and Clock Feedthrough in SC Filters," *IEEE J. Solid-State Circ.*, 23(8), 997, 1988.

[91] Bienstman, L. A. and de Man, H. J., "An Eight-Channel 8 Bit Microprocessor Compatible NMOS D/A Converter with Programmable Scaling," *IEEE J. Solid-State Circ.*, 15(6), 1051, 1980.

[92] Yen. R., Robert, C., and Gray, P. R., "A MOS Switched-Capacitor Instrumentation Amplifier," *IEEE J. Solid-State Circ.*, 17(6), 1008, 1982.

[93] Lewis, S. and Gray, P. R., "A Pipelined 5-MSamples/s 9-Bit Analog-to-Digital Converter," *IEEE J. Solid-State Circ.*, 22(12), 954, 1987.

[94] Willingham, S. D. and Martin, K. W., "Effective Clock-Feedthrough Reduction in Switched Capacitor Circuits," *Proc. of the 1990 IEEE Int. Symp. on Circuits and Systems*, pp. 2821-2824, 1990.

[95] Toumazou, C., Lidgey, F. J., and Haigh, D. G. (eds.),*Analog IC Design: The Current-Mode Approach*, Peter Peregrinus Ltd., London, 1990.

[96] Chua, L. O. and Kang, S. M., "Section-Wise Piecewise-Linear Functions: Canonical Representation, Properties and Applications," *Proc. of the IEEE*, 67 (6), 915, 1977.

[97] Chua, L. O. and Deng, A. C., "Canonical Piecewise Linear Representation," *IEEE Trans. on Circ. Sys.*, 35 (1), 101, 1988.

[98] Huertas, J. L., Chua, L. O., Rodríguez-Vázquez, A., and Rueda, A., "Nonlinear Switched-Capacitor Networks: Basic Principles and Piecewise-Linear Design," *IEEE Trans. on Circ. Sys.*, Vol. 32 (April), pp. 305-319, 1985

[99] Rodríguez-Vázquez, A., Delgado-Restituto, M., Huertas, J. L., and Vidal, F., "Synthesis and Design of Nonlinear Circuits," Chapter 32 in *The Circuits and Filters Handbook*, Chen, W.-K., ed., CRC Press, Boca Raton (FL), 1995.

[100] Delgado-Restituto, M., Rodríguez-Vázquez, A., Espejo, S., and Huertas, J. L., "A Chaotic Switched-Capacitor Circuit for Noise Generation," *IEEE Transactions on Circuits and Systems-I*, Vol. 39 (April), pp. 325-328, 1992.

[101] Domínguez-Castro, R., Rodríguez-Vázquez, A., Medeiro, F., and Huertas, J. L., "High Resolution CMOS Current Comparators," *Proc. 1992 Euro. Solid-State Circ. Conf.*, 242, 1992.

[102] Rodríguez-Vázquez, A., Domínguez-Castro, R., Medeiro, F., and Delgado-Restituto, M., "High Resolution CMOS Current Comparators: Design and Applications to Current-Mode Function Generation," *A. Int. Circ. and Signal Processing*, 7(2), 149, 1995.

[103] Laker, K. R. and Sansen, W. M. C., *Design of Analog Integrated Circuits and Systems*, McGraw-Hill, New York, 1994.

[104] Rodríguez-Vázquez, A. and M. Delgado-Restituto, M., 1994."Generation of Chaotic Signals using Current-Mode Techniques," *Journal of Intelligent and Fuzzy Systems*, Vol. 2, No. 1 (January), pp. 15-37.

[105] Gotz, M., Kilias, T., Kutzer, K. and Schwarz, W., "Design of Broadband Generators Using Chaotic Electronic Circuits," *Proc. 1995 Europ. Conf. Circ. Theory and Design*, 111, 1995.

[106] Kilias, T., Leimer, F., and Schwarz, W., "Chaos and Pseudorandomness" in *Nonlinear Dynamics of Electronic Systems*, Davies, A. C. and Schwarz, W. (eds.), World Scientific, Singapore, 1994.

[107] Espejo, S., Rodríguez-Vázquez, A., Huertas, J. L., and Quintana, J. M., "Application of Chaotic Switched-Capacitor Circuits for Random Number Generation," *Proc. 1989 Euro. Conf. Circ. Theory and Design*, 440, 1989.

[108] Rodríguez-Vázquez, A., Espejo, S., Huertas, J. L., and Martin, J. D., "Analog Building Blocks for Noise and Truly Random Number Generation in CMOS VLSI," *Proc. 1990 Euro. Conf. Solid-State Circ.*, 225, 1990.

Chapter 12

Robustness of Chaos in Analog Implementations

Sergio Callegari and Riccardo Rovatti
DEIS
University of Bologna
viale Risorgimento 2, 40136 Bologna, Italy
s.callegari@chaos.cc, r.rovatti@chaos.cc

Gianluca Setti
DI
University of Ferrara
via Saragat 1, 44100 Ferrara, Italy
g.setti@chaos.cc

12.1 Introduction

As the interest in chaos grows, the research focus progressively expands from speculative analysis to embrace practical exploitation. At the same pace at which new applications [1–9] are discovered, the already difficult matter of analysing chaotic behaviour becomes paired with synthesis problems and the need to implement chaotic systems in hardware. Designers are confronted with the need to produce circuits which operate in an unconventional way and do so *reliably*.

Notwithstanding the fact that the comprehension of chaotic dynamics is still partial and that only a small fraction of this knowledge is directly applicable to electronic design, useful circuits have already been proposed for several years [10–14], often to be employed as chaotic noise sources [5, 13, 15, 16]. Restricting the scope to those systems for which sufficient mathematical instruments are ready available (e.g., discrete-time Markov systems, etc.) can help keep the design manageable, still leaving enough degrees of freedom to fit the needs of many applications. A certain level of understanding of the chaotic phenomena is in fact indispensable not just to characterize the behaviour, but also to guarantee that the implementation ensures proper robustness and low environmental sensitivity. As we well know, these are measures of how a circuit can cope with adverse operating conditions, either induced externally (fluctuation in the power supply, interference, etc.) or internally (deviation from nominal component values, non-ideal matching between the component behaviour and the models adopted at design time, etc.).

Even though sample circuits can be built with an *a posteriori* verification of their robustness, such an approach is not advisable for production tasks. Furthermore, a trade-off is often involved in the reduction of sensitivity to non-ideal component behaviour, so that to improve accuracy one may have to compromise on other performance indexes or on cost. Explicitly tackling robustness/sensitivity issues from the beginning helps identify such ties and may prove useful for the evaluation of different design approaches. Clearly, one would yearn for *formal, methodic strategies*, but unfortunately, a settled, general theory for estimating the sensitivity of chaotic circuits does not yet exist. Nonetheless, there are some useful theorems and many common issues have already been identified and addressed.

12.1.1 Chaos and Analog Design

Many design areas allow us to dramatically simplify the task of dealing with robustness and sensitivity issues by the well-known technique of *data restoration*. Typically, this is done in digital circuits where signals can assume only two values: whenever some non-ideal condition brings a signal away from the allowed states, data restoration is employed to reset it to the closest allowed value, thus normally guaranteeing proper behaviour. From a strictly theoretical point of view, chaos does not allow the use of data restoration. In fact, quantization annihilates the *sensitivity to initial conditions* which is a key feature of chaos. Furthermore, limiting the allowed system states to a finite set prevents the possibility of *aperiodic trajectories* which are another cardinal feature of chaos. From a practical point of view, it is possible to mimic the behaviour of chaotic systems on digital hardware, but this approach is not suitable for every application. Thus, herein we shall deal only with analog implementations.

The design of analog circuits is far more delicate than that of digital ones, particularly regarding the point of implementation robustness: no analog electronic circuit can reflect *exactly* a given model and, from a rigorously formal point of view, differences can be striking. Just to mention a few issues:

1. An analog circuit cannot be *autonomous* because of noise and external interference;

2. An analog circuit cannot be *stationary* because of parameter drift;

3. An analog circuit has a *higher dimensional state space* than its originating model because of parasitic reactive components and internal couplings;

4. An analog circuit *deviates* from its originating model because of explicit approximations introduced by the designer and implicit approximations introduced by the use of simplified device models for describing its components;

5. No analog circuit is equal to another, because of *parameter spread*: this compels the designer to deal with an *infinite continuous set of systems* all of which must operate within specifications.

Nonetheless, from an engineering point of view, the importance of all this is relative. No one expects to match exactly a given model. What is required is to get sufficiently close to it. Formally, it is possible to think of a *cost* function which measures the *distance* of a given *physical* implementation from the ideal performance. Then, the application environment suggests a threshold and all the chips for which the cost function evaluates to less than the threshold are acceptable. In this way the concept of *production yield* is operatively defined. Obviously, the higher the yield, the better the design.

In order to produce good designs, one cannot resort to an *a posteriori* verification of the yield. Throughout the whole design process, it is necessary to try to estimate and maximize it. It is exactly this estimation process that can be hard, if not impossible, for the case of chaotic circuits, making it difficult to deal with robustness and sensitivity in the chaotic circuit design arena.

12.1.2 Estimation of Yield in Chaotic Circuits

Estimation of yield can be performed at various levels of abstraction. At least two of them are worth mentioning: the first one is the *symbolic* or *analytical* level.

At this level of abstraction, the analog designer employs the statistical information about device parameter deviation and device matching [17] together with an analytical model of the circuit (possibly based on simplified device equations). By symbolic manipulation the designer obtains statistical information about the circuit deviations from ideal performance. Since everything is done analytically, this approach is inherently hard and its complexity grows quickly with the circuit dimensions. Often it cannot be applied, or it can be applied only to selected circuit blocks.

Another important level of abstraction is the *numeric analysis level.* In this case, numeric tools are adopted and a much more sophisticated model of the circuit can be employed. The typical situation is a *Spice* simulation of the circuit with *Montecarlo analysis.* This procedure is very frequently applied.

Unfortunately, in the case of chaotic circuits, both of these approaches can become unmanageable.

The main difficulty with chaotic circuits and particularly with chaotic noise sources is that *performance has to be estimated in a statistical and functional domain.*

For instance, an application may require a chaotic source producing an output sequence characterized by a uniform distribution of its values, or by a δ-like autocorrelation. In the case of a physical chip, the process of evaluating its conformance to design specifications would consist of testing the chip in a suitable environment and in *collecting output statistics from the generated time-series.* In this way, functions such as output distributions or autocorrelation can be estimated. Then, the evaluation of a function norm allows us to compute the distance from the ideal statistical behaviour.

When one tries to migrate this process at a symbolic manipulation level, statistics is brought into action in two distinct ways: first of all, as mentioned above, device non-idealities (deviations from nominal values, mismatch, etc.) are described to the designer by means of a statistical characterization. Secondly, the system behaviour itself is characterized through statistics collected over time at the system output. The problem of determining how an input statistic (space of device parameters) influences an output statistic (deviations from ideal behaviour) over something which is *itself* characterized statistically (the system behaviour) is in most cases too difficult to be solved analytically.

Spice Montecarlo analysis, on the contrary, is mere number-crunching, so that no problems arise from a theoretical point of view. Nonetheless, the fact that behaviour must be characterized through statistics computed over time can be an important source of trouble. Collecting time-series means computation intensive *transient analysis.* For the estimation of functions such as output distributions or correlations, these must be conducted over long periods (e.g., for a discrete time system, one might have to consider 10^4–10^6 periods or even more). In order to make the Montecarlo analysis meaningful, each of these transient analyses must be repeated hundreds of times. Altogether the process can become much too time consuming, even on today's computation machines.

Hence, for the design-time estimation of yield of chaotic circuits there are two distinct obstacles: the first one is the lack of mathematical tools (for the elegant, analytical approach), and the second one is the lack of computational power (for the Montecarlo, brute-force approach). In spite of the present impossibility of foreseeing production yields, many partial results exist about how to improve them, and particularly on how to avoid their collapse.

12.1.3 The Continuity Assumption

There is a property which almost always characterizes analog circuits, when *conventional* design areas are considered. This consists of a tie in between implementation errors and behavioural errors which is locally continuous and monotonic. For instance, in a filter, varying the value of resistors and capacitors may affect the tuning. Nonetheless, one can comfortably rely on the certainty that tuning accuracy can be improved by improving the accuracy of resistors and capacitors.

This property derives from years of refinement of design practices and is so deeply rooted in the experience of designers that it is often taken for granted. The breakage of this property is synonymous with lack of robustness in the resulting circuits. In fact, if an infinitesimally small change in some parameter can drastically change the system performance, then there is a good chance that some chips will show an unsatisfactory or totally incorrect behaviour.

In areas where design practices are well established, it is clear how to avoid these situations. For instance, in linear dynamical systems the respect of suitable *phase* and *gain* margins has precisely the purpose of avoiding local discontinuities in the implementation error/behavioural error relationship. In a field such as chaotic system design, taking for granted the *continuity assumption* is surely not advisable: *bifurcations* are a fundamental concept in the study of chaotic systems, and a bifurcation corresponds exactly to a parameter set which sits on the borderline between two different dynamical behaviours. Looking for some interesting output statistics, there is a risk to tune the parameters too close to a bifurcation point, if not precisely on it. This is of course a major source of trouble. Nonetheless, circuits have been presented in the past, showing problems of this sort. Fortunately, these situations are often easy to identify, as will be clarified further on.

12.1.4 Aim of This Chapter

As mentioned above, at the present time it is not possible to deal with the problem of robustness and sensitivity of *generic* chaotic circuits. We shall then abandon a general vision, to focus on particular cases which still have importance from an applicative point of view. Herein only the synthesis of discrete-time chaotic sources will be considered, mostly focusing on those based on *piecewise affine maps*. A particular emphasis will be on circuits meant to produce uniformly distributed chaotic streams with rapidly decaying autocorrelations, because these are common requirements in many application fields. Furthermore, the hypothesis of uniform sample distribution allows some important simplifications.

Systems will be considered not only by means of mathematical models, but also by the analysis of circuits suitable for their implementation. Once again, not every possible design approach will be taken into account, but only those which — to date — have proven to be the most effective.

The opening of this chapter is dedicated to providing a brief mathematical background so that the following themes can be introduced without misunderstandings about namings and conventions. After this, the architecture of the chaotic systems under examination will be reviewed and some building blocks introduced. The core of the chapter will cover robustness and methodologies to avoid operation on the boundary of two different dynamical behaviours. For what concerns sensitivity, an example will be provided, showing how analysis conducted via numerical simulation can be useful to its improvement. Finally, in the last part of the chapter, some of the results will be generalized by means of mathematical theorems.

12.2 Robustness and Chaotic Maps

12.2.1 System Definition

The systems under examination are autonomous, stationary, and discrete-time, as in:

$$x_{n+1} = M(x_n) \tag{12.1}$$

where x is the state variable and M, defined from some subset of $\mathbb{R}$ to some subset of $\mathbb{R}$, is the state evolution function, i.e., the *system map*. From an engineering point of view, it makes sense to restrict the domain and co-domain of M to an interval $\mathcal{A}$. At this point, we consider no output variable since we need to observe the system state for considering robustness issues. Note that a first-order system like the one in (12.1) is enough to exhibit chaotic behaviour. Therefore, although some of the results proposed in the following sections can be extended to higher dimensional systems, we shall base our discussion on the simplest case only.

In (12.1), the subscripts n and $n+1$ are used to represent evolution and in a real circuit a fixed amount of time T (the *time-step* or *sampling period*) must be allowed before the system can pass from state x_n to state x_{n+1}. Hence, a *delay element* is needed in the circuit and any hardware chaotic noise source of this kind has the form schematized in Figure 12.1. The *delay element* operation is generally synchronized to an external clock source.

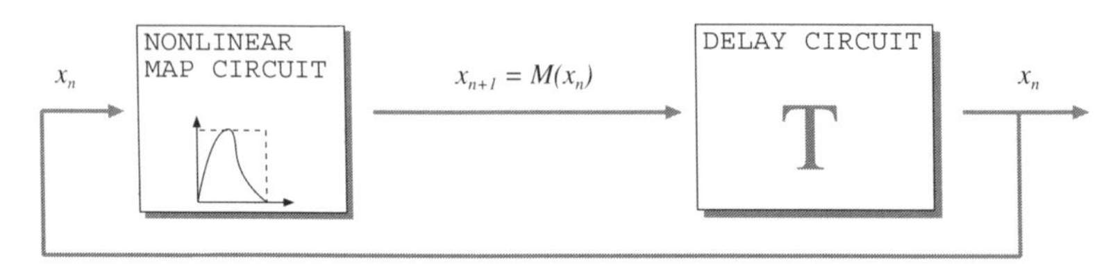

Figure 12.1: Basic structure of a chaotic noise generator based on the iteration of a chaotic map.

The maps that we are going to consider are piecewise affine (PWA). There are three main reasons for this choice:

1. PWA maps have proven able to provide a large variety of statistical behaviours;

2. Non-PWA maps are more difficult to implement accurately;

3. Many significant mathematical results exist for a subclass of PWA maps called PWA-Markov maps (PWAM).

Furthermore, the maps that we are going to consider are *eventually everywhere expanding*, meaning that a $\overline{k}$ exists so that the magnitude of the slope of M^k (k^{th} iteration of the map, as described in the following section) is always greater than 1 for all $k > \overline{k}$.

12.2.2 A Few Mathematical Definitions

In the following, we shall often refer to the definitions at the beginning of Chapter 3 and, as much as possible, we shall maintain the same notation. We also suggest other references [18–20]. One of the concepts that will be needed is that of kth iteration of a map, M^k, defined as the composition of $M(M(\ldots M(x)))$, k times if $k > 0$ and as the identity map if $k = 0$.

If we consider the maps as operators applicable to sets rather than to single values, then we have immediately the concept of *image of a set $\mathcal{X}$ through a map*, $M(\mathcal{X})$, and its extension to $M^k(\mathcal{X})$. Similarly, we can think of the counter-image of a set through a map and of its generalization $M^{-k}(\mathcal{X})$. Given these concepts, the idea of invariant set follows intuitively, as a set which is mapped into itself by M. Often, more rigorously, the invariant set is defined as a set $\mathcal{X}$ for which $M^{-1}(\mathcal{X}) = \mathcal{X}$.

For an invariant set, a *basin of attraction* exists, defined as the set of all the points which eventually fall in the invariant set through the iteration of the map. Clearly, in the domain $\mathcal{A}$ of M, many invariant sets may exist, each one having a basin of attraction larger than or equal to itself. From an engineering point of view, invariant sets which are intervals are the most interesting.

If the state of a chaotic system remains bounded within an invariant set which is an interval, for analysis purposes it is common to *rescale* the map so that the invariant set is transformed into the interval $[0, 1]$. This process is called *normalization*. In the following we shall often consider normalized maps, unless otherwise stated.

In order to define mathematical tools to analyze chaotic systems, it is necessary to operate on sets on which a *measure* is defined. Given that we operate on subsets of $\mathbb{R}$, the obvious choice is to adopt the *Borel-measure*. The measure of a set $\mathcal{X}$ is generally indicated as $\mu(\mathcal{X})$, and we shall indicate the Borel-measure as $\mu_B(\mathcal{X})$.

Most properties which have been derived for chaotic maps are valid for maps which are *non-singular*. A map is non-singular if the counter-image of every possible set of null measure has null measure.

Apart from Borel-measure, another class of measures is extremely important in the study of chaotic maps. These are measures which derive from *probability density functions* (PDFs). Given a map with an invariant interval $\mathcal{I}$, suppose that values are drawn from $\mathcal{I}$ according to a given PDF $\rho : \mathcal{I} \to \mathbb{R}^+ \cup \{0\}$. Given a subset $\mathcal{J}$ of $\mathcal{I}$ its measure can be expressed as the *probability* of drawing values from $\mathcal{J}$, i.e., $\mu_\rho(\mathcal{J}) = \int_\mathcal{J} \rho(x)\, dx$. It is important to notice that not only can a map transform *points* into *points*, indirectly it can also be thought of as a means to transform *(probability) densities* into *densities*. If a value x_0 is drawn in $\mathcal{I}$ according to a PDF ρ_0, then the PDF associated with $M(x_0)$ is defined through the well-known *Perron-Frobenius* operator associated with M (see Chapter 3).

With regard to measures derived from probability densities, a few more definitions have great importance. First of all, among the many measures one can think of, only a few can be *conserved by a map M*, so that for all $\mathcal{X} \subseteq \mathcal{A}$,

$\mu(M^{-1}(X)) = \mu(X)$. If a measure conserved by M derives from a PDF ρ, then ρ is invariant for the Perron-Frobenius operator associated with M.

Invariant densities are fundamental for defining what is necessary to have complex dynamics exploitable from an engineering point of view.

A map having $\mathcal{I} = [0, 1]$ as an invariant interval is said to be *ergodic* on $\mathcal{I}$ if given a measure μ, for every invariant subset $\mathcal{J} \subseteq \mathcal{I}$ either $\mu(\mathcal{J}) = 0$ or $\mu(\mathcal{I} \setminus \mathcal{J}) = 0$.

Ergodicity has many implications. If M is ergodic on $\mathcal{I}$, then the behaviour of the dynamical system defined by M has to be studied over the whole $\mathcal{I}$: almost all the trajectories do repeatedly pass arbitrarily close to any point in the set at arbitrarily large intervals. Furthermore, there is a unique invariant PDF and the *Birkhoff ergodic theorem* [19] states that statistics acquired from a system time-series are the same as those one would obtain from the invariant PDF.

A map for which $\mathcal{I} = [0, 1]$ is said to be *mixing* if given a measure which is conserved by the map $\overline{\mu}$ (and hence an invariant density $\overline{\rho}$ associated with $\overline{\mu}$), the following holds:

$$\lim_{n\to\infty} \overline{\mu}(Y \cap M^{-n}(Z)) = \overline{\mu}(Y)\overline{\mu}(Z) \qquad \forall\, Y, Z \subseteq \mathcal{I}$$

Even stronger is the concept of exactness which requires:

$$\lim_{n\to\infty} \overline{\mu}(M^{n}(Z)) = 1 \qquad \forall Z \subseteq \mathcal{I} \text{ so that } \overline{\mu}(Z) > 0$$

The ideas of ergodicity, mixing, and exactness are extremely important when considering the implementation of chaotic noise sources. Ergodicity implies the ability of the system to have trajectories which wander through *all* the bounded space in which the system is meant to operate. Mixing (and exactness) assure that the system is effectively *sensitive to initial conditions.* In fact, from the mixing property it derives that points which are very close together cannot evolve into trajectories contained by a flux-pipe. Trajectories will diverge apart, leading to an *unpredictable* system. Note that exactness implies mixing which in turn implies ergodicity.

Although many definitions of chaos exist, it is generally accepted that for a system to be *chaotic*, mixing is a minimal requirement. From an engineering point of view, it is more common to look for exactness, which is characterized by a definition which is more practical to handle. Hence, in a robustness analysis, *exactness* is a property to watch for. A system losing its exactness under moderate perturbation is to be treated as a non-robust system.

12.2.3 Some Robustness Issues to Consider

From what has been defined above, it should be clear that when a map is perturbed and the system dynamics consequently changes, many different properties are affected. The way in which this happens may or may not lead to qualitative changes in the system behaviour. In the following we shall observe a few qualitative changes which may easily be encountered. The list is not fully exhaustive, and the most trivial cases will not be considered. Nor does the classification of potential qualitative changes purport to be the only possible one. The illustration of robustness issues will be proposed by means of examples, still trying to keep the discussion as general as possible.

In order to know whether and how the qualitative behaviour is affected if small modifications are applied to a map, it is often necessary to know how the map is defined *outside* its ideal principal invariant set $\mathcal{I}$.

A very typical situation in the implementation of chaotic systems is to have a characteristic which is initially specified only inside the principal invariant set (because this defines the statistics). Then, in the implementation, the map is prolonged outside the IS, normally conserving the slope that it shows in the vicinity of the boundaries. However, at a certain distance from the boundaries of the invariant set, the physical limits of the system come into action and the characteristic saturates (Figure 12.2). Saturation may happen in many ways:

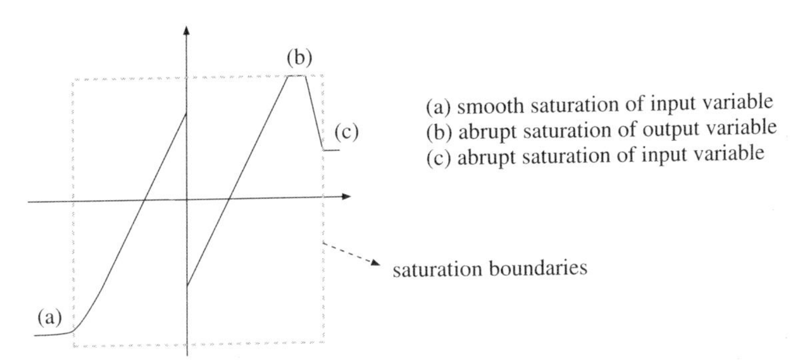

Figure 12.2: An example of the effects of saturation when a map is implemented on a circuit.

abruptly or smoothly, because of saturation in the circuit that manages the input of the functional block or because of the circuit that manages the output of it, etc. Generally, one can express saturation as:

$$M_s(x) = S_o(M(S_i(x))) \tag{12.2}$$

where S_i and S_o are input and output saturation characteristics. S_i and S_o define a bounding box around the map as shown in Figure 12.2. In some cases saturation may even happen at some internal circuit processing block leading to weird saturation characteristics. All this has marginal importance, though. The important thing is that eventually the map characteristics flatten.

12.2.3.1 Singularity

The first robustness issue we shall look at is a rather elementary one. It is the introduction of singularity, in the sense presented in the previous section. Singularity may arise if a map is implemented so that the circuit saturation points are too close to the principal invariant set $\mathcal{I}$.

If the saturation points are on the boundary of $\mathcal{I}$, then it is possible for a perturbation to bring the saturation points *inside* the invariant set. In this case, the map becomes *singular.*

Consider the following example, in which a symmetric tent map is used:

$$M(x) = \alpha x - 2\alpha \left(x - \frac{1}{2} \right)^{+} \tag{12.3}$$

where α is a parameter in $(1, 2]$ and the operator $(\cdot)^{+}$ returns its argument if this is positive, or zero otherwise. The principal invariant set for (12.3) is $[\alpha(1-\alpha/2), \alpha/2]$. Hence the normalized map can be expressed as:

$$M_n(x) = 2 - \alpha + \alpha x - 2\alpha \left(x - (\alpha - 1)/\alpha \right)^{+} \tag{12.4}$$

(Figure 12.3A). Since the invariant set is now $[0, 1]$, one can possibly think of

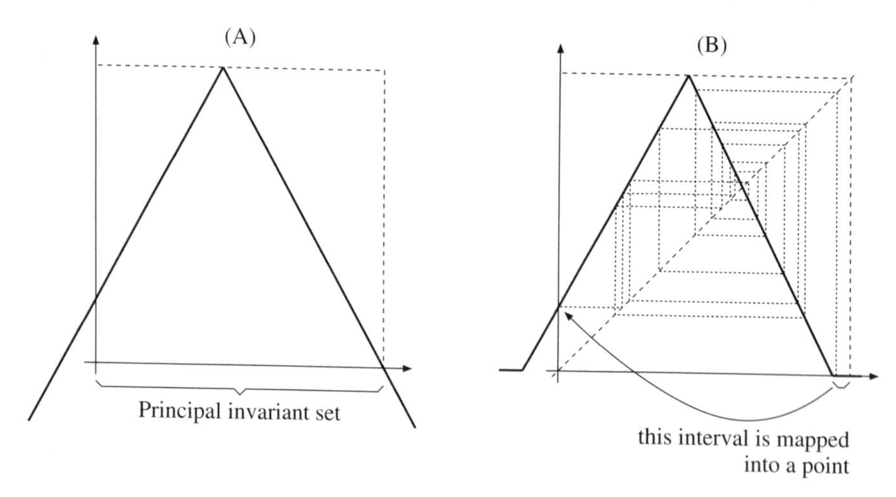

Figure 12.3: Generalized Tent Map: (A) ideal implementation; (B) implementation with slope error and saturation. (B) is singular and results in a periodic stable orbit (dotted line).

implementing this map in a circuit in which every variable is bounded to be positive. Hence we have the following saturation functions:

$$S_o(x) = \max(0, \min(K_o, x)) \qquad\qquad S_i(x) = \max(0, \min(K_i, x))$$

where K_o and K_i are constants greater than one, and the implemented map is effectively $S_o(M_n(S_i(x)))$. Now consider what happens if the map M_n is non-ideally shaped so that it results as:

$$\hat{M}_n(x) = 2 - \alpha + \alpha x - (2\alpha + \epsilon) \left(x - (\alpha - 1)/\alpha \right)^{+} \tag{12.5}$$

with $\epsilon > 0$. One gets a map such as the one shown in Figure 12.3B, which is singular, as it transforms a set of non-zero measure into a set of null measure.

The result of singularity is that the ideal principal invariant set becomes the basin of attraction of another invariant set, which is a (stable) limit cycle (dotted line in Figure 12.3). Hence the system becomes periodical and highly predictable.

Apart from pathological cases, the golden rule to avoid singularity is to assure that a certain clearance is left from the edges of the ideal principal invariant set to the points in which saturation intervenes.

Another case in which singularity can be introduced is when breakpoints are not implemented correctly. At a breakpoint, the map slope should suddenly change, and it is natural that such a behaviour can be reproduced only with approximation by an electronic circuit. If a circuit implementing a breakpoint introduces a flat map region, as shown in Figure 12.4C, then the map becomes singular.

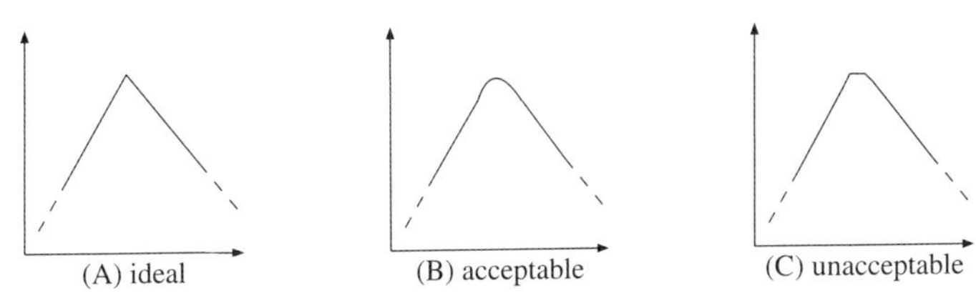

Figure 12.4: Breakpoint implementation.

12.2.3.2 Traps (loss of ergodicity)

The second robustness issue we shall consider is the emergence of *traps*. A trap appears when an invariant set $\mathcal{T}$, subset of the ideal principal invariant set $\mathcal{I}$, becomes attractive. If the measure of the basin of attraction of $\mathcal{T}$ is non-null, then there is a non-null probability (i.e., eventual certainty) of remaining trapped.

From the definition of ergodicity that we have given before, it is obvious that a system with a trap is non-ergodic.

The following example illustrates the creation of a trap due to the non-ideal implementation of a map. Consider the following map (in the family of the 'W' maps):

$$\begin{aligned} M(x) = 1 - \frac{1}{\alpha}x + \frac{1-\alpha}{\alpha(1-2\alpha)}(x-\alpha)^+ + \\ -\frac{2}{\alpha(1-2\alpha)}\left(x-\frac{1}{2}\right)^+ + \frac{1-\alpha}{\alpha(1-2\alpha)}(x+\alpha-1)^+ \end{aligned} \tag{12.6}$$

with $\alpha \in (0, 1/2)$ (also plotted in Figure 12.5A).

If, due to an implementation error the map receives an offset, so that one has:

$$\begin{aligned} M(x) = 1 - \frac{1}{\alpha}x + \frac{1-\alpha}{\alpha(1-2\alpha)}(x-\alpha)^+ + \\ -\frac{2}{1-2\alpha}\left(x-\frac{1}{2}\right)^+ + \frac{1-\alpha}{\alpha(1-2\alpha)}(x+\alpha-1)^+ + \epsilon \end{aligned} \tag{12.7}$$

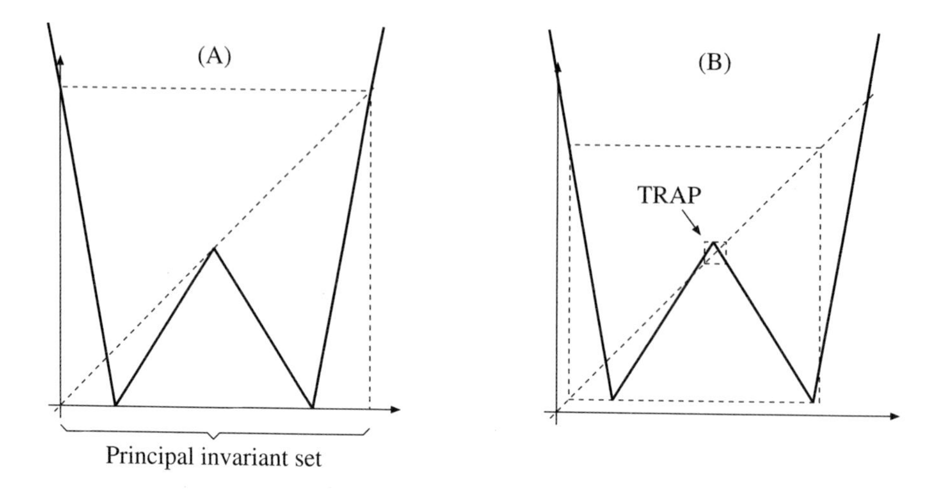

Figure 12.5: A map in the 'W' family: (A) ideal implementation; (B) implementation with offset error. (B) contains a trap.

with $\epsilon > 0$, and α is less than 1/4 then a trap appears, because from the unstable equilibrium point $P = 1/2$ a whole attracting invariant set emerges (Figure 12.5B). More precisely, at the trap the system can be locally simplified to:

$$\frac{1}{1-2\alpha}x - \frac{2}{1-2\alpha}\left(x - \frac{1}{2}\right)^{+} + \left(\epsilon - \frac{1}{2-4\alpha}\right) \tag{12.8}$$

which is the equation of a *shifted and stretched* tent map such as the one in Equation (12.3).

The resulting trapping behaviour is exemplified in the time-series shown in Figure 12.6.

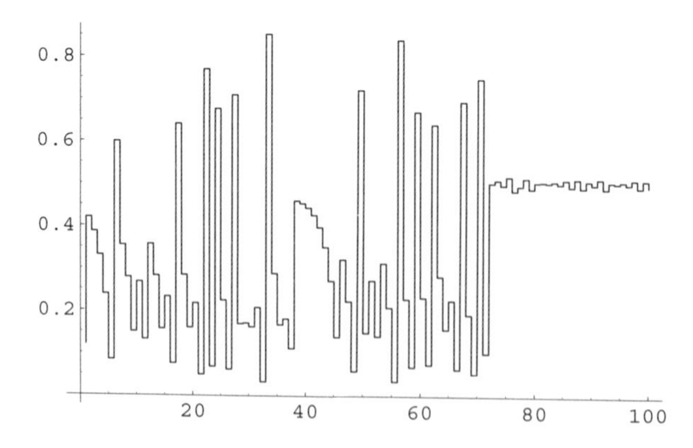

Figure 12.6: Time-series showing a trapping behaviour.

It can be observed that problems of this kind can arise in every map where a breakpoint inverting the map slope is placed on the identity axis $y = x$. Unfortunately, this is a situation that can occur very frequently if one decides to implement maps taken from textbooks. This is because maps shown in books make a heavy use of symmetry to keep the analysis simple and do often have breakpoints which are also fixed points. It can be shown that non-robust trapping behaviours can surely arise if the map slopes on the sides of the fixed breakpoint are μ_1 and μ_2, and $|1/\mu_1 - 1/\mu_2| < 1$.

Another example of trapping occurs if a map has a branch whose slope magnitude is one (or less than one) passing close to the axis $y = x$. In this case it is trivial to show that offset errors can create a stable equilibrium point.

Apart from the two examples mentioned above, it is possible to have more complex trapping behaviours. In these cases, more than two map branches came into action to define the trap. Obviously, the analysis is far more intricate. However, criteria may exist even for these situations. The criteria given in Section 12.5 are obviously relevant. For now, intuitively accept that the two trapping behaviours exemplified above account for the large majority of trapping-like non-robust behaviours that can be found in practical cases.

12.2.3.3 Weak traps (intermittence)

Traps, as described in the previous section, do completely lock the system behaviour at some small set.

However, to influence disruptively the system statistical behaviour, it is not necessary to completely lock the state. It is in fact sufficient to have a region characterized by an unfortunate distribution of *get-out* times.

Given a system which is chaotic on a principal invariant set $\mathcal{I}$ and a subset $\mathcal{T}$ of $\mathcal{I}$, the get-out time distribution for $\mathcal{T}$ can be defined pretending that at a time n_0 the system state x enters $\mathcal{T}$. Then one can build the series:

$$(\Pr(x_{n0} \notin \mathcal{T}), \Pr(x_{n0+1} \notin \mathcal{T}), \Pr(x_{n0+2} \notin \mathcal{T}), \dots)$$

where Pr expresses the probability of an event. Clearly, the series starts at zero and converges to one. Slow convergence means that once entered in $\mathcal{T}$ the system state tends to remain blocked there for long times. In this case, it is common to talk of *intermittent* behaviour, because the system shows regular dynamics for most time and then, intermittently, has its state blocked inside the region $\mathcal{T}$.

An example of an intermittent system is offered by the same 'W' map used in Section 12.2.3.2. Consider Equation (12.7) when $\epsilon > 0$ and $\alpha > 1/4$: the resulting behaviour is as shown in Figure 12.7. The intermittent behaviour is self evident: for most time the system state wanders through the whole invariant set, but at some times it remains locked inside a neighbourhood of 1/2.

The conditions which favour the emergence of weak traps are very similar to those which cause the emergence of regular traps. Once again common problems are caused by breakpoints which invert the map slope placed on the axis $y = x$. If the slope condition $1/\mu_1 + 1/\mu_2 < 1$ mentioned in Section 12.2.3.2 is not satisfied, then there might not be a regular trap. Nonetheless, there is still a chance for intermittence, as pictured in the example above.

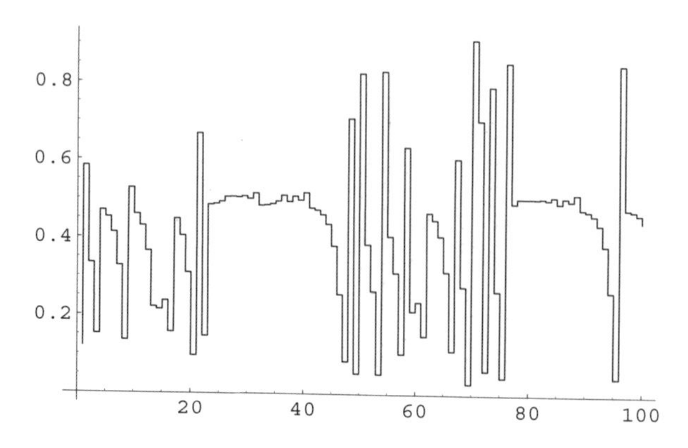

Figure 12.7: Time-series showing an intermittent behaviour.

Another case in which intermittence emerges is if a map branch having slope magnitude only slightly greater than one passes close to $y = x$.

As for regular traps, there are many more situations in which intermittence can appear and which are not so trivial. However, once again, consider that the two situations described above (breakpoints on $y = x$, low slope branches in the vicinity of $y = x$) do account for most problems in practical situations.

12.2.3.4 Effects of noise on traps and singularity

It is well known that there are situations in which noise can be beneficial to chaotic systems. It is common to be offered examples in which noise makes a system highly unpredictable, because sensitivity to initial conditions amplifies its effects at every map iteration.

However, noise can be beneficial also to systems subject to traps or weak traps. In fact, if the noise floor is comparable to the Borel measure of the trap, then it can create a get-out path from the trap or influence positively the distribution of get-out times for a weak trap. A region which appears to be trapping or weakly trapping in absence of noise can have negligible effects on the output time-series in the presence of noise.

Also maps which are made singular by non-ideal implementations can benefit from noise. In fact, singularity causes a whole region of the invariant set (with non-null measure) $\mathcal{X}$ to be mapped into a single point or set of points having measure zero $M(\mathcal{X})$. The effect of noise is however to *spread* each mapping: if a point a is meant to be mapped into $M(a)$, then the effect of noise is to have a mapped into a point that can be anywhere in a whole region surrounding $M(a)$. Consequently the set $\mathcal{X}$ corresponding to the singular region on the map will be mapped *within* some neighborhood of $M(\mathcal{X})$ having non-null measure. Of course, the beneficial effects of noise can be appreciated only if the singular regions on the map have an extension which is not much larger than the noise floor.

12.2.3.5 Maps with discontinuities: splitting of the invariant set and other issues

Another robustness issue emerges in maps which are non-continuous. In fact, from a engineering point of view, it is generally a requirement to have a behaviour which is ergodic on an interval. Errors in the implementation of the map can lead to the splitting of the invariant set in a union of disjoint intervals. For instance, consider the following map:

$$M(x) = -k\left(x - \frac{1}{4}\right)\chi_{(-\infty,1/4]} + \left(1 + k\left(x - \frac{1}{2}\right)\right)\chi_{(1/4,1/2]} - \left(1 - k\left(x - \frac{1}{2}\right)\right)\chi_{(1/2,3/4]} + k\left(x - \frac{3}{4}\right)\chi_{(3/4,+\infty]} \quad (12.9)$$

where $\chi_{\mathcal{A}}$ is the characteristic function of the set $\mathcal{A}$ which evaluates to one if $x \in \mathcal{A}$ and zero otherwise. The constant k is meant to be set to 2 to give the map shape shown in Figure 12.8A.

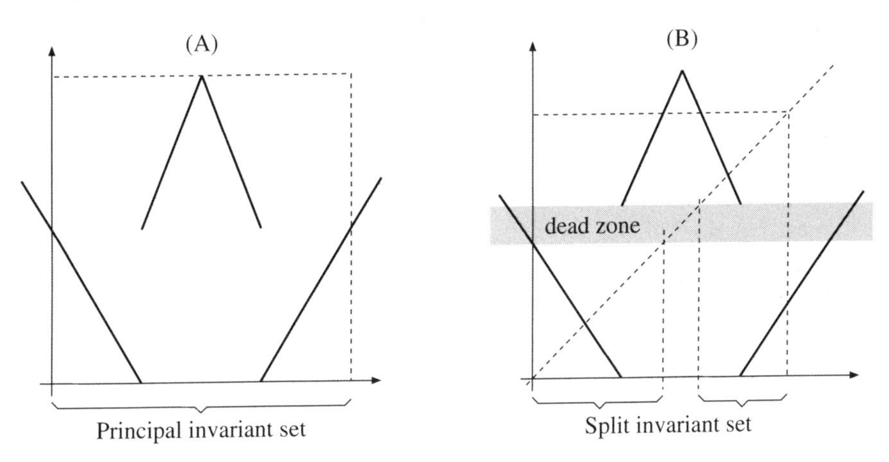

Figure 12.8: The map in Equation (12.9). Ideal case, with $k = 2$ (A) and a possible implementation with $k = 2 - \epsilon$ (B).

If, due to an implementation error, k is set to $2 - \epsilon$, with $\epsilon > 0$ then the map is transformed as shown in Figure 12.8B. The consequence is that the state is prevented from entering a region $(1/2 - \epsilon/4, 1/2 + \epsilon/4)$, as in Figure 12.9. The system is not ergodic on the interval $[0, 1]$, however it is ergodic on $[0, 1/2 - \epsilon/4] \cup [1/2 + \epsilon/4, 1 - \epsilon/2]$.

In some cases, there is not a real separation of the invariant set as illustrated above, but simply a sharp fall or a sharp increase in the probability of finding the state in some region.

From a mathematical point of view, it must be noticed that maps having discontinuities are often not as well known as continuous maps. However, it should also be considered that the physical implementation of a discontinuity is generally not possible. From a practical point of view, maps with discontinuities

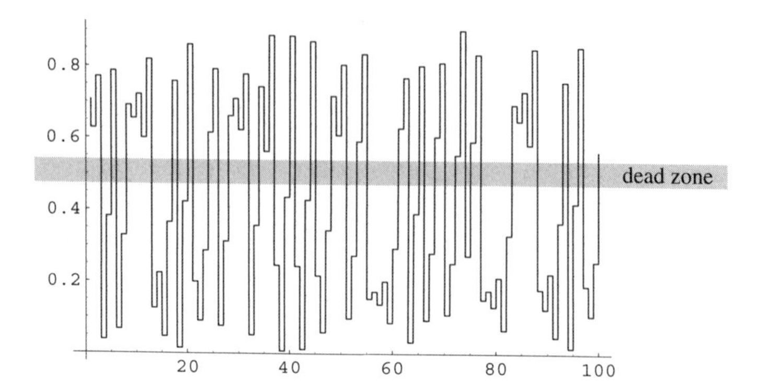

Figure 12.9: Time-series relative to the map in Figure 12.8B. The splitting of the invariant set and the appearance of a dead zone can be easily spotted.

can (and should) often be studied as maps in which branches with very large slopes are substituted for the discontinuities.

12.2.3.6 Escape from the principal invariant set

A typical result of a map perturbation is obviously a change in the system invariant set. This means that if $\mathcal{I}$ is the principal invariant set for a map M, a badly implemented version of M, say $\hat{M}$ is likely not to have $\mathcal{I}$ among its invariant sets. We say *the invariant set changes* because of an underlying assumption of continuity. When moving from M to an $\hat{M}$ very close to it, we expect to be able to find an $\hat{\mathcal{I}}$ invariant set for $\hat{M}$ very close to $\mathcal{I}$. However, this might not be the case, and is obviously a major robustness issue to consider.

The saturation introduced with the implementation does create parasitic stable equilibrium points and/or parasitic stable limit cycles *outside* the principal invariant set $\mathcal{I}$ (Figure 12.10). This is of course unavoidable and due to the fact that saturation lowers the map slope towards zero, either progressively or sharply. The parasitic stable equilibrium is normally not a problem, because it remains unreachable outside the system invariant set. However, in some cases, misimplementation and/or noise can make the parasitic equilibrium reachable.

A typical example is offered by tent map systems (12.3). For tent maps, saturation does usually introduce a parasitic equilibrium point at some x_E lower than the values in the ideal principal invariant set (Figure 12.10A). In order to obtain a uniform distribution of the output samples, it is common to set the map slope α to 2. In this case, however, a slope error rising the breakpoint over $y = 1$ does make the basin of attraction of the parasitic equilibrium point x_E intersect the ideal principal chaotic invariant set of the system. The consequence is that both the principal invariant set and the chaotic behaviour are dissolved and that the system settles at the parasitic equilibrium point (Figure 12.11).

Apart from tent-map systems, many other systems can have problems of this

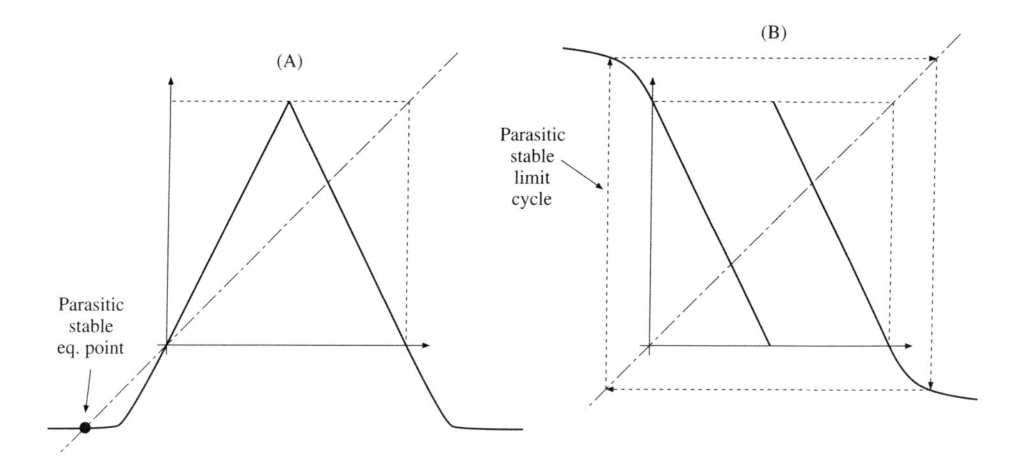

Figure 12.10: Parasitic stable equilibrium points (A) and limit cycles (B).

kind. For instance, also the Bernoulli shift:

$$M(x) = 2x \bmod 1$$

is non-robust in this sense. A situation in which problems of this sort do always occur is when a map crosses the axis $y = x$ at (one of) the boundaries of the principal invariant set.

This phenomenon is a particularly subtle one, because many common formal robustness criteria (for example refer to [34]) do not catch this kind of non-robust behaviour. This is because, for the sake of simplicity, many robustness criteria make very strong assumptions on the perturbations to be considered. If a normalized map is considered, perturbations which make $M(x)$ lay outside $[0, 1]$ for $x \in [0, 1]$ are generally excluded. Unfortunately, this kind of non-robust behaviour is very common. Indeed, it is so common that when circuit designers started tackling the problem of implementing chaotic maps in hardware, it was the first robustness issue to be identified [10, 14].

In order to avoid escape from the ideal principal invariant set and to guarantee smooth changes in the invariant set in presence of perturbations, it is generally enough to assure that the basin of attraction $\mathcal{B}$ of the principal invariant set $\mathcal{I}$ is larger than $\mathcal{I}$ and *contains* $\mathcal{I}$ *with some clearance* (it is also common to refer to this situation speaking of *into* vs. *onto* maps [21]). We say that $\mathcal{B}$ contains $\mathcal{I}$ with clearance if for every couple of points (a, b) with $a \in \mathcal{I}$ and $b \notin \mathcal{B}$ we have that the set $\mathcal{X} = \{x | x = \lambda a + (1 - \lambda)b, \lambda \in [0, 1]\}$ is defined so that the measure of $\mathcal{X} \cap (\mathcal{B} \setminus \mathcal{I})$ is not null (Figure 12.12).

If such a condition is not respected, as for the tent map case illustrated above, then there are techniques for enlarging the basin of attraction of the ideal principal invariant set with regard to the invariant set itself. Some of these techniques require modifications of the map *inside* the ideal principal invariant set, while others require modifications to be applied *outside*. The main difference in between the two approaches is that if we accept to modify the map inside its ideal

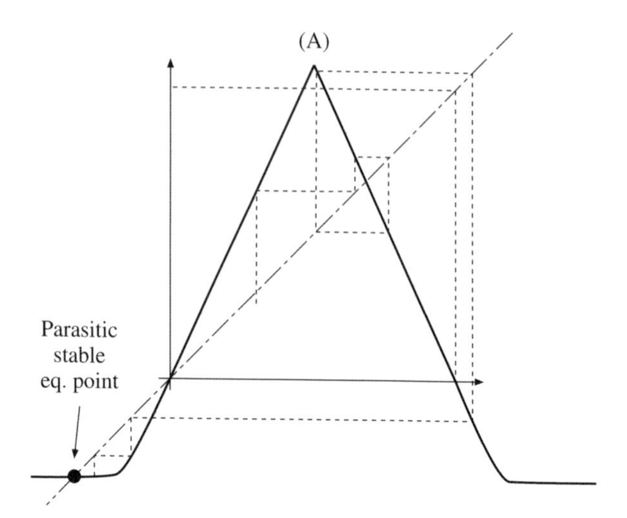

Figure 12.11: Escape from the ideal principal invariant set towards an attractive parasitic stable point.

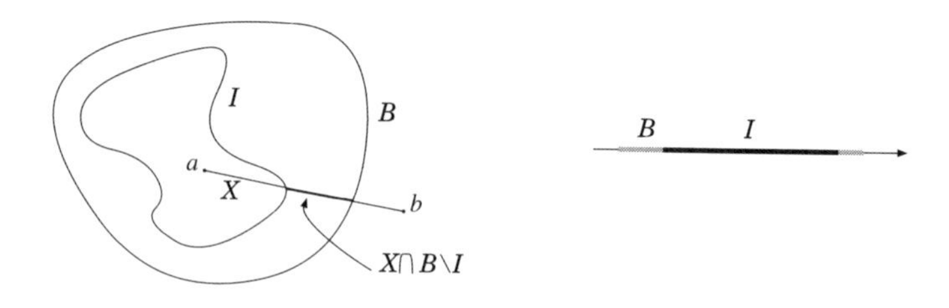

Figure 12.12: Invariant sets contained with clearance in their basin of attraction. Bi-dimensional and unidimensional cases.

principal invariant set, then we accept an *a priori* modification of the system statistical behaviour. In some cases, it is however possible to affect some statistical properties which are not application critical, leaving some others untouched. In the following paragraph a few of these techniques are illustrated by means of examples. As a sample case, the tent map will be considered, but it is possible to imagine extensions of the proposed techniques to many other maps.

Decrease of the map slope. This is the first strategy that has historically been adopted in order to make tent-map and Bernoulli-shift systems robust. It requires us to modify the map within its principal invariant set and it affects *all* the system statistical properties. Hence, it is not very effective and it is reported here mainly because of its simplicity.

If the slope of a tent map system is globally diminished (Figure 12.13A), then the system principal invariant set is made smaller and some clearance is created with regard to its basin of attraction (Figure 12.13B). At this point, a perturbation of the map taking the breakpoint to a higher value does not harm the system behaviour, because the system state is always re-injected into the set where it should belong (Figure 12.13C). As mentioned above, this technique is suitable also for Bernoulli-shift systems (Figure 12.13D).

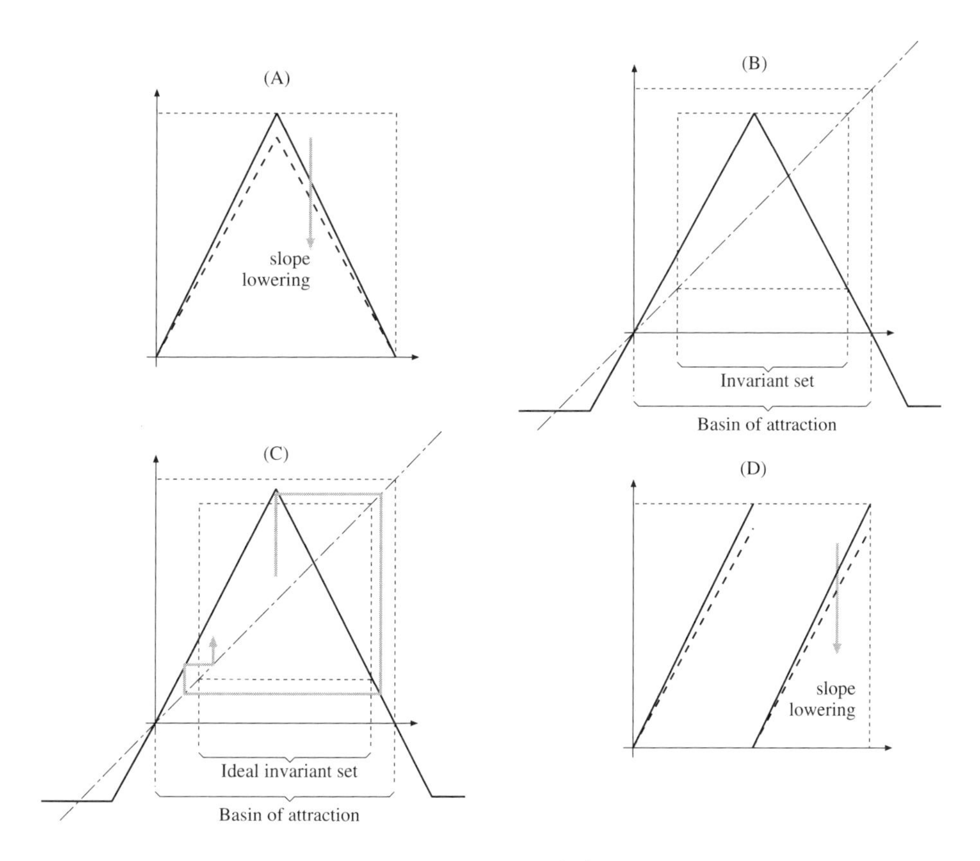

Figure 12.13: The slope decrease technique: (A) application to a tent map system; (B) reduction of the invariant set and creation of a clearance with respect of its basin of attraction; (C) state escape from the unperturbed invariant set and re-injection; (D) application of the same technique to a Bernoulli-shift system (the state re-injection mechanism is slightly different).

The most important advantage of this technique is that no additional breakpoint is required to make a system robust. A system which, by unfortunate design, is subject to the escape of the system state from the specified invariant set can be made robust with minimal changes. The disadvantage is that the statistical properties of the system are all affected. Symmetric tent map systems are usually meant to produce output samples characterized by uniform distribution, δ-like autocorrelation and thus a flat power density spectrum. This is achieved by setting:

$$M(x) = 1 - 2\left|x - \frac{1}{2}\right|$$

However, robustness can be achieved by setting:

$$M(x) = 1 - 2(1-\mu)\left|x - \frac{1}{2}\right|$$

where $\mu \in (0, 1/2)$ and close to zero. The larger μ the greater the robustness of the system, but the further the sample distribution, the autocorrelation and the spectrum from the ideal ones. As an example, Figure 12.14 shows the probability density function and the autocorrelation of a lowered slope tent map system where μ is set to 0.025 (2.5% slope decrease). The variation in the IS boundaries is obviously not a concern, as it is very easy to compensate for it by output rescaling, but the change in shape in the PDF and in the autocorrelation can be critical to many applications.

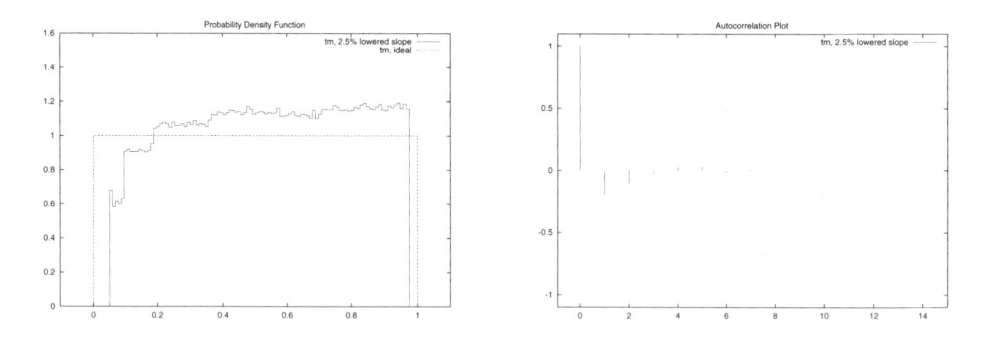

Figure 12.14: Properties of a lowered slope tent map system.

Tailing. Tailing is another technique which modifies the map inside the invariant set. However, when applied to systems having a uniform probability distribution of the output samples, tailing preserves this statistical property, and only second-order statistics (such as autocorrelation and power density spectrum) are affected.

Tailing consists in joining a scaled down version of the map with a "tail" (a linear branch having slope -1), as shown in Figure 12.15. Also in this case, clearance is created in between the invariant set and its basin of attraction, as shown in Figure 12.16A, for the case of a tent map. Figure 12.16B shows the mechanism of state re-injection when the map is perturbed.

Once again, the stronger the action with regard to the original map (i.e., the larger the tail), the more robust the resulting system. However, second-order statistics are affected. A very nice property of tailed-tent-map systems (and tailed-shift systems) is that by modulating the tail size, both the system level of sensitivity to initial conditions (as expressed by the Lyapunov exponent) and the output sequence autocorrelation decay can be tuned. For example, Figure 12.17 shows the autocorrelation decay for two tailed tent-map systems characterized

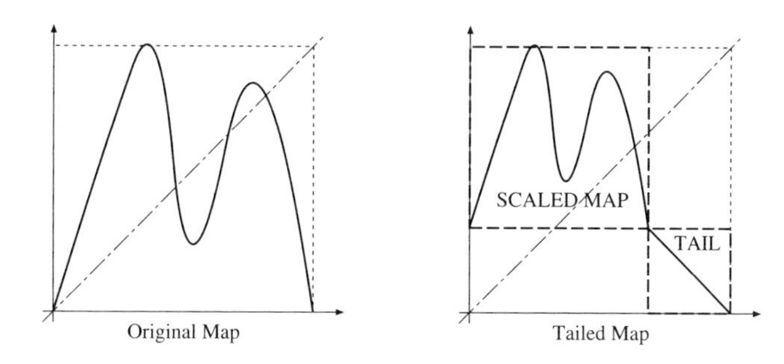

Figure 12.15: The tailing technique.

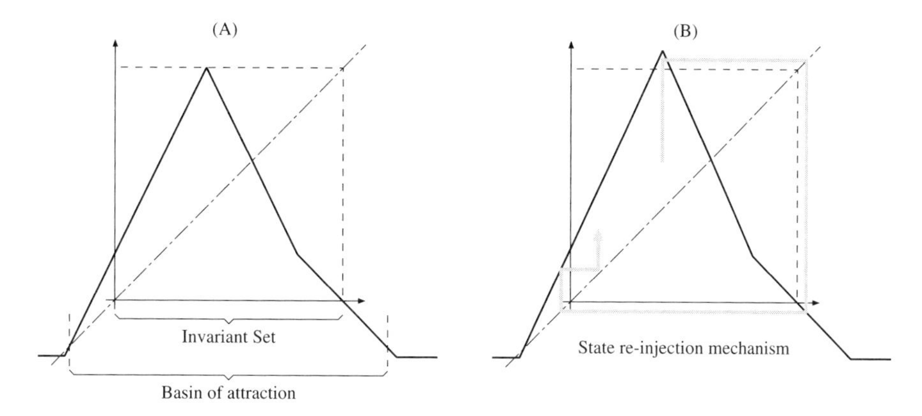

Figure 12.16: Tailed Tent Map: invariant set and basin of attraction (A); state re-injection mechanism when slope errors occur (B).

by a 1% and a 5% tail, as in:

$$M(x) = 1 - \left| x - \frac{1-\Phi}{2} \right| + (x + \Phi - 1)^{+}$$

where the tail coefficient Φ is set to 0.01 first and to 0.05 later. This tuning feature is analogous to the one used in Chapter 3 to optimize the performance of certain telecommunication systems.

Cutting and Flipping. This technique also involves modifying the map shape inside the invariant set and can be applied to systems having a uniform invariant probability density function, just like the tailing technique.

It consists in literally *cutting* the original map into pieces and *composing* the pieces in a different way. Figure 12.18 shows an example, with regard to a tent map system. If the operation is performed with care, the chaotic behaviour is preserved, and a different system is obtained, still characterized by a uniform invariant probability density function. Robustness can be achieved because this

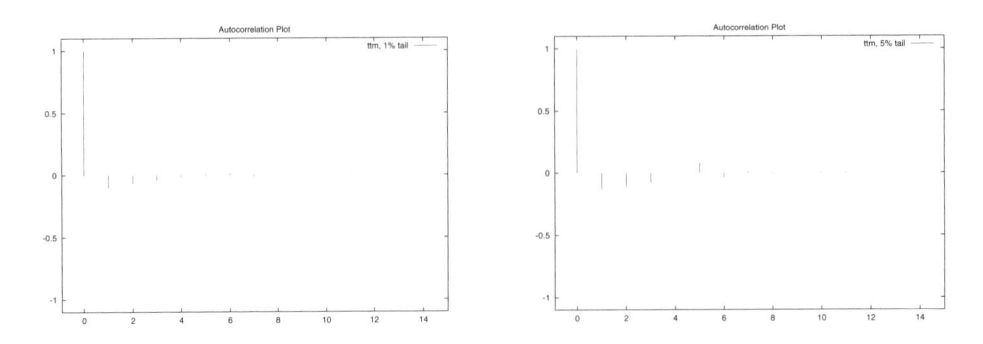

Figure 12.17: Tuning of the auto-correlation in tailed tent map systems.

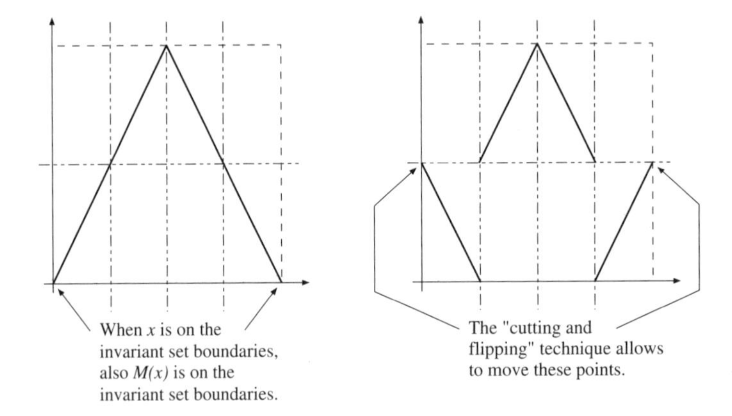

Figure 12.18: Cutting and flipping technique applied to the tent map.

technique allows us to enlarge the basin of attraction of the principal invariant set, without affecting the dimensions of the principal invariant set itself.

Actually, this technique, that was used in [22] is sort of middle-way in between techniques which alter the map within the invariant set and techniques which alter it outside. In fact, the map is changed within the invariant set just in order to move its value at the invariant set boundaries away from the boundaries themselves (Figure 12.18). In this way, the extrapolation of the map terminal branches is allowed to enlarge the IS basin of attraction. Figure 12.19 shows the enlargement and the resulting state re-injection mechanism.

Map cutting and flipping must be performed with care. In fact it is possible to relieve one robustness issue and to introduce other problems. For example, the map shown in Figure 12.20A, derived from a tent map using cutting and flipping is robust with regard to the escape from the invariant set, but is non-robust with regard to trapping behaviours. The map shown in Figure 12.20B, once again obtained from a tent map, is not exact.

It is obvious that cutting and flipping can preserve a uniform probability density function, but that it will not preserve other statistical properties. However, maps showing a uniform invariant PDF and which are *doubly symmetric* [22], are

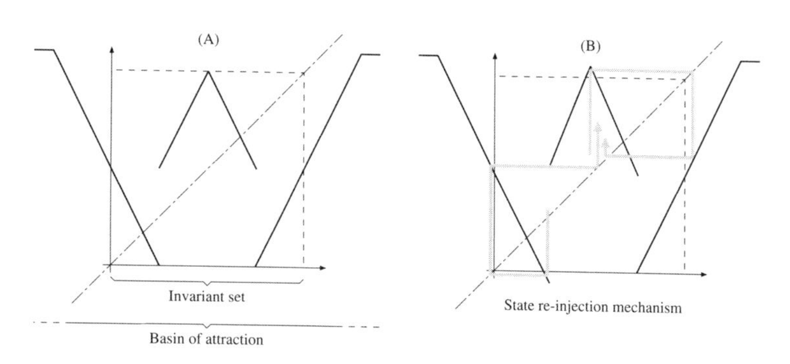

Figure 12.19: Enlargement of the basin of the attraction of the principal invariant set by means of the cutting and flipping technique (A). State re-injection mechanism (B).

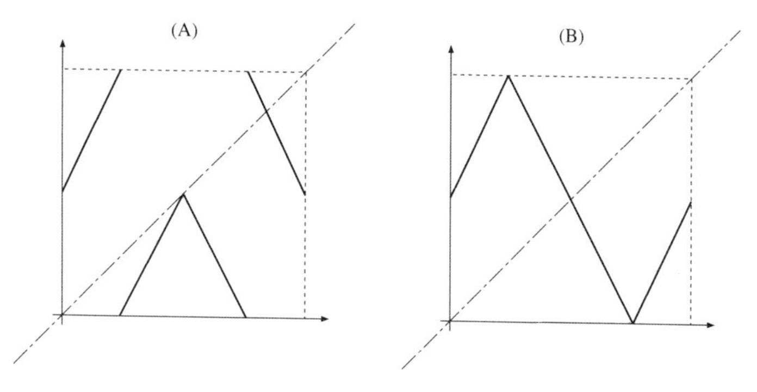

Figure 12.20: Other maps obtained from the tent map by means of the cutting and flipping technique. These maps solve one robustness issue, but introduce other problems, such as the trapping behaviours (A) and the loss of exactness (B).

characterized by δ-like autocorrelations. Hence, if one can think of a cutting and flipping strategy which preserves this symmetry, then also δ-like autocorrelation and power spectrum properties can be conserved. The map in Figure 12.18 is characterized by a δ-like autocorrelation and by a uniform power density spectrum.

Hooks. The last technique that shall be presented in order to prevent escape from the ideal principal invariant set involves changes in the map to be applied *only* outside the map principal invariant set. Clearly this technique does not affect *a priori* any statistical property of the system.

The idea underlying this technique is quite simple. Branches are added to the map outside the principal invariant set just in order to *bounce* the system state inside the region it should belong to if, for some reason, the state is found to be outside. These additional branches shall be referred to as *hooks*, just like those routines in computer programs which activate when some event triggers them.

Although simple, to the best of our knowledge, there are no examples of the application of the hook technique in the literature. The main problem is that it appears expensive, since some breakpoints must be introduced to define the hook branches. However, in many cases a single additional breakpoint can be enough. For example, Figure 12.21 shows a system based on the tent map made robust by the addition of a single hook.

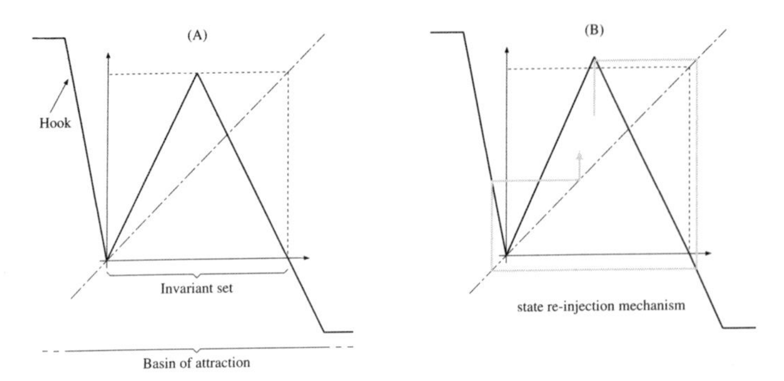

Figure 12.21: Hooked tent map. Enlargement of the basin of attraction of the invariant set (A). Re-injection mechanism for the system state in case of slope errors in the implementation (B).

When adding hooks, care must be put in avoiding the introduction of different robustness problems. In the case of Figure 12.21, for example, a hook having slope less or equal to 2 cannot be introduced without introducing trapping mechanisms. If the hook slope is only slightly higher than 2 then intermittence becomes possible. A sufficiently high slope or a discontinuity in the origin are enough to prevent these problems.

12.2.3.7 System startup failure

The problem of system startup is strictly connected with the problem of assuring a sufficiently large basin of attraction to the system principal invariant set, as illustrated in the previous section.

In order to assure proper system start-up, special circuitry can be added so that at the system switch-on, the state is forced inside the principal invariant set. Alternatively, it is enough to force the system state inside the basin of attraction of the principal invariant set, since this assures that the state will eventually reach the region where it is meant to remain bounded.

Although in specific cases some assumptions can be made, if no special circuitry is adopted the system state at the startup is generally undetermined. In this case, the system startup behaviour is left to chance and may result in a severe robustness issue. For example, in a certain lot of systems, some may start up correctly and others may fail, depending just on some parasitic capacitance value. However, it can be intuitively accepted that the larger the basin of attraction of the principal invariant set, the higher the chances of having the switch-on state within it and to start up correctly.

If the basin of attraction of the principal invariant set is large enough to cover all the possible values the state can assume in the physical system, then the system will always start up correctly. Here is where saturation can be actively exploited. Saturation assures that the system state remains bounded. If the basin of attraction is larger than the saturation boundaries, then the system startup will have no chances to fail.

The previous section illustrates various techniques which have the effect of enlarging the basin of attraction of the principal invariant set with regard to the principal invariant set itself. Some of this techniques do introduce only some limited clearance (decrease in the map slope, tailing) and hence are not sufficient to substitute for a proper startup circuitry. Other techniques allow arbitrary expansions of the basin of attraction of the system invariant set (cutting and flipping, hooks) and can hence effectively avoid the burden of a dedicated start-up circuit.

12.3 Robust Implementation of Chaotic Maps

In this section, considerations about robustness and sensitivity will be directly tied to circuits and circuit blocks that can be used for the implementation of discrete time chaotic systems based on piece-wise affine maps. The purpose of this section is not to describe implementation strategies, which can be conveniently found in Chapter 11 of this book, but rather to look at the unavoidable implementation errors and at their effects on the global system behaviour. Henceforth, only the minimal details will be given about architectures, alternative design paradigms, and circuit blocks.

Any analog discrete-time chaotic sample generator, as shown in Figure 12.1 can be split in two major elements: a map circuit and an analog memory, which is meant to operate as the analog counterpart of a digital edge-triggered register. From each of these two blocks, a high level of accuracy is expected. It is necessary to remark from the very beginning that circuits operating in a chaotic loop must be *very* accurate, and that the bounds in which errors must be constricted are usually much tighter than those that can be accepted in other fields of analog design. Gain errors, for instance, are generally translated into slope errors in the map and even very low errors can result in large variations of the output statistics from the expected behaviour. For instance, Figure 12.14 shows the output sample distribution for a tent map system in which the map slope, meant to be exactly 2, has been lowered by 2.5%. It is self-evident that even such a small error is enough for the distribution to largely drift away from uniformity.

Hence, a rule in the design of chaotic systems is never to have gains or references expressed as absolute physical quantities, but to design circuits where gains and references are expressed as ratios of homogeneous physical quantities. This allows us to rely on relative parameter accuracy which in integrated circuits can be extremely good. Following this elementary consideration, the design strategies which are best established so far, exploit transistor replication or capacitor replication to express all the fundamental system parameters, and to avoid the use of gain blocks based on operational amplifiers and resistors.

One of the problems connected with the high accuracy levels that are required is that in analog circuits accuracy is often paid in terms of other performance indexes. In the case of a discrete-time circuit, accuracy is generally paid in terms of cycle-time. Hence every effort should be put into determining the required accuracy levels and going no further. Unfortunately, this task is unmanageable in general analytical terms, so that extensive simulation is always needed. In any case, as the system is composed by the cascade of a map circuit and a memory circuit, the global accuracy of the map that is iterated is dominated by the worst of these two elements and there is not much point in pushing one subsystem at accuracy levels much higher than the other.

Contrary to its role in other analog design fields, noise is generally not of much concern in the design of chaotic systems. Signal-to-noise ratios as low as 30 dB can be easily tolerated, as shown in Figure 12.22. However, it is very important

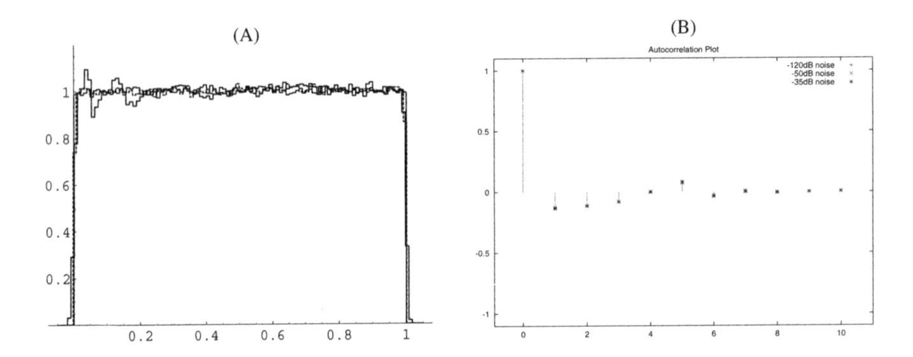

Figure 12.22: Effects of noise on a tailed tent map system (tail=5%). Noise levels equal to -30 dB, -50 dB, and -120 dB with regard to the maximum signal have been simulated, obtaining plots for PDF (A) and autocorrelation (B). Smaller spikes in the PDF plot are effects of the limited sequence length.

that systematic errors are not misinterpreted for noise. Another important point is that a side effect of noise is to enlarge the set in which the system state is allowed to wander (note the extremes of the plots in Figure 12.22A). As a consequence, the system may become more sensitive to "escape from the invariant set" robustness problems. For instance, the system simulated in Figure 12.22 cannot tolerate noise-to-signal levels worse than ~ 27 dB.[1]

All the examples that will be proposed in this section follow the current mode design approach to chaotic signal sources. In this approach, the system

[1] The noise levels are expressed considering Gaussian noise and calculating the maximum noise level ($\pm 3\sigma \Rightarrow 6\sigma$) to the invariant-set-width ratio.

state is represented by a branch current in the circuit. This is believed to be the best established approach so far, although strategies which make use of switched capacitors are being investigated.

12.3.1 Architecture

In the current-mode implementation of discrete-time chaotic systems based on piecewise linear maps, it is possible to identify a minimal set of building blocks. These comprise:

1. current comparators, which are necessary to introduce discontinuities and can be used also to introduce breakpoints;

2. current rectifiers, which can be used to introduce breakpoints;

3. current mirrors, which are used as accurate gain blocks to set the branch slopes;

4. current mode sample-and-hold or track-and-hold circuits, which are used to build up the current mode memory (delay) element.

The idea is to use an array of current rectifiers/comparators to build up a set of functions characterized by a single breakpoint/discontinuity, and then to compose them into the desired map by using a linear-combiner made up of current mirrors and summing nodes (Figure 12.23). Then a master-slave cascade of two S/H or T/H blocks can realize the required memory element.

Obviously, more effective strategies can be devised, in order to distribute more efficiently the parasitic capacitances (maybe in tree-structures), or to reduce the total current (and hence power-consumption) which is required by the PWA signal processing. Such strategies may become necessary especially when implementing maps with a large number of breakpoints. However, they do not change the set of required building blocks nor they do invalidate the considerations that will follow in the next sections, where some of the building blocks will be examined in order to obtain design guidelines.

12.3.2 Design Blocks

12.3.2.1 Current mirrors

Current mirrors are very basic and well-known building blocks. Considerations about the need of providing redundant gate areas for improving matching [17] and introducing cascode arrangement to improve the overall linearity and accuracy are henceforth elementary. What is worth noticing is that when current mirrors are used in a discrete-time environment, their step response time becomes an important design factor.

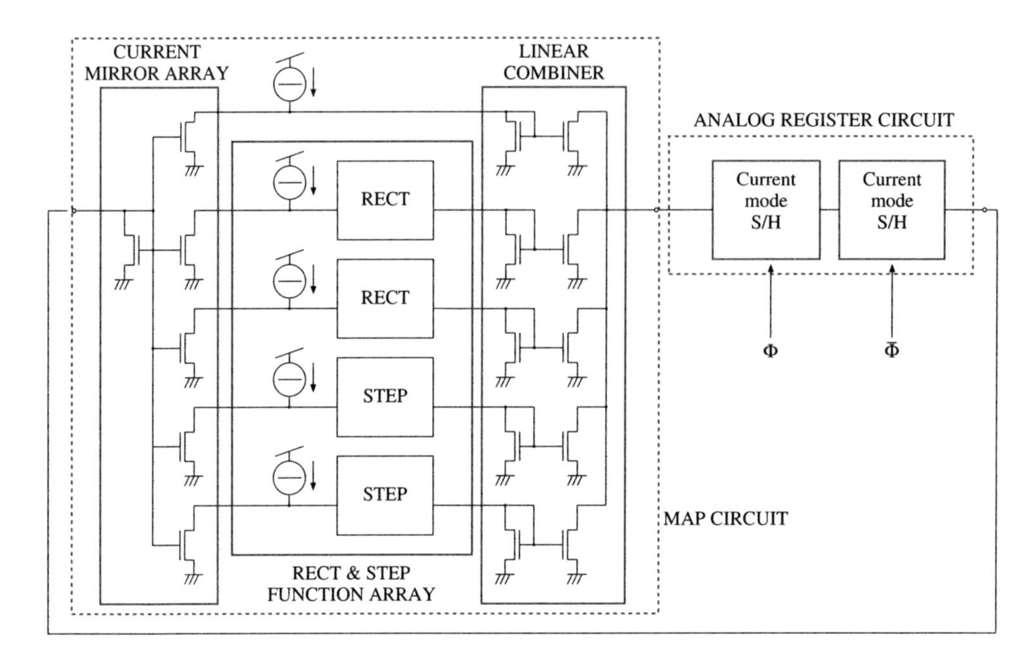

Figure 12.23: Basic current mode implementation of a chaotic discrete time system based on the iteration of a mono-dimensional PWA map. The evaluation of discontinuities and breakpoints occurs in parallel. "STEP" blocks are comparators (in charge for discontinuities), while "RECT" blocks are current rectifiers (in charge of slope discontinuities). The reference currents and the current mirror ratios are the *parameters* that select the map shape.

The step response of current mirrors has been studied in [23], obtaining the model:

$$T_{settle} = \frac{C_G}{\sqrt{2I_f\beta}} \ln\left(\frac{\sqrt{I_i}-\sqrt{I_f}}{\sqrt{I_i}+\sqrt{I_f}} \frac{\sqrt{I_f - \text{sign}(I_f - I_i)I_{settle}}+\sqrt{I_f}}{\sqrt{I_f - \text{sign}(I_f - I_i)I_{settle}}-\sqrt{I_f}}\right) \tag{12.10}$$

where C_G is the total capacitance at the common gate of the current mirror MOS transistors, I_i and I_f are respectively the current at the beginning and at the end of the input step, $\beta = \beta' W/L$ is the MOSFET transconductance parameter and I_{settle} is the output settling band in which the output is assumed to have reached its final value. The plot in Figure 12.24 shows the settling times for a current mirror designed on a well-established 0.8μm process, with $W = L = 4\mu$m and currents in the range 0–10μA.

As can be seen, the settling time grows very rapidly when current I_f is diminished, and then saturates for I_f values as large as the settling band. Note also that this model is valid only for transistors operating in the strong inversion region and becomes *too optimistic* when the transistors are driven in the weak inversion region.

In a typical operating condition, the settling band is defined as a fraction of the range of currents on which the mirror is meant to operate, say $[I_{min}, I_{max}]$,

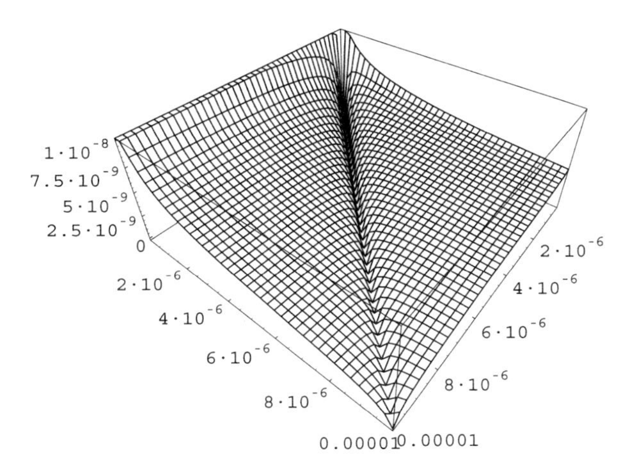

Figure 12.24: Response time of a current mirror implemented in a 0.8μm technology, with $W = L = 4\mu$m and stimulated with currents in the range 0–10μA. The settling band is assumed to be 0.5%. On the x axis (to the right) the initial current is reported, while the y axis (to the left) shows the step final current.

so that I_{settle} can be set to $(I_{max} - I_{min})\alpha$, with α in the order of some percent or fraction of percent. Once I_{min} and I_{settle} are defined, the formula in Equation (12.10) allows calculation of a minimal bound for the system cycle time.

If I_{min} is set to zero, the minimal cycle time is defined only by αI_{max}. Hence, to have satisfactory cycle times, it is necessary to operate with large I_{max} or large α. In the current trend of low power design, rising I_{max} cannot always be feasible. However, it can be seen that the best power·delay products, once the range $I_{max} - I_{min}$ is defined, are obtained if $I_{min} > 0$. Hence, when operating with low currents it can be convenient to add positive offsets to all the current-mode signals.

More relevant to our considerations about robustness is the choice of α. Clearly, if α is taken too large, in an effort to minimize the system cycle time and to improve throughput, some current mirrors might introduce unacceptable errors. Hence, as a rule of thumb, α should be set so that the errors introduced by the current mirror settling times are no larger than the other *dominant* errors in the system signal processing chain. Actually, things can be a bit more subtle than that. In fact, the error introduced by current mirrors settling times is of a very systematic sort and hence can affect the system statistical behaviour.

By considering the model in Equation (12.10) and simplifying it, assuming the initial current as far apart as possible from the final one (worst case), it is possible to draw a plot of the mirror output as a function of the step final input current I_f, once the cycle time (and hence the settling time) has been set. The qualitative result is shown in Figure 12.25.

It can be seen that the mirror operates with negligible errors for all the

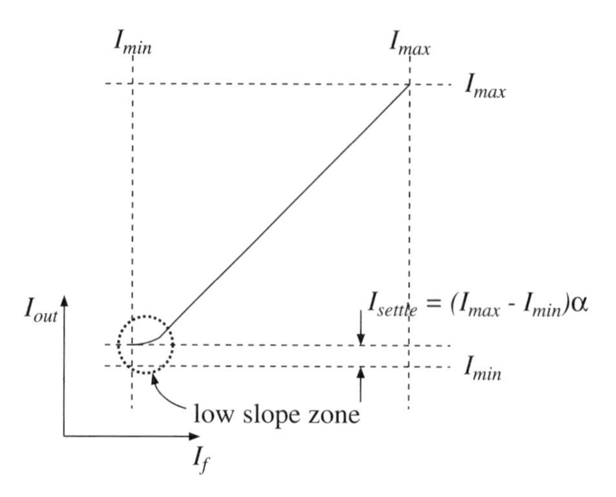

Figure 12.25: Qualitative response of a current mirror once the cycle time has been set.

currents higher than a certain level, but that it lags on low currents. The consequence is that the input-output characteristics are not straight, as desired, but show a smooth-saturation profile. If a mirror is used for setting a branch slope in a PWA map, then it is possible to obtain a map which, at some points, is "dynamically" characterized by a slope much lower than expected. As a consequence, in some (though rare) cases, systems may show unexpected (weak) trapping-like behaviour which disappears if the cycle time is increased.

12.3.2.2 Rectifiers (and comparators)

In the following we shall consider only rectifiers, but the considerations can be very easily extended to comparators. The simplest current rectifier one can think of is the current mirror, which is a unidirectional device. In the literature there are several examples of chaotic PWA systems based on current mirrors which are exploited as rectifiers (e.g., in [13]).

For example, consider the circuit in Figure 12.26 where the mirror MIR_2 prevents the current from flowing when $I_{IN} < I_{REF}$, by causing the switching-off of the up-hill mirror MIR_1.

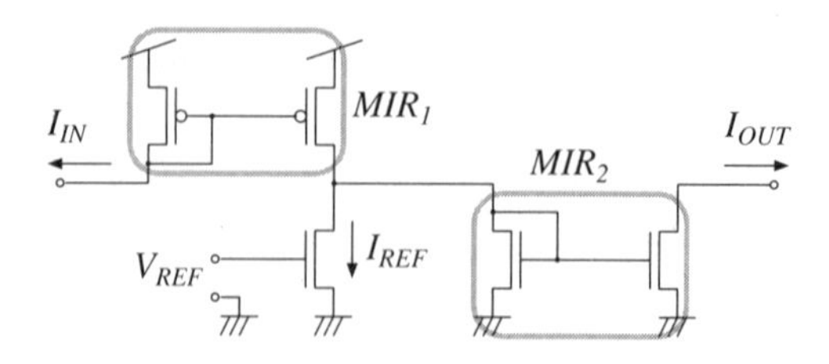

Figure 12.26: Current mirror used as a rectification block: MIR_2 acts as a current rectifier.

The problem with this circuit is that it is not very fast. Using a current mirror as a rectification circuit implies compelling the mirror to operate with a current range which comprises the value zero. As shown in the previous section, in this operating condition the mirror response time can become rather long, particularly for what concerns the mirror switch-off.

The model shown in Equation (12.10) is not rigorously valid when one exploits a current mirror as a rectification element. In fact, with reference to Figure 12.26, when I_{REF} is greater than I_{IN}, the source I_{REF} contributes to the discharge of the gate capacitance of MIR_2, which can happen very rapidly. The worst case is clearly when I_{IN} is equal to I_{REF}. In this particular case, the switch-off time goes exactly as described by the model (12.10). Another difference with that model is that in order to obtain (12.10) the gate voltage has been assumed never to go below the MOST threshold voltage. In the current mirror rectifier, on the contrary, the gate voltage of MIR_2 can go below threshold due to I_{REF}. Hence there is also a mirror *switch-on* time to be considered. It corresponds to the situation when the rectifier has to pass from $I_{IN} < I_{REF}$ to $I_{IN} > I_{REF}$: this requires the gate voltage to climb a whole MOS threshold before the mirror can start conduction. The process can be relatively slow if I_{IN} is made only slightly greater than I_{REF}. If we define a tolerance band I_{settle}, then we can approximate the rectifier worst response time as:

$$T_{rect} \approx \max\left(\frac{\sqrt{2}C_G\left(\sqrt{I_{max}} - \sqrt{I_{settle}}\right)}{\sqrt{I_{max} I_{settle} \beta}}, \frac{C_G V_{th}}{I_{settle}} + \frac{C_G}{\sqrt{I_{settle}\beta}}\right) \tag{12.11}$$

where the "max" operator takes into account the fact that we have to consider the worst case among the rectifier switch-off and switch-on times.

Once again, the time shown in Equation (12.11) sets a bound to the minimal cycle time that can be adopted if the rectifier is used in a discrete-time system.

Also in this case, it is possible to verify what happens when the rectifier is used as a part of a PWA map circuit. As an example, consider a functional block described by:

$$I_{OUT} = 2I_{in} - 2\left(I_{IN} - I_{BK}\right)^{+} \tag{12.12}$$

which realizes a tent map. If a current step is applied to the input of the functional block, from a certain I_0 to I_{IN}, and a plot is drawn putting I_{IN} on the x axis and I_{OUT} on the y axis, then the behaviour shown in Figure 12.27 is obtained. Leaving more time to the rectifier, the output converges to the ideal one. What results is a behaviour which is similar to hysteresis, in the sense that two different maps are obtained, depending on the initial current value I_0.

If a block of this sort is used for implementing a chaotic circuit, then misbehaviours can arise, disappearing as the cycle time is sufficiently increased. Misbehaviours may range from "escape from the invariant set" problems (due to the overshot shown in the left-hand side characteristics), to weak traps (due to the low slope characteristics in the right-hand side plot).

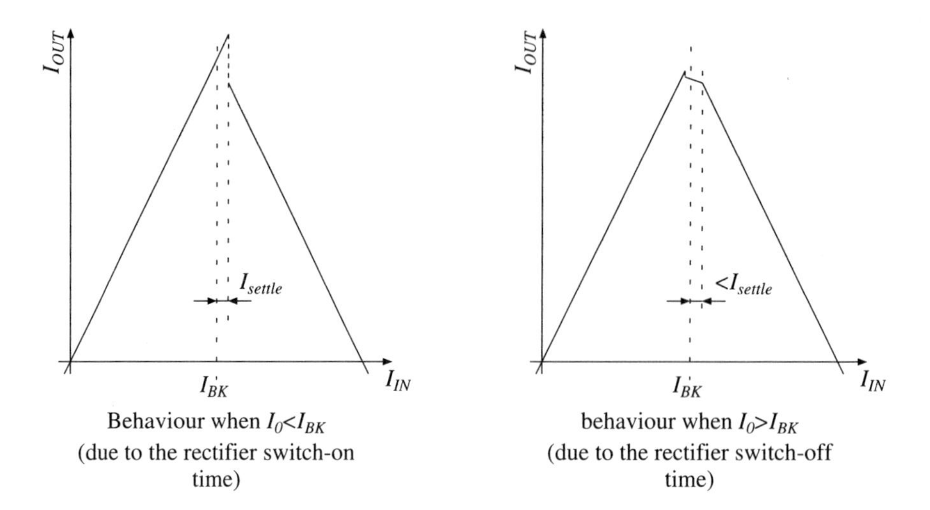

Figure 12.27: Behaviour obtained from Equation (12.12) when rectifiers based on current mirrors are adopted.

In order to improve the behaviour and to achieve better cycle times, it is convenient to adopt more sophisticated, active rectifying circuits, such as the one shown in Figure 12.28. In this circuit, an active amplifier operates in a feedback

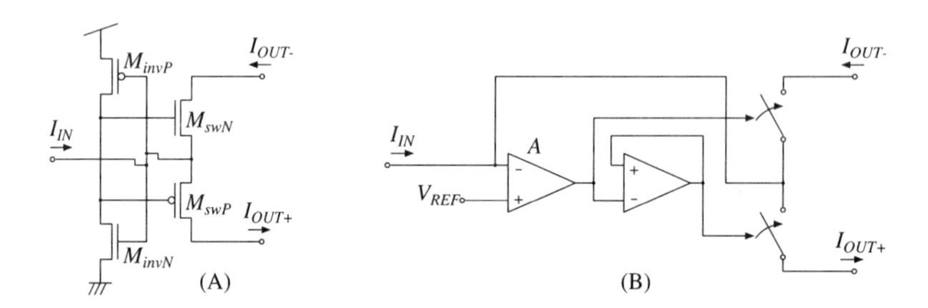

Figure 12.28: Active rectifier: circuit (A) and simplified schematic (B).

loop, in order to reduce the input node swing. This is the same as to say that the input impedence of the circuit is made very low. Since the circuit response time is dominated by the input capacitance to input resistance product, such a low input impedance makes the system very responsive. For a detailed analysis consult [24] and also Chapter 11 in this same book. We regard as important that the qualitative response of an active rectifier is very similar to that of a current mirror rectifier, apart from a much faster response.

It can be noticed also that connecting an inverter at the amplifier output node, the circuit can be used also as a comparator. Hence it is understandable how the inaccuracies characteristic of current rectifiers are reflected also in current comparators.

12.3.2.3 Sample-and-hold elements

In order to syncronize the operation of a discrete time system with an external clock, registers are required. In the case of an analog circuit, an *analog register* is required, which can be built as the cascade of two sample-and-hold (or track-and-hold) elements. In a current mode circuit, sample-and-hold elements store on a capacitor a voltage which is the result of a V to I conversion performed at the input. Then the opposite conversion is done at the output [25]. Although many different architectures can be used, this basic operating principle remains always the same, so that every analog sample-and-hold circuit is subject to clock feedthrough errors [26].

In a chaotic circuit, it is not uncommon for the feedthrough errors introduced by the analog memory element to be the dominant errors in the signal processing chain. In order to deal with these errors, it is often convenient to think of a system in which the errors introduced by the sample-and-hold circuitry are referred to the map. This is so to say that if an analog register is characterized by an input-output relationship given by $I_{OUT} = \Phi(I_{IN})$ and if this register is meant to be introduced in a chaotic circuit whose map is M, then one can think of an ideal memory element used to iterate the map $\hat{M}(x) = \Phi(M(x))$. In this way, once the input-output characteristics of a given memory element are measured or simulated, the usual tools can be used to study the behaviour of the chaotic system.

When it comes to consider the input-output characteristics Φ, it is common to split its deviation from the identity function in two components: a *signal independent* error, i.e., an offset, and a *signal dependent* error, so that:

$$\Phi(x) = x + \Phi'(x) + \sigma$$

where the term $\phi'(x)$ represents the signal dependent error, and σ is the signal independent one. For most architectures, at least under certain operating conditions, $\Phi'(x)$ can be approximated as a linear function of x with negligible error. In this case, one is allowed to write:

$$\Phi(x) = x + \Delta\mu x + \sigma - \frac{1}{2}\Delta\mu$$

for a normalized Φ defined over the set $[0, 1]$.

Once one has an order one model such as the one above, a couple of points are worth noting.

First of all, many different architectures exist for analog memory elements, setting many different possible combinations for $\Delta\mu$ and σ. For example, by proper cascading two sample-and-hold blocks, it is possible to reduce the value of σ (because reciprocal cancellation of the signal independent errors introduced by the two S/H stages can be achieved); furthermore, sample-and-hold circuits based on replication techniques [27] exist which allow us to improve both $\Delta\mu$ and σ as long as a very good matching can be achieved between the two signal paths that are required for the error cancellation; alternatively, S^2I sample-and-hold circuits can be used, which produce relatively large σ values, but very low $\Delta\mu$ [28, 29].

Secondly, once the architecture is known, one is usually able to say *a priori* what will be the sign of $\Delta\mu$. This is relevant to robustness, since for $\Delta\mu < 0$ the clearance among the principal invariant set and its basin of attraction is normally increased. Obviously, it can be convenient to choose a memory element architecture which enhances the system robustness, rather than the contrary.

For instance, consider the following situation. A system is to be based on a tent map circuit, whose slope is set to 2. As seen in Section 12.2.3.6, a system of this sort is not robust. However, if one adopts a naive analog register, such as the one shown in Figure 12.29, then one can be sure that the analog register will be characterized by a signal dependent error with $\Delta\mu < 0$, because feedthrough errors in this circuit are larger when the gate voltages are larger. The effect of such an error on the map $\hat{M}(x) = \Phi(M(x))$ is to produce a lowered slope tent map, which is, to a certain extent, a robust map.

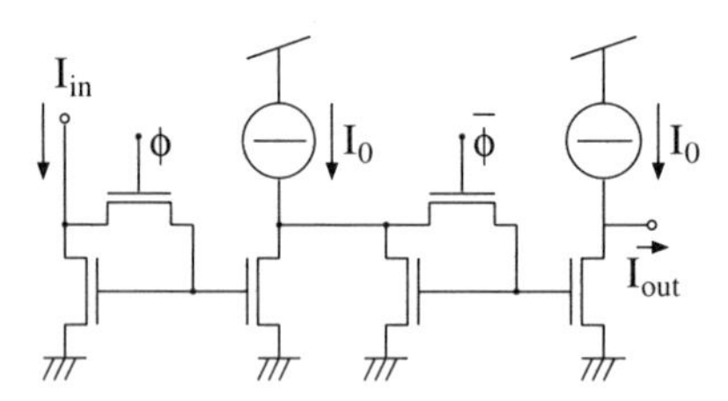

Figure 12.29: A naive analog register architecture. This circuit guarantees a signal dependent error whose slope $\Delta\mu$ is negative.

12.4 An Example of Sensitivity Analysis

In those cases in which a dominant error source can be identified, it is possible to perform a sensitivity analysis on that error source, by means of simulations. In fact, the identification of a single dominant error source allows us to introduce simplified models and to execute time-efficient simulations running at a behavioural rather than at a circuit level. This allows us to overcome the computation intensive tasks mentioned in Section 12.1.2.

Since the feedthrough errors in the analog register are often the dominant errors, it seems reasonable to try to estimate the effects of those errors on the output statistics of some important systems. Operating on errors introduced by the analog memory is generally easier than dealing with other error sources, since S/H errors are applied globally, while branch slope errors and breakpoint positioning errors are applied locally. This simplification, together with the approximation of the analog register behaviour with an order one model in which only a linear signal dependent error factor $\Delta\mu$ and an offset σ are considered, allows us to analyze the system sensitivity to its major cause of error, by simply running simulations for all possible $(\Delta\mu, \sigma)$ couples on a certain mesh.

For example, in [29], an analysis of this sort has been performed on Bernoulli shift, tent map, and tailed tent map systems, defining suitable norms for reading the deviations from the ideal output statistics. Figure 12.30 shows the result of the simulations for what concerns the deviations from the ideal cumulative probability density function (CPDF, also called Probability Distribution Function, i.e., the integral of the PDF) for tent map and tailed tent map systems. Note that for determining these deviations, an integral norm has been used. The white areas in the plot represent those regions in which the system behaviour has been broken by the error magnitudes which have been taken into account.

The results obtained from this analysis are the following:

1. For all the maps that have been tested there is a much more pronounced sensitivity to signal dependent errors than to offsets. This means that S/H architectures which are optimal for what concerns signal dependent error should be considered. As a consequence of this analysis, some chaotic circuits based on S^2I S/H circuits have been presented in the literature [30].

2. Different maps behave differently with respect to identical errors. For example, a tailed tent map system is generally less influenced by slope errors than a tent map or a hooked tent map system. Based on experience, designers should determine which maps are best used when some particular statistical feature has to be obtained with a good level of accuracy.

12.5 Robustness and Topological Conjugation

The aim of this section is to complement the considerations of the first part, providing formal results about map robustness.

Formal robustness issues exist [31, 32, 34, 36] and some of them, though quite technical in nature, will be recalled in the following. Their scope is limited by the need for a precise definition of robustness (which may not coincide with the one dictated by application-specific constraints), of perturbation (which may not model the disturbance to which the circuit will actually be exposed), and possibly by more technical assumptions that are needed to obtain mathematical proofs.

We here concentrate on the relationship between robustness and topological conjugation, which is a means of studying wide classes of maps by thoroughly investigating the behaviour of just one of them.

In particular, we can show that for a set of maps the equivalence classes induced by topological conjugation [39] are such that if one of the elements in the class is robust then all the elements in that class are robust. In this way, some maps which would not satisfy classical sufficient conditions like those in [31, 32] can be shown to be robust if another element in their equivalence class meet those requirements.

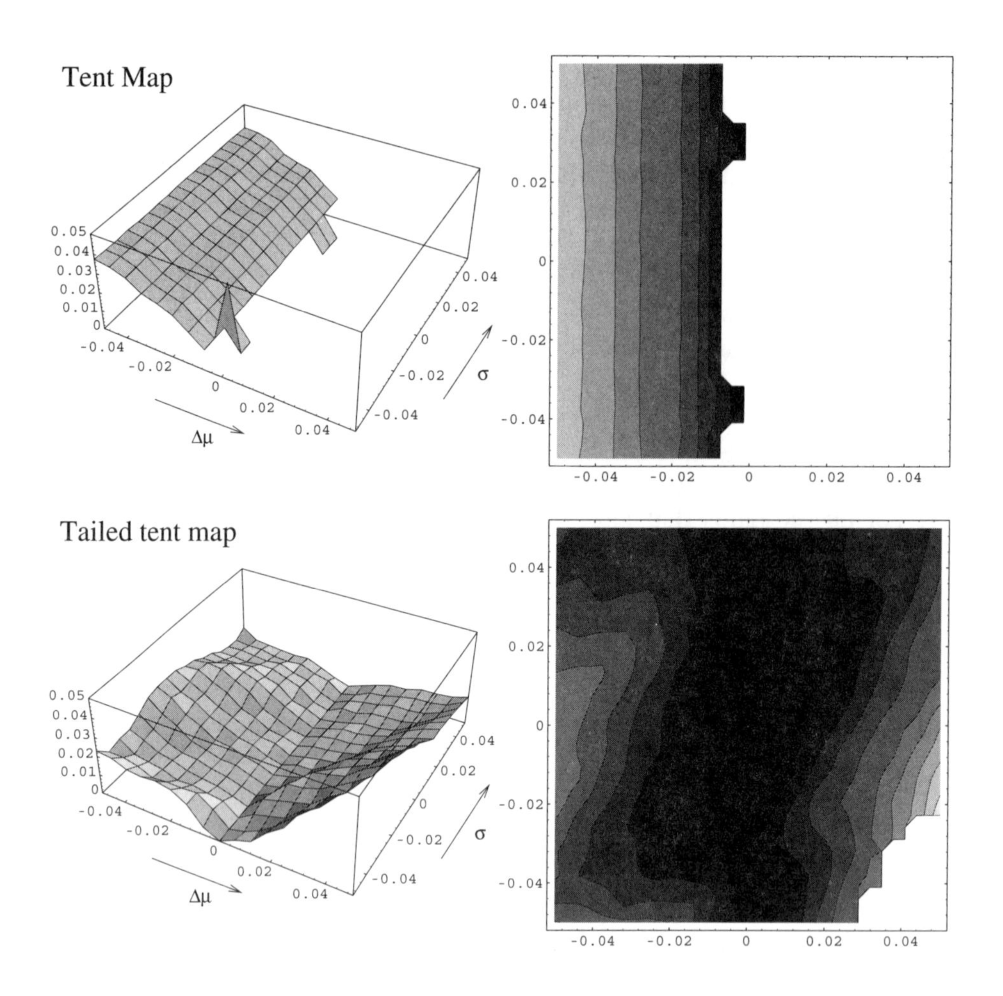

Figure 12.30: Results of a sensitivity analysis showing the effects of signal dependent/signal independent S/H errors on chaotic systems based on the iteration of a tent map and a tailed tent map. On the x axis, signal dependent (i.e., $\Delta\mu$) errors are reported, while on the y axis signal independent (offsets, σ in the model) are reported. The deviation from the prescribed CDF is then calculated using an integral norm. The density plots on the right help read the trends.

To formalize this point we assume that the statistical features we want to preserve are completely contained in the spectrum of the Perron-Frobenius operator $\mathbf{P}_M$ associated with the map M, i.e., in the collection of its eigenvalues which are those complex numbers λ_M for which at least a bounded variation density function ρ_M exists such that $\mathbf{P}_M \rho_M = \lambda_M \, \rho_M$.

Since the Perron-Frobenius operator is infinite-dimensional, one may reasonably expect it possesses an infinite collection of eigenvalues. This is actually the case and we know from [31, 33, 37, 38] that, for any map of interest, $\mathbf{P}_M$ is a quasi-compact operator, i.e., it has only a finite number of isolated eigenvalues with finite-dimensional eigenspaces. Moreover, if one sets

$$r_{\text{ess}} = \lim_{k \to \infty} \left[\sup \frac{1}{|(d/dx)M^k|} \right]^{1/k}, \tag{12.13}$$

then any $\gamma_M \in \mathbb{C}$ such that $|\gamma_M| < r_{\text{ess}}$ is an eigenvalue of $\mathbf{P}_M$. As no real number $r > r_{\text{ess}}$ has the same property, r_{ess} is called the essential spectral radius of $\mathbf{P}_M$. Nevertheless, $\mathbf{P}_M$ has a unit spectral radius $r_{\text{sp}} = 1$ and there are $l > 0$ eigenvalues $\lambda_M^1 = 1$ and $\lambda_M^2, \ldots, \lambda_M^l$ of modulus 1 which correspond to the finite-dimensional eigenspaces $\mathbb{V}_1, \ldots, \mathbb{V}_l$. Finally, $\mathbf{P}_M$ may feature other isolated eigenvalues γ_M^i whose modulus is between r_{ess} and r_{sp}.

As already clarified in other chapters in this volume, the iteration of $\mathbf{P}_M$ accounts for the evolution of the set of trajectories of the discrete time system $x_n = M^n(x_0)$, whose initial conditions x_0 are drawn according to an initial probability density. An additional important characteristic of the spectrum of $\mathbf{P}_M$ is related to the complexity of such a dynamical system, from ergodicity to mixing and exactness [19, chap. 4].

For example, one may show that if f is at least mixing, and has an everywhere expanding iterate, the spectral decomposition of $\mathbf{P}_M$ holds with $l = 1$, $\lambda_M^1 = 1$ and $\dim \mathbb{V}_1 = 1$ so that there is a unique invariant probability density $\bar{\rho}_M$ which is the fixed point of the Perron-Frobenius operator, i.e., its eigenfunction with unit eigenvalue. For the same kind of chaotic systems, another important quantity can be defined which is related to the rate of decay of the correlation of the generated signals. Such quantity is called the *rate of mixing* r_{mix} and is the modulus of the second largest eigenvalue of $\mathbf{P}_M$. In particular, it can be shown [33] that, if an exact map is adopted, after an exponentially vanishing transient, the points x_n will distribute according to the probability density $\bar{\rho}_M$.

12.5.1 The Perturbation Model

One of the most general models [31, 32] of perturbation relies on the following:

Definition 1. *A random perturbation of a map $M : [0,1] \mapsto [0,1]$ is a Markov chain M^ϵ ($\epsilon > 0$) with transition probability $\Pi_M^\epsilon : [0,1] \times 2^{[0,1]} \mapsto [0,1]$ where $2^{[0,1]}$ is the collection of all the subsets of $[0,1]$. Moreover, the following requirements must be fulfilled.*

1. *The map $x \mapsto \Pi_M^\epsilon(x, \cdot)$ is continuous for each ϵ.*
2. *Each $\Pi_M^\epsilon(x, \cdot)$ is absolutely continuous with respect to the Lebesgue measure m.*
3. *The Koopman operator $\mathbf{K}_{M^\epsilon}$ of M^ϵ is such that $\lim_{\epsilon \to 0} \mathbf{K}_{M^\epsilon} T = \mathbf{K}_M T$ for any continuous test function $T : [0,1] \mapsto \mathbb{R}$, where*

$$\mathbf{K}_{M^\epsilon}T(x) = \int_{[0,1]} T(y)\Pi^\epsilon_M(x, dy)$$

and $\mathbf{K}_M T(x) = T(M(x))$ *is the unperturbed Koopman operator.*

In other words, the last assumption simply ensures that the perturbation vanishes for $\epsilon \to 0$ as the transition probability Π^ϵ_M gives 1 for any set containing $M(x)$ and 0 for any other subset of $[0, 1]$.

Interesting enough, it can be seen [31] that for small ϵ this model accounts for both implementation inaccuracies and noise-like disturbances.

To get an intuitive idea of the significance of Definition 1, let us refer to Figure 12.31. While the unperturbed function M would map a point x at time n into the corresponding point $y = M(x)$ at time $n+1$ (continuous-line), due to the random perturbation suffered by the map, a transition between point x and $\xi \in [y - \Delta y/2, y + \Delta y/2]$ occurs with probability $\Pi^\epsilon_M(x, \Delta y)$ (dashed-line).

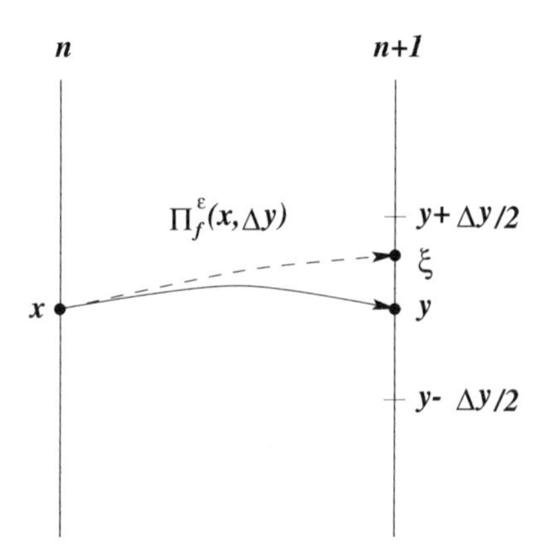

Figure 12.31: Schematic interpretation of the effect of a random perturbation on the evolution of the system trajectory: transition for the unperturbed map (continuous-line) and for the perturbed one (dashed-line).

If one defines the Perron-Frobenius operator $\mathbf{P}_{M^\epsilon}$ for M^ϵ simply as the adjoint of the perturbed Koopman operator $\mathbf{K}_{M^\epsilon}$, the concepts of spectral decomposition and of rate of mixing r^ϵ_{mix}, previously introduced may be extended to the perturbed case. In general, since $[0, 1]$ is a compact set, from the first two points of Definition 1 it follows [31] that M^ϵ admits an invariant density $\bar{\rho}_{M^\epsilon}$ such that $\mathbf{P}_{M^\epsilon}\bar{\rho}_{M^\epsilon} = \bar{\rho}_{M^\epsilon}$. Other eigenvalues satisfy the equation $\mathbf{P}_{M^\epsilon}\rho_{M^\epsilon} = \lambda_{M^\epsilon}\rho_{M^\epsilon}$, and if 1 is the only point of the spectrum of $\mathbf{P}_{M^\epsilon}$ on the unit circle, the interpretation of r^ϵ_{mix} carries over as in the unperturbed case [31].

Several results are available in the special case of random perturbation by convolution, namely where the transition probabilities $\Pi^\epsilon_M(x, dy)$ have densities $\pi^\epsilon_M(y - M(x))$, where $\pi^\epsilon_M : \mathbb{R} \mapsto \mathbb{R}^+$ with $\text{supp}\pi^\epsilon_M \subset [-\epsilon, \epsilon]$.

12.5.2 A Formalization of Stochastic Robustness

In order to introduce the concept of stochastic robustness, the key point is that the statistical features of a chaotic, perturbed or unperturbed, map are linked to the spectral characteristics of the associated Perron-Frobenius operator. Therefore, the notion of stochastic robustness with respect to the perturbations defined above is given comparing an eigenvalue-eigenfunction couple (λ_M, ρ_M) of the Perron-Frobenius operator corresponding to the unperturbed map and the corresponding couple $(\lambda_{M^\epsilon}, \rho_{M^\epsilon})$ for the statistically perturbed one. The map M is said to be robust if both $|\lambda_{M^\epsilon} - \lambda_M| \to 0$ and $\|\rho_{M^\epsilon} - \rho_M\| \to 0$ as $\epsilon \to 0$ for a certain norm $\|\cdot\|$.

Different concepts of stochastic robustness may be generated by different choices of the eigenvalue-eigenfunction pair and of the convergence norm $\|\cdot\|$. For example in [35,36] $\lambda_M = 1$ is chosen and weak convergence of $\bar{\rho}_{M^\epsilon}$ to $\bar{\rho}_M$ is sought while in [31] a stronger norm $\|\cdot\|$ is adopted to define robustness ($\|\cdot\|_1 = \int|\cdot|$ for maps on the circle and the bounded variation norm for maps on $[0,1]$ into itself) and convergence of r^ϵ_{mix} to r_{mix} is also investigated.

For piecewise $\mathcal{C}^2$, piecewise expanding maps, several results guaranteeing robustness are available [31]. More precisely, assume that a partition $0 = a_0 < a_1 < \cdots < a_{N-1} < a_N = 1$ of $[0,1]$ exists such that, for any i, the restriction $M|_{[a_i,a_{i+1}]}$ can be extended to a $\mathcal{C}^2$ map with $\min|M'| \geq \lambda > 1$. The points a_i are called the turning points and we say that M has no periodic turning point if $M_k(a_i) \neq a_i$ for all $k > 0$ and $i \neq 0, N$. The following theorem holds.

Theorem 1 ([31] [32]). *Let f be an exact piecewise $\mathcal{C}^2$, piecewise expanding map and indicate with M^ϵ a random perturbation by convolution (see Section 12.5.1) with invariant probability density $\bar{\rho}_{M^\epsilon}$ and rate of mixing $r^\epsilon_{\text{mix}} < 1$. Assume also that f has no periodic turning point. Then*

1. $\|\bar{\rho}_{M^\epsilon} - \bar{\rho}_M\|_1$ *tends to 0 as* $\epsilon \mapsto 0$
2. $\limsup_{\epsilon \mapsto 0} r^\epsilon_{\text{mix}} \leq \sqrt{r_{\text{mix}}}$

In the following discussion we will address a generic perturbation corresponding to Definition 1, as well as a generic eigenvalue-eigenfunction pair allowing a wide class of convergence norms (namely those not stricter than the bounded variation norm) so that, no matter how it is defined, stochastic robustness can be shown to propagate through topological conjugation.

12.5.3 Topological Conjugation Propagates Robustness

Two maps M and N both mapping $[0,1]$ into itself are said to be topologically conjugate if a homeomorphism (i.e. a continuous invertible function with continuous inverse) ϕ exists such that $N = \phi \circ M \circ \phi^{-1}$ [39, chap. 2]. Let us first prove the following [40]:

Theorem 2. *Let M and N be two topologically conjugate maps. Then topological conjugation maps perturbations of M into perturbations of N.*

Proof. Note first that, since ϕ is a bijection, the equality $N = \phi \circ M \circ \phi^{-1}$ translates the perturbation M^ϵ with transition probability Π_M^ϵ into N^ϵ with transition probability $\Pi_N^\epsilon(x, X) = \Pi_M^\epsilon(\phi^{-1}(x), \phi^{-1}(X))$. Figure 12.32 clarifies such a relationship highlighting the different paths that can be followed to go from x to $g(x)$ and its perturbed versions.

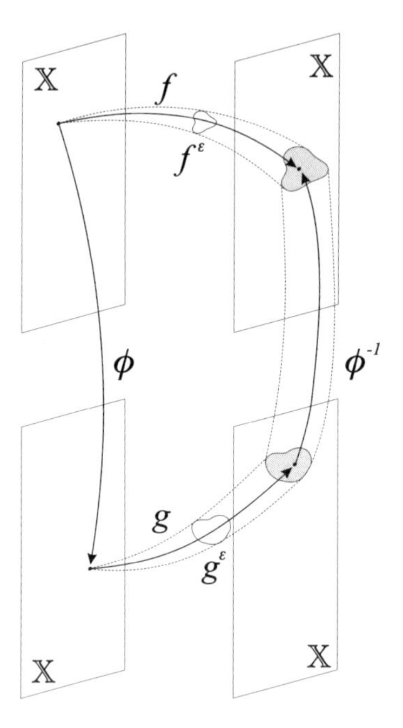

Figure 12.32: Topological conjugation acting on perturbed maps.

As long as ϕ is continuous, the continuity of Π_M^ϵ implies the continuity of Π_N^ϵ. Hence, to prove that N^ϵ is a true perturbation of N we only need to show that for any continuous test function $T : [0,1] \mapsto \mathbb{R}$ and for any $x \in [0,1]$

$$\lim_{\epsilon \to 0} \int_{[} 0,1] T(y) \Pi_N^\epsilon(x, dy) = T(N(x)) \tag{12.14}$$

To do so let us rewrite the left-hand limit of (12.14) as

$$\begin{aligned} &\lim_{\epsilon \to 0} \int_{[} 0,1] T(y) \Pi_M^\epsilon(\phi^{-1}(x), \phi^{-1}(dy)) = \\ &= \lim_{\epsilon \to 0} \int_{[} 0,1] T'(\phi^{-1}(y)) \Pi_M^\epsilon(\phi^{-1}(x), \phi^{-1}(dy)) \\ &= T'(M(\phi^{-1}(x))) \end{aligned}$$

where the continuous test function $T' = T \circ \phi$ has been introduced and the definition of perturbation has been exploited for M^ϵ. Thus, we may finally note that $T'(M(\phi^{-1}(x))) = T(\phi(M(\phi^{-1}(x)))) = T(N(x))$ and verify (12.14). $\square$

This is the first step in proving that stochastic robustness propagates through conjugation. To formalize the last step let us prove the following [40]

Theorem 3. *Assume that an eigenvalue-eigenfunction pair (λ_N, ρ_N) of the Perron-Frobenius operator $\mathbf{P}_N$ has been chosen and that for any perturbation N^ϵ a solution of the perturbed eigenequation $\mathbf{P}_{N^\epsilon}\rho_{N^\epsilon} = \lambda_{N^\epsilon}\rho_{N^\epsilon}$ exists such that*

$$\begin{aligned}\lim_{\epsilon\to 0}|\lambda_{N^\epsilon} - \lambda_N| &= 0\\ \lim_{\epsilon\to 0}\|\rho_{N^\epsilon} - \rho_N\| &= 0\end{aligned}$$

for a norm $\|\cdot\|$ not stricter that the bounded variation norm. If M is topologically conjugate to N we also have

$$\begin{aligned}\lim_{\epsilon\to 0}|\lambda_{M^\epsilon} - \lambda_M| &= 0\\ \lim_{\epsilon\to 0}\|\rho_{M^\epsilon} - \rho_M\| &= 0\end{aligned}$$

Proof. By applying the very definition (3.3) of the Perron-Frobenius operator for an unperturbed map, from the conjugation relationship, one easily gets the equivalent tie

$$\mathbf{P}_N = \mathbf{P}_\phi \mathbf{P}_M \mathbf{P}_{\phi^{-1}} = \mathbf{P}_\phi \mathbf{P}_M \mathbf{P}_\phi^{-1}.$$

in terms of Perron-Frobenius operators.

A similar result also holds for perturbations. In fact, $\mathbf{P}_{N^\epsilon}$ is the adjoint operator of $\mathbf{K}_{N^\epsilon}$. Yet one has that, for any continuous test function $T : [0,1] \mapsto \mathbb{R}$

$$\begin{aligned}&\mathbf{K}_{N^\epsilon}T(x) =\\ &= \int_{[0,1]} T(y)\Pi_M^\epsilon(\phi^{-1}(x), \phi^{-1}(dy)) =\\ &= \int_{[0,1]} \mathbf{K}_\phi T(\phi^{-1}(y))\Pi_M^\epsilon(\phi^{-1}(x), \phi^{-1}(dy)) =\\ &= \mathbf{K}_{\phi^{-1}}\mathbf{K}_{M^\epsilon}\mathbf{K}_\phi T(x)\end{aligned}$$

which is the adjoint of $\mathbf{P}_\phi \mathbf{P}_{M^\epsilon}\mathbf{P}_{\phi^{-1}}$.

Note now that the eigenequation $\mathbf{P}_N\rho_N = \lambda_N\rho_N$ can be written as

$$\mathbf{P}_\phi \mathbf{P}_M \mathbf{P}_\phi^{-1}\rho_N = \lambda_N\rho_N$$

i.e., $\mathbf{P}_M(\mathbf{P}_\phi^{-1}\rho_N) = \lambda_N(\mathbf{P}_\phi^{-1}\rho_N)$. Thus, the spectrum of the Perron-Frobenius operator is invariant by conjugation while each eigenfunction ρ_N is changed into an eigenfunction $\rho_M = \mathbf{P}_\phi^{-1}\rho_N$. Hence, when ϵ vanishes we get from $\lambda_{N^\epsilon} \to \lambda_N$ that also $\lambda_{M^\epsilon} = \lambda_{N^\epsilon} \to \lambda_M = \lambda_N$.

The paths allowing the analysis of perturbed eigenfunctions are summarized in the following diagram.

$$\begin{array}{ccc}
N^\epsilon & \xrightarrow{\phi^{-1}} & M^\epsilon \\
{\scriptstyle \mathbf{P}_{N^\epsilon}\rho_{N^\epsilon}=\lambda_{N^\epsilon}\rho_{N^\epsilon}}\big\downarrow & & \big\downarrow{\scriptstyle \mathbf{P}_{M^\epsilon}\rho_{M^\epsilon}=\lambda_{M^\epsilon}\rho_{M^\epsilon}} \\
\rho_{N^\epsilon} & \xleftarrow{\mathbf{P}_\phi} & \rho_{M^\epsilon} \\
{\scriptstyle \epsilon\to 0}\big\downarrow & & \big\downarrow{\scriptstyle \epsilon\to 0} \\
\rho_N & \xleftarrow[\mathbf{P}_\phi]{} & \rho_M
\end{array}$$

The key property here is the linearity of the $\mathbf{P}_\phi^{-1}$ operator. In fact, we know that the Perron-Frobenius is bounded in the space of bounded variation function with respect to any norm $||\cdot||$ not stricter than the bounded variation norm. With this we may write

$$\begin{aligned}
||\rho_{M^\epsilon} - \rho_M|| &= ||\mathbf{P}_\phi^{-1}\rho_{N^\epsilon} - \mathbf{P}_\phi^{-1}\rho_N|| \leq \\
&\leq ||\mathbf{P}_\phi^{-1}||||\rho_{N^\epsilon} - \rho_N||
\end{aligned}$$

which is enough to say that $\lim_{\epsilon\to 0} ||\rho_{M^\epsilon} - \rho_M|| = 0$, i.e., that also f is robust. □

As an example of application of the above theorem, consider the logistic map $N(x) = 4x(1-x)$. It does not satisfy the classical requirements for robustness [31,32] as it is not piecewise expanding. Nevertheless successful implementations are reported in the literature (see, e.g., [41]) and this suggests that it must feature a certain degree of robustness.

In fact, the logistic map is topologically conjugate to the tent map $M(x) = 1 - 2|x - 1/2|$ by means of $\phi(x) = \sin^2(\pi x/2)$. For this last map it is well known that as long as its breakpoints lay within $[0, 1]$, small perturbations lead to a graceful deviation from the nominal statistical properties. This robustness is propagated to the logistic map by Theorems 2 and 3.

12.6 Conclusions

In this chapter, we have dealt with the problem of robust implementation of chaotic maps. Using an empirical approach, some typical robustness issues have been shown for systems based on PWA maps. Then these considerations have been applied to circuits, taking into account the current mode approach to the design of current mode discrete time chaotic systems. An example of a very specialized sensitivity analysis run by simulation has also been provided, in order to show how even a very rough examination of the factors which influence the adherence of an implemented system to its ideal statistical behaviour can be useful for extrapolating design guidelines.

A formal criterion to propagate robustness from the design of one map to another which is topologically conjugated has been also provided. This was possible since it has been recognized that robustness of the statistical features of a map is equivalent to low sensitivity to perturbation of the spectral properties of the associated Perron-Frobenius operator.

Altogether, we hope that this chapter will help those concerned with the design of discrete time chaotic circuits, if not by directly providing solutions, at least by calling attention to a few situations which always need to be double checked.

References

[1] Mazzini, G., Setti, G., and Rovatti, R., Chaotic complex spreading sequences for asynchronous DS-CDMA. I. System modeling and results. *IEEE Transactions on Circuits and Systems, Part I*, 44(10):937–947, October 1997.

[2] Rovatti, R., Setti, G., and Mazzini, G., Chaotic complex spreading sequences for asynchronous DS-CDMA. II. Some theoretical performance bounds. *IEEE Transactions on Circuits and Systems, Part I*, 45(4):496–506, April 1998.

[3] Kennedy, M. P., Chaotic communications: State of the art. In *Proceedings of ECCTD'99*, volume 1:437-440, Stresa, September 1999.

[4] Setti, G., Rovatti, R., and Mazzini, G., Syncronization mechanism and optimization of spreading sequences in chaos based DS-CDMA systems. *IEICE Transactions on Fundamentals*, E82A(9):1737–1746, September 1999.

[5] Rovatti, R., Setti, G., and Graffi, S., Chaos based FM of clock signals for EMI reduction. In *Proceedings of ECCTD'99*, volume 1:373-376, Stresa, September 1999.

[6] Rovatti, R., Setti, G., and Graffi, S., Chaotic FM and AM of clock signals for improved EMI compliance. In *Proceedings of NOLTA'99*, volume 1:219–222, Hawaii, December 1999.

[7] Rovatti, R. and Setti, G., Chaos based generation of $1/f^{]0,1[}$ noise for circuit simulation. In *Proceedings of NOLTA'99*, volume 2:565–568, Hawaii, December 1999.

[8] Giovanardi, A., Mazzini, G., Rovatti, R., and Setti, G., Features of chaotic maps producing self-similar processes. In *Proceedings of NOLTA'98*, volume 1:203-207, 1998.

[9] Clarkson, T. G., Ng, C. K., Bean, J., Review of hardware pRAMs, *Neural Networks World*, (5):551–564, 1993.

[10] Delgado-Restituto, M., Medeiro, F., and A. Rodríguez-Vázquez, Nonlinear, switched current CMOS IC for random signal generation. *Electronics Letters*, (25):2190–2191, 1993.

[11] Langlois, P. J., Bergmann, G., and Bean, J. T., A current mode tent map with electrically controllable skew for chaos applications. In *Proceedings of NDES'95 (Nonlinear Dynamics of Electronics Systems Conference)*, Dublin, 1995.

[12] Rovatti, R., Manaresi, N., Setti, G., and Franchi, E., A current-mode circuit implementing chaotic continuous piecewise-affine Markov maps. In *Proceedings of MICRONEURO'99*, p. 781, Granada, April 1999.

[13] Bean, J. T. and Langlois, P. J., A noise generator based on chaos for a neural network application. In *Proceedings of NDES'93 (Nonlinear Dynamics of Electronics Systems Conference)*, Dresden, July 1993.

[14] Callegari, S., Setti, G., and Langlois, P. J., A CMOS Tailed Tent Map for the generation of uniformly distributed chaotic sequences. In *Proceedings of IEEE ISCAS'97*, volume 2, p. 781, Hong Kong, June 1997.

[15] Delgado-Restituto, M., Rodríguez-Vázquez, A., Porra, V., Design considerations for a linear modulation DCSK chaos-based radio. In *Proceedings of ICECS'98*, volume 1, p. 239, Lisboa, September 1998.

[16] Bernstain, G. M. and Lieberman, M. A., Secure random number generation using chaotic circuit. *IEEE Transactions on Circuits and Systems*, 37(9):1157–1164, January 1990.

[17] Pelgrom, M. J. M., Duinmaijer, A. C. J., and Welbers, A. P. G., Matching properties of MOS transistors. *IEEE Journal of Solid State Circuits*, 24(5):1433–1440, October 1989.

[18] Devaney, R., *An Introduction to Chaotic Dynamical Systems.* Second Edition, Addison-Wesley, Menlo Park (CA), 1989.

[19] Lasota, A. and Mackey, M. C., *Chaos, Fractals and Noise. Stochastic Aspects of Dynamics.* Second Edition, Springer-Verlag, New York, 1995.

[20] Nayfeh, A. H. and Balachandran, B., *Applied Nonlinear Dynamics.* John Wiley & Sons, New York, 1995.

[21] Tsuneda, A., Eguchi, K., and Inoue, T., Design of chaotic binary sequences with good statistical properties based on piecewise linear into maps. In *Proceedings of MICRONEURO'99*, p. 261, Granada, April 1999.

[22] Delgado Restituto, M. and Rodríguez-Vásquez, A., Piecewise affine Markov maps for chaos generation in chaotic communication. In *Proceedings of ECCTD'99*, volume 2, Stresa, September 1999.

[23] Nairn, D. G., Analytic step response of MOS current mirrors. *IEEE Transactions on Circuits and Systems, Part I*, 40(2):133–135, February 1993.

[24] Rodríguez Vásquez, A., Domínguez Castro, R., Medeiro, F., and Delgado Restituto, M., High resolution CMOS current comparators: Design and applications to current-mode function generation. *Analog Integrated Circuits and Signal Processing*, (7):149–165, 1995.

[25] Fiez, T. S., Liang, G., and Allstot, D. J., Switched-current circuit design issues. *IEEE Journal of Solid State Circuits*, 26(3):192–201, March 1991.

[26] Sheu, B. J. and Hu, C., Switch-induced error voltage on a switched capacitor. *IEEE Journal of Solid State Circuits*, SC-19(4):519–525, August 1984.

[27] Yang, H. C., Fiez, T. S., and Allstot, D. J., Current feedthrough effects and cancellation techniques in switched-current circuits. In *Proceedings of IEEE ISCAS'90*, pp. 3186–3188, 1990.

[28] Hughes, J. B. and Moulding, K. W., S^2I: a two step approach to switched currents. In *Proceedings of IEEE ISCAS'93*, volume 2, pages 1235–1238, 1993.

[29] Callegari, S. and Rovatti, R., Analog chaotic maps with sample-and-hold errors. *IEICE Transactions on Fundamentals*, E82A(9):1754-1761, September 1999.

[30] Callegari, S., Rovatti, R., and Setti, G., A tailed tent map chaotic circuit exploiting S^2I memory elements. In *Proceedings of ECCTD'99*, volume 1, pp. 193-196, Stresa, September 1999.

[31] Baladi, V. and Young, L.-S., On the Spectra of Randomly Perturbed Expanding Maps. *Communications in Mathematical Physics*, vol. 156, pp. 355-385, 1993.

[32] Baladi, V. and Young, L.-S., *Erratum* - On the Spectra of Randomly Perturbed Expanding Maps. *Communications in Mathematical Physics*, vol. 166, pp. 219-220, 1994.

[33] Keller, G., On the Rate of Convergence to Equilibrium in One-Dimensional Systems. *Communications in Mathematical Physics*, vol. 96, pp. 181-193, 1984.

[34] Rovatti, R. and Setti, G., Statistical Features and Robustness of Chaotic Fuzzy Maps. *IEEE International Conference on Fuzzy Systems*, pp. 793-799, 1997.

[35] Kifer, Y., *Ergodic Theory of Random Transformations.* Birkhäuser, Boston, 1986.

[36] Kifer, Y., *Random Perturbation of Dynamical Systems.* Birkhäuser, Boston, 1988.

[37] Hofbauer, F. and Keller, G., Ergodic Properties of Invariant Measures for Piecewise Monotonic Transformations. *Mathematische Zeitschrift*, vol. 180, pp. 119-140, 1982.

[38] Baladi, V. and Keller, G., Zeta Functions and Transfer Operators for Piecewise Monotone Transformations. *Communications in Mathematical Physics*, vol. 127, pp. 459-477, 1990.

[39] Kuznetsov, Y. A., *Elements of Applied Bifurcation Theory.* Springer-Verlag, New York, 1995.

[40] Rovatti, R. and Setti, G., Topological Conjugacy Propagates Stochastic Robustness of Chaotic Maps. *IEICE Transactions on Fundamentals*, vol. 81, no. 9, 1998.

[41] McGonical, G. C. and Elmasry, M. I., Generation of Noise by Electronic Iteration of the Logistic Map. *IEEE Transactions on Circuits and Systems*, vol. 34, pp. 981-983, 1987.

Index